# Modern Derivatization Methods for Separation Sciences

# Modern Derivatization Methods for Separation Sciences

Edited by

**Toshimasa Toyo'oka**
*School of Pharmaceutical Sciences, University of Shizuoka, Japan*

**JOHN WILEY AND SONS**
Chichester • New York • Weinheim • Brisbane • Singapore • Toronto

Baffins Lane, Chichester,
West Sussex PO19 IUD, England

National 01243 779777
International (+44) 1243 779777
e-mail (for orders and customer service enquiries): cs-books@wiley.co.uk
Visit our Home Page on http://www/wiley.co.uk
or http://www.wiley.com



*Other Wiley Editorial Offices*

John Wiley & Sons, Inc., 605 Third Avenue,
New York, NY 10158-0012, USA

WILEY-VCH Verlag GmbH, Pappelallee 3,
D-69469 Weinheim, Germany

Jacaranda Wiley Ltd, 33 Park Road, Milton,
Queensland 4064, Australia

John Wiley & Sons (Asia) Pte Ltd, Clementi Loop #02-01,
Jin Xing Distripark, Singapore 129809

John Wiley & Sons (Canada) Ltd, 22 Worcester Road,
Rexdale, Ontario M9W IL1, Canada

***Library of Congress Cataloging-in-Publication Data***

Modern derivatization methods for separation sciences/edited by
Toshimasa Toyo'oka.
p. cm.
Includes bibliographical references and index.
ISBN 0-471-98364-0 (alk. paper)
1. Chromatographic analysis—Methodology. 2. Derivatization.
I. Toyo'oka, Toshimasa.
QD79.C4M63 1999 98-27197
543′.089—dc21 CIP

***British Library Cataloging in Publication Data***

A catalogue record for this book is available from the British Library

ISBN 0 471 98364 0

Typeset in 10/12pt Times by Laser Words, Madras, India
Printed and bound in Great Britain by Bookcraft (Bath) Ltd
This book is printed on acid-free paper responsibly manufactured from sustainable forestry, in which at least two trees are planted for each one used for paper production

# Contents

List of Contributors xi

Preface xiii

1. Pre-treatment for Real Sample Analysis and Choice of Suitable Reagent 1
    1.1. Derivatization for Drugs
        *Akihiko Nakagawa, and Yukinori Kawahara* 1
        1.1.1. Introduction 2
        1.1.2. Application of Derivatization of Drugs 2
            1.1.2.1. Determination of Pravastatin Sodium in Plasma by HPLC with Laser-induced Fluorescence Detection after Immobilized Antibody Extraction 2
            1.1.2.2. Enantiospecific Determination of Ibuprofen in Rat Plasma Using Chiral Fluorescence Derivatization Reagent, (−)-2-[4-(1-aminoethyl)phenyl]-6-methoxybenzoxazole 4
            1.1.2.3. Specific Fluorogenic Derivatization of Ivermectin Using Non-fluorescent Reagents 6
            1.1.2.4. Other Methods 8
        1.1.3. Conclusion 10
    References 11

    1.2. Derivatization for Agrochemicals
        *Hiroyuki Nakazawa and Koichi Saito* 13
        1.2.1. Introduction 13
            1.2.1.1. Sample Preparation 14
            1.2.1.2. Derivatization in Food Analysis 15
        1.2.2. Analysis of Nutrients 16
            1.2.2.1. Carbohydrates 16
            1.2.2.2. Amino Acids and Peptides 17
            1.2.2.3. Fatty Acids 22
            1.2.2.4. Organic Acids 23
            1.2.2.5. Vitamins 24
            1.2.2.6. Minerals 26

1.2.3. Analysis of Food Additives 27
1.2.3.1. Preservatives 27
1.2.3.2. Antioxidants 27
1.2.3.3. Sweeteners 28
1.2.3.4. Emulsifiers 28
1.2.3.5. Other Food Additives 29
1.2.4. Analysis of Veterinary Medicine 29
1.2.4.1. Sulfonamides 30
1.2.4.2. Antibiotics 30
1.2.4.3. Antiparasitic Agents 32
1.2.5. Analysis of Pesticides 34
1.2.6. Analysis of Natural Toxins 35
1.2.6.1. Mycotoxins 35
1.2.6.2. Marine Toxins 39
1.2.6.3. Plant Toxin 42
1.2.7. Analysis of Food Contaminants 42
1.2.7.1. Amines 42
1.2.7.2. Organotin 43
1.2.7.3. Hydroperoxides 45
1.2.7.4. Isocyanate Monomers 45
References 45

1.3. Derivatization for Environmental Contaminants
*Kazuo Iwaki* 51
1.3.1. Introduction 51
1.3.2. Environmental Samples 51
1.3.2.1. Gas Sample 51
1.3.2.2. Soil and Various Water Samples 53
1.3.3. Application 55
1.3.3.1. Aldehydes and Ketones in Air 55
1.3.3.2. Aliphatic Amines in Air 57
1.3.3.3. Residual Pesticides and Herbicides in Soil and Water 59
References 62

2. Reagent for UV-VIS Detection
*Kiyoshi Zaitsu, Masaaki Kai and Kenji Hamase* 63
2.1. Introduction 64
2.2. Label of Amines and Amino Acids ($-NH_2$, NH) 65
2.2.1. Nitrobenzenes 65
2.2.1.1. 1-Fluoro-2,4-dinitrobenzene (FDNB, 2,4-dinitro-1-fluorobenzene, Sanger Reagent) 65
2.2.1.2. 4-Fluoro-3-nitrotrifluoromethylbenzene (4-fluoro-3-nitrobenzotrifluoride) (FNBT) 66
2.2.1.3. 2,4,6-Trinitrobenzene-1-sulfonic Acid (TNBS) 67
2.2.1.4. Na-(2,4-Dinitro-5-fluorophenyl)-L-alanine amide (FDAA, 1-Fluoro-2,4-dinitrophenyl-5-L-alanine amide, Marfey's Reagent) 67

2.2.2. Acetic Anhydride 67
2.2.3. Isocyanates and Isothiocyanates 67
2.2.3.1. Isocyanates 67
2.2.3.2. Phenyl Isothiocyanate (PITC, Edman Reagent) 68
2.2.3.3. Butylisothiocyanate (BITC) 70
2.2.3.4. 4-*N*,*N*-Dimethylaminoazobenzene-4′-isothiocyanate (DABITC) 70
2.2.4. 1,2-Naphthoquinone-4-sulfonate (NQS) 70
2.2.5. Acyl Chlorides 71
2.2.5.1. Benzoyl Chloride 71
2.2.5.2. Dansyl Chloride (DNS-Cl) 72
2.2.5.3. 4-*N*,*N*-Dimethylaminoazobenzene-4′-sulfonyl Chloride (DABS-Cl, Dabsyl-Cl) 73
2.2.6. 9-Fluorenylmethyl Chloroformate (9-Fluorenylmethoxycarbonyl Chloride, FMOC-CL, Fluorescence and UV Detection) 73
2.2.7. Ninhydrin 74
2.2.8. Diethylethoxymethylenemalonate (DEMM) 75
2.2.9. 6-aminoquinolyl-*N*-hydroxyuccinimidyl Carbamate (AQC) 76
2.2.10. Disuccinimido Carbonate (DSC) 76
2.2.11. Solid-phase Reagent with UV or VIS Light Absorbing Moiety 76
2.2.11.1. Polymeric 3,5-Dinitrobenzoyl Tagged Derivatization Reagent 77
2.2.11.2. Polymeric 6-Aminoquinoline (6-AQ) Tagged Derivatization Reagent 78
2.2.11.3. Polymeric Benzotriazole Activated Reagent Containing FMOC Group 78
2.3. Label of Carboxyl (−COOH) 78
2.3.1. Alkyl Halides 78
2.3.2. Aromatic Amines 79
2.3.3. Hydrazines 81
2.3.4. Hydroxylamine 82
2.4. Label of Hydroxyl (−OH) 84
2.4.1. Acyl Halides 84
2.4.2. Acid Anhydrides 86
2.4.3. Isocyanates 87
2.4.4. Other Reagents 88
2.5. Label of Reducing Carbohydrate 89
2.5.1. Reductive Amination 89
2.5.2. 1-Phenyl-3-methyl-5-pyrazolone 91
2.5.3. Post-column Derivatizing Reagent 92
2.6. Label of Thiol (−SH) 92
2.6.1. 2-Halopyridinium Salt 92
2.6.2. Disulfide Reagent 93
2.6.3. Other Reagents 94
2.7. Labelling of Other Compounds 94
2.7.1. 1-(2-Pyridyl)-piperazine (PYP) 94
2.7.2. Diethyldithiocarbamate (DDTC) 94
2.7.3. 9-Methylamino-methylanthracene (MAMA) 95
References 95

3. Reagent for FL Detection
*Masatoshi Yamaguchi, and Junichi Ishida* 99
Abbreviations 100
3.1. Introduction 102
3.2. Reagents for Amines and Amino Acids 103
3.2.1. General Amino Compounds 103
3.2.1.1. Primary Amines and Amino Acids 103
3.2.1.2. Primary and Secondary Amines and Amino Acids 106
3.2.2. Particular Amines and Amino Acids 116
3.2.2.1. Catecholamines 116
3.2.2.2. Tryptophan and Indolamines 117
3.2.2.3. 5-hydroxyindoleamines (Serotonin Related Compounds) 117
3.2.2.4. Guanidino Compounds 119
3.2.3. Peptides 121
3.2.3.1. General Peptides 121
3.2.3.2. Arginine-containing Peptides 121
3.2.4. Fluorescence Derivatization for CE 121
3.3. Reagents for Organic Acids 121
3.3.1. Carboxylic Acids 121
3.3.1.1. Fatty Acids 129
3.3.1.2. Prostaglandins 130
3.3.1.3. Steroids 131
3.3.1.4. Glucuronic Acid Conjugates 132
3.3.2. $\alpha$-Keto Acids 133
3.3.3. Sialic Acids and Dehydroascorbic Acid 133
3.4. Reagents for Alcohols 135
3.5. Reagents for Phenols 139
3.6. Reagents for Thiols 141
3.7. Reagents for Aldehydes and Ketones 146
3.8. Reagents for Carbohydrates 150
3.9. Reagents for Dienes 152
3.10. Reagents for Nucleic Acid Related Compounds 153
References 158

4. Reagent for CL Detection
*Naotaka Kuroda and Kenichiro Nakashima* 167
4.1. Introduction 167
4.2. Label of Amines ($-NH_2$, $-NH$) 171
4.3. Label of Carboxyl ($-COOH$) 176
4.4. Label of Hydroxyl ($-OH$) and Thiol ($-SH$) 178
4.5. Label of Other Functional Groups 179
4.6. Application 183
References 187

5. Reagents for Electrochemical Detection
*Kenji Shimada, Tomokazu Matsue and Kazutake Shimada* 191
5.1. Introduction 192
5.1.1. Amperometric Analysis in Flowing Streams 192

5.1.2. Electrochemical Detectors 192
5.1.3. Electrode Material 193
5.1.4. Electrode Configuration 194
5.1.5. Mobile Phase 195
5.1.6. Chemical Derivatization 196
5.2. Labeling of Primary and Secondary Amines 198
5.2.1. *o*-phthalaldehyde 198
5.2.2. Naphthalene-2,3-dicarbaldehyde 199
5.2.3. Ferrocene 200
5.2.4. Isocyanate and Isothiocyanate 202
5.2.5. Salicylic Acid Chloride 203
5.2.6. 2,4-Dinitrofluorobenzene 204
5.2.7. 1,2-Diphenylethylenediamine 204
5.2.8. Bolton and Hunter Type Reagent 204
5.2.9. Others 205
5.3. Labeling of Carboxy Groups 205
5.4. Labeling of Hydroxy Groups 207
5.5. Labeling of Thiol Groups 208
5.6. Labeling of Other Functional Groups 208
5.7. Applications 211
References 213

6. Derivatization for Resolution of Chiral Compounds
*Toshimasa Toyo'oka* 217
6.1. Introduction 218
6.1.1. Fundamentals of Stereochemistry 218
6.1.2. Chirality and Biological Activity 220
6.1.3. Chirality Application as Single Isomers 222
6.2. Resolution of Racemates 223
6.2.1. Direct Resolution 223
6.2.2. Indirect Resolution 223
6.3. Reactions of Various Functional Groups 225
6.3.1. Amines 225
6.3.2. Carboxyls 226
6.3.3. Hydroxyls 227
6.3.4. Thiols and Others 227
6.4. Derivatization for GC Analysis 228
6.4.1. Label for Alcohols 228
6.4.2. Label for Carboxylic Acids 230
6.4.3. Label for Aldehydes and Ketones 231
6.4.4. Label for Other Functional Groups 233
6.5. Derivatization for LC Analysis 233
6.5.1. Label for UV-VIS Detection 235
6.5.1.1. Label for Primary and Secondary Amines 235
6.5.1.2. Label for Carboxylic Acids 240
6.5.1.3. Label for Alcohols 241
6.5.1.4. Label for Other Functional Groups 241

6.5.2. Label for Fluorescence (FL), Laser-Induced Fluorescence (LIF) and Chemiluminescence (CL) Detection 243
6.5.2.1. Label for Primary and Secondary Amines Including Amino Acids 249
6.5.2.2. Label for Carboxylic Acids 257
6.5.2.3. Label for Alcohols 262
6.5.2.4. Label for Aldehydes and Ketones 266
6.5.2.5. Label for Thiols 266
6.5.2.6. Label for Other Functional Groups 267
6.5.3. Derivatization for Electrochemical (EC) Detection 267
6.6. Derivatization for Capillary Electrophoresis (CE) Analysis 267
6.7. Conclusion and Further Perspective 284
References 284

Index 291

# Contributors

Kenji Hamase, Ph.D.,
Department of Analytical Chemistry,
Faculty of Pharmaceutical Sciences,
Kyushu University,
3-1-1 Maidashi, Higashi-ku,
Fukuoka 812-8582, Japan,
Phone: +81-92 642-6598,
FAX: +81-92 642-6601,
E-mail: hamase@analysis.phar. kyushu-u.ac.jp

Junichi Ishida, Ph.D.,
Department of Analytical Chemistry,
Faculty of Pharmaceutical Sciences,
Fukuoka University, Nakakuma,
Johnan-ku, Fukuoka 814-0180, Japan,
Phone: +81-92-871-6631

Kazuo Iwaki, Ph.D.,
Deputy General Manager,
Chemical Analysis Department,
Ebara Research Co. Ltd.,
2-1 Honfujisawa 4-chome,
Fujisawa-shi Kanagawa 251-8502, Japan,
Phone: +81-466-83-8276,
Fax: +81-466-83-3160,
E-mail: IwakiL0034@erc.ebara.co.jp

Prof. Masaaki Kai, Ph.D.,
Chemistry of Functional Molecules,
School of Pharmaceutical Sciences,
Nagasaki University,
1-14 Bunkyo-Machi,
Nagasaki 852-8521, Japan,
Phone: +81-95 847-1111,
Fax: +81-95 843-1742,
E-mail: ms-kai@net.nagasaki-u.ac.jp

Yukinori Kawahara, Ph.D.,
Director of Analytical and Metabolic Research Laboratories,
Sankyo Co. Ltd., Hiromachi 1-2-58,
Shinagawa-ku, Tokyo 140, Japan,
Phone: +81-3-3492-3131

Naotaka Kuroda, Ph.D.,
Department of Hygienic Chemistry,
School of Pharmaceutical Sciences,
Nagasaki University,
1-14 Bunkyo-Machi, Nagasaki 852, Japan,
Phone: +81-958-47-1111,
Fax: +81-958-48-4219,
E-mail: anal-kuro@cc.nagasaki-u.ac.jp

Prof. Tomokazu Matsue, Ph.D.,
Graduate School of Engineering,
Tohoku University,
Aramaki, Aoba, Sendai 980-77, Japan,
Phone: +81-22-217-7209,
Fax: +81-22-217-7293,
E-mail: matsue@est.che.tohoku.ac.jp

Akihiko Nakagawa, Ph.D.,
Analytical and Metabolic Research Laboratories,
Sankyo Co. Ltd., Hiromachi 1-2-58,
Shinagawa-ku, Tokyo 140, Japan,
Phone: +81-3-3492-3131,
E-mail: akihik@shina.sankyo.co.jp

Prof. Kenichiro Nakashima, Ph.D.,
Department of Hygienic Chemistry,
School of Pharmaceutical Sciences,
Nagasaki University, 1-14 Bunkyo-machi,
Nagasaki 852, Japan,
Phone: +81-958-47-1111 (Ex. 2526),
Fax: +81-958-42-3549,
E-mail: naka-ken@net.nagasaki-u.ac.jp

Prof. Hiroyuki Nakazawa, Ph.D.,
Department of Analytical Chemistry,
Hoshi University, 2-4-41 Ebara,
Shinagawa-ku, Tokyo 142, Japan,
Phone: +81-3-5498-5763,
Fax: +81-3-5498-5765,
E-mail: nakazawa@hoshi.ac.jp

Koichi Saito, Ph.D.,
Senior researcher,
Department of Food Chemistry,
Saitama Institute of Public Health,
639-1, Kamiokubo, Urawa, Saitama 338, Japan,
Phone: +81-48-853-6121,
Fax: +81-48-840-1041,
E-mail: saitok@interlink.or.jp

Prof. Kazutake Shimada, Ph.D.,
Department of Analytical Chemistry,
Faculty of Pharmaceutical Sciences,
Kanazawa University,
Takaramachi 13-1, Kanazawa 920-0934, Japan,
Phone: +81-76-234-4459,
Fax: +81-76-234-4459,
E-mail: shimada@dbs.p.kanazawa-u.ac.jp

Prof. Kenji Shimada, Ph.D.,
Department of Analytical Chemistry,
Niigata College of Pharmacy,
5-13-2 Kamishinei-cho, Niigata 950-2081, Japan
Phone: +81-25-268-1172,
Fax: +81-25-268-1177,
E-mail: shimada@niigata-pharm.ac.jp

Prof. Toshimasa Toyo'oka, Ph.D.,
Department of Analytical Chemistry,
School of Pharmaceutical Sciences,
University of Shizuoka,
52-1 Yada, Shizuoka 422-8526, Japan,
Phone: +81-54-264-5656,
Fax: +81-54-264-5593,
E-mail: toyooka@ys2.u-shizuoka-ken.ac.jp

Prof. Masatoshi Yamaguchi, Ph.D.,
Department of Analytical Chemistry,
Faculty of Pharmaceutical Sciences,
Fukuoka University, Nakakuma,
Johnan-ku, Fukuoka 814-0180, Japan,
Phone: +81-92-871-6631 (Ex. 6618),
E-mail: pp034545@psat.fukuoka-u.ac.jp

Prof. Kiyoshi Zaitsu, Ph.D.,
Department of Analytical Chemistry,
Faculty of Pharmaceutical Sciences,
Kyushu University, 3-1-1 Maidashi,
Higashi-ku, Fukuoka 812-8582, Japan,
Phone: +81-92 642-6596,
FAX: +81-92 642-6601,
E-mail: zaitsu@analysis. phar. kyushu-u.ac.jp

# Preface

Biologically active compounds are ubiquitous in our lives. They are used for medicines, agrochemicals (e.g. pesticides and harbicides), food additives, biogenic amines and flavors. It is fairly difficult to determine them with accuracy and precision, because they usually exist in munite amounts, especially in real biological and environmental samples. The choice of a suitable method that provides good reproducibility is essential to obtain correct results. Separation analysis represented by various chromatography is recommended for the quantitation of analytes in complex matrices.

Derivatization was an important technique for analysis using gas chromatography in the early stages. The main purpose of the derivatization was to add volatility to saccarides and amino acids. In this derivatization, selectivity and sensitivity were not considered. However, derivatization is the essential technique in separation sciences using thin-layer chromatography (TLC), liquid chromatography (LC) and capillary electrophoresis (CE), as well as gas chromatography (GC). For analysis by high-performance liquid chromatography (HPLC), various reagents have been developed to increase separability, selectivity and sensitivity. This is due to the development of various types of detection instruments such as ultraviolet-visible (UV-VIS), fluorescence (FL), chemiluminescence (CL) and electrochemical (EC). The use of derivatization in separation sciences is mainly to improve the chromatographic properties and detection sensitivity.

The major aim of this book is to provide an easy-to-read overview of various derivatization methods that are available for minute analyses of biological importance. Emphasis is placed on practical use, and the characteristics (merits and demerits) of the various approachs are critically discussed. The derivatization listed in this book is a reaction which produces covalent binding between the analyte and the reagent. This book describes recent advances in chemical derivatization for the separation sciences mainly by GC, LC and CE.

The first chapter presents a general introduction of the pre-treatment of real samples such as biological, food and environmental. The pre-treatment is the clean-up method for derivatization to obtain a trace amount of analyte without contamination. This part is most important because the accuracy and precision of the result obtained is dependent on the pre-treatment method, especially in trace analysis. In Chapters 2 and 3, homogeneous reactions suitable for the derivatization of various functional groups of trace analytes with UV-VIS and FL labels are described in detail. Pre- and post-column applications and typical derivatization procedures are given for each functional group. Chapter 4 deals with the chemiluminescence (CL) detection of fluorophores derived from the fluorogenic reaction and fluorescence labelling reaction. This is one of the most sensitive detection methods, capable of detecting fmol-amol levels. However, this detection system is not applicable to all fluorescent materials, a notable disadvantage. In Chapter 5, the theory of electrochemical reactions and derivatization for elctrochemical detection are described, together with some examples. Detection is based on a redox reaction of the electrodes and is suitable for compounds easily oxidized and reduced with low potential. Finally, chiral resolution continues with the derivatization for effective separation and high-sensitivity detection.

There are some excellent books on derivatization, but they do not address the specifics of the topic. Each chapter includes sufficient references to the literature to serve as a valuable starting point for more detailed investigations.

As shown in the bibliography, scientists in Japan are very active in these fields. The contributors and the authors selected for this book are outstanding research chemists. This book should be useful to many investigators in various fields, including clinical, pharmaceutical, biological and environmental.

The editor would like to express sincere thanks to the authors for their contributions, as well as to their colleagues for providing stimulating discussions. Thanks are also due to the entire publishing staff at John Wiley & Sons for their continued support and contribution towards the completion of this book.

Toshimasa Toyo'oka
Shizuoka, Japan

# 1

# Pre-treatment for Real Sample Analysis and Choice of Suitable Reagent

## 1.1 DERIVATIZATION FOR DRUGS

**Akihiko Nakagawa, and Yukinori Kawahara**
Analytical and Metabolic Research Laboratories, Sankyo Co. Ltd., Tokyo, Japan

1.1.1. Introduction 2
1.1.2. Application of Derivatization of Drugs 2
1.1.2.1. Determination of Pravastatin Sodium in Plasma by HPLC with Laser-induced Fluorescence Detection after Immobilized Antibody Extraction 2
1.1.2.2. Enantiospecific Determination of Ibuprofen in Rat Plasma Using Chiral Fluorescence Derivatization Reagent, (−)-2-[4-(1-aminoethyl)phenyl]-6-methoxybenzoxazole 4
1.1.2.3. Specific Fluorogenic Derivatization of Ivermectin Using Non-fluorescent Reagents 6
1.1.2.4. Other Methods 8
1.1.3 Conclusion 10
References 11

Edited by Toshimasa Toyo'oka: *Modern Derivatization Methods for Separation Sciences* 

## 1.1.1. Introduction

Drug analysis is mainly divided into three fields; materials, formulations and bioanalysis of specimens obtained *in vitro* and *in vivo*. The former two are controlled by regulations such as UPS in the USA, and the physico-chemical properties of drugs are summarized in a series of books entitled 'Analytical Profiles of Drug Substances', published periodically by Academic Press Inc. from 1972. In this section, methods for derivatization for drugs, in order to monitor their levels in biological samples, are described.

Drug level monitoring in biological fluids such as blood and urine, and in tissues is essential to elucidate its disposition in the body regarding pharmacological and toxicological properties. From the early 70's, chromatographic methods, especially using HPLC, have played important roles in trace level drug monitoring. In this field, sensitivity and selectivity of the target drug in a complex matrix, are the most important parameters.

UV detection methods in HPLC, which is the most widely used one, sometimes lack sensitivity or selectivity for trace level drug analysis. Chemical derivatization can modify drugs to give efficient absorption in UV or visible wavelength and luminescent properties such as fluorescence and chemi- or bio-luminescence, or electrochemical activity can attain highly sensitive and selective determination of drugs using HPLC.

A vast number of drugs exhibit the property of chirality and some of them are used therapeutically as the racemates. When racemic drugs are administered, individual enantiomers often have different activities, toxicities, and pharmacokinetic properties. For enantioseparation, HPLC is one of the most powerful techniques. There are two principles in chiral separation using HPLC; direct separation using a chiral stationary phase column, and diastereomer formation with a suitable chiral reagent. The diastereomeric derivatization method has the advantage for trace analysis of enantiomers in biological matrix because of the utilization of the reagent with high sensitivity in detection.

Papers published recently on the derivatization of drugs using HPLC are mostly classified into the two categories mentioned above. In the following section, our recent publications on the above categories are mainly described.

## 1.1.2. Application of Derivatization of Drugs

### 1.1.2.1. Determination of Pravastatin Sodium in Plasma by HPLC with Laser-induced Fluorescence Detection after Immobilized Antibody Extraction

Pravastatin sodium, Fig. 1.1.1, is an inhibitor of 3-hydroxy-3-methylglutaryl-coenzyme A (HMG-CoA) reductase, a key enzyme in cholesterol biosynthesis [1]. Since plasma concentration in rats and humans is quite low due to specific uptake [2], the development of a highly sensitive and specific assay in biological matrices has been required to study of the pharmacokinetics of

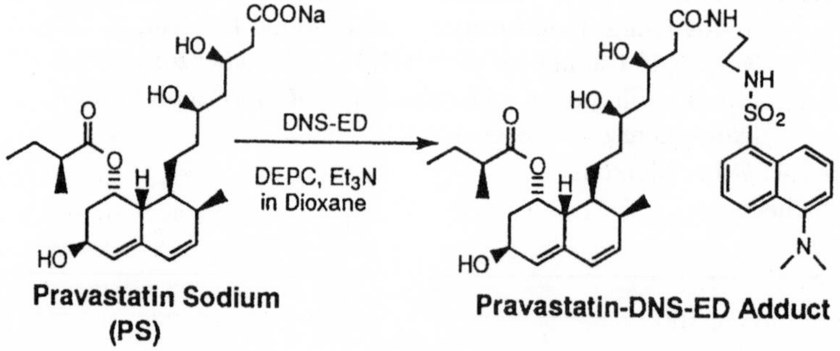

COONa
HO
HO
O
O
H
HO
Pravastatin Sodium
(PS)
DNS-ED
DEPC, $Et_3N$
in Dioxane
CO-NH
NH
$SO_2$
N
Pravastatin-DNS-ED Adduct

**Fig. 1.1.1.** Chemical structures of pravastatin sodium and its *N*-dansyl-ethylenediamine derivative. [Reproduced from ref. 3, p. 1631, Chart 1.].

pravastatin sodium. HPLC with UV detection was not sensitive enough to measure reliably the low ng/mL level of quantification needed for biological samples.

Dumousseaux *et al.* [3] developed a method for the determination of pravastatin in plasma by using an immobilized antibody column extraction followed by HPLC with a laser-induced fluorescence detector after fluorogenic derivatization. For the extraction of pravastatin in plasma samples, in which 100 μg/mL levels of organic acids and/or fatty acids exist, a simple yet specific immobilized-antibody-mediated clean-up was performed. The immobilized-antibody-mediated extraction method was first introduced by Glencross *et al.* [4] for the determination of 17$\beta$-estradiol. A plasma sample was applied to the column and washed with water, and the drug was eluted with methanol.

After evaporation of methanol, pravastatin was derivatized with *N*-dansyl-ethylenediamine (DNS-ED) at the carboxyl end in the presence of diethyl phosphorocyanidate (DEPC) and triethylamine (TEA) in dioxane (Fig. 1.1.1). The optimal DNS-ED derivatization condition was extensively examined. The best solvent for the derivatization was found to be dioxane, amongst DMF, tetrahydrofurane, ethyl acetate, and dioxane. Pravastatin sodium was first dissolved in a small amount of DMF and then diluted with dioxane (2 to 4-fold, v/v) because of its solubility. Fig. 1.1.2 shows the effect of DNS-ED concentration on the derivatization. The peak area of the pravastatin-DNS-ED adduct was increased with an increase of DNS-ED and became constant in the range of DNS-ED concentrations of more than 100 μg/mL. The effect of all the reagent concentrations on the derivatization and HPLC determination was studied, as shown in Fig. 1.1.3. The reagents were prepared as dioxane solutions. The highest peak area was obtained when pravastatin was derivatized under molar concentrations of DNS-ED:DEPC:TEA = 0.001:1:1. A concentration of 100 μg/mL for DNS-ED was chosen as the best compromise between the reaction yield and the chromatographic noise due to a high concentration of the reagent.

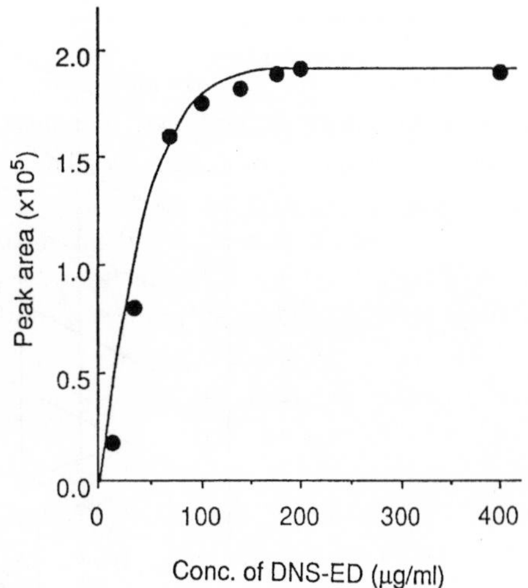


**Fig. 1.1.2.** Effect of *N*-dansyl-ethylenediamine concentration on the derivatization. [Reproduced from ref. 3, p. 1633, Fig. 2.].

From these results, the concentration of reagents was fixed as follows: DNS-ED, 100 μg/ml (= 0.34 mM); DEPC, 51.8 μl/ml (0.34 M); and TEA, 47.6 μl/ml (0.34 M). For this condition, the derivatization process was completed within 5 min at room temperature to yield a maximum and a constant peak area.

The DNS-ED derivatization method was designed to make good use of a highly sensitive He-Cd laser-induced fluorescence detector. The detection limit was 2 pg/injection of pravastatin with a He-Cd laser-induced fluorescence detector, which was 20 times more sensitive than the conventional fluorescence detection (Fig. 1.1.4).

For HPLC separation of DNS-ED-derivatized pravastatin, a column-switching technique was used to remove excess reagents and by-products. The HPLC system was described elsewhere [5]. In this case, a combination of two reversed-phase columns of different lipophilicity were employed: A C4 column was used as a preseparation column (first column) to delete the major peaks derived from the reagents and the plasma components, and a fraction containing derivatized pravastatin was introduced into the analytical C18 column (second column). A comparison between the chromatograms obtained with the ODS column

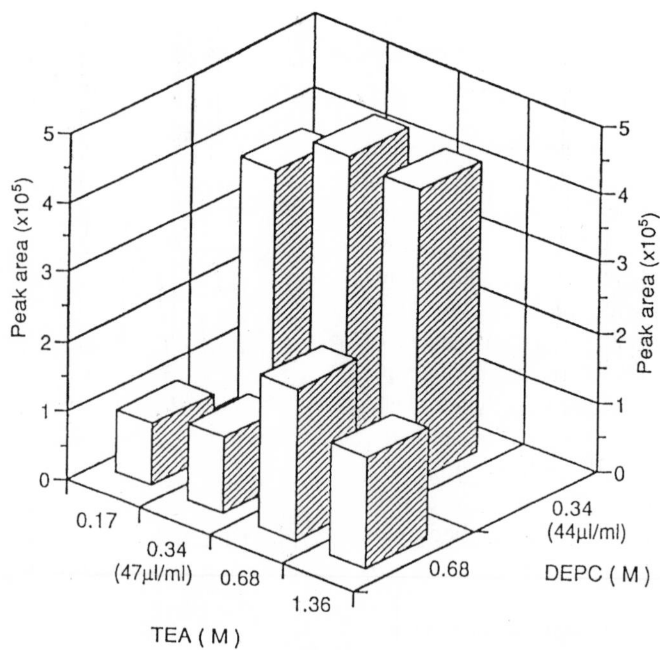


**Fig. 1.1.3.** Influence of diethyl phosphorocynidate (DEPC) and triethylamine (TEA) concentrations on the derivatization. [Reproduced from ref. 3, p. 1633, Fig. 3.].

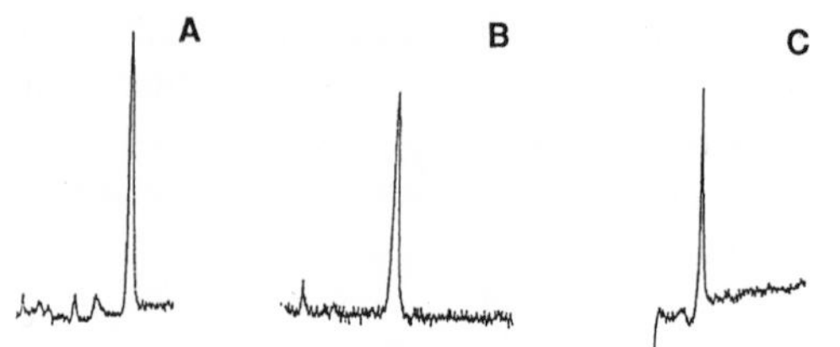


**Fig. 1.1.4.** Comparison of the detection limit of *N*-dansyl-ethylenediamine derivative of pravastatin sodium by three different detectors in HPLC. A; UV (239 nm) detection with 10 ng/injection, B; conventional fluorescence detection (Ex. 350 nm and Em. 530 nm) with 50 pg/injection and C; laser-induced fluorescence detection with 15 pg/injection. [Reproduced from ref. 3, p. 1634, Fig. 5.].

alone and with the two columns by the column-switching technique is shown in Fig. 1.1.5. The limit of quantitation of the method was 100 pg/ml when 1 ml of plasma sample was available. An average coefficient of variations of the overall method were less than 8% at the concentration range of 1–100 ng/ml.

### 1.1.2.2. Enantiospecific Determination of Ibuprofen in Rat Plasma Using Chiral Fluorescence Derivatization Reagent, (−)-2-[4-(1-aminoethyl)phenyl]-6-methoxybenzoxazole

Ibuprofen, a well known non-steroidal anti-inflammatory agent (NSAID), is a derivative of 2-phenylpropionate. This type of NSAID has a chiral center in its 2-phenylpropionate moiety and is administered clinically as a racemic mixture except for *S*-(+)-naproxane. Because these compounds undergo metabolic chiral inversion from the inactive *R*-enantiomers to their pharmacologically active *S*-enantiomers [6], it is essential to know their enantiospecific disposition.

For this purpose, many optically active amines have been applied as chiral derivatization reagents because NSAIDs have a carboxyl functional group, which is easily transformed into the amide group, on a chiral carbon center, i.e., L-leucinamide [7], 1-phenylethylamine [8], 1-(1-naphthyl)ethylamine [9], 1-(4-dimethylaminonaphthalen-1-yl)ethylamine [10].

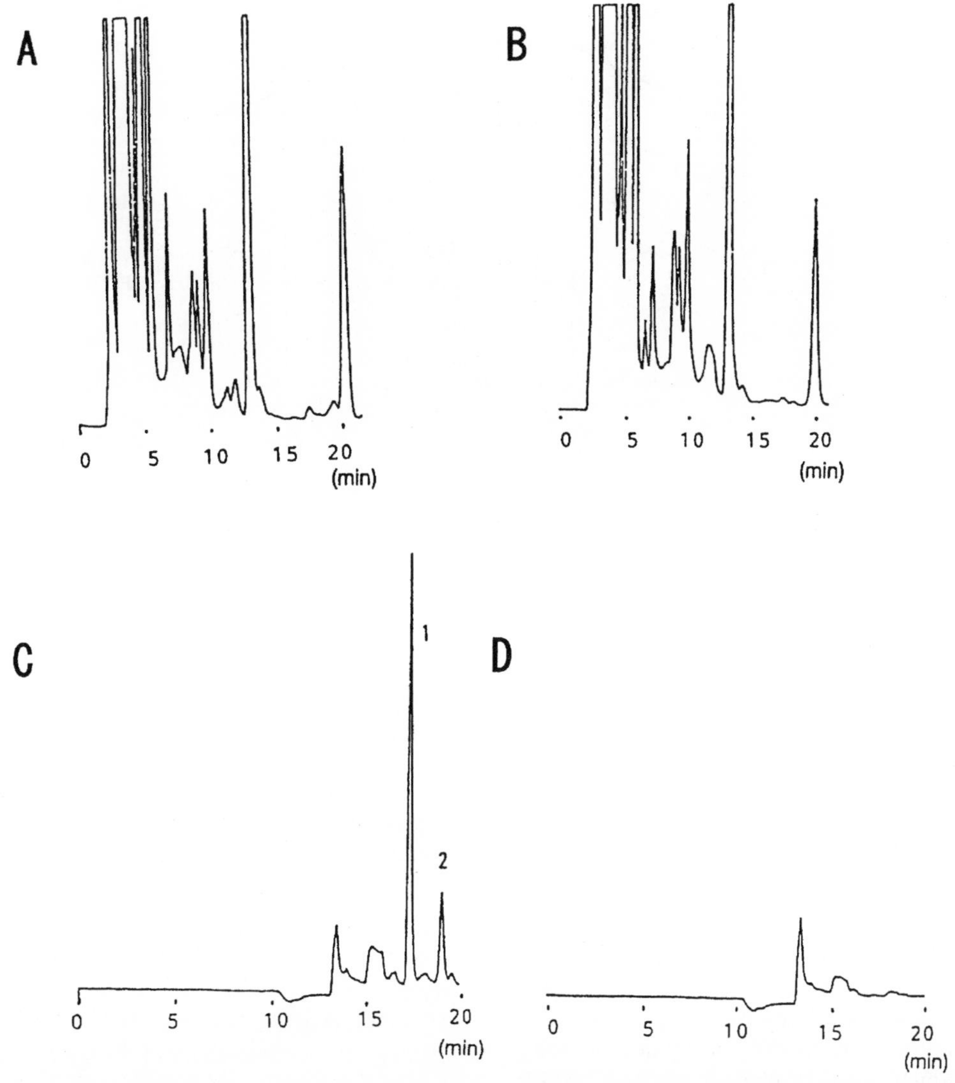


**Fig. 1.1.5.** Typical chromatograms of reagents, plasma blank, and samples spiked with pravastatin sodium generated by a C18 column alone (A, B) and by the column-switching technique (C, D). A; standard pravastatin sodium derivatized with *N*-dansyl-ethylenediamine (1 μg/ml), B; reagent blank for A, C; extracted and derivatized plasma sample spiked with pravastatin sodium (50 ng) and D; blank for C. 1; pravastatin sodium and 2; internal standard. [Reproduced from ref. 3, p. 1634, Fig. 4.].

Kondo *et al.* [10,11] have developed (−)-2-[4(1-aminoethyl)phenyl]-6-methoxybenzoxazole((−)-APMB), 2-[4-(L-leucyl)amino-phenyl]-6-methoxybenzoxazole (L-LeuBOX), 2-[4-(D-phenylglycyl)amino-phenyl]-6-methoxybenzoxazole (D-PgBOX), 2-[4-(L-phenylalanyl)amino-phenyl]-6-methoxybenzoxazole(L-PheBOX) as highly sensitive chiral fluorescence derivatization reagents for the resolution of carboxylic acid enantiomers. Using (−)-APMB (Fig. 1.1.6) as a chiral derivatization reagent, the diastereomeric amides formed were separated on both a normal and a reversed-phase column. Using L-LeuBOX, D-PgBOX, and L-PheBOX, the diastereomeric amides were

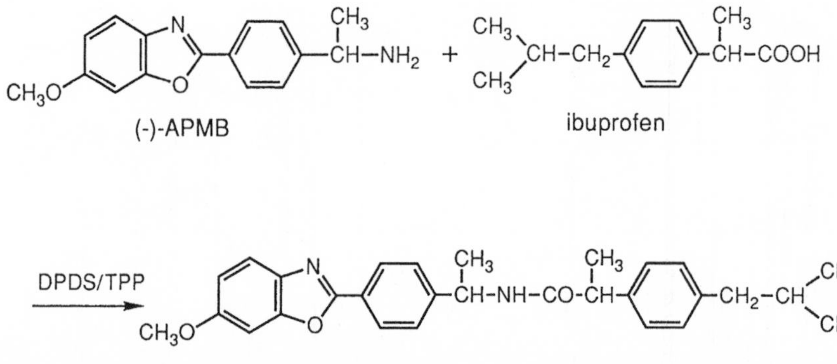

**Fig. 1.1.6.** Derivatization reaction of ibuprofen with the chiral derivatization reagent, (−)-2-[4(1-aminoethyl)phenyl]-6-methoxybenzoxazole ((−)-APMB). DPDS; 2,2′-dipyridyl disulphide and TPP; triphenylphosphine. [Reproduced from ref. 13, p. 172, Fig. 1.].

separated on a normal-phase column, but no satisfactory separation of the diastereomeric amides was obtained on a reversed-phase column. In this section, a method for enantiospecific determination of ibuprofen in rat plasma by reversed-phase HPLC with (−)-APMB is described [13].

An aliquot of rat plasma sample was spiked with internal standard, dichlofenac sodium, and then acidified with 1N HCl. Samples were applied to solid phase extraction column, Chem Elut, and ibuprofen and the internal standard were eluted with *n*-hexane-diethyl ether-isopropyl alcohol (50:50:1). The eluent was evaporated to dryness under nitrogen gas, and the resultant residue was added with derivatization reagent; 200 μL of a dichloromethane solution of (−)-APMB (2 μmole/mL), 100 μL each of dichloromethane solution of 2,2′-dipyridyl disulphide (DPDS, 20 μmole/mL) and triphenylphosphine (TPP, 20 μmole/mL). The reaction mixture was allowed to stand for 5 min at room temperature, and was evaporated to dryness under nitrogen gas. The residue was dissolved in the mobile phase and an aliquot of the solution was injected into HPLC.

The analytical column was a 5 μm TSK gel ODS-80 (150 mm × 4.6 mm i.d., Tosoh) and the mobile phase was acetonitrile: water: acetic acid (700:300:1). The fluorescence detector was operated at wavelengths of 320 nm and 380 nm for excitation and emission, respectively. Although many chiral derivatization reagents for the amide formation reaction of 2-arylpropionic acid require a long derivatization time, the formation of the fluorescence amide with (−)-APMB in the presence of DPDS and TPP in dichloromethane was completed almost quantitatively by evaporation with a stream of nitrogen within 5 min. No racemization of ibuprofen occurred during the derivatization reaction, even when the reaction time was increased up to 60 min. The effect of all the reagent concentrations on the derivatization was carefully examined beforehand and fixed as mentioned above.

The *S*- and *R*-diastereoisomeric amides of ibuprofen and the amide of dichlofenac were well resolved using reversed-phase liquid chromatography with retention times of 11.0, 12.2 and 14.0 min, respectively, as shown in Fig. 1.1.7. The excess of the derivatization reagent and degradation products of the reagent were eluted with the solvent front. Good linear correlations were obtained between the peak-area ratios of each isomer to the internal standard with a concentration range of 0.2–15 μg/ml in plasma when 100 μl plasma samples were used.

### 1.1.2.3. Specific Fluorogenic Derivatization of Ivermectin Using Non-fluorescent Reagents

Ivermectin is a parasite control agent for sheep, cattle and swine. Tolan *et al.* [14] first reported the determination method of ivermectin in an animal specimen using high-performance liquid

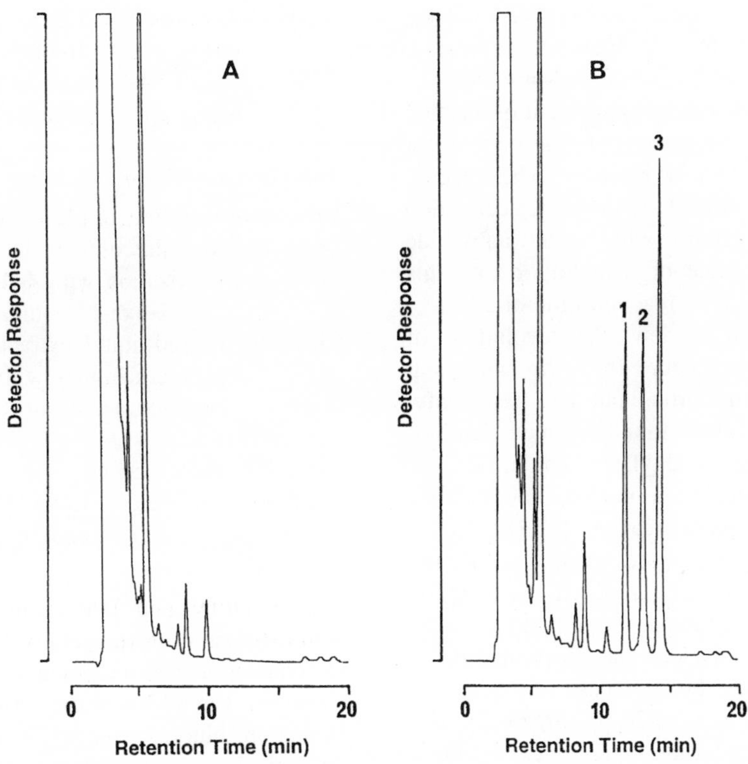


**Fig. 1.1.7.** Typical chromatograms obtained from A; blank rat plasma and B; spiked plasma containing 10 µg/ml racemic ibuprofen. Peaks are (−)APMB amides of 1; *S*-ibuprofen, 2; *R*-ibuprofen and 3; internal standard. [Reproduced from ref. 13, p. 173, Fig. 2.].

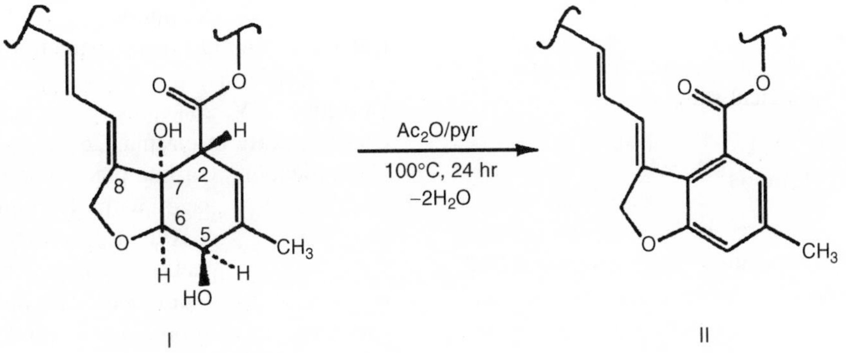

$Ac_2O$/pyr

100°C, 24 hr

$-2H_2O$

I → II

**Fig. 1.1.8.** Derivatization reaction of ivermectin. [Reproduced from ref. 16, p. 74, Scheme.].

chromatography with fluorescence detection. The method involves dehydrative aromatization using non-fluorescent reagents.

Ivermectin has two hydroxyl groups on the dihydroxycyclohexene ring (Fig. 1.1.8). The hydroxyl groups are sterically positioned to allow for the easy elimination of water to produce a delocalized electron system: i.e., heating with acetic anhydride in pyridine causes the elimination of two moles of water, producing a fluorophore consisting of an aromatic ring [15] conjugated with a diene system (Fig. 1.1.8; excitation 375 nm; emission 475 nm). This reaction does not require a separation of excess fluorescent reagent or reaction by-products from the analytical fluorophore for application to biological samples. The method reported by Tolan *et al.* [14] involves acetylation of the hydroxyl groups with acetic anhydride in pyridine in advance of dehydration and this derivatization required a reaction time of 24 h.

Fink *et al.* [16] review the evolution of the derivatization procedure from the standpoint of improvements in throughput and sensitivity. Connors and co-workers found 4-dimethylamino-pyridine and some *N*-alkylated imidazoles are superior to pyridine as a nucleophilic catalyst for aetylation by acetic reagent [17,18]. When 1-methylimidazole was used as an acetylation catalyst, the derivatization reduced the reaction time to 1 h [19]. Trifluoroacetic anhydride as the acetylation reagent reduced the reaction time to less than 30 seconds. For this case, the detection limit of ivermectin was ca. 20 pg (S/N = 2) [20]. The use of laser-induced fluorescence detection further reduced the detection limit [21].

In this case, knowledge based on organic chemistry contributes to establish a sophisticated derivatization method without using fluorescent reagents. The reaction does not require a separation of excess fluorescence reagent or reaction by-products from the analytical fluorophore for application to biological samples.

### 1.1.2.4. Other Methods

In this section, some recent publications (1996 or later) on the derivatization of drugs using HPLC are summarized.

*Antiviral Agents*

**Cidofovir** and cytosine-containing compounds in plasma [22].
Derivatization: Precolumn derivatization with phenacyl bromide to form 2-phenyl-3,*N*4-ethenocytosine derivative
Detection: Fluorescence
Analytical parameters: LOQ, 5 ng/ml (ca. 100 fmol on column); between-day and within-day precision, <17%.

*Antitubercular Agents*

**Ethambutol** in human plasma and urine [23].
Derivatization: Precolumn fluorescence derivatization with 4-fluoro-7-nitrobenzo-2-oxa-1, [3-diazole]
Clean-up: Liquid-liquid extraction, from basified plasma samples with diethyl ether and back-extracted into 0.01 M phosphoric acid
Detection: Fluorescence
Analytical parameters: LOQ, 10 ng/ml in plasma and 10 μg/ml in urine.

*Antineoplastic Agents*

**5-Fluorouracil** in hair [24].
Derivatization: Fluorescence
Clean-up: Liquid-liquid extraction
Detection: Fluorescence
Analytical parameters: Detection limit of 5-FU in hair, 0.01 ppm.
**Cisplatin** in human blood [25].
Derivatization: Precolumn derivatization with *bis* (salicylaldehyde)tetramethylethylenediimine
Clean-up: Liquid-liquid extraction with chloroform
Detection: UV, 254 nm.
**Methotrexate** in plasma [26].
Derivatization: On-line precolumn oxidative cleavage with Cerium (IV) trihydroxyhydroperoxide (CTH) as a packed oxidant to form 2,4-diaminopteridine derivatives
Clean-up: Protein precipitation and solvent purification
Detection: Fluorescence detection, Ex. 367 nm and Em. 463 nm
Analytical parameters: Coefficients of variance, <4%; mean relative errors, 1.11%–7.83%; recovery, 93.74%–98.11%.
**Cremophor** EL in plasma [27].

Derivatization: Saponification of CrEL in alcoholic KOH and derivatization with 1-naphthylamine
Clean-up: Liquid-liquid extraction
Detection: UV 280 nm
Analytical parameters: LOQ, 0.01% (v/v); percentage deviation, <8%; precision, <7%.

**Busulfan** in human plasma [28].
Derivatization: Precolumn derivatization with sodium diethyldithiocarbamate
Clean-up: Liquid-liquid extraction
Detection: UV detection, 251 nm
Analytical parameters: Range, 60–3000 ng/ml; limit of detection, 20 ng/ml (signal-to-noise ratio of 6); coefficients of variance, 4.41–13.5%; mean derivatization and extraction yield, 93.4–107%.

*Anticholesteremic Agents*

**Simvastatin** and its active metabolite in human plasma [29].
Derivatization: Esterification with 1-bromoacetylpyrene in the presence of 18-crown-6
Clean-up: C8 and phenylboronic acid solid-phase extraction and column-switching HPLC
Detection: Fluorescence
Analytical parameters: Range, 0.1–10 ng/ml; C. V. % (Intra-day), <11.0%; accuracies, 91.7–117%.

*Antihypertensive Agents*

**Captopril** in human plasma and urine [30].
Derivatization: The first method, precolumn derivatization of captopril with the fluorescent label monobromobimane (MBB); the second method, postcolumn reaction with the fluorescent reagent *o*-phthaldialdehyde (OPA)
Clean-up: Protein precipitation and reducing disulfide bond with tributylphosphine for total captopril analysis
Detection: Fluorescence
Analytical parameters: In plasma, LOQ, 12.5 ng/ml; for the MBB method (limit of detection 30 pg) and 25 ng/ml; for the OPA method (limit of detection 50 pg); in urine, LOQ, 250 ng/ml (limit of detection 50 pg).

**Propranolol** and 4-hydroxypropranolol in human plasma [31].
Derivatization: Diastereomeric derivatization with 2,3,4,6-tetra-*O*-acetyl-beta-glucopyranosyl isothiocyanate
Clean-up: Liquid-liquid extraction
Detection: Fluorescence
Analytical parameters: Range, 1–100 ng/ml for propranolol; 2–50 ng/ml for 4-hydroxypropranolol enantiomers, using 0.5 ml of human plasma.

*Anti Arrhythmia Agents*

Enantioselective determination of **diprafenone** in human plasma [32].
Derivatization: Precolumn derivatization with homochiral *R*(−)-1-(1-naphthyl)ethyl isocyanate
Clean-up: Liquid-liquid extraction
Detection: UV, 220 nm
Analytical parameters: LOQ, 10 ng/ml for *S*(−)- and *R*(+)-diprafenone in plasma.

*Cardiotonic Agents*

**Digoxin** and its metabolities in human serum [33].
Derivatization: Precolumn fluorescence derivatization with 1-naphthoyl chloride in the presence of 4-dimethylaminopyridine
Clean-up: Cyclodextrin and C1 solid-phase extraction
Detection: Fluorescence

Analytical parameters: Range, 0.25–4.0 ng/ml; recoveries of digoxin and its metabolites from serum ranged from 62–86%; coefficients of variance, 5.8%–20.9%.

*GABA-Agonists*

**Baclofen** in human plasma [34].
Derivatization: Precolumn derivatization with *o*-phthalaldehyde-tert.-butanethiol
Clean-up: SCX solid-phase extraction
Detection: Electrochemical
Analytical parameters: LOQ, 10 ng/ml; within and between-day R.S.D., <10%; inaccuracy, <7%.

*Excitatory Amino Acid Antagonists*

Enantioselective determination of **eliprodil** in human plasma and urine [35].
Derivatization: Precolumn derivatization with (*S*)-(+)-naphthylethyl isocyanate
Clean-up: Liquid-liquid extraction and column-switching HPLC
Detection: Fluorescence, Ex. 275 nm and Em. 336 nm
Analytical parameters: Range, 0.15–10 ng/ml for unchanged eliprodil enantiomers in plasma; 50–25 000 ng/ml for total (conjugated) eliprodil enantiomers in urine; LOQ, 0.15 ng/ml for each unchanged enantiomer in plasma; 50 ng/ml for each unchanged enantiomer in urine.

*Serotonin Uptake Inhibitors*

Simultaneous determination of **D-fenfluramine** (FEN), **D-norfenfluramine** (NF) and **fluoxetine** (FLX) in plasma, brain tissue and brain microdialysate [36].
Derivatization: Precolumn derivatization with dansylchloride
Analytical parameters: LOQ, 200 fmol for FEN and NF; 500 fmol for FLX in brain microdialysate, and 1 pmol for NF and FEN and 2 pmol for FLX in plasma; inter-assay variability (relative standard deviation), 6.6%, 6.9% and 9.3% for FEN, 4.6%, 3.7% and 7.9% for NF and 10.4%, 4.9% and 12.2% for FLX; for brain microdialysate (2 pmol/μL), plasma (2 pmol/μL) and brain tissue (50 pmol /mg); extraction recovery, 108% and 48% for FEN, 105% and 78% for NF and 94% and 45% for FLX; in plasma (2 pmol/μL) and brain tissue (5 pmol/mg).

*Analgesics*

Enantioselective determination of **eliprodil** in human plasma [37].
Derivatization: Precolumn derivatization with *R*-(+)-1-(1-naphthyl)ethylamine
Detection: Fluorescence
Analytical parameters: Range, 20–2000 ng/ml for each enantiomer; LOQ, 20 ng/ml for each enantiomer; precision, (% relative standard deviations), <11.

### 1.1.3. Conclusion

The recent progress in liquid chromatography-mass spectrometry (LC-MS), especially LC-MS/MS, drastically changes the situation of trace analysis of drugs in a biological matrix. Electrospray ionization (ESI) and atmospheric pressure chemical ionization (APCI) techniques permit highly sensitive and selective measurement of analytes in biological samples without tedious clean-up or derivatization steps.

On an economical basis, although LC-MS/MS instruments are quite expensive, the total analytical cost is almost similar or less compared with conventional HPLC methods owing its high sample throughput with less man-power. In addition, guidelines for new drug applications world-wide, require high reliability of analytical methods together with Good Laboratory Practice standards. From this point of view, HPLC analytical methods with derivatization and/or clean-up steps have disadvantages not only in sensitivity and selectivity, but also sample throughput and reliability. On reliability, LC-MS and LC-MS/MS can utilize the ultimate internal standard, i.e. stable isotope labeled analogue.

In LC-MS/MS operations, APCI covers lower to middle polar compounds and ESI covers middle to high polar ones. However, highly water-soluble compounds sometimes prove difficult in sensitive measurement even if ESI is employed. These compounds are extracted with difficulty and concentrated from the biological matrix. Demands for novel derivatization reagents/methods for these types of compounds are increasing, especially for those which are reactive in a water matrix.

For enantiospecific determination of drugs, HPLC methods using chiral derivatization reagents still has advantages over other techniques. However, it is difficult for selective determination of a drug that has two or more chiral centers in the molecule. We have encountered a compound that has three chiral centers, i.e. eight optical isomers. The compound is an active metabolite of CS-670, a non-steroidal anti-inflammatory agent, that is a derivative of 2-phenylpropionate, the same group as ibuprofen. In this case, one chiral center has separated using immunoaffinity purification followed by enantioselective derivatization to trace the active metabolite level in plasma [38].

There are many racemate drugs in market and some of those are half active and half inactive. Regulations require clarification of the precise disposition of an individual enantiomer in a body including its toxicity and pharmacokinetic properties. See the recent reviews in this field [39,40]. In addition, derivatized enantiomers should be detected using high sensitivity and specificity methods, such as LC-MS/MS.

## REFERENCES


[1] Tsujita, Y., Kuroda, M., Shimada, Y., Tanzawa, K., Arai, M., Kaneko, I., Tanaka, M., Masuda, H., Tarumi, C., Watanabe, Y. and Fujii, S. (1986) *Biochim. Biophys. Acta*, **877**, 50.

[2] Arai, M., Serizawa, N., Terahara, A., Tsujita, Y., Tanaka, M., Masuda, H. and Ishikawa, S. (1988) *Ann. Rep. Sankyo Res. Labs*, **40**, 1.

[3] Dumousseaux, C., Muramatsu, S., Takasaki, W. and Takahagi, H. (1994) *J. Pharm. Sci.* **83**, 1630.

[4] Glencross, R.G., Abeywardene, S.A., Corney, S.J. and Morris, H.S. (1981) *J. Chromatogr.*, **223**, 193.

[5] Takahagi, H., Inoue, K. and Horiguchi, M. (1986) *J. Chromatogr.*, **352**, 369.

[6] Caldwell, J., Hutt, A.J. and Fournel-Gigleux, S. (1988) *Biochem. Pharmacol.* **37**, 105.

[7] Foster, R.T. and Jamali, F. (1987) *J. Chromatogr.* **416**, 388.

[8] Wright, M.R., Sattari, S., Brocks, D.R. and Jamali, F. (1992) *J. Chromatogr.* **583**, 259.

[9] Lemko, C.H., Caille, G. and Foster, R.T. (1993) *J. Chromatogr.* **619**, 330.

[10] Goto, J., Goto, N. and Nambara, T. (1982) *J. Chromatogr.* **239**, 559.

[11] Kondo, J., Imaoka, T., Kawasaki, T., Nakanishi, A. and Kawahara, Y. (1993) *J. Chromatogr.* **645**, 75.

[12] Kondo, J., Suzuki, N., Imaoka, T., Kawasaki, T., Nakanishi, A. and Kawahara, Y. (1994) *Anal. Sci.* **10**, 17.

[13] Kondo, J., Suzuki, N., Naganuma, H., Imaoka, T., Kawasaki, T., Nakanishi, A. and Kawahara, Y. (1994) *Biomed. Chromatogr.* **8**, 170.

[14] Tolan, J.W., Eskola, P., Fink, D.W., Mrozik, H. and Zimmerman, L.A. (1980) *J. Chromatogr.* **190**, 367.

[15] Mrozik, H., Eskola, P., Fisher, M.H., Egerton, J.R., Cifelli, S. and Ostlind, D.A. (1982) *J. Med. Chem.* **25**, 658.

[16] Fink, D.W., de Montigny, P. and Shim, J.-S.K. (1997) *Biomed. Chromatogr.* **11**, 74.

[17] Pandit, N.K. and Connors, K.A. (1982) *J. Pharm. Sci.* **71**, 485.

[18] Connors, K.A. and Albert, K.S. (1973) *J. Pharm. Sci.* **62**, 845.

[19] Tway, P.C., Wood, J.S. and Downing, G.V. (1981) *J. Agric. Food Chem.* **29**, 1059.

[20] de Montigny, P., Shim, J.-S.K. and Pivnichny, J.V. (1990) *J. Pharm. Biomed. Anal.* **8**, 507.

[21] Rabel, S.R., Stobaugh, J.F., Heinig, R. and Bostick (1993) *J. Chromatogr.* **617**, 79.
[22] Eisenberg, E.J. and Cundy, K.C. (1996) *J. Chromatogr.* **679**, 119.
[23] Breda, M., Marrari, P., Pianezzola, E. and Strolin Benedetti, M. (1996) *J. Chromatogr.* **729**, 301.
[24] Uematsu, T., Nakashima, M., Fujii, M., Hamano, K., Yasutomi, M., Kodaira, S., Kato, T., Kotake, K., Oka, H. and Masuike, T. (1996) *Eur. J. Clin. Pharmacol.* **50**, 109.
[25] Khuhawar, M.Y., Lanjwani, S.N. and Memon, S.A. (1997) *J. Chromatogr.* **693**, 175.
[26] Emara, S., Razee, S., Khedr, A. and Masujima, T. (1997) *Biomed. Chromatogr.* **11**, 42.
[27] Sparreboom, A., van Tellingen, O., Huizing, M.T., Nooijen, W.J. and Beijnen, J.H. (1996) *J. Chromatogr.* **681**, 355.
[28] Heggie, J.R., Wu, M., Burns, R.B., Ng, C.S., Fung, H.C., Knight, G., Barnett, M.J., Spinelli, J.J. and Embree, L. (1997) *J. Chromatogr.* **692**, 437.
[29] Ochiai, H., Uchiyama, N., Imagaki, K., Hata, S. and Kamei, T. (1997) *J. Chromatogr. B Biomed. Appl.* **694**, 211.
[30] Kok, R.J., Visser, J., Moolenaar, F., de-Zeeuw, D. and Meijer, D.K. (1997) *J. Chromatogr. B. Biomed. Appl.* **693**, 181.
[31] Wu, S.T., Chang, Y.P., Gee, W.L., Benet, L.Z. and Lin, E.T. (1997) *J. Chromatogr. B Biomed. Appl.* **692**, 133.
[32] Zhong, D., Chen, R., Fieger Buschges, H. and Blume, H. (1997) *Pharmazie* **52**, 106.
[33] Tzou, M.C., Sams, R.A. and Reuning, R.H. (1995) *J. Pharm. Biomed. Anal.* **13**, 1531.
[34] Millerioux, L., Brault, M., Gualano, V. and Mignot, A. (1996) *J. Chromatogr.* **729**, 309.
[35] Malavasi, B., Ripamonti, M., Rouchouse, A. and Ascalone, V. (1996) *J. Chromatogr.* **729**, 323.
[36] Clausing, P., Rushing, L.G., Newport, G.D. and Bowyer, J.F. (1997) *J. Chromatogr.* **692**, 419.
[37] Tsina, I., Tam, Y.L., Boyd, A., Rocha, C., Massey, I. and Tarnowski, T. (1996) *J. Pharm. Biomed. Anal.* **15**, 403.
[38] Takasaki, W., Asami, M., Muramatsu, S., Hayashi, R., Tanaka, Y., Kawabata, K. and Hoshiyama, K. (1993) *J. Chromatogr.* **613**, 67.
[39] Srinivas, N.R. and Igwemezie, L.N. (1992) *Biomed. Chromatogr.* **6**, 163.
[40] Toyo'oka, T. (1996) *Biomed. Chromatogr.* **10**, 265.

# 1.2 DERIVATIZATION FOR AGROCHEMICALS

**Hiroyuki Nakazawa,* and Koichi Saito†**
*Department of Analytical Chemistry, Hoshi University Tokyo 142–8501, Japan and
†Department of Food Chemistry, Saitama Institute of Public Health, Saitama 338–0824, Japan

1.2.1 Introduction 13
1.2.1.1 Sample Preparation 14
1.2.1.2 Derivatization in Food Analysis 15
1.2.2 Analysis of Nutrients 16
1.2.2.1 Carbohydrates 16
1.2.2.2 Amino Acids and Peptides 17
1.2.2.3 Fatty Acids 22
1.2.2.4 Organic Acids 23
1.2.2.5 Vitamins 24
1.2.2.6 Minerals 26
1.2.3 Analysis of Food Additives 27
1.2.3.1 Preservatives 27
1.2.3.2 Antioxidants 27
1.2.3.3 Sweeteners 28
1.2.3.4 Emulsifiers 28
1.2.3.5 Other Food Additives 29
1.2.4 Analysis of Veterinary Medicine 29
1.2.4.1 Sulfonamindes 30
1.2.4.2 Antibiotics 30
1.2.4.3 Antiparasitic Agents 32
1.2.5 Analysis of Pesticides 34
1.2.6 Analysis of Natural Toxins 35
1.2.6.1 Mycotoxins 35
1.2.6.2 Marine Toxins 39
1.2.6.3 Plant Toxin 42
1.2.7 Analysis of Food Contaminants 42
1.2.7.1 Amines 42
1.2.7.2 Organotin 43
1.2.7.3 Hydroperoxides 45
1.2.7.4 Isocyanate Monomers 45
References 45

## 1.2.1 Introduction

Chemical compounds found in foods can be divided into two categories: essential nutrients that maintain human lives and substances which have unfavorable effects on man. The former includes saccharides, fats, amino acids, proteins, vitamins, and minerals. The latter contains environmental hazardous compounds such as organic mercury and PCB. Organochlorine and

Edited by Toshimasa Toyo'oka: *Modern Derivatization Methods for Separation Sciences* 

organophosphorous pesticides, food additives such as antioxidants and preservatives, synthetic antibacterials and antibiotics used for drugs in feeds and animals, have been used for purposes of insecticide, preservation, disease control in the process of culturing, harvesting, and food preparation. Criteria of usage and residue levels are set on these compounds. Controlled by the Food Sanitation Act, some of these compounds are subjected to administrative examination. To analyze these compounds under the food regulation, accurate, easy, quick and sensitive methods are required.

In food analysis, the most significant and major part of the procedure involves how effectively to extract trace analytes of interest from the complicated matrix composed of fats, proteins and minerals. Sample pretreatment, i.e. separation, clean-up, concentration of analytes from food materials is inevitable for HPLC and CE measurement with derivatization. It is no exaggeration to say that the performance of the method is evaluated by the sample pretreatment stages.

### 1.2.1.1 Sample Preparation

Organic solvents such as acetone, acetonitrile and chloroform have been traditionally used for the extraction of analytes of interest, followed by a clean-up procedure. It incorporates liquid-liquid extraction with *n*-hexane/methanol and *n*-hexane/acetonitrile, solid-phase extraction(SPE) with ODS cartridges containing silica gel-based, chemically bonded ODS phase packings, and column chromatography to eliminate interferences. Prior to describing derivatization, this chapter discusses up-to-date extraction and clean-up procedures in food analysis.

*Matrix Solid-Phase Dispersion (MSPD)*

Simultaneous extraction and clean-up pretreatment, MSPD, has been developed in recent years. The procedure includes blending of the sample and C18 packing, followed by packing into a disposable syringe. Fat is removed with *n*-hexane and the analytes of interest are eluted with dichloromethane or methanol. The method is widely accepted by the residual analysis of synthetic antibacterials such as sulfonamides [1–7] and furazolidone [8,9] in fish, meat and milk, nicarbazin [10] in chicken meat, oxolinic acid [11] in fish; anthelmintics such as benzimidazoles [12], ivermectin [13] and hygromycin B [14] in meat. The method is also suitable for residual analysis of pesticides such as oxamyl and methomyl [15] in fruits. To analyze organochlorine pesticides [16,17] in oysters and beef fat, sample-combined C18 packing stacked on Florisil is carried out. MSPD distinctly yields target analytes without loss during the extraction process.

*Gel Permeation Chromatography (GPC)*

A preliminary separation column for GPC is used for the clean-up procedure of pesticide residue analysis. GPC employs column packings based on styrene-divinyl benzene cross-linked macromolecular polymers and mixed solvent of dichloromethane, cyclohexane or ethyl acetate for eluent. High molecular weight compounds such as phosphoric lipids, chlorophyll and polymers travel with the solvent front. On the other hand, low molecular weight compounds such as pesticides, are PCB retarded in their progress down the column and exit. The low molecular weight compounds are collected and concentrated and then subjected to instrumental measurement of HPLC and GC-MS. Exemption from degradation and absorption loss in pesticide residue analysis is a distinguishing feature of GPC, which is adopted as an official method by FDA, AOAC, USDA, and EPA. GPC columns are available from Bio Beads SX-3, Optima GPC, Shodex CLNpak and EnviroSep-ABC. Application of residue analysis for food materials has been reported for organotin [18] in fish, organochlorine pesticides [19–23] and carbamates [24–25] in brown rice, vegetables, fruits, dairy products, fish and seafood, peanut butter and animal fat.

*Supercritical Fluid Extraction (SFE)*

Extraction procedure based on the distribution of solutes among immiscible solvents has been customarily carried out to extract analytes of interest in foods. SFE, using supercritical $CO_2$ or $CHF_3$ as a fluid, has been developed in recent years. SFE can extract thermally unstable compounds and easily control the solubility of solutes by temperature and/or pressure, and consequently the selective extraction of analytes can be easily adjusted. $CO_2$ is less polar like *n*-hexane, so a polar solvent like methanol is added in part as a modifier. Application of the method has been reported on a number of food samples; residue analysis of atrazine and fluazifop-*P*-butyl [26] in onion; organochlorine pesticides such as BHC, DDT, HCE [27,28] in chicken fat, lard, vegetables, hamburgers and peanut butter; carbamates of carbofuran [29] in potatoes, methomyl, methiocarb and eptam [30] in apples. This method is also used for the analysis of PCB [31] in fish, sulfonamides [32] in chicken liver and *N*-nitrosamines [33] in hams. Granted to be a clean analysis, this method uses no dichloromethane and chloroform for the extraction of analytes of interest. In addition, SFE has served not only for extraction but also, more recently, in chromatography to separate optical isomers [34].

*Solid Phase Microextraction (SPME)*

SPME, a new conception different from traditional solvent extraction, has been reported for food analysis. Extraction of the analytes of interest from the sample matrix is carried out without using a solvent. Immersed in sample solution, fused silica fiber coated with liquid phase (polydimethylsiloxane, or polyacrylate or an equivalent) adsorbs analytes of interest based on the principle of partition equilibrium. The analytes would pass through the SPME/HPLC interface, and then be injected into HPLC [35]. Not many reports have been published for SPME on food samples so far. Using this method, analysts have less chance to use solvents. In addition, exhaust fumes and drainage from the laboratory should seldom contain solvents. Thus, this method is preferable from the standpoint of environmental and industrial hygiene. SPME fiber and instrument kits are offered by Spelco Co.

#### 1.2.1.2 Derivatization in Food Analysis

Coupled with a high sensitivity detector, fluorescent derivatization is feasible for the detection trace of analytes in food materials and has recently been developed with a wide variety of applications available. The bulk of fluorescent derivatization reactions fall into three general reaction types: (a) a derivative reagent itself is non-fluorescent or weakly fluorescent, however, derivative compounds of the reagent and the analyte of concern fluoresce; (b) a fluorogenic reagent reacts with the analyte of interest (fluorescent labeling); (c) a reagent itself does not react with the analyte, but modifies the moiety of the chemical structure of the analyte with the consequence of fluorescence or fluorescent intensification (including oxidation or reduction). Use of (a) or (b) is popular for derivatization in food analysis. Application of (c) was reported for the analysis of aflatoxin [36], a kind of mycotoxin, and avermectin [37], an insecticide for hops, using trifluoroacetic acid (TFA) as a reaction reagent. In both cases, the analytical technique was derivatization, but TFA is not a commonly used derivatization reagent. On the other hand, a non-fluorescent reagent which has an intermolecular fluorescent site, on being interfered with by some intermolecular group, is modified by the analyte of interest, and fluoresces. For example, polysaccharide bound to 2-pyridylamino group is non-fluorescent. When combined with $\alpha$-amylase causing severe intermolecular $\alpha$-1,4-glycoside bonding from polysaccharide, the 2-pyridylamino group become fluorescent. This technique is applied for post-column derivatization [38]. The HPLC and CE methods customarily employ pre-column and post-column derivatization and this chapter describes an outline of newly developed on-column derivatization.

The principle of on-column derivatization is as follows: after the HPLC column is equilibrated with the mobile phase in the presence of the derivative reagent, the analyte of interest, introduced from an injector, is subject to derivatization with the reagent to its derivative at the inlet site of the HPLC column and the derivative is subsequently chromatographed and detected. This method was designed to overcome some of the drawbacks of the current derivatization methods used for pre-column and post-column techniques. Pre-column derivatization often needs complicated procedures and cannot handle unstable reaction compounds. In addition, post-column derivatization requires complicated instrumentation. However, on-column derivatization also has disadvantages. Available derivatization reagents are limited because on-column derivatization must proceed promptly at the inlet site of the HPLC column. Optimum analytical conditions of analytes are difficult to set up due to interference with reaction conditions and HPLC mobile phase conditions. On-column derivatization has been developed for the analysis of amines in foods and its applications are discussed in the following chapter.

## 1.2.2 Analysis of Nutrients

Saccharides, amino acids, fatty acids, organic acids and vitamins are the main constituents of foods and used as food additives such as nutrition supplements and flavoring. This chapter describes the use of derivatization methods for these compounds without distinguishing the origin on natural or added.

### 1.2.2.1 Carbohydrates

Carbohydrates are structurally classified into monosaccharides, oligosaccharides and polysaccharides. Monosaccharides and some oligosaccharides taste sweet. Compounds whose aldehyde group in aldose is substituted by primary alcohol are called sugar alcohol. This chapter describes the analysis of carbohydrates including sugar alcohol.

*Analysis by HPLC*

Many liquid chromatographic systems have been described for the analysis of major sugars and polyhydric alcohols in foods. These systems have usually incorporated refractive index (RI) detection coupled with ion-exchange or amino-phase separation. However, insufficient sensitivity and specificity (the detection limit is at the µg level) of the RI detector has promoted the improvement of the detection system or use of the derivatization method. The former includes pulsed amperometric detection (PAD), the use of an electrochemical detector (ECD) whose glassy carbon electrode surface was coated with Cu/Cl-containing film using $CuCl_2$ [39] or ECD detection with an ion-paired reagent in the alkaline mobile phase regarding saccharides as weak acid [40].

The use of post-column derivatization with *p*-aminobenzoic acid hydrazide followed by detection at visual-absorption (VIS) 410 nm was reported [41]. The detection limit of the method was 5 ppb. A variant post-column method with immobilized enzyme reactors which detect indirectly produced hydrogen peroxide with high sensitivity, was also reported. A fundamental system which determines glucose uses immobilized glucose oxidase (GOD) to produce hydrogen peroxide followed by ECD detection [42]. This method was applied for the analysis of such disaccharides as maltose [43] and sucrose [44,45]. The analysis of maltose uses immobilized glucoamylase (GAM) combined with a glucose detection system. Maltose was hydrolysed to $\alpha$-D-glucose and $\beta$-D-glucose. $\beta$-D-glucose was transferred to gluconolacton and hydrogen peroxide by GOD and then hydrogen peroxide was detected with ECD. As for the analysis of sucrose, immobilized reactors of invertase (INV) and pyranose oxidase (PyOD) were connected and the end product of hydrogen peroxide was determined by chemiluminescence method with luminol [44]. In this case, sucrose was hydrolysed to glucose and fructose by INV, the glucose produced in the first reactor was then used to produce hydrogen peroxide by GOD. The PyOD, which is also referred to as glucose 2-oxidase, oxidizes the hydroxyl group at the C-2 position of the pyranose ring of hexoses and pentoses, thus differing from the extensively employed glucose oxidase which oxidizes at the C-1 position. On the other hand, after hydrolysis of sucrose with INV, the fructose produced was used for the transformation of hexacyanoferrate(III) to hexacyanoferrate(II) using fructosedehydrogenase (FDH) followed by ECD detection [45].

$$\text{D-Fructose} + 2\text{Fe(CN)}_6{}^{3-} = \text{5-keto-D-fructose} + 2\text{Fe(CM)}_6{}^{4-}$$

To analyse glucosides, immobilized $\beta$-glucosidase was used for the hydrolysis of $\beta$-D-glucose, followed by luminol chemiluminescence detection of hydrogen peroxide produced by GOD reactor [46] and immobilized GAM was used to transform maltooligosaccharide to glucose followed by PAD detection [47]. The detection limit of these HPLC methods with immobilized enzyme reactors for glucose is 2–10 ng, and the sensitivity is 1000 times higher than that of RI detector (10 μg). However, exchange of reactors is required because the lifetime of these immobilized enzyme reactors is reported to be three months for GOD [42,43], one month for GAM [43] and eight days for INV [45].

The use of the pre-column derivatization method is reported for the reaction with *p*-nitorobenzoyl chloride (PNB-Cl) followed by UV detection at 260 nm [48,49], with phenyl isocyanate (PHI) by UV detection at 240 nm [50,51], and with isatoic anhydride to form fluorescent anthraniloyl derivatives by fluorescent detection (λex360 nm, λem420 nm) [52]. PNB-Cl was used for the analysis of biological samples with excellent sensitivity, but the washing procedure after the reaction is complicated. On the other hand, PHI reacts highly with the free hydroxyl groups of carbohydrates and sugar alcohols. The resulting derivatives are very stable and show excellent sensitivity. UV monitoring at 240 nm permits detection down to the nanogram level. However, PHI derivatives of reducing sugars gave a peak for each enantiomer while non-reducing sugar alcohols gave a single peak. PNB-Cl derivatives were analysed with normal-phase HPLC, PHI derivatives with reversed-phase HPLC and anthraniloyl derivatives with reversed-phase or normal-phase HPLC.

*Analysis by Capillary Electrophoresis*

Capillary electrophoresis (CE) for the analysis of saccharides commonly employs UV detection, and the drawback is sensitivity. Introduction of boron in the alkaline mobile phase allows 2–20 fold higher sensitivity [53], which may be explained by anion electrification of saccharides which results in the reaction of the hydroxy group of saccharides with boron ion. To improve sensitivity, the addition of sorbic acid which acts as both an electrolyte and a chromophore with detection at 265 nm was developed (detection limit, 2 pmol) [54].

A derivatization method with 1-phenyl-3-methyl-5-pyrazolone [55], 2-aminopyridine (AP) [56], *N*-2-pyridylglycamine [57] and ethyl-*p*-aminobenzoate [58] with detection at UV 240 nm–305 nm was reported. The detection limit of these CE ranges from 10 fmol to 10 pmol [54,57,58]. Pre-column fluorophore derivatization with AP [59], 5-aminonaphthalene-2-sulfonate (ANA) [60], 8-aminonaphthalene-1,3,6-trisulfonic acid(ANTS) [61] and 9-aminopyrene-1,4,6-trisulfonate (APTS) [62], as reductive amination, is reported (Fig. 1.2.1). The sugars were derivatized through the Shiff base formation between the aromatic amine of a reagent and the aldehyde form of a sugar, followed by reduction of the Shiff base to a stable product. Other fluorescence derivatization methods use aminated reduced sugar and aminosaccharides with 3-(4-carboxybenzoyl)-2-quinolinecarboxaldehyde (CBQCA) [63–65]. These fluorescent derivatives are excited by laser ray, AP, ANA and ANTS derivatives by helium-cadmium laser (λex325 nm, λem375 nm for AP, 475 nm for ANA, 514 nm for ANTS), and APTS by argon-ion laser (λex457 nm, λem550 nm for CBQCA; λex488 nm, λex512 nm for APTS). The limit of laser-induced fluorescence detection for APTS derivatives of sugars is 2 pmol, for CBQCA is at atto-mole level.

The CE analysis of saccharides in this section is limited to standards of sugars and saccharides and hydrolysis products. The application of CE for food materials which contain a complex matrix would require such preliminary separation as clean-up.

#### 1.2.2.2 Amino Acids and Peptides

Proteins are the main constituents of organisms, and about 20 amino acids build polypeptide chains with peptide bonding. Molecular weights of proteins range from 4000 (protamine etc.) to some ten millions (virus proteins etc.). Proteins which are made up of amino acids only are called simple proteins, and those made up of other than

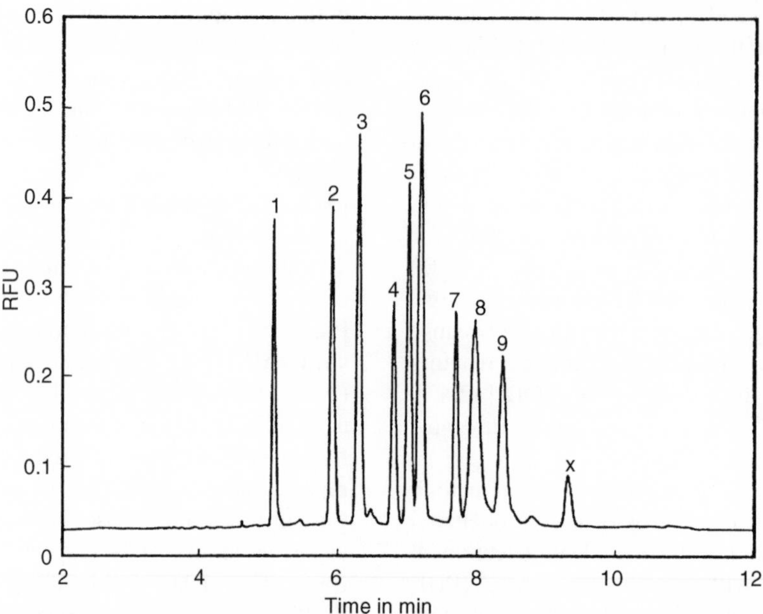


**Fig. 1.2.1.** Electropherogram of nine APTS-derivatized monosaccharides at 1.0 μM each. conditions: untreated fused-silica capillary, 20 μm (i.d.) × 27 cm; light source, 488 nm argon-ion laser; emission filter, 520 ± 20 nm (Oriel, Startford, CT) and a notch filter at 488 nm (Barr Associates); applied potential, 20 kV/19 μA; buffer, 100 mM borate, pH 10.2; sample injection, 20 s, 3.5 kPa pressure; outlet, cathode. Peak identification: (1) *N*-acetylgalactosamine; (2) *N*-acetylglucosamine; (3) rhamnose; (4) mannose; (5) glucose; (6) fructose; (7) xylose; (8) fucose; and (9) galactose; (x) impurity peak derived from APTS. (Reproduced from ref. 62: *Anal. Chem.*, (1995) **67**, p. 2244, Fig. 5.).

amino acids are called conjugated proteins. The representative conjugated proteins are glycoprotein and lipoproteins. Peptides include oligopeptide (the number of amino acids is 2–10), and polypeptides (the number of amino acids is 10–50). Peptides and proteins are either analysed by the high molecule itself or by amino acids, the hydrolysates of peptides and proteins. Most HPLC methods may employ the latter case because the amino acid composition of proteins is often studied.

### *Amino Acids*

The analysis of amino acids has traditionally been based on ion-exchange chromatography followed by post-column derivatization with ninhydrin. This approach is reliable and the resolution of the amino acids is reasonable but the analysis time is rather long with limited detectability. Although the method is still the main method in use, the following HPLC methods tend to supersede the classical method. Due to the use of simpler instrumentation in pre-column derivatization and the lower cost of such systems compared with post-column derivatization, the pre-column derivatization is generally preferred. Typical reagents for pre-column derivatization are phenyl isothiocyanate (PITC) [66–68]. *o*-phthalaldehyde (OPA) [69–71], 9-fluorenylmethyl chloroformate (FMOC) [72–75], 2,4-dinitrofluorobenzene (DNFB) [76], dansyl chloride (DNS) [77–79], 6-aminoquinolyl-*N*-hydroxysuccinimidyl carbamate (AQC) [80,81], PHI [82], *N*-(acridinyl) maleimide [83], and fluorescamine [84,85]. Some of the features of these methods for amino acid analysis of protein hydrolysates were summarized by Sawar *et al.* [86].

PITC has been used for the Edman degradation method to determine the amino acid sequence. Derivatization with PITC which employs UV

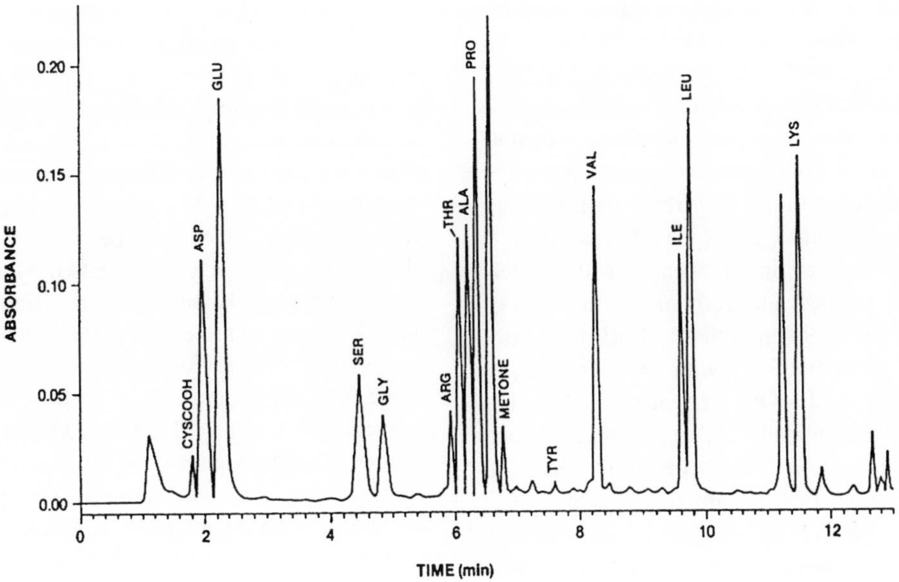


**Fig. 1.2.2.** Chromatogram of PTC cysteic acid (CYSCOOH) and methionine sulfone (METONE) in performic acid plus 6 M HCl hydrolysate of milk-based infant formula. (Reproduced from ref. 86: *J. Chromatogr.*, (1993) **615**, p. 12, Fig. 2.).

detection (254 nm) is less sensitive than fluorescent derivatization and also the method requires the evaporative elimination of excess PITC under reduced pressure after the reaction. The PITC derivatives are fairly stable at room temperature, and the method has been used to determine the amino acid composition of a variety of foods. Fig. 1.2.2 shows the chromatogram of amino acids (except tryptophan) in hydrolysate of milk-based infant formula. This method determines sulfur-containing amino acids as stable cysteic acid and methionine sulfone using oxidation with performic acid and hydrolysis with HCl. The analysis of tryptophan includes alkaline hydrolysis with barium hydroxide followed by PITC derivatization, and applied for the analysis of infant formula [68].

OPA reacts with primary amino group in the presence of a thiol reagent to form highly fluorescent isoindoles. Showing no stray fluorescence and high reactability, OPA/thiol has been widely used for the analysis of amino acids in biological samples as a pre-column and post-column derivatization reagent. The use of amino acid analysis in food was reported for the determination of sulfur-containing amino acid such as alliin, which was derivatized with OPA/*tert.*-butylthiol reagent and detected with both UV (337 nm) and fluorescence (λex230 nm, λem420 nm) [69,70]. To determine free amino acids, this method does not apply hydrolysis but extraction with formic acid containing aqueous methanol followed by SPE (C18) clean-up. Formic acid was added to deactivate the enzyme, alliinase, in garlic. A derivatization method with fluorescamine for the determination of alliin in garlic was also reported (λex405 nm, λem480 nm) [84].

The main disadvantage of the OPA derivatization method lies in the fact that OPA does not react with the secondary amino acids, such as proline and hydroxyproline. The analysis of free amino acids in shrimp reported the employment of OPA derivatives of primary amino acids and 4-chloro-7-nitrobenzofurazan (NBD-Cl) derivatives of secondary amino acids (proline and hydroxyproline) [71].

Recently, an automated pre-column derivatization of amino acids in cheese hydrolysate was reported using OPA/3-mercaptopropionic acid (3-MP) for the primary amino acids, followed by FMOC for the secondary amino acids [87]. FMOC was introduced as a protective reagent for the amino group during peptide synthesis, FMOC derivatives show excellent reactivity, sensitivity and stability, furthermore they react with primary and secondary amines. Thus FMOC has been a widely used derivatizing reagent for the determination of amines and fumonisins, a kind of mycotoxin. The problem is that FMOC itself fluoresces, and then interferes with the fluorescence determination of FMOC derivatives of amino acids. Thus the elimination of excess reagent such as pentane extraction [72], heptylamine addition [74], and alkaline hydroxylamine addition [73] is required after derivatization. Amino acid analysis with FMOC has another problem. Derivatization of aspartic acid and glutamic acid with FMOC proceeds slower than that of other amino acids. A prolonged reaction time to complete derivatization of these two amino acids will cause the formation of more hydrophobic disubstituted derivatives (*di*-FMOC-Tyr, and *di*-FMOC-His) of tyrosine and histidine. To solve this problem, alkaline treatment of these FMOC disubstituted derivatives was carried out to form each monosubstituted derivative. As the peak area of monosubstituted derivatives of Tyr and His are 2.3–2.5 fold that of disubstituted derivatives, this transformation will be favorable for the analysis of Tyr and His. After FMOC derivatization, some FMOC derivatives of other amino acids might be subjected to decomposition by alkaline treatment. A mixture of hydroxylamine and sodium hydroxide proved effectiveness in converting both *di*-FMOC-Tyr and *di*-FMOC-His to the monosubstituted derivatives, because it had little effect on normal *N*-FMOC groups. The method was applied for the analysis of white lysozome hydrolysate of eggs with fluorescence detection (λex263 nm, λem313 nm) [73], and of free amino acids and amines in wine, fruit or vegetable juice, fish, and cheese [74].

Dansylation of amino acids has been widely studied and appeared to be suitable for UV (254 nm) or fluorescence (λex330 nm, λem530 nm) measurement. However, the reaction was complicated by a side reaction whereby the amino acid was degraded. DNS is a electrophilic reagent and reacts with the unprotonated form of the amino group, and thus a pH of 9 or greater is required. However, at high pH this reaction competes with hydrolysis of DNS. Moreover, a second competing reaction takes place between excess DNS and dansylated amino acid, leading to the loss of dansylated amino acids and the formation of dansylamide. To minimize this side reaction, it was necessary to closely control reaction pH, temperature, and duration, as well as the ratio of DNS reagent to amino acids. The addition of methylamine and ethylamine is also suggested [88]. The use of DNS was applied for the analysis of hydrolysates of feedstuff [77] and casein [78], taurine in milk and infant formula [79]. For the analysis of taurine, the pre-column method with DNFB and UV detection at 350 nm was reported [76]. Other than infant formula, this method was also applied for the measurement of taurine in meat, tuna, and squid (detection limit 10 pmol).

AQC, a novel derivatization reagent, can react in seconds with all primary and secondary amino acids without appreciable matrix interference to form single, quantitative, and very stable derivatives. In addition, the excess reagent was hydrolyzed to 6-aminoquinoline(AMQ) in less than 2 min, thus preventing any unwanted side reactions and the florescence emission maxima of AMQ and AQC-derivatized amines are approximately 100 nm apart, allowing for selective detection of the desired analytes without significant reagent interference. This method was applied for the analysis of hydrolysate of grain and corn powder [80,81]. All data were generated by fluorescence detection except for the analysis of tryptophan, whose fluorescence response is weak due to internal quenching, which used UV detection.

Little is known about the post-column derivatization of amino acids in foods. Ninhydrin was used to measure taurine in infant formula and milk(VIS 570 nm) [89]. To enhance the sensitivity, the use of OPA/3-MP instead of ninhydrin was reported for post-column derivatization with fluorescence detection of amino acids, hydrolysate of collagen-containing meat [90]. In this case, OPA does not react with secondary amino acids. Before the reaction with OPA reagent, oxidation with sodium hypochlorite (NaOCl) of secondary amino acid is required for the first step. The drawback of NaOCl oxidation is the decomposition of tryptophan.

Recently developed on-column fluorescence derivatization is applied to the analysis of amino acids [91] and taurine [92]. Derivatizing reagents for on-column methods are used in pairs such as OPA and an active thiol, *N*-acetyl-L-cysteine(NAC), Thiofluor and 2-ME. The analysis of taurine [92] employs SPE (Bond Elut SCX) clean-up as sample pretreatment was used for infant formula and milk products, eggs, honey, fish powder, shellfish and so on. High sensitivity of OPA derivatives and improvement of clean-up, enabled trace analysis, and taurine in egg yolk and honey to be detected for the first time (Fig. 1.2.3).

On-column methods use OPA mixed with mobile phase, thus simultaneous analysis of multi-components with gradient elution is difficult. Combining the column-switching technique and step-gradient elution, the simultaneous analysis of polyamines and their precursors, amino acids, was carried out. This method was applied to the investigation of the arginine degradation pathway of *Photobacterium phosphreum*, a putrefactive bacteria [93] and the measurement of citrulline, which is produced stoichiometrically from arginine when nitric oxide (NO) is produced, in order to study the effect of food stuffs on the NO productivity of macrophage cultured cell [94].

The use of CE with derivatization of amino acids for the analysis of foods is hardly reported. For derivatization with CE, the pre-column method is the most common, and the reaction system is fundamentally similar to the HPLC method. A recent report shows that free amino acids in fish and lysine, tyrosine, histidine, and ornithine in soya flour hydrolysates, are derivatized to *di*-dansyl amino acids followed by separation and detection with micellar electrokinetic chromatography (MEKC) [95].

### *Peptides*

Peptides in food exhibit multiple functional properties (surfactant, antimicrobial, antioxidant,

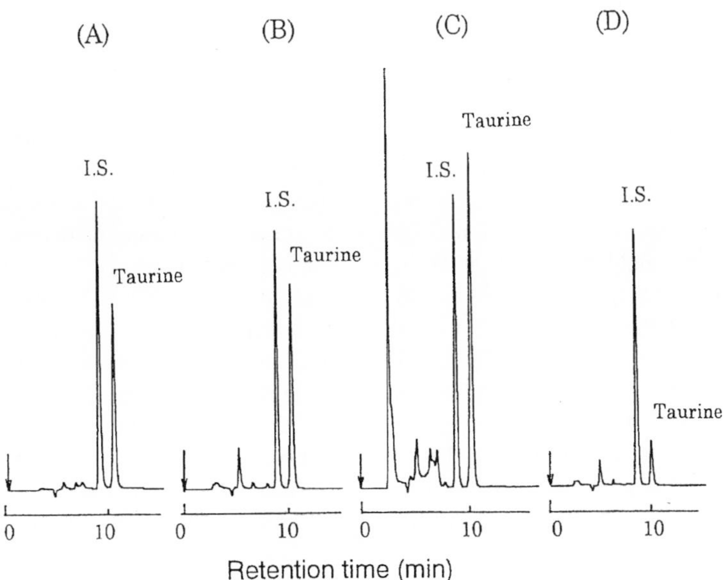


**Fig. 1.2.3.** Typical chromatograms of taurine from: (A) clam; (B) Yolk; (C) salmon reo; and (D) honey. (Reproduced from ref. 92: *J Food Hyg. Soc. Japan*, (1997) **38**, 403, Fig. 2.).

etc.) and contribute to bitter and sweet tastes. Separation of peptides is mainly performed by HPLC or CE. However, identification of peaks is, undoubtedly, the most difficult task of peptide analysis due to the wide range of peptide structures (amino acid composition and sequence).

The following methods have been reported recently; a small hydrophilic peptide in cheese was analysed, and extracted samples are subject to gel permeation (Sephadex G-25) followed by FMOC derivatization with the pre-column method and separation on HPLC. The peptide derivatives were deprotected with piperidine followed by sequencing by the Edman degradation method [96]; small peptides in wine are identified after separation with Sephadex G-10 followed by post-column OPA derivatization and photodiode array detection [97].

### 1.2.2.3 Fatty Acids

Fatty acids are naturally present in foods and are an important energy source. They are also functional groups of many surface active agents presently used as food additives. Naturally occurring fatty acids are the ester compounds of glycerol and higher alcohol, free fatty acids are not commonly found. Fatty acid analysis has been used to determine fatty acid composition. Highly unsaturated fatty acids such as linoleic acid and arachidonic acid are known to be essential fatty acids, and the determination of their composition in foods is important. Fatty acids often need to be quantified as indicators of rancidity, freshness or adulteration of fats. The analysis of fatty acids in food includes extraction of fat, hydrolysis of fats to free fatty acids, GC or HPLC determination after derivatization or GC determination after direct transesterification (methyl esterification) under acid or alkaline catalyst. Traditional fatty acid analysis is based on the GC method, while HPLC method has been seldom reported.

The analysis of fatty acids with HPLC employs UV detection (215 or 192 nm) or RI detector, however, the method is not suitable for ppb level trace analysis of food due to the interference of coexistents. As for the HPLC method with derivatization the post-column ion-pair extraction technique is reported [98]. The method is based on the principle that fatty acids are extracted as ion-pairs with chloroform from the aqueous acetonitrile mobile phase after the post-column addition of aqueous methylene blue solution. The chloroform phase containing the ion-pairs is monitored with an absorbance detector at 651 nm. This method was applied for the analysis of orange juice, margarine and butter. The detection limits ranged from 26 to 83 ng, depending upon the acid. One problem associated specifically with the system as set up for fatty acids was the need routinely to backflush the detector cell with acetonitrile to remove deposits which caused fatty acid adsorption and thus loss of resolution and sensitivity.

This same approach was applied to the direct determination of the artificial sweeteners cyclamate, saccharin and acesulfame K in diet beverages using the dye methyl violet 2B [99]. An additive, sodium dioctylsulfosuccinate, was also detected in beverage powder by applying a similar approach [100].

The use of pre-column HPLC method for the analysis of fatty acid with 2-nitrophenylhydrazine hydrochloride (2-NPH) was recently reported [101,102]. After alkaline hydrolysis of vegetable oils such as coconut oil, olive oil, and margarine, and fat in milk, yogurt, ice cream, cheese, butter, beef tallow and lard, and sardine oil, free fatty acids are reacted with 2-NPH and then derivatized to corresponding fatty acid hydrazides. Each of the derivatives were separated on reversed-phase HPLC with isocratic elution and detected at VIS 400 nm. This method was able to determine 29 saturated and mono and polyunsaturated fatty acids (C8:0–C22:6), including cis-trans isomers and double-bond positional isomers.

For the analysis of fatty acids (C12–C18) in beer, after liquid–liquid extraction with ether/pentane under acidic conditions with sulfuric acid, the pre-column method with 9-chloromethylanthа-racene(9-CA) followed by fluorescent detection ($\lambda$ex365 nm, $\lambda$em412 nm) is reported (Fig. 1.2.4) [103]. The detection limit of this method is 0.2 g/l.

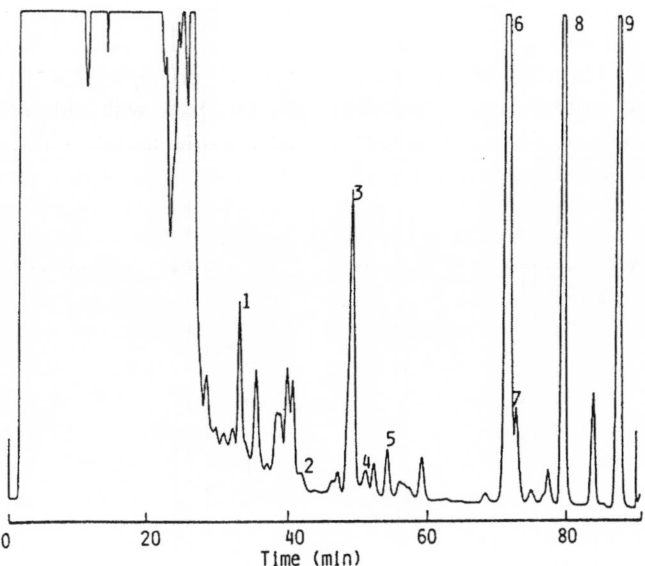


**Fig. 1.2.4.** HPLC chromatogram of fatty acids in beer. (1) Lauric acid (C12:0); (2) linolenic acid(C18:3); (3) myristic acid (C14:0); (4) palmitoleic acid (C16:1); (5) linoleic acid (C18:2); (6) palmitic acid (C16:0); (7) oleic acid (C18:1); (8) *n*-heptadecanoic acid (C17:0); (9) stearic acid (C18:0). (Reproduced from ref. 103: *J. Agric. Food Chem.*, (1990) **38**, p. 1365, Fig. 5.).

### 1.2.2.4 Organic Acids

Organic acids are compounds which have a carboxylic group in the molecule, including fatty acids and amino acids. Organic acids in foods are often called fruit acids, mainly contained in fruit and vegetable and tasting sour. Organic acids are also intermediate or end products in biological metabolism. As for foods, such agricultural products as fruit, vegetable and grains, such fermented products as Japanese sake and bean paste(miso), meat and dairy products of milk, and such sea foods as fish, shellfish and seaweeds are good sources of organic acids. These organic acids are not only associated with food taste but are used for additives for some purposes of pH adjustment, nutrition fortification, sour agents and preservatives.

Recently the use of HPLC or CE for the analysis of organic acid, has been applied, but derivatization methods have not been well reported. The pre-column method includes esterification with phenacyl bromide [104,105] and *O*-(4-nitrobenzyl)-*N*,*N*′-diisopropylisourea (PNBDI) [106] followed by separation with reversed-phase HPLC and detection at UV (254 or 265 nm) has been traditionally applied to the analysis of organic acids in fruit juice (Fig. 1.2.5). However, PNB derivatization could not produce characterizable derivatives of oxalic acid and pyruvic acid.

Fluorescent derivatization with pre-column method uses the reaction of lactic acid with ceric sulfate to form acetaldehyde followed by condensation of 1,3-cyclohexadione to produce fluorescent 9-methyl-1,8-dioxaoctahydroacridine($\lambda$ex366 nm, $\lambda$em455 nm). This method was applied to the analysis of lactic acid in wine [107]. Pretreatment includes 10-fold dilution of the sample with water, boiling down to half the volume of the diluted sample to eliminate native acetaldehyde and ethyl acetate in wine, followed by clean-up with Bio-Rex 5 anion-exchange column. Fluorescent derivatives were separated by ion-exchange partition chromatography (Aminex HPX-87H).

As for the post-column method, after separation on the ion-exchange column, organic acids were reacted with 1-ethyl-3-(3-dimethylaminopropyl)

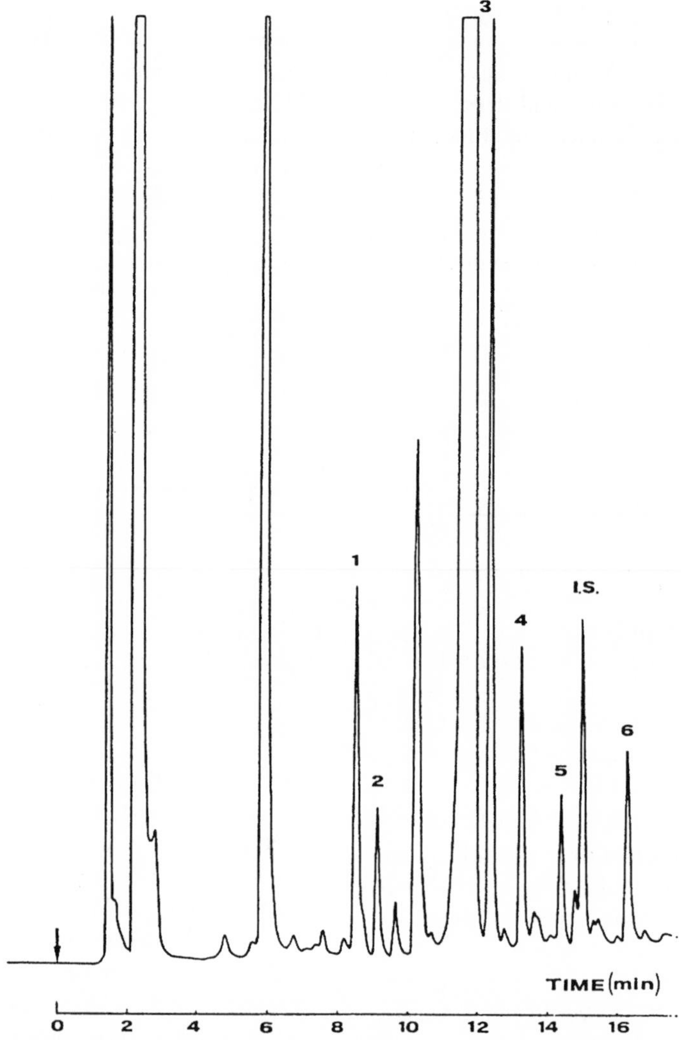


**Fig. 1.2.5.** Chromatogram of a commercial sample of Italian wine (Asti Spumante). Peaks: 1 = lactate (1.54 g/l); 2 = acetate (0.4 g/l); 3 = tartrate (2.83 g/l); 4 = malate (0.74 g/l); 5 = succinate (0.30 g/l); 6 = citrate (0.50 g/l). I.S. = Internal standard. (Reproduced from ref. 105: *J. Chromatogr.*, (1986) **362**, p. 51, Fig. 1.).

carbodiimide hydrochloride and 2-nitrophenylhydrazine to form hydrazide, which is subject to heat treatment to produce purple under alkaline conditions followed by detection at VIS 530 nm [108]. Simple pretreatment and specificity of this method was applied to the commercially available carbonic acid analyser [109].

The use of a devised post-column method is reported. After separation of organic acids on the HPLC column, the method includes enzyme reaction in immobilized enzyme reactor and ECD detection of produced hydrogen peroxide [110]. This method was applied to the analysis of lactic acid in yogurt and concentrated tomatoes using reversed-phase HPLC and L-lactate oxidase as a immobilized enzyme reactor.

### 1.2.2.5 Vitamins

Vitamins are essential nutrients for human and animal organisms and extremely small quantities of vitamins regulate metabolism. Vitamins must be obtained from food because they cannot be synthesized in the body. Vitamins have widely different functions in the body, and deficiency of vitamins may upset the balance and function in the body. Vitamins are used as food additives for nutrient fortification. Although HPLC is the main method in use for vitamin analysis in recent years, a few have been reported for the derivatized method.

*Vitamin $B_1$ (Thiamine)*

The Post-column and pre-column method with cyanogen bromide or potassium ferricyanide is used to form thiochrome fluorescence (λex375 nm, λem440 nm) after the separation of reversed-phase column, used for the analysis of water soluble vitamin $B_1$ [111–114]. The drawback of the pre-column method is the formation of by-products. Although thiochrome fluoresces strongly under alkaline conditions, the fluorescence intensity of thiochrome decreases when the ODS column is used within the recommended pH region of the mobile phase. Therefore, preference may be give to using an alkaline-tolerant column or the introduction of an alkaline solution to the ODS column using the post-column technique [115]. This method is applied for the simultaneous determination of thiamine and thiamine phosphoric ester in a biological sample.

The post-column method can separate each thiamine phosphoric ester and interfering peaks are seldom observed. The use of oxidizer for the post-column reaction solution requires such maintenance as washing in flow. The post-column method with potassium ferricyanide was applied for food analysis and a peak of hydroxyethylamine, an active form of thiamine, was observed, but thiochrome from hydroxyethylamine was not recognized when cyanogen bromide was used [111,116].

*Vitamin $B_6$ (Pyridoxine)*

Three organic compounds of pyridoxine (PN), pyridoxal (PL) and pyridoxamine (PM) act as vitamin $B_6$ and the most common forms of these compounds in animals and plans are phosphoric esters. They are widely found in animal and plant foods, and in particular cereals contain large quantities of vitamin $B_6$. Phosphoric derivatives of PL and PM exist in nature and work as co-enzymes. PN was extracted with perchloric acid and detected with fluorescence HPLC from juice, fruit, vegetable, grains and dairy products [117]. This method employs derivatization with potassium cyanide to obtain more than 10 times stronger fluorescence intensity because the sensitivity of the PL ester is low. The use of fluorescence post-column method with sulfurous acid after extraction with perchloric acid and separation with ion-paired reversed-phase HPLC was applied to the analysis of PN (6 derivatives) in pork liver and milk [118]. This method was seldom interfered with from unknown peaks, showing high sensitivity.

*Vitamin C (Ascorbic Acid)*

Vitamin C has long been recognized as an important nutrient in several food products. The reduced and oxidized form of the vitamin are referred to as ascorbic acid (AA) and dehydroascorbic acid (DHAA), respectively. In humans, both forms are biologically active. The total vitamin C activity is the sum of both forms.

As for analysis for vitamin C, being electrochemically inactive, DHAA cannot be detected with UV absorbance. Therefore DHAA is reduced to AA with dithiothreitol [119], DL-homocysteine [120] or L-cysteine [121] and separated with ion-exchange or reversed-phase HPLC followed by ECD detection. This method is applied for the analysis of fruit juice and green tea [121]. As the method measures total vitamin C, the difference between total vitamin C and AA is calculated in order to measure DHAA by measurement of AA solely without reduction. Simultaneous determination of 3-(1,2-dihydroxyethyl)furo[3,4-b]quinoxaline-1-one (DFQ) and AA was developed using the reaction of *o*-phenylenediamine (OPDA) and DHAA to form fluorescence derivatives, DFQ [122]. A variant method includes oxidization of AA to DHAA with Norit, fluorescent derivatization to DFQ (λex350 nm, λem430 nm) and separation on reversed-phase column and detection as total vitamin C [123].

The method was applied for the analysis of canned corn, potatoes, green beans, potato chips, cereals and infant formula and favorable chromatograms with the absence of interfering peaks were observed. In the pre-column derivatization with OPDA, problems were encountered with the stability of the derivative in aqueous solutions.

The use of the post-column method is reported; after separation of AA and DHAA by ion-exclusion HPLC, DHAA was reduced with sodium borohydride under alkaline conditions and detected at UV300 nm [124]; after separation by anion-exchange HPLC, AA was oxidized to DHAA with the addition of $HgCl_2$ or $CuSO_4$ reagent followed by DFQ derivatization with OPDA [125,126]. This method was used for the analysis of fruit juice and meat-based food.

Vitamin C is the most unstable among vitamins and destroyed by air oxidation and heat. Caution is required for extraction and clean-up procedure. Extracting solvents of vitamin C are pyrogallol added citric acid [125], metaphosporic acid [127], EDTA containing sulfuric acid [119], trichloroacetic acid [128] and acetic acid containing metaphosporic acid [123,126], suppressing oxidase and copper ion.

*Folacin (Pteroylglutamic acid)*

Folacin is a glycoside of *p*-aminobenzoic acid and glutamic acid with pteridine skeleton. The pre-column method is used to separate folic acid by oxidation with potassium permanganate and hydrogenperoxide to form fluorescence product (2-amino-4-hydroxypteridine-6-carboxylic acid) followed by HPLC [112]. Sample pre-treatment of oral nutrients for the analysis of folic acid employs extraction under alkaline conditions and clean-up with SPE (Bond Elut SAX).

*Biotin*

Biotin is a growth factor of yeast and some bacteria and anti-albumin disorder factor. Taking too much albumin causes nutritious disorder. Biotin occurs in most foods and yeast, liver, egg yolk and tomatoes are rich sources of this vitamin.

As for biotin analysis in foods, microbiological assay with *Lactobacillus plantarum* has been traditionally used. The use of the HPLC method using fluorescence pre-column technique with 4-bromomethyl-7-methoxycoumarin (Br-MMC) [129], 9-anthryldiazomethane (ADAM) [130], 1-pyrenyldiazomethane (PDAM) [131], panacyl bromide [132] was developed and applied to the analysis of biological samples.

The post-column method includes chlorination of the amide group of biotin, introduction of thiamine to form thiochrome and detection with fluorescence [133]. The analysis of pharmaceutical preparations employs this method and the detection limit is 10 ng.

*Vitamin K*

Most 2-methyl-1-naphtoquinone derivatives show vitamin K activity to some extent. Naturally occurring vitamin K is vitamin $K_1$ (phytonadione) found in plants and vitamin $K_2$ (menaquinones) in bacteria. The reduction fluorescence method is developed for the analysis of vitamin K. The method includes post-column reduction of vitamin K with zinc or platina column to form fluorescence product (corresponding hydroquinones) and fluorescent detection. Application was reported for the analysis of cereals, fats and vegetables using the post-column method with sodium tetrahydroboric acid [134], for infant formula with zinc column [135], and for breast milk with platina column [136]. To measure 2,3-epoxide of vitamin $K_1$, an electrode of ECD detector was used for the reduction to form a fluorescence product [137], but the electrochemical reduction method suffers from incomplete reduction of the injected K vitamins.

### 1.2.2.6 Minerals

Minerals, or trace elements, are essential in the body and seven minerals such as calcium, magnesium, sodium, potassium, phosphorus, sulfur and chlorine consist of 60–80% of the total minerals. Iron, copper, fluorine, iodine, manganese, cobalt, zinc and molybdenum are also needed.

Molybdenum is required to activate such specific enzymes as the flavoprotein enzyme which catalyses aldehyde oxidase and xanthine oxidase. Good sources are bovine kidney, some cereals and beans. The use of the pre-column

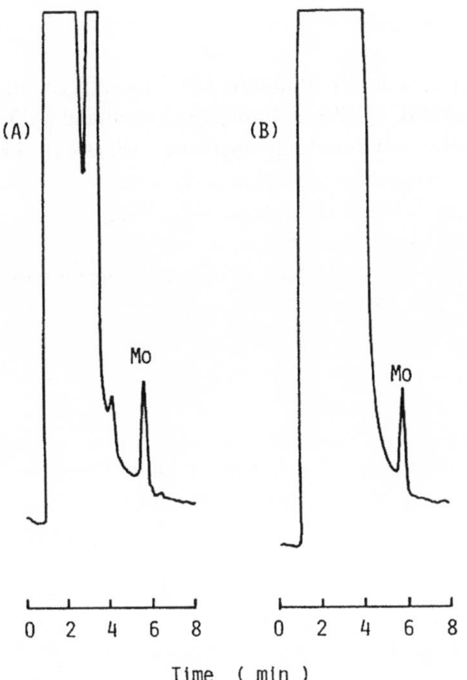


**Fig. 1.2.6.** Determination of Mo in seawater (A) and bovine liver (B) by on-column chelation. (Reproduced from ref. 140: *J. AOAC Int.*, (1995) **78**, p. 1310, Fig. 4.).

method with chelating agent such as 2-(5-bromopyridylazo)-5-diethylaminophenol [138] and tetracycline [139] was reported for the analysis of molybdenum. As the reaction with these chelating agents was slow, on-column and pre-column methods with 8-hydroxyquinoline (8-HQ) were developed, and applied to sea water and bovine liver samples [140]. The detection was conducted spectrophotometrically at VIS 390 nm. The on-column chelation method showed relatively less interference from manganese and greater simplicity in operation compared with the pre-column chelation method (Fig. 1.2.6).

## 1.2.3 Analysis of Food Additives

### 1.2.3.1 Preservatives

Preservatives are added to prevent growth of bacteria associated with putrefaction and decomposition. Benzoic acid, sorbic acid, dehydroxy acetic acid, *p*-oxybenzoic acid esters, and propionic acid are representative preservatives.

Analysis of preservatives with HPLC commonly employs UV detection (210–270 nm). A method using the derivatization of the carboxylic group with UV or fluorescent labeling was reported. Propionic acid was phenacyl esterified with *p*-bromophenacyl bromide catalysed by 18-crown-6, separated on reversed phase-HPLC, and detected at UV 254 nm [141], and propionic acid in bread was determined by this method.

Analysis of sorbic acid employs the pre-column method with 4-bromomethyl-6,7-dimethoxycoumarin (Br-DMC) to produce fluorophore derivatives followed by separation on reversed-phase HPLC and fluorescence detection ($\lambda$ex355 nm, $\lambda$em420 nm) [142]. After extraction with steam distillation, margarine and butter samples were analysed by this method.

Lysozyme has been used for a drug and a food preservative lysing the cell wall of gram positive bacteria to destroy growth of bacteria. Measurement of lysozyme activity in food has recently been developed using the fluorescence technique with synthesis substrate, 4-methylumbelliferyl tri-*N*-acetyl-$\beta$-chitotrioside (4-MU-$(GlcNAc)_3$) [143]. This substrate is a chemical compound of fluorescent substance, 4-methylumbelliferone (4-MU) and trimer acetylglucosamine. The whole molecule shows non-fluorescence because of bonding with acetylglucosamine. 4-methylumbelliferone (4-MU) fluoresces when liberated by hydrolysis by lysozyme. This principle was applied for the pre-column HPLC method, which includes separation of 4-MU produced by enzyme reaction and 4-MU-$(GlcNAc)_3$, substrate, and the use of polystyrene reversed-phase column with pH 12 alkaline buffer as the mobile phase to yield sufficient fluorescence intensity. The method is employed for the measurement of lysozyme activity in Japanese sake, Mirin (sweet sake), cheese and custard cream.

### 1.2.3.2 Antioxidants

Unsaturated fatty acids in fats and oils are subject to oxidation to produce hydroperoxide, and toxic oxides are further produced by chain reactions.

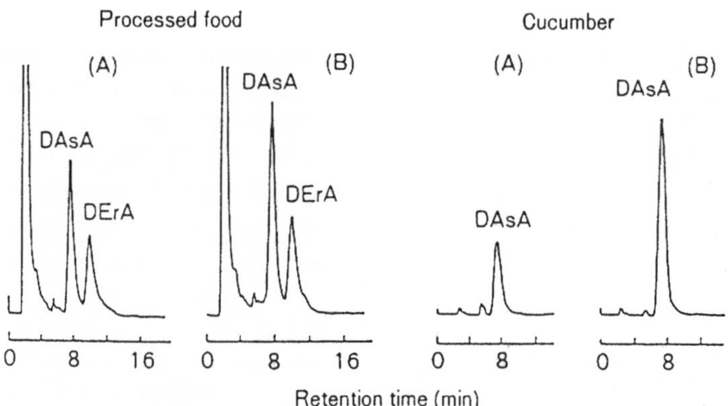


**Fig. 1.2.7.** Typical high-performance liquid chromatograms of fluorescent derivatives. Processed food: mixture of fishes, shellfishes and meats: (A) without oxidation; (B) after oxidation. (Reproduced from ref. 144: *J. Food Hyg. Soc Japan*, (1990) **31**, p. 49, Fig. 4.).

Antioxidants have been used to prevent these reactions.

Ascorbic acid (AA) and erysorbic acid (ErA) are stereoisomers, and used for nutritional enrichment and antioxidants. AA exists in nature as a food component, while ErA does not exist. Each has a reduction form and an oxidation form. The oxidation forms of AA and ErA react with OPDA to form fluorescent derivatives. The use of derivatization with OPDA after oxidation of the reduction form with indophenol is followed by measurement of total AA and total ErA [144]. This method applied to processed food includes extraction with metaphosphoric acid, oxidation with indophenol, derivatization with OPDA and clean-up with SPE (Sep-pak C18). Separation with HPLC employs the alkaline mobile phase and polymer column regarding separation and detection sensitivity. Detection limits for AA and ErA are both 0.5 ng (Fig. 1.2.7).

### 1.2.3.3 Sweeteners

Typical sweeteners are saccharin, glycyrrhizic acid, aspartame and sorbitol.

Analysis of aspartame (*N*-L-$\alpha$-aspartyl-L-phenylalanine 1-methyl ester), dipeptide, uses precolumn method with fluorescamine to form fluorescent derivatives followed by reversed-phase HPLC and fluorescent detection ($\lambda$ex390 nm, $\lambda$em480 nm) [145], and applied for such liquid food as soft drinks and soy sauce, chewing gum, bean paste(miso) and chocolate. The detection limit for liquid food is 2 ppm, for solid food is 4 ppm. The aromatic phenylalanine moiety of aspartame molecule is transformed to electrochemically active derivatives with UV (254 nm) irradiation. This reaction is applied for post-column photolytic derivatization with HPLC separation [146], and the method is used for diet colas and pharmaceutical products. Preliminary operation requires only dilution with water and the detection limit is 0.5 ppm as the standard concentration.

The derivatized HPLC of such sweeteners as acesulfame K, cyclamate and saccharin is reported using post-column ion-pair extraction as described in analysis of fatty acids in '1.2.2 Analysis of food nutrients' [99]. The principle involves mixing the analyte, as it elutes from the HPLC column, with an appropriate counter-ion. The resulting ion-pair is dynamically extracted into an organic phase, which enters the detector for detection. This method uses methyl violet 2B or crystal violet for dyes of counter-ion, applied for the analysis of beverages.

### 1.2.3.4 Emulsifiers

Emulsifiers serve to improve storage and quality of processed food. Widely used emulsifiers are monoglycerides (MGL), sucrose esters of fatty

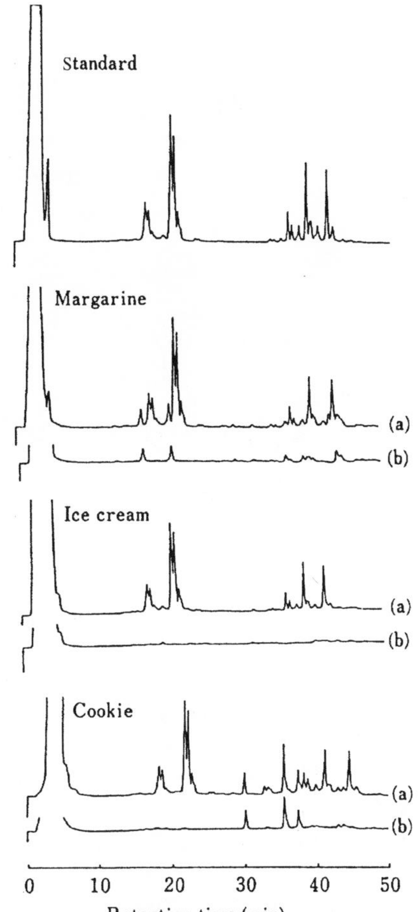


**Fig. 1.2.8.** Liquid chromatograms obtained from several kinds of foods added to a mixture of sucrose mono- and di-fatty acid esters or not by the proposed procedure: (a) added to the mixture; (b) no addition. HPLC condition: column, Inertsil C8 (4.6 mm i.d. × 250 mm); mobile phase, 15% $H_2O$ in $CH_3CN$ (A) and 100% $CH_3CN$ (B); Gradient (A) → (B) for 30 min, following to (B) for 20 min. Flowrate: 1.0 ml/min; detection: UV230 nm; injection: 20 µl. (Reproduced from ref. 148: *J. Food Hyg. Soc Japan*, (1989) **30**, p. 311, Fig. 5.).

acids (SuE), propylene glycol esters of fatty acids (PGE) and sorbitan esters of fatty acids. Analysis of these emulsifiers (MGL, SuE, PGE) uses 3,5-dinitrobenzoil chloride (DNBC) with the pre-column method to form DNBC derivatives followed by separation with reversed-phase HPLC and UV detection at 230 nm [147–149] and the method is applied for margarine, shortening, mayonnaise and ice cream (Fig. 1.2.8). The detection limits of MG, SuE, PGE are 10 ppm, respectively.

### 1.2.3.5 Other Food Additives

Phytic acid used as a sour agent is widely distributed in foods of plant origin. In most seeds phytic acid is found as magnesium phytate. Analysis of phytic acid is reported for the post-column method with sulfosalicylic acid and iron compound [150]. This reagent reacts with phosphates. The iron in the reagent is bound by the phosphate and thus removed, resulting in a decrease in color and is measured at VIS 500 nm. Phytic acid extracted with trichloroacetic acid from soy beans, grain sorghum, and tofu was separated on anion-exchange HPLC.

Polyphosphates are found in soft drinks and in livestock products to prevent discoloration and precipitation, and to improve binding capacity. Analysis of polyphosphates employs the post-column method with sulfosalicylic acid and iron compound similar to analysis of phytic acid [151]. The method applied for soft drinks uses separation on anion-exchange HPLC with gradient elution.

Sodium dioctylsulfosuccinate (DSS) is a wetting agent permitted as a food additive in a variety of products in the United States and in Canada. The use of post-column ion-pair extraction is reported for the analysis of DSS similar to cyclamate, a sweetener [100], and methylene blue is used as an ion-pair dye for the analysis of DSS. The method is applied for soft drinks.

## 1.2.4 Analysis of Veterinary Medicine

In the livestock and marine products industry, antibiotics, antibacterial agents and hormones have been used for prevention and treatment of animal diseases or as growth stimulants or for improvement in efficiency of feed utilization. The use of these drugs can leave residues in animal-derived foods that can pose a risk to public health. Therefore, sensitive methods for monitoring these residues are required to ensure that the animal tissues are safe for human consumption.

### 1.2.4.1 Sulfonamides

Among widely used chemotherapeutic agents, sulfonamides have a long history and are the most common veterinary medicine. As for antibacterial residue analysis in foods, considerable research papers for sulfonamides have been published and most of them use HPLC with UV detection. An HPLC method with fluorescence derivatization with OPA using pre-column labeling was applied to the analysis of sulfadimidine and 12 sulfonamides in pork ($\lambda$ex285 nm, $\lambda$em445 nm) [152,153]. In this method, the amino group was derivatized under acidic conditions at 60 °C for 30 min without using active thiol. However, fluorescence derivatization of the amino group with OPA is usually carried out in alkaline conditions with active thiol. Regarding derivatization with fluorescamine, the use of the pre-column method for the analysis of 5–8 sulfonamides in meat and milk [154–156], and of the post-column method for 12 sulfonamides in pork tissues [157] and 14 sulfonamides in salmon tissues [158] were reported. Also *p*-dimethylaminobenzaldehyde was used for the analysis of six sulfonamides in liver and kidney of chicken and pig with post-column derivatization monitoring at VIS 450 nm [159]. Detection limits, which differ in samples, for pre-column derivatization with OPA were 0.1–1 ppb and 2–10 ppb with fluorescamine, proving that derivatization with OPA was more sensitive.

### 1.2.4.2 Antibiotics

*Penicillin*

Penicillins are non-toxic, and used widely in veterinary medicine. However, improper use of the antibiotics in livestock farming may lead to human hazards. Trace amounts in the livestock food supply may cause allergic reactions in some individuals.

Analysis of penicillins with derivatization was reported using the post-column technique with 1,2,4-triazole/mercuric chloride to form UV detectable derivatives. This method was applied to the analysis of penicillin G [160,161] in dairy products such as milk, cheese and yogurt, and cloxacillin [162] in bovine tissues followed by separation with reversed-phase HPLC and detection at UV 325 nm after SPE (Bond Elute C18) clean-up. The detection limit of this method was 3–5 ppb. The pre-column method with formaldehyde/trichloroacetic acid was another fluorescence derivatization and applied for the analysis of amoxicillin [163] in catfish and salmon (Fig. 1.2.9) and ampicillin [164] in bovine milk. Fluorescent derivatives with this reagent were assumed to be 2-hydroxy-3-phenol-6-methylpyrazine for amoxicillin (Fig. 1.2.10) and 2-hydroxy-3-phenyl-6-methylpyrazine for ampicillin. These detection limits were 0.31–0.8 ppb, and this method has achieved enhanced sensitivity and detectability compared with 1,2,4-triazole/mercuric chloride in UV detection.

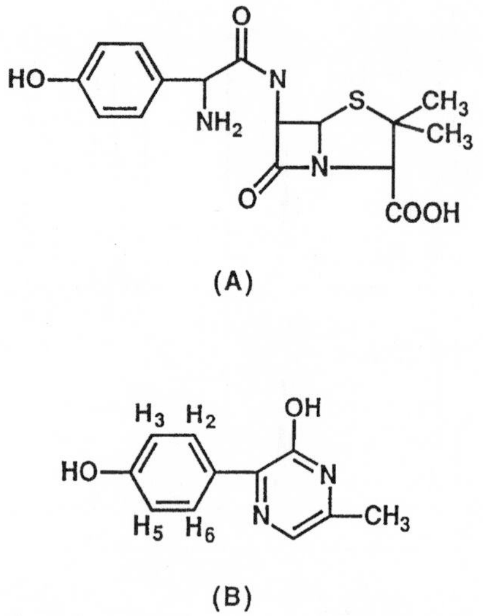

(A)

(B)

**Fig. 1.2.9.** Structures of: (A) amoxicillin; and (B) its major derivative, 2-hydroxy-3-phenol-6-methylpyrazine. (Reproduced from ref. 163: *J. AOAC Int.*,(1996) **79**, p. 395, Fig. 7.).

*Aminoglycoside Antibiotics*

Streptomycin (SM) and dihydrostreptomycin (DSM), representative aminoglycoside antibiotics, form the streptamine group of aminoglycoside antibiotics that consist of a disaccharide molecule with a streptidine (diguano derivative)

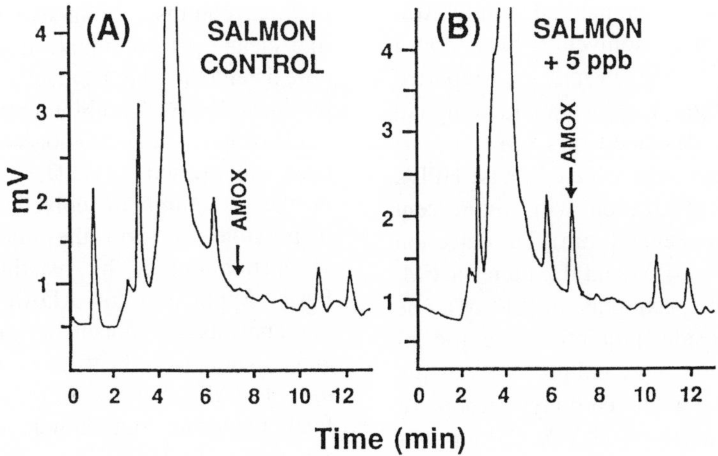


**Fig. 1.2.10.** Chromatograms of: (A) salmon muscle tissue control; and (B) salmon muscle tissue spiked with amoxicillin at 5 ng/g. (Reproduced from ref. 163: *J. AOAC Int.*, (1996) **79**, p. 394, Fig. 6.).

aminocyclitol moiety. Both drugs are approved for use in food-producing animals, and are available in various formulations for the treatment of a broad spectrum of local and systemic infections. However, both drugs are potentially toxic by causing damage in vestibular and auditory functions.

Derivatization of SM have been reported using fluorescence detection (λex400 nm, λem495 nm) to react with ninhydrin in an alkaline medium because SM possesses guanidino group. In this case, the derivatization reaction was completed with the introduction of sodium hydroxide using the post-column technique in the presence of ninhydrin in the mobile phase. The method applied for the analysis of beef, chicken and pork employs extraction with perchloric acid followed by SPE (C8) clean-up and ion-paired HPLC (detection limit 0.5 ppm) [165]. 1,2-Naphtoquinone-4-sulfonic acid (NQS) reacts under alkaline conditions with guanidino group to give fluorescent derivatives like ninhydrin, thus post-column method in the presence of NQS in the mobile phase was reported for the fluorescent detection of SM and DSM (λex347 nm, λem418 nm) [166–168]. The use of on-line enrichment of column-switching with the pre-column technique improved the detection sensitivity of SM and DSM in beef and milk samples [167,168]. These detection limits were 10 ppb for SM and 20 ppb for DSM. Ninhydrin and NQA specifically react with guanidino compounds to form fluorescent derivatives, but not with primary and secondary amines. Therefore, no peaks of fluorescent derivatives are observed on the chromatograms for aminoglycoside antibiotics without the guanidino group.

As for the analysis of aminoglycoside antibiotics of SM analogs, the use of the pre-column fluorescent derivatization method (λex400 nm, λem495 nm) with OPA/2-ME for kanamycin (KM), destomycin A (DM) and hygromycin B (HM) were reported [169,170]. KM, DM and HM in cattle kidney and pork were extracted with 10% trichloroacetic acid, and the extract was purified with Amberlite CG-50 and/or Dowex 1-X8. The detection limits were 0.1 ppm for DM and 0.3 unit/g for HM.

The use of the pre-column method with OPA/2-ME in the solid phase has been reported. This method was applied to the analysis of gentamycin (GM) in beef tissues and milk [171,172]. Retained on the Sep-pak silica, GM was derivatized on the silica phase with OPA/2-ME solution and the reaction compound was eluted with ethanol. The detection limit was 0.2 ppm.

The use of the post-column method with OPA/2-ME for the analysis of fradiomycin (FM) and DM in beef and neomycin (NM) in milk was reported [173,174]. The method employs extraction and clean-up procedure described in KM analysis followed by separation with reversed-phase HPLC and post-column derivatization with fluorescent detection (λex340 nm, λem450 nm). The detection limits were 0.1 ppm for DM and 0.2 ppm for FM.

For the analysis of spectinomycin (SPCM), one of other aminoglycoside antibiotics, the use of *p*-nitrophenylhydrazine, a pre-column derivatizing agent, which reacts with carbonyl group of SPCM followed by detection at VIS 420 nm was reported [175]. Sample reparation for chicken and pork tissues employs deproteinization with trichloroacetic acid and SPE (C18) clean-up. The detection limit was 0.05 ppm.

*Polyether Antibiotics*

Monensin, salinomycin, naracin, and lasalocid are used primarily as coccidiostats. It has been reported that narasin can be used as a growth stimulant, and monensin and lasalocid improve the efficiency of feed utilization. These antibiotics are a new class of drug known as carboxylic polyethers and have similar chemical structures.

Derivatization of monensin, salinomycin and naracin employs post-column method with vanilline followed by detection at VIS 522 nm. This method has been applied for the analysis of bovine tissues, milk, chicken tissues and feed [176–178]. Sample pretreatment employs extraction with aqueous methanol, followed by liquid-liquid extraction with dichloromethane, and clean-up with SPE (Sep-pak silica). The detection limits were 5–25 ppb. The post-column method was also applied for the analysis of semduramicin, a relatively new polyether antibiotic [179].

On the other hand, a method using pre-column technique was developed for the analysis of polyether antibiotics, i.e. allylic hydroxyl group of salinomycin was acidified with pyridinium dichromate to give $\alpha,\beta$-unsaturated keton and detected at UV 225 nm [180]. However, this method requires column-switching as well as off-line clean-up to eliminate interfering peaks from co-existents in the residue analysis of livestock products. Derivatization with ADAM (λex365 nm. λem418 nm) and 1-bromoacetylpyrene (BAP) has been widely used [181]. The carboxylic acid group in the structures of polyether antibiotics made it possible to form the fluorescent derivatives of each antibiotic by reacting with ADAM or BAP. Application of ADAM was reported for the analysis of monensin, salinomycin, naracin and lasalocid in beef liver [182,183]. For the use of BAP, analysis of polyether antibiotics in feed, monensin and salinomycin in chicken was reported [184,185]. These polyether antibiotics have ionophore action, and carboxylic and hydroxyl group form complex compounds with alkaline metal-ion, thus direct derivatization by ADAM proceeds with difficulty. Therefore, the extraction with chloroform under weak acidic condition in order to obtain free monensin (acid form) [186], acetylation of hydroxyl group with acetic anhydride to release ionophore state [182,183], or addition of crown ether such as Kryptofix 222 as a catalyst of ADAM has been tried [184,185]. The crown ether reacts with alkaline metal-ion to form complex compounds and the reaction is supposed to be the ionization of carboxyl group resulted in catalyzation of the reaction.

### 1.2.4.3 Antiparasitic Agents

*Avermectins*

Avermectins, which are isolated from the myceria of *Streptmyces avermitilis*, are a family of antiparasitic drugs which are potent broad-spectrum agents at very low dosage levels. Avermectins have been used for veterinary medicine, especially for antiparasitic agents for livestocks or pesticides for agricultural insect and mites. The former includes ivermectin, doramectin and moxidectin, while the latter includes avermectin $B_1$ and milbemectin. The usage of these drugs differs, but these avermectins have similar chemical structures consisting of a 16-membered macrocyclic lactone ring. Thus common derivatization method has been adopted, i.e., a fluorescence product (λex360 nm, λem475 nm) is formed through a chemical reaction of the drug with acetic anhydride which results in the formation of a conjugated dehydration product. This dehydration produced an aromatic system which is conjugated with the 8–11 diene to furnish the fluorophore. Being effective for end- and ectparasite, ivermectin has been used for veterinary medicine and the wide range of residue analysis in livestock products were reported using the pre-column HPLC method with fluorescent derivatization mentioned earlier. Samples of animal tissues and bovine milk were extracted with aqueous acetone followed by liquid-liquid extraction with isooctane, acetonitrile, and hexane in order [188–193]. MSPD combined with SPE (Sep-pak alumina) as simultaneous extraction and clean-up was applied for the analysis of bovine liver [13]. The derivatization procedure was common among the reports. A reaction with 1-methylimidazole/acetic anhydride/*N*-*N*-dimethylformamide $(2 + 3 + 9)$, a derivatizing agent, was carried out at 95 °C for 1 hour followed by silica gel column clean-up such as Sep-pak silica and the detection limit was 1–2 ppb. The analysis of moxidectin in cattle tissues uses the same derivatization as ivermectin [194].

Among the avermectins, avermectin $B_1$ was found to be the most effective against agricultural insects and mites. Avermectin residues degrade rapidly by both oxidative and photochemical pathways to various products when applied to various crops. However, the only residues of toxicological significance are avermectin $B_1$ and 8,9-*Z*-avermectin $B_1$ which are an isomer of avermectin $B_1$ isomerized by UV light below 280 nm. Analytical methods of these compounds were similar to those for ivermectin, but acetic anhydride used for the derivatization reagent was replaced by trifluoroacetic anhydride. The method included extraction with aqueous methanol or acetonitrile followed by SPE (aminopropyl) clean-up was applied for the analysis of dried hops [38], wine [195] and apple [196] and the detection limit was 1–2 ppb.

*Malachite Green*

Malachite green (MG) is a triarylmethane dye used for aquaculture as a parasiticide, commonly applied to newly laid fish eggs to inhibit fungal and protozoal infections. Some triphenylmethane dyes, including MG, are recognized as animal carcinogens.

Determination of MG employs detection with visible absorption at 610 nm and the reduction of malachite green to its colorless leuco base (LMG) is a facile process, which has been reported to occur in fish flesh. The 90% of the total malachite green species present in fish were in the leuco form. To determine MG and LMG simultaneously, post-column reactor packed with lead dioxide ($PbO_2$) was used to transform LMG to MG with detection at VIS 618 nm after HPLC separation [197]. Application for rainbow trout or cat fish used acetonitrile-buffer extraction, liquid-liquid extraction with dichloromethane, and SPE (alumina, propylsulfonic acid) clean-up (detection limit 1–2 ppb) [198,199]. An another method for acidification of LMG to MG employs post-column electrochemical oxidation [200]. This HPLC system consists of coulometricall efficient electrochemical cell (ESA Coulochem model 5100A; a potential of 0.45 V). Analysis for rainbow trout included extraction with dichloromethane/acetonitrile/perchloric acid and clean-up with SPE (Bakerbond C18). The limit of detection for the method is 3 ppb for MG and 6 ppb for LMG.

The oxidation efficiency from MG to LMG was reported to be 100.5% for $PbO_2$ column reactor [199] and 57% for electrochemical oxidation [200]. The use of $PbO_2$ column reactor requires replenishment for every 30–40 injection, while the use of the electrochemical cell needs occasional generation with 5 M nitric acid.

*Hormones*

For livestock and the marine industry, hormones have been used as a growth stimulant, for the improvement of efficiency of feed utilization and production of protein-rich lean meat, but some synthetic hormones such as diethylstilbestrol are

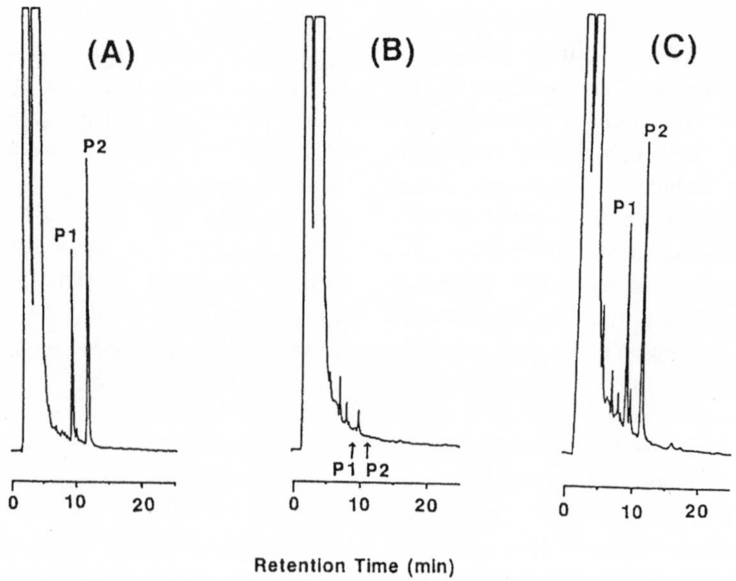


**Fig. 1.2.11.** High-performance liquid chromatograms of progesterone standard 2 ng (A), beef (B) and fortified beef as 20 ppb progesterone (C). (Reproduced from ref. 203: *J. Food Hyg. Soc. Japan*, (1995) **36**, p. 599, Fig. 6.).

known as carcinogens. Both natural and synthetic hormones have been regulated and their usage and residue levels are different in each country. Determination levels of hormones are quite low as little as ppb–ppt. Because the analytes of interest are animal tissues, analysis of these samples is easily affected by contaminants and requires a complicated procedure. Therefore, the use of GC/MS is common, and immunoassay which is a feasible technique for detection of trace analytes with a high sensitivity has been recently applied for the residue analysis. The HPLC method with derivatization was employed for the analysis of clenbuterol and progesterone. Belonging to a $\beta$-blocker with chemical structure of phenylethylamine, Clenbuterol is a basic molecule and consists of an aromatic ring bonded with the amino group. An analytical method for animal tissues such as liver includes extraction with diluted hydrochloride followed by clean-up with SPE (C18) [201] or enzymic degradation with subtilisine A followed by liquid-liquid extraction clean-up with Chem Elute (CE1020) [202], post-column derivatization with diazo-coupling reaction and determination with VIS (500 nm)-HPLC.

As for the analysis of progesterone, the carbonyl group reacts with dansylhydrazine (DNS-H) to form fluorescent derivatives under acidic catalyst ($\lambda$ex340 nm, $\lambda$em520 nm) [203]. In this case, progesterone reacts to form dansylhydrazon on the steroid framework with two positions of the second and twentieth carbon site, while reaction with the carbonyl group of the third carbon site produces both *anti-* and *syn-* stereoisomers represented by two peaks on the chromatogram (Fig. 1.2.11). Sample pretreatment includes extraction with acetonitrile, liquid-liquid extraction with dichloromethane, and clean-up with SPE (Bond Elute DEA), moreover Sephadex LH-20. This method was feasible to detect as low as 2 ppb progesterone.

## 1.2.5 Analysis of Pesticides

Among the analytical methods for pesticides found in foods, GC and GC/MS have been the main methods in use and the use of HPLC and CE has been seldom reported. However, pesticides which have large polarity are not easily analysed by GC and some pesticides are unstable to the high temperatures used by GC. For the analysis of these pesticides, the HPLC method has been developed.

*N*-methylcarbamates are carbamates with *N*-methyl substituents including carbamide acid ester and its salts. These carbamates were analysed by following technique, which involved a reversed-phase separation followed by a post-column base hydrolysis that liberated methylamine, which further reacted with OPA/thiol reagents to form a highly fluorescent isoindole (λex340 nm, λem455 nm). The general system includes separation of several *N*-methylcarbamates on a reversed-phase column with gradient elution followed by hydrolysis (ca 100 °C) with sodium hydroxide to form methylamine and derivatization with OPA/2-ME (or 3-MP). Recently, hydrolysis without an alkaline solution was developed, and the use of catalytic solid-phase hydrolysis such an anion-exchange resin as Aminex A-27 and magnesium oxide (Fig. 1.2.12) [204,205], UV irradiation, decomposition with a photolysis reactor [30] and a single-stage reaction of hydrolysis and derivatization with a mixed solution of alkaline solution and OPA/thiol reagent was reported [15]. Preliminary treatment of GPC [25], SFE [30], MSPD [15], and SPE [205,206] for several vegetables and fruits was investigated. Simultaneous multi-component analysis of *N*-methylcarbamates and their metabolites with gradient elution on reversed-phase HPLC is customarily carried out. This post-column method for *N*-methylcarbamates offers simplicity and sensitivity, but numerous interference peaks from co-existents appeared when the method was applied for the analysis of such citrus fruit as grapefruit. Therefore, this method is not suitable for identification but screening.

The approach with liberation of methylamine from *N*-methylcarbamates using a post-column photolysis reactor mentioned earlier has been examined for the analysis of other carbamates, nitorogenous pesticides and organophosphates [207,208]. Depending on analyte pesticides, detection with fluorescent or ECD sensitive products after photolysis [208], and with fluorescent OPA derivatives with a liberated allylamine such as methylamine was reported [207]. These methods were applied to standards. For food application, bentazon (BEN), pyrazoxyfen (PYR) and chinomethionate (CIN) in brown rice was analysed by post-column photolysis with UV irradiation followed by fluorescent detection (λex329 nm, λem415 nm for BEN; λex333 nm, λem405 nm for PYR; λex377 nm, λem450 nm for CIN) (Fig. 1.2.13) [209]. In this case, disappearance of interfering peaks and improvement of sensitivity was observed compared to detection with UV 215 nm (for BEN), UV 251 nm (for PYR), and UV 260 nm (for CIN) (Fig. 1.2.14). The detection limit was 10 ppb, respectively.

## 1.2.6 Analysis of Natural Toxins

### 1.2.6.1 Mycotoxins

*Aflatoxins*

Mycotoxins are toxic metabolites produced by fungi and more than 300 mycotoxins are known. Among them, aflatoxins yielded by *Aspergillus flavas* are the most potent carcinogens in nature, and the legislative level of aflatoxins is set in most countries. More than 10 analogues of aflatoxins $B_1$, $B_2$, $G_1$, $G_2$, $M_1$, $M_2$, $P_1$, $Q_1$ and aflatoxicol are recognized. Strong carcinogenicity and contamination prevalence of aflatoxin $B_1$, $B_2$, $G_1$, $G_2$ and $M_1$ in food necessitates the monitoring of these compounds. These aflatoxins have native fluorescence (λex365 nm, λem450 nm), but the fluorescence intensity of $B_1$ and $G_1$ is weaker than that of $B_2$ and $G_2$. Therefore, the use of fluorescence enhancement by hydration of $B_1$ and $G_1$ with trifluoroacetic acid (TFA) to hemiacetals, i.e., reaction of double bonding of terminal furan ring in $B_1$ and $G_1$ with TFA is widely accepted [37]. In this case, $B_2$ and $G_2$ do not react with TFA. The method applied for the analysis of wine and fruit juice used reversed-phase HPLC with pre-column technique, and its detection limit was 0.02 ppb.

A technique has been developed using post-column derivatization with aqueous iodine to enhance the fluorescence of aflatoxins $B_1$ and $G_1$, i.e., after separation by reversed-phase HPLC [210]. Aflatoxins $B_1$, $B_2$, $G_1$ and $G_2$ were

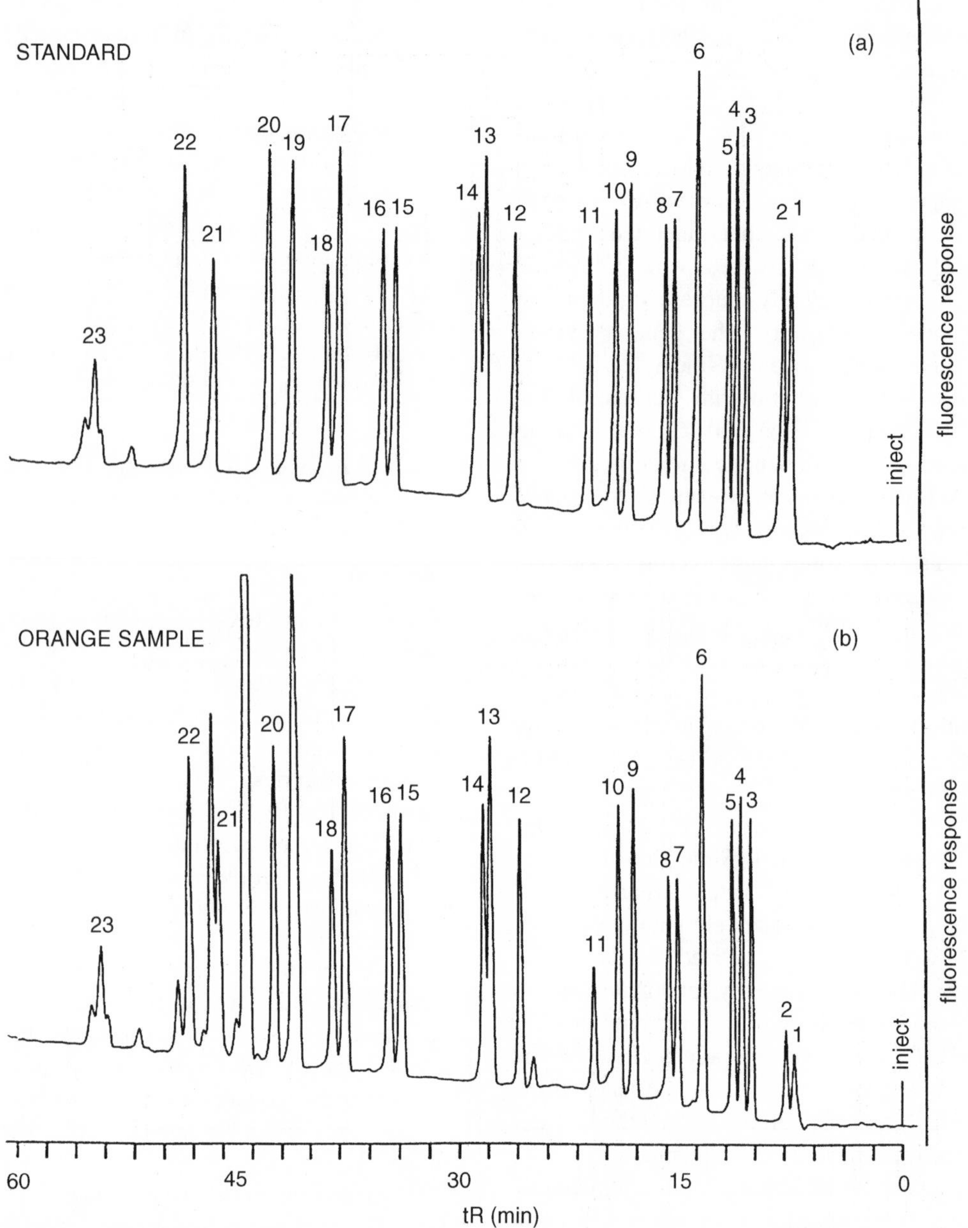


**Fig. 1.2.12.** Liquid chromatograms of: (a) a standard mixture of 23 *N*-methylcarbamates; and (b) an orange sample fortified with 23 *N*-methylcarbamates at 0.13 mg/kg level after extraction and clean-up on aminopropyl-bonded silica SPE column (by ASPEC): 1, butocarboxim sulfoxide; 2, aldicarb sulfoxide; 3, butoxycarboxim; 4, aldicalb sulfone; 5, oxamyl, 6, methomyl; 7 ethiofencarb sulfoxide; 8, thiofanox sulfoxide; 9, 3-hydroxycarbofuran ethiofencarb sulfone; 10, methiocarb sulfoxide; 11, dioxacarb, thiofanox sulfone, tranid; 12, methiocarb sulfone, butocarboxim; 13, aldicarb; 14, 3-ketocarbofuran, cloethocarb; 15, propoxur, 16, carbofuran, bendiocarb; 17, carbaryl, thiofanox; 18, ethiofencarb; 19, isoprocarb, landrin; 20, carbanolate; 21, methiocarb; 22, promecarb; 23, bufencarb, mexacarbate, aminocarb. (Reproduced from ref. 205; *J. AOAC Int.*, (1992) **75**, p. 1067, Fig. 1.).

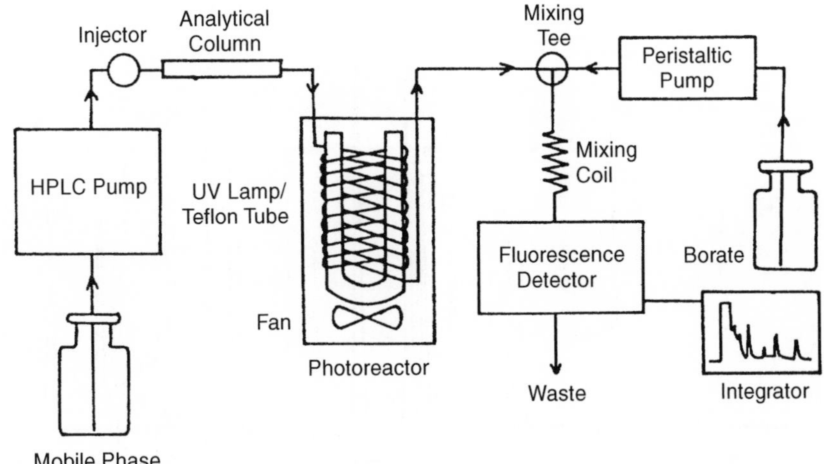


**Fig. 1.2.13.** Schematic diagram of HPLC post-column UV photolysis fluorescence detection system. (Reproduced from ref. 209; *J. Food Hyg. Soc. Japan*, (1995) **36**, p. 602, Fig. 1.).

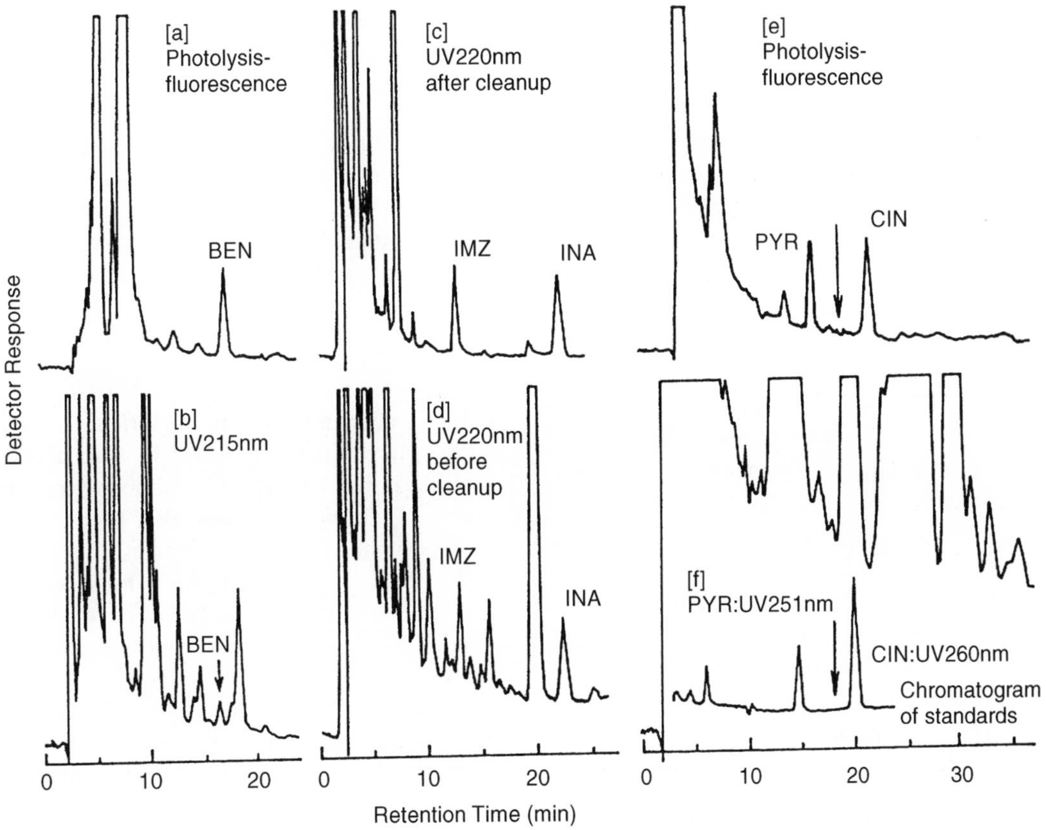

**Fig. 1.2.14.** Chromatograms of BEN [a,b], IMZ/INA [c,d] and PYR/CIN [e,f] in rice fortified at 0.025 μg/g. Arrows in [e,f] indicate changing wavelength. (Reproduced from ref. 209; *J. Food Hyg. Soc. Japan*, (1995) **36**, p. 604, Fig. 2.).

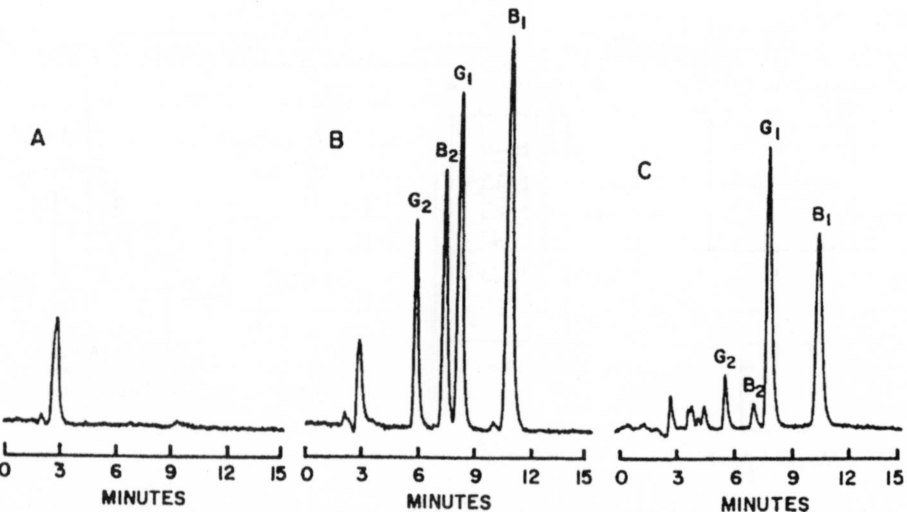


**Fig. 1.2.15.** A, Chromatogram of aflatoxin-free peanut sample prepared as described in Method; B, chromatogram of aflatoxin-free peanut sample that was spiked with standard solution to give 20 ng $B_1$ and $G_1$/g and 6 ng $B_2$ and $G_2$/g before extraction and clean-up; C, chromatogram of naturally contaminated peanut sample diluted 1:10 before LC injection based on minicolumn analysis. (Reproduced from ref. 210: *J. Assoc. Off. Anal. Chem.*, (1988) **71**, p. 46, Fig. 2.).

derivatized to produce fluorescence with the introduction of 10 mg% iodine solution to the reaction coil which is thermostated at 75 °C. Fig. 1.2.15 shows the chromatogram for peanuts extracted with chloroform followed by laminated clean-up with Florisil and neutral alumina. The detection limit using the system was 0.1 ppb for aflatoxins $B_1$ and $G_1$.

A unique HPLC system was reported, which incorporates HPLC in-line derivatization of aflatoxins $B_1$ and $G_1$ by post-column introduction of $\beta$-cyclodextrin solution [211]. Cyclodextrins are chiral, toroidal-shaped formed by the action of the enzyme cyclodextrin transglycosylase on starch. These cyclic oligomers contain from 6 to 12 glucose units bonded through alpha linkage. The three smallest homologues ($\alpha$, $\beta$, and $\gamma$) are available commercially. In this case, $\beta$-cyclodextrin provides an excellent cavity for the aflatoxins, thus creating an inclusion complex formation interaction. The transparency of $\beta$-cyclodextrin allows the partially encased aflatoxin to be sufficiently excited by UV irradiation, and concomitantly minimizes the intramolecular motion so that fluorescence enhancement occurs ($\lambda$ex365 nm, $\lambda$em428 nm). The enhancement is similar to that observed when aflatoxins are in contact with solid supports, such as silica gel. The method was applied to the analysis of corn.

### *Fumonisins*

Fumonisins (fumonisin $B_1$, $B_2$ and $B_3$), a group of toxins produced mainly by *Fusarium moniliforme* strains, have been implicated in leukoencephalomalacia in horses and with pulmonary dedema in swine. Their presence has also been associated with oesophagal cancer in humans.

To detect fumonisins with high sensitivity, various derivatizations for HPLC have been developed. In the early stages under alkaline conditions, maleyl derivatives reacted with maleic anhydride with UV detection (230 nm) [212], fluorescent detection with fluorescamine derivatives ($\lambda$ex390 nm, $\lambda$em475 nm) [212] and OPA/2-ME derivatives ($\lambda$ex335 nm, $\lambda$em440 nm) [213] were reported. However, maleyl derivatization applied to the analysis of corn was unsuccessful. Base-line separation of the fumonisin

$B_1$ peak was not possible due to the presence of substrate matrix interference. Disadvantages of the latter two fluorescence derivatizing reagents for HPLC are formation of dual peaks with fluorescamine and instability of the derivatives with OPA/2-ME (half-life period 1–2 min). The dual peaks from fluorescamine resulted from the formation of the acid alcohol and the lactone derivatives of the fluorescent complex both exhibiting identical fluorescent characteristics. To solve these problems, the use of derivatizing reagents such as 4-fluoro-7-nitrobenzofurazan (NBD-F) (λex460 nm, λem500 nm) [214], 4-(*N*,*N*-dimethylaminosulfonyl)-7-fluoro-2,1,3-benzoxadiazole (DBD-F) (λex450 nm, λem590 nm) [215], naphthalene-2,3-dicarboxaldehyde (NDA) (λex420 nm, λem500 nm) [214,216], FMOC (λex263 nm, λem313 nm) [217], AQC (λex395 nm, λem418 nm) [218] has been considered. Determing the usefulness of each reagent is not easy, but reacting conditions with pre-column method are: 1 min at 60 °C for NBD-F; 60 min at 60 °C for DBD-F; 30s at room temperature for FMOC; and 10 min at 55 °C for AQC, and the reacting conditions for FMOC appear to be the best. The stability of fumonisin derivatives at room temperature is reported as: 30 min for NBD-F; 48 hours for DBD-F; 24 hours for NDA; 72 hours for FMOC and 48 hours for AQC, the use of FMOC is also excellent for stability. However, the derivatives with FMOC require removal of excess reagents after reaction and extraction with pentane is carried out.

The use of the post-column method with OPA/NAC was developed [219]. A derivative with OPA/NAC is more stable than that with OPA/2-ME and the reagent has been used for the analysis of amines with the on-column method.

Thus, various derivatization methods for fumonisins have been investigated, extraction and clean-up procedures for samples such as corn are almost the same; after extracted with 50% acetonitrile or 75% methanol, samples are subject to clean-up with SPE (C18 and/or SAX). A clean-up using immunoaffinity column (Fumonitest column) specific to fumonisin $B_1$ after 80% methanol extraction was also reported [220].

### 1.2.6.2 Marine Toxins

*Paralytic Shellfish Poisoning Toxins*

Poisoning by seafoods contaminated with marine toxins are posing serious problems to public health and fishery industries worldwide. Paralytic shellfish poisoning (PSP) is a potentially deadly illness caused by ingestion of shellfish that have accumulated potent neurotoxins produced by marine algae such as those of the genus *Alexandrium*. The group of compounds known as PSP toxins is composed of saxitoxin and its derivatives. For toxin detection, derivatization of the toxins to fluorescent compounds by oxidation under alkaline condition yields methods with high sensitivity and fairy good specificity for the toxins [221]. This chemical assay is 100 times more sensitive than the existing bioassay, and pre-column and post-column fluorescent derivatization has been applied to HPLC. The representative pre-column method employs peroxide or periodate as an oxidizing agent. PSP toxins in mussel were thermally extracted with diluted hydrochloride followed by clean-up with SPE (C18 and SAX), oxidization with peroxide or periodate to form PSP fluorescent derivatives, separation with reversed-phase gradient HPLC and detection (λex330 nm, λem400 nm) [222,223]. The main disadvantage of the pre-column method is that some toxins (PSP analogues) cannot be separated (e.g., neosaxitoxin and $B_2$, GTX-1 and GTX-4, GTX-2 and GTX-3) on HPLC due to the similar chemical structures of PSP oxidized products.

On the other hand, available post-column methods employ periodate as a reaction reagent [224], and electrochemical oxidation (a potential of 0.75V) using a electrochemical detector as post-column reactor (Fig. 1.2.16) [225]. As the fluorescence intensity of PSP oxidized products is the strongest at pH 5 not under alkaline conditions, either method requires neutralization with introduction of diluted acetic acid in flow after post-column oxidization. The major drawback of basic electrochemical oxidation was the noisy baseline signal, caused by fluctuations in flow from the post-column pumps and thus variation in the effluent composition.

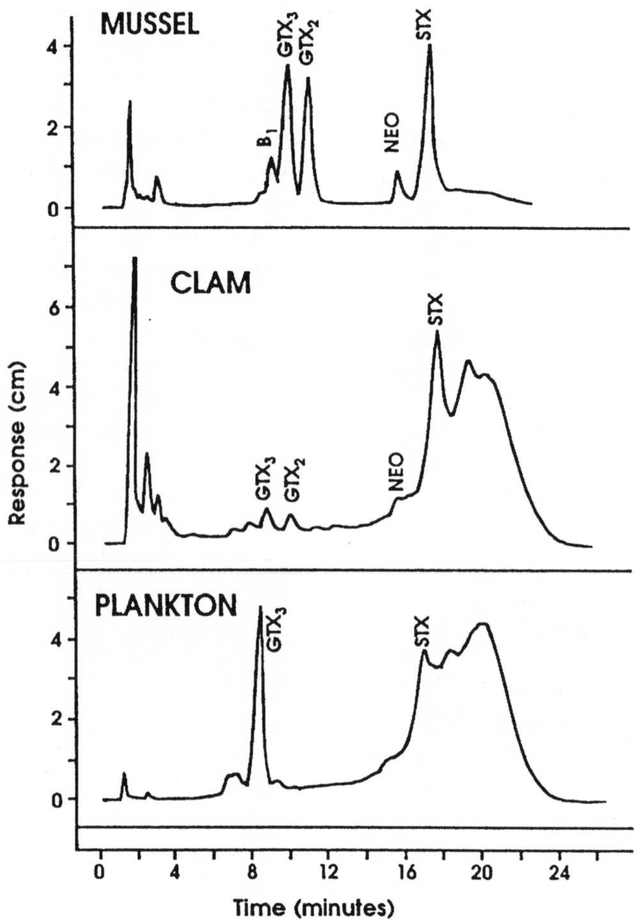


**Fig. 1.2.16.** Chromatograms depicting separation of PSP toxins in various extracts. The mussel extract contained total PSP toxin at 25 µg/g, and the calm extract was contaminated at 1 µg/g (0.5 g tissue ml; dilution factor, 1/30 for both samples). Plankton chromatogram is qualitative only. Integrator attenuation was set at 8 for the mussel sample and at 5 for the clam and plankton samples. (Reproduced from ref. 225: *J. AOAC Int.*, (1995) **78**, p. 703, Fig. 9.).

*Diarrhetic Shellfish Poisoning Toxins*

Diarrhetic shellfish poisoning (DSP) is a severe gastrointestinal illness caused by consumption of shellfish contaminated with toxigenic dinoflagellates. The main toxins responsible for DSP are okadaic acid (OA), dinophysistoxin-1 (DTX-1), DTX-2, and DTX-3 which are lipophilic polyether compounds.

HPLC with pre-column fluorescent derivatization in the presence of ADAM has been used in many laboratories worldwide [226]. A representative new method for DSP toxins analysis on mussels employs aqueous-methanolic extraction and partitioning into chloroform followed by SPE ($NH_2$) clean-up and derivatization with ADAM (Fig. 1.2.17) [227]. A problem with ADAM is its instability, which leads to decomposition products that interfere in the analysis. To solve this problem, the method for *in situ* formation of ADAM from the stable 9-anthraldehyde hydrazone has been used for the derivatization of DSP toxins [228].

A silica or Florisil column clean-up after the ADAM reaction is necessary to eliminate decomposition products of the reagent itself or reaction by-products [226]. A method using

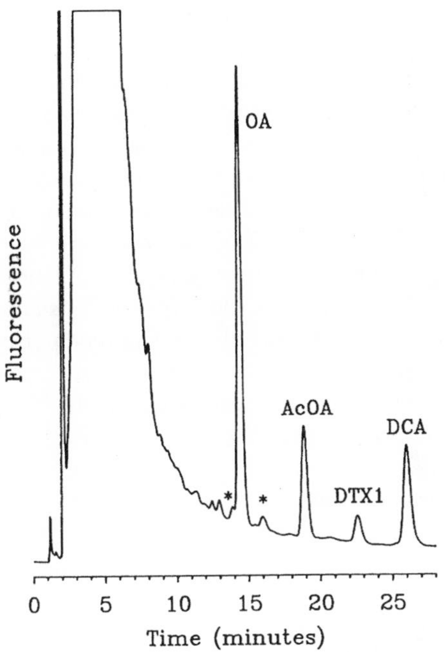


**Fig. 1.2.17.** LC-FLD chromatogram for MUS-2, a mussel tissue reference material, after extraction, derivatization with ADAM, and clean-up by optimized silica SPE. Isomers of OA are marked with asterisks. (Reproduced from ref. 227: *J. AOAC Int.*, (1995) **78**, P. 568, Fig. 11.).

a column-switching system was introduced instead of the clean-up procedure [229], but chromatograms of samples were interfered with by co-existences. Thus SPE is preferred. The toxicity of ADAM has not been reported, but it is a potential carcinogen and skin irritant.

Other derivative reagents for DSP toxins are BAP [230]. 9-CA [231,232], 2,3-(anthracenedicarboximido)ethyl trifluoromethanesulfonate (AE-OTf) [233]. According to the report comparing BAP and ADAM, a clear baseline signal was observed when BAP was used, but the sensitivity was as low as 1/4 that of ADAM [230]. Being stable, 9-CA replaces ADAM, but its reactability is inferior to ADAM. Reaction with ADAM proceeds at room temperature (or 37 °C), while the reaction with 9-CA requires 75 °C or 90 °C. A similar derivatization reaction occurs when 9-CA is used instead of ADAM; 9-CA yields the same fluorescent products (9-anthrylmethyl ester) of OA and DTX-1 as the ADAM reaction, and the similar clean-up was used. The big advantage of 9-CA is that it is relatively inexpensive and it is shelf-stable. Even the reagent solutions once prepared are stable for at least a week if refrigerated.

On the other hand, being developed as a labeling reagent for fatty acid [234], AE-OTf easily reacts with OA and DTX-1 and the derivatization reaction completes within 10 min at room temperature. Moreover, the sensitivity of fluorescent derivatives is high ($\lambda$ex298 nm, $\lambda$em462 nm), the detection limit for mussel and scallop was 20–30 ng/g [233].

*Tetrodotoxin*

In some Asian countries, pufferfish is a highly esteemed food. Most incidents of intoxication by tetrodotoxin (TTX) are caused by accidental intake of toxic parts of the pufferfish such as the liver. TTX is known to give a fluorescent 2-amino-quinazoline derivative when heated in a strong alkaline solution [235]. This reaction is applied for post-column HPLC method [236,237]. The analytical procedure includes extraction of TTX with diluted acetic acid, defattening with diethyl ether, clean-up with Amberlite CG-50 and separation with cation-exchange or reversed-phase HPLC. Post-column reaction employs 4N-NaOH solution and fluorescent derivatization is carried out in boiling water ($\lambda$ex357 nm, $\lambda$em510 nm). The detection limit is 5 ng/injection and the HPLC method is far more sensitive and specific than mouse bioassay. This method was useful for investigating naturally occurring TTX analogues in puffers and other animals [237,238].

*Ciguatoxin*

Ciguatera is a term given to a peculiar form of poisoning caused by ingestion of fish that live mainly in coral reef areas. Ciguratoxin (CTX) has a primary alcohol group reactive to fluorescent reagents [239]. Reaction with 1-anthrylcarbocyanide produced a fluorescent ester of CTX, which was successfully determined by

HPLC [240]. While the method was useful in identifying CTX as the principal toxin in carnivorous fish, establishing a proper clean-up procedure is difficult due to the extremely low concentration of CTX required in fish flesh to cause human illness. Furthermore, fluorometric methods have not been reported for the detection of CTX precursors and several congeners of CTX which lack primary hydroxyls. Further exploratory efforts are needed in the development of highly sensitive HPLC methods for the determination of CTXs.

#### 1.2.6.3 Plant Toxin

Hypoglycin A (HG-A; 2-amino-4,5-methylenehex-5-enoic acid) is a water-soluble toxic compound found in unripe fruits and seeds of ackee (*Blighia sapida*). The unripe ackee aril (freshly edible material) contains HG-A at 100–111 mg/100g. When the fruit is ripe, the HG-A levels decrease to less than 10 mg/100 g. Cooked ripe ackee fruit is non-toxic and has been a staple part of the Jamaican diet for centuries. The ingestion of unripe ackee fruit products, however, is associated with Jamaican 'vomiting disease'. HG-A, the toxic component of ackee fruit, was identified as the causative agent of Jamaican vomiting sickness, characterized by repeated vomiting, severe acidosis, and hypoglycemia accompanied by depletion of liver glycogen. As canned ackee fruit is exported from Jamaica, the concern does exist that the high levels of HG-A may be present in these products if immature fruit is packed.

The use of HPLC method for the analysis of HG-A in canned ackee fruit was reported and the method included 80% ethanol extraction followed by pre-column derivatization in the presence of PITC and UV detection [241].

### 1.2.7 Analysis of Food Contaminants

#### 1.2.7.1 Amines

Amines in foods are produced by enzymatic degradation associated with fermentation or putrefaction, and amine levels are shown to be an index of food decomposition. These amines include volatile amines such as methylamine and trimethylamine, and non-volatile amines such as histamine, putrescine and spermidine. Mucosa stimulation is a generally accepted physiological activity of amines, but the amines usually do not cause any danger to people unless large amounts are ingested. However, some of the amines like spermidine and spermine may be nitrosated in the presence of nitrite and act as precursors of mutagenic or carcinogenic nitrosamines. Ingestion of large amounts of amines such as tyramine causes a blood pressure rise and migraine. Histamine is known to be a cause of allergy-like food poisoning. Some polyamines such as putrescine, speridine and spermine are known to potentiate hitamine-toxicity and to act synergistically in allergy-like food poisoning. Therefore, determination of amine content is necessary to control the quality and evaluate the safety of a variety of foods. A number of methods have been reported for the analysis of amines.

As amines have no prominent functional groups except the amine group, some derivatization is required for the analysis of amines and its applications are reported. Amines in canned fish were detected by UV absorption using pre-column derivatization with benzoylchloride [242]. Pre-column technique with diazonium coupling reagent was used for the determination of histamine in fish with VIS 420 nm detection [243]. The use of pre-column derivatization with DNS was reported for the analysis of amines in cheese, fish, wine, soy-sauce and dried sausage with UV or fluorescent detection (the detection limit is 1–20 ppm) [244–246]. Pre-column derivatization with FMOC was applied to the analysis of tyramine, $\beta$-phenylethylamine and tryptamine in chicken with UV detection [247] and amines in wine, fruit or vegetable juice, fish, cheese and amino acids with fluorescent detection ($\lambda$ex265 nm, $\lambda$em315 nm) [74]. These fluorescent derivatization reagents themselves and their hydrolysis products such as 9-fluorenylmethanol have strong fluorescence, and thus the reagents are used for only pre-column derivatization in the HPLC method. After the derivatization reaction, excess reagents should be removed by addition of proline for DNS, heptylamine for FMOC or extraction by pentane. Bis-substituted derivatives of histamine and tyramine are sometimes found when the analyte is derivatized with DNS and FMOC.

On the other hand, OPA, which does not fluoresce by itself, can be used either pre-column and post-column, or for on-column derivatization. OPA, generally used in pairs with active thiol compounds such as 2-ME, 3-MP or NAC, solely reacts with histamine to form a specific fluorescent derivative. This pre-column derivatization was applied for the analysis of histamine in tuna and grapefruit juice [249]. The use of pre-column derivatization with OPA/2-ME was reported for the analysis of amines such as histamine, tyramine and polyamines in wine, followed by reversed-phase column chromatography with gradient elution and detection with ECD [250]. Post-column derivatization was also used for the analysis of amines in dried herring, vegetables, fish and meat products, beers, malts, and hops [251–254]. Detection limits of these samples were 0.3–1 ppm.

The use of on-column derivatization with OPA/NAC was applied to the determination of histamine and its metabolite, 1-methylhistamine, in soy-bean products such as soy-sauce and bean paste (miso), and fish [255], and for the simultaneous determination of histamine and polyamines with isocratic elution [256]. These methods involve pretreatment of simultaneous deproteinization and extraction with trichloroacetic acid, followed by clean-up with Amberlite CG-50, and formation of OPA fluorescent derivatives using on-column derivatization, and chromatography on the reversed-phase column (detection limit 0.5–20 ppm). On-column derivatization with OPA/NAC uses an alkaline eluent, therefore the HPLC column should be tolerant towards alkaline, and a polymer based reversed-phase column such as Asahipak ODP-50 was selected. Moreover, on-column derivatization with OPA/NAC was applied for the analysis of amines in wine [257]. This application simplifies pretreatment procedure using SPE (C18 or SAX) for clean-up. The detection limits were 0.1–0.3 ppm.

Several automated on-line derivatization of amines were reported. For example, controlled by computerized autosampler, derivatization was automatically carried out in the pre-column [258]; after HPLC separation in the presence of ninhydrin in the mobile phase, derivatization was completed with heating followed by VIS 546 nm detection [259]. This method might be categorized into a post-column derivatization.

To perform on-line derivatization, HPLC instrumentation equipped with column-switching has been developed. The reaction compounds of FMOC and amines in an autosampler syringe are introduced into pre-column while excess reagent and co-existent materials are removed by the end-cut technique, and then FMOC derivatives are forwarded to chromatographic separation and detection [260]; after being purified on a cation-exchange polymer pre-column, amines were back-flashed from the pre-column and labeled with OPA fluorophore in the presence of OPA/NAC in the mobile phase using on-column derivatization, followed by chromatographic separation on a reversed-phase C18 column and detection (Fig. 1.2.18). The method was applied for the analysis of amines in anchovy and dry sausage (Fig. 1.2.19) [261].

#### 1.2.7.2 Organotin

Organotin, a generic name for a tin compound whose tin and carbon are directly combined, is represented by the chemical formula of $RnSnX_4$-n ($n \leqq 4$). The value of Sn is four, and four forms of tetra-, tri-, di- and mono compounds exist. R shows alkyl groups such as the butyl group, phenyl or allyl groups. X reveals a halogen represented by chlorine, oxide, hydroxyl group or an anion. Organotin has been used for fungicide, insecticide and wood preservatives, and also antifouling paints for ships and fishing nets. During 1980 highly concentrated pollution of organotin on cultured fish and seafoods was revealed, and the malgrowth and malbreeding of cultured fish was feared. As for the effects on organisms of organotin, the degree of acute toxicity against mammals is weakened by the following order of tri- $\geqq$ tetra- $>$ di- $>$ mono compounds, and the toxity- of alkyl tin is stronger than that of allyl tin. Amongst several alkyl tins, an alkyl group consisting of shorter carbon chains

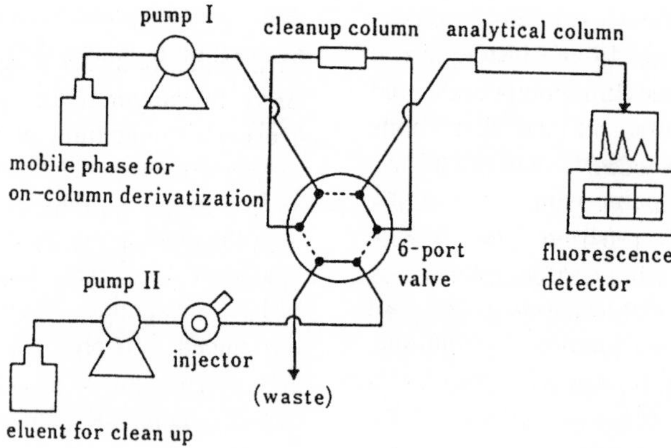


**Fig. 1.2.18.** Schematic diagram of on-column derivatization HPLC with automated sample clean-up system.

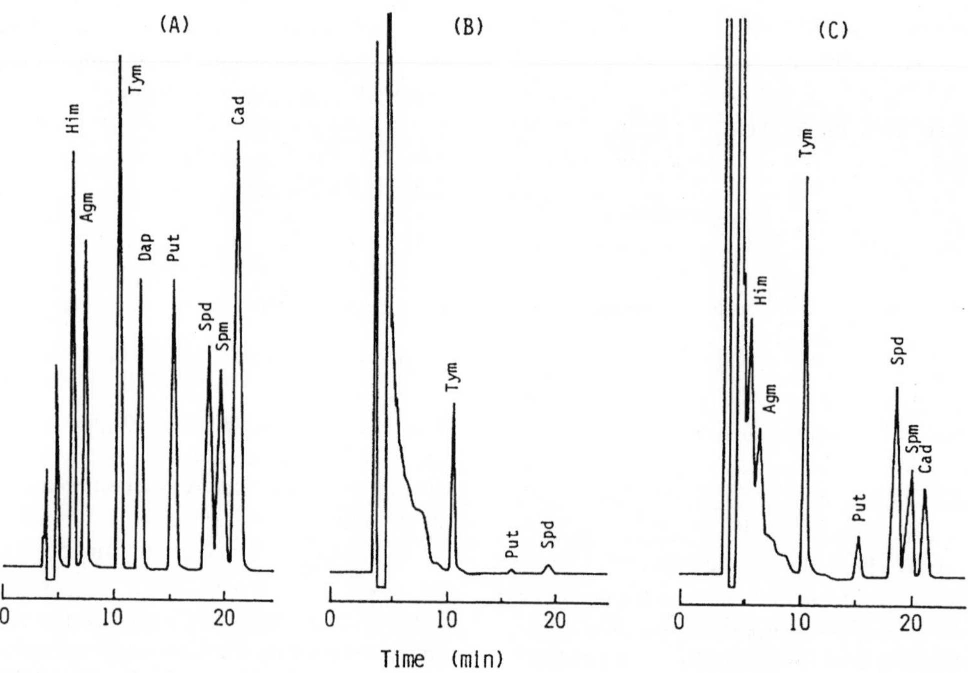


**Fig. 1.2.19.** Typical chromatograms of amines from: (A) standards; (B) anchovy; and (C) dry sausage. Standards: Him (4 ng), Agm (4 ng), Tym (10 ng), Dap (40 ng), Put (20 ng), Spd (20 ng), Spm (80 ng), Cad (20 ng). (Reproduced from ref. 261: *J. Food Hyg. Soc. Japan*, (1995) **36**, p. 640, Fig. 1.).

shows stronger toxicity, however, carcinogenicity and teratogenicity are not known at present.

For the analysis of alkyltin, ECD-GC and FPD-GC are the main methods in use and HPLC is less used. Sample preparation for HPLC is simpler than that for the GC method. The use of HPLC for the analysis of alkyltin has been reported after alkali decomposition. Bis(tri-*n*-butyltin) oxide (TBTO) in fish was converted to tri-*n*-butyltin chloride(TBTC) with hydrochloric acid, followed by clean-up with an ion-exchange column [262]. A post-column fluorescent derivatization of octyltin compounds, dodecyltin compounds and dibutyltin (DBT) in peanut oil and diet foods to form the fluorescent complex (λex420 nm, λem495 nm) with molin reagent (2′,3,4′,5,7-pentahydroxyflavon) was reported [263]. DBT and dioctyltin compounds in food containers and packings made of polyvinylchloride were analysed following purification in the pre-column equipped with column-switching, followed by post-column derivatization with molin reagent and fluorescent detection [264].

#### 1.2.7.3 Hydroperoxides

Lipid peroxidation has received much attention because of its toxicity, bitter taste and off-flavors. To prevent lipid peroxidation of foods, some efforts have been made such as packing, adding antioxidation reagents, and removing oxygen. Lipid peroxidation *in vivo* is also speculated to relate to some diseases or aging. Therefore, it is necessary to measure lipid hydroperoxides.

The most widely used method for the study of lipid peroxidation in biological materials is the thiobarbituric acid method. The use of the post-column HPLC method with luminol and isoluminol has been popular [265–267]. Recently, post-column HPLC with diphenyl-1-pyrenylphosphine (DPPP) was developed for the analysis of lipid peroxides in edible oil [268]. In this method, hydroperoxides in vegetable oil, butter and margarine are dissolved in chloroform or extracted with chloroform-methanol (2:1), separated on the reversed-phase HPLC and reacted with DPPP using the post-column method. Hydroperoxides oxidize DPPP quantitatively to a strongly fluorescent oxide (λex352 nm, λem380 nm). DPPP itself has almost no fluorescence but its oxidized product, DPPP oxide, has strong fluorescence. DPPP was less reactive to dialkyl peroxides, and not reactive to unoxidized fatty acids, hydroxy acids and their esters [269]. This method detected 2 pmol triacylglycerol monohydroperoxides.

#### 1.2.7.4 Isocyanate Monomers

Within the food packaging industry isocyanates are used in polyurethane polymers and adhesives. During manufacture, residual unpolymerized isocyanate monomer can remain in the polymer and may migrate into food that subsequently comes into contact with the polymer. Analytical methods for isocyanate have been mainly employed for the analysis of air, using tryptamine [270] and 9-(*N*-methylaminomethyl) anthracene (MAMA) as derivatizing reagents. The use of the pre-column method with MAMA was reported for the analysis of polyurethane used for food containers and packaging and isocyanate in laminate samples [271]. This method determines 10 kinds of isocyanate as MAMA derivatives followed by separation on reversed-phase HPLC and fluorescent detection (λex254 nm, λem412 nm). The detection limit is 0.03 ppm.

## REFERENCES


[1] Long, A.R., Hsieh, L.C., Malbrough, M.S., Short, C.R. and Barker, S.A. (1990) *J. Assoc. Off. Anal. Chem.*, **73**, 868.

[2] Reimer, G.J. and Suarez, A. (1991) *J. Chromotogr.*, **555**, 315.

[3] Reimer, G.J. and Suarez, A. (1992) *J. AOAC Int.*, **75**, 979.

[4] Walker, L.V., Walsh, J.R. and Webber, J.J. (1992) *J. Chromatogr.*, **595**, 179.

[5] Long, A.R., Short, C.R. and Barker, S.A. (1990) *J. Chromatogr.*, **502**, 87.

[6] Long, A.R., Hsieh, L.C., Malbrough, M.S., Short, C.R. and Barker, S.A. (1990) *J. Agric. Food Chem.*, **38**, 423.

[7] Poucke, L.S.G., Depourcq, G.C.I. (1991) *J. Chromatogr. Sci.*, **29**, 423.

[8] Long, A.R., Hsieh, L.C., Malbrough, M.S., Short, C.R. and Barker, S.A. (1991) *J. Assoc. Off. Anal. Chem.*, **74**, 292.

[9] Long, A.R., Hsieh, L.C., Malbrough, M.S., Short, C.R. and Barker, S.A. (1990) *J. Agric. Food Chem.*, **38**, 430.

[10] Schenck, F.J., Barker, S.A. and Long, A.R. (1992) *J. AOAC Int.*, **75**, 659.

[11] Jarboe, H.H. (1992) *J. AOAC Int.*, **75**, 428.

[12] Long, A.R., Malbrough, M.S., Hsieh, L.C., Short, C.R. and Barker, S.A. (1990) *J. Assoc. Off. Anal. Chem.*, **73**, 860.

[13] Scheneck, F.J., Barker, S.A. and Long, A.R. (1992) *J. AOAC Int.*, **75**, 655.
[14] McLaughlin, L.G. and Henion, J.D. (1992) *J. Chromatogr.*, **591**, 195.
[15] Stafford, S.C. and Lin, W. (1992) *J. Agric. Food Chem.* **40**, 1026.
[16] Lott, H.M. and Barker, S.A. (1993) *J. AOAC Int.*, **76**, 67.
[17] Long, A.R., Soliman, H.M. and Barker, S.A. (1991) *J. Assoc. Off. Anal. Chem.*, **74**, 493.
[18] Sasaki, K., Ishizaka, T., Suzuki, T. and Saito, Y. (1988) *J. Assoc. Off. Anal. Chem.*, **71**, 360.
[19] Ishii, Y., Adachi, N., Taniuchi, J., and Sakamoto, T. (1990) *Nippon Noyaku Gakkaishi*, **15**, 225.
[20] Hopper, M.L. (1991) *J. Assoc. Off. Anal. Chem.*, **74**, 974.
[21] Hopper, M.L. McMahon, B., Griffitt, K.R., Cline, K., Fleming-Jones, M.E. and Kendall, D.C. (1992) *J. AOAC Int.*, **75**, 707.
[22] Ishii, Y., Taniuchi, J. and Sakamoto, T. (1990) *Nippon Noyaku Gakkaishi.* **15**, 231.
[23] Goodspeed, D.P. and Chestnut, L.I. (1991) *J. Assoc. Off. Anal. Chem.*, **74**, 388.
[24] Chaput, D. (1988) *J. Assoc. Off. Anal. Chem.*, **71**, 542.
[25] Hong, J., Eo, Y., Rhee, J., Kim, T. and Kim, K. (1993) *J. Chromatogr.*, **639**, 261.
[26] Wigfield, Y.Y. and Lanouette, M. (1993) *J. Agric. Food Chem.*, **41**, 84.
[27] France, J.E., King, J.W. and Snyder, J.M. (1991) *J. Agric. Food Chem.* **39**, 1871.
[28] Hopper, M.L. and King, J.W. (1991) *J. Assoc. Off. Anal. Chem.*, **74**, 661.
[29] King, J.W., Hopper, M.L., Luchtefeld, R.G., Taylor S.L. and Orton, W.L. (1993) *J. AOAC Int.*, **76**, 857.
[30] Howard, A.L., Brauce, C. and Taylor, L.T. (1993) *J. Chromatogr. Sci.*, **31**, 323.
[31] Lee, H.-B., Peart, T.E., Niimi, A.J. and Knipe, C.R. (1995) *J. AOAC Int.*, **78**, 437.
[32] A-Khorassani, M., Taylor, L.T. and Schweighardt, F.K. (1996) *J. AOAC Int.*, **79**, 1043.
[33] Pensabene, J.W., Fiddler, W., Maxwell, R.J., Lightfield, A.R. and Hampson, J.W. (1995) *J. AOAC Int.*, **78**, 744.
[34] Macaudiere, P., Caude, M., Rosset, R. and Tambute, A. (1989) *J. Chromatogr. Sci.*, **27**, 583.
[35] Chen, J. and Pawliszyn, J.B. (1995) *Anal. Chem.*, **67**, 2530.
[36] Takahashi, D.M. (1977) *J. Assoc. Off. Anal. Chem.*, **60**, 799.
[37] Cobin, J.A., and Johnson, N.A. (1995) *J. AOAC Int.*, **79**, 503.
[38] Omichi, K. and Ikenaka, T. (1988) *Anal. Biochem.*, **168**, 332.
[39] Prabhu, S.V. and Baldwin, R.P. (1990) *J. Chromatogr.*, **503**, 227.
[40] Stefansson, M. and Lu, B. (1993) *Chromatographia*, **35**, 61.
[41] Merber, N.C., Lingeman, H. and Brinkman, U.A. (1993) *Anal. Chim. Acta*, **279**, 39.
[42] Murakami, K., Kakemoto, M., Harada, T. and Yamada, Y. (1991) *Bunseki Kagaku*, **40**, 125.
[43] Murakami, K., Kakemoto, M., Harada, T., Yamada, Y. and Ogawa, H. (1992) *Bunseki Kagaku*, **41**, 343.
[44] Kiba, N., Saegusa, K. and Furusawa, M. (1993) *Bunseki Kagaku*, **42**, 649.
[45] Kiba, N., Shitara, K., Fuse, H., Furusawa, M. and Takata, Y. (1991) *J. Chromatogr.*, **549**, 127.
[46] Koerner, P.J. Jr. and Nieman, T.A. (1988) *J. Chromatogr.*, **449**, 217.
[47] Lare, L.A. and Johnson, D.C. (1988) *Anal. Chem.*, **60**, 1867.
[48] Lloyd, P. and Crabbe, J.C. (1985) *J. Chromatogr.*, **343**, 402.
[49] Petchey and Crabbe, J.C. (1984) *J. Chromatogr.*, **307**, 180.
[50] Bjorkqvist, B. (1981) *J. Chromatogr.*, **218**, 65.
[51] Indyk, H.E. and Woollard, D.C. (1994) *Analyst*, **119**, 397.
[52] Kargacin, M.E., Bassell, G., Ryan, P.J. and Honeyman, T.W. (1987) *J. Chromatogr.*, **393**, 454.
[53] Hoffstetter-Kuhn, S., Paulus, A., Gassmann, E. and Widmer, H.M. (1991) *Anal. Chem.*, **63**, 1541.
[54] Vorndran, A.E., Oefner, P.J., Scherz, H. and Bonn, G.K. (1992) *Chromatographia*, **33**, 163.
[55] Honda, S., Yamamoto, K., Suzuki, S., Ueda, M. and Kakehi, K. (1991) *J. Chromatogr.*, **588**, 327.
[56] Delgado, C., Talou, T. and Gaset, A. (1993) *Analusis*, **21**, 281.
[57] Honda, S., Iwase, S., Makino, A. and Fujiwara, S. (1989) *Anal. Biochem.*, **176**, 72.
[58] Vorndran, A.E., Grill, E., Huber, C., Oefner, P.J. and Bonn, G.K. (1992) *Chromatographia*, **34**, 109.
[59] Suzuki, S., Kakehi, K. and Honda, S. (1992) *Anal. Biochem.*, **205**, 227.
[60] Stefansson, M. and Novotny, M. (1994) *Anal. Chem.*, **66**, 1134.
[61] Chiesa, C. and Horvath, C. (1993) *J. Chromatogr.*, **645**, 337.
[62] Evangelista, R.A., Liu, M.-S. and Chen, F.-T.A. (1995) *Anal. Chem.*, **67**, 2239.
[63] Liu, J., Shirota, O. and Novotny, M. (1991) *J. Chromatogr.*, **559**, 223.
[64] Liu, J., Shirota, O. and Novotny, M. (1991) *Anal. Chem.*, **63**, 413.
[65] Liu, J., Shirota, O. and Novotny, M. (1992) *Anal. Chem.*, **64**, 973.
[66] Bidlingmeyer, B.A., Cohen, S.A., Tarvin, T.L. and Frost, B. (1987) *J. Assoc. Off. Anal. Chem.*, **70**, 241.

[67] Sarwar, G., Botting, H.G. and Peace, R.W. (1988) *J. Assoc. Off. Anal. Chem.*, **71**, 1172.

[68] Alegria, A., Barbera, R., Farre, R., Ferrers, M., Lagarda, M.J. and Lopez, J.C. (1996) *J. Chromatogr. A.*, **721**, 83.

[69] Muetsch-Eckner, M., Sticher, O. and Meier, B. (1992) *J. Chromatogr.*, **625**, 183.

[70] Auger, J., Mellouki, F., Vannereau, A., Boscher, J., Cosson, L. and Mandon, N. (1993) *Chromatographia*, **36**, 347.

[71] Vazquez-Ortiz, F.A., Caire, G., Higuera-Ciapara, I. and Hernandez, G. (1995) *J. Liquid Chromatogr.*, **18**, 2059.

[72] Einarsson, S., Josefsson, B. and Lagerkvist, S. (1983) *J. Chromatogr.*, **282**, 609.

[73] Haynes, P.A., Sheumack, D., Kibby, J. and Redmond, J.W. (1991) *J. Chromatogr.*, **540**, 177.

[74] Kirschbaum, J., Luckas, B. and Beinert, W.-D. (1994) *J. Chromatogr.*, **661**, 193.

[75] Thomas, D.J. and Parkin, K.L. (1994) *J. Agric. Food Chem.*, **42**, 1632.

[76] Polanuer, B., Ivanov, S. and Sholin, A. (1994) *J. Chromatogr. B.*, **656**, 81.

[77] Thio, A.P. and Tompkins, D.H. (1989) *J. Assoc. Off. Anal. Chem.*, **72**, 609.

[78] Sanz, M.A., Castillo, G. and Hernandez, A. (1996) *J. Chromatogr. A.*, **719**, 195.

[79] Woollard, D.C. and Indyk, H.E. (1993) *Food Chem.*, **46**, 429.

[80] Cohen, S.A. and Antonis, K.M. (1994) *J. Chromatogr. A.*, **661**, 25.

[81] Liu, H.J., Chang, B.Y., Yan, H.W., Yu, F.H. and Liu, X.X. (1995) *J. AOAC Int.*, **78**, 736.

[82] Calull, M., Fabregas, J., Marce, R.M. and Borrull, F. (1991) *Chromatographia*, **31**, 272.

[83] Hashimoto, M., Iwatsuki, S., Kuwata, G. and Imai, M. (1992) *Nippon Eiyo Shokuryo Gakkaishi*, **45**, 363.

[84] Mochizuki, E., Nakayama, A., Kitada, Y., Saito, K., Nakazawa, H., Suzuki, S. and Fujita, M. (1988) *J. Chromatogr.*, **455**, 271.

[85] Sanchez, F.G. and Gallardo, A.A. (1992) *Anal. Chim. Acta*, **270**, 45.

[86] Sarwar, G. and Botting, H.G. (1993) *J. Chromatogr.*, **615**, 1.

[87] Butikofer, U., Fuchs, D., Bosset, J.O. and Gmur, W. (1991) **31**, 441.

[88] Tapuhi, Y., Schmidt, D.E., Lindner, W. and Karger, B.L. (1981) *Anal. Biochem.*, **115**, 123.

[89] Nicolas, E.C., Phender, K.A., Aoun, M.A. and Hemmer, J.E. (1990) *J. Assoc. Off. Anal. Chem.*, **73**, 627.

[90] Ashworth, R.B. (1987) *J. Assoc. Off. Anal. Chem.*, **70**, 248.

[91] Yoneda, Y., Takahashi, M. and Kitamura, T. (1997) *Bunseki Kagaku*, **46**, 89.

[92] Saito, K., Horie, M., Tokumaru, Y. and Nakazawa, H. (1997) *J. Food Hyg. Soc. Japan*, **38**, 400.

[93] Saito, K., Itaya, T., Horie, M., Nakazawa, H. and Imanari, T. (1994) *Jpn J. Toxicol. Environ. Health*, **40**, 140.

[94] Ishii, R., Saito, K., Takahashi, K., Hoshino, Y., Suzuki, S. and Nakazawa, H. (1995) *Bunseki Kagaku*, **44**, 829.

[95] Skocir, E. Prosek, M. and Oskomic, M. (1997) *Chromatographia*, **44**, 267.

[96] Roturier, J.M., Bars, D. and Gripon, J.C. (1995) *J. Chromatogr. A.*, **696**, 209.

[97] Bartolome, B., Moreno-Arribas, V., Pueyo, E. and Polo, M.C. (1997) *J. Agric. Food Chem.*, **45**, 3374.

[98] Lawrence, J.F. and Charbonneau, C.F. (1988) *J. Chromatogr.*, **445**, 189.

[99] Lawrence, J.F. (1989) *Analyst*, **112**, 879.

[100] Lawrence, J.F. (1987) *J. Assoc. Off. Anal. Chem.*, **70**, 15.

[101] Miwa, H. and Yamamoto, M. (1996) *J. AOAC Int.*, **79**, 493.

[102] Miwa, H. and Yamamoto, M. (1990) *J. Chromatogr.*, **523**, 235.

[103] Kaneda, H., Kano, Y., Kamimura, M., Osawa, T. and Kawakishi, S. (1990) *J. Agric. Food Chem.*, **38**, 1363.

[104] Mentasti, E., Gennaro, M.C., Sarzanini, C., Baiocchi, C. and Savigliano, M. (1985) *J. Chromatogr.*, **322**, 177.

[105] Caccamo, F., Carfagnini, G., Corcia, A. and Samperi, R. (1986) *J. Chromatogr.*, **362**, 47.

[106] Badoud, R. and Pratz, G. (1986) *J. Chromatogr.*, **360**, 119.

[107] Burini, G. (1993) *J. AOAC Int.*, **76**, 1017.

[108] Horikawa, R. and Tanimura, T. (1982) *Anal. Lett.*, **15**, 1629.

[109] Kidouchi, K., Sugiyama, N., Morishita, H., Wada, Y., Nagai, S. and Sakakibara, J. (1987) *J. Chromatogr.*, **423**, 297.

[110] Moegele, R., Pabel, B. and Galensa, R. (1992) *J. Chromatogr.*, **591**, 165.

[111] Ujiie, T., Tsutake, Y., Morita, K., Matsuno, M. and Kodaka, K. (1991) *Vitamins (Japan)*, **65**, 249.

[112] Hasegawa, H., Arima, h., Suzuki, Y., Isshiki, H., Iida, K. and Yamamoto, Y. (1992) *Vitamins (Japan)*, **66**, 513.

[113] Ishiwatari, Y., Shimada, S., Suzuki, K., Ikeda, R. and Yasuda, K. (1992) *Vitamins (Japan)*, **66**, 271.

[114] Nishimune, T., Ito, S., Abe, M., Kimoto, M., and Hayashi, R. (1988) *J. Nutr. Sci. Vitaminol.*, **34**, 543.

[115] Iwata, H., Matsuda, T. and Tonomura, H. (1988) *J. Chromatogr.*, **450**, 317.

[116] Ujiie, T., Tsutake, Y., Morita, K., Tamura, M. and Kodaka, K. (1990) *Vitamins (Japan)*, **64**, 379.

[117] Tsuge, H. and Hirose, N. (1989) *Vitamins (Japan)*, **63**, 349.
[118] Bitsch, R. and Moeller, J. (1989) *J. Chromatogr.*, **463**, 207.
[119] Kim, H.-J. (1989) *J. Assoc. Off. Anal. Chem.*, **72**, 681.
[120] Behrens, W.A. and Madere, R. (1990) *J. Food Comp. Anal.*, **3**, 3.
[121] Iwase, H. and Ono, I. (1993) *J. Chromatogr. A.*, **654**, 215.
[122] Zapata, S. and Dufour, J.-P. (1992) *J. Food Sci.*, **57**, 506.
[123] Dodson, K.Y., Young, E.R. and Soliman, A.-G. (1992) *J. AOAC Int.*, **75**, 887.
[124] Yadui, Y. and Hayashi, M. (1991) *Anal. Sci.*, **7**, 125.
[125] Vanderslice, J.T. and Higgs, D.J. (1984) *J. Chromatogr. Sci.*, **22**, 485.
[126] Ali, M.S. and Philippo, E.T. (1996) *J. AOAC Int.*, **79**, 803.
[127] Oderiz, M.L.V., Blanco, M.E.V., Hernandez, J.L., Lozano, J.S. and Rodriguez, M.A.R. (1994) *J. AOAC Int.*, **77**, 1056.
[128] Chiari, M., Nesi, M., Carrea, G. and Righetti, P.G. (1993) *J. Chromatogr.*, **645**, 197.
[129] Desbene, P.L., Coustal, S. and Frappier, F. (1983) *Anal. Biochem.*, **128**, 359.
[130] Hayakawa, K. and Oizumi, J. (1987) *J. Chromatogr.*, **413**, 247.
[131] Yoshida, T., Uetake, A., Nakai, C., Nimura, N. and Kinoshita, T. (1988) *J. Chromatogr.*, **456**, 421.
[132] Stein, J., Hahn, A., Lembcke, B. and Rehner, G. (1992) *Anal. Biochem.*, **200**, 89.
[133] Yokoyama, T. and Kinoshita, T. (1991) *J. Chromatogr.*, **542**, 363.
[134] Sakano, T., Nozumoto, S., Nagaoka, T., Morimoto, A., Fujimoto, K., Masuda, S., Suzuki, Y. and Hirauchi, M. (1988) *Vitamins (Japan)*, **62**, 393.
[135] Indyk, H.E. and Woollard, D.C. (1997) *Analyst*, **122**, 465.
[136] Hiraike, H., Kimura, M. and Itokawa, Y. (1988) *Vitamins (Japan)*, **63**, 25.
[137] Hirauchi, M., Sakano, T., Nagaoka, T. and Morimoto, A. (1988) *J. Chromatogr.*, **430**, 21.
[138] Lin, C. and Zhang, X. (1987) *Analyst*, **112**, 1659.
[139] Ding, X. and Fu, C. (1993) *Talanta*, **40**, 641.
[140] Nagaosa, Y. and Kobayashi, T. (1995) *J. AOAC Int.*, **78**, 1307.
[141] Yabe, Y., Tan, S., Ninomiya, T. and Okada, T. (1983) *J. Food Hyg. Soc. Japan*, **24**, 329.
[142] Burini, G. and Damiani, P. (1991) *J. Chromatogr.* **543**, 69.
[143] Shibata, T., Tsuji, S., Kobayashi, K., Asai, Y., Honda, T., Noda, K., Iwaida, M., Mochizuki, E., Sugahara, O. and Ito, Y. (1992) *Jpn J. Toxicol. Environ. Health.* **38**, 437.
[144] Hayashi, H. and Miyagawa, A. (1990) *J. Food Hyg. Soc. Japan.* **31**, 44.
[145] Tamase, K., Kitada, Y., Sasaki, M., Ueda, Y. and Takeshita, R. (1985) *J. Food Hyg. Soc. Japan.* **26**, 515.
[146] Galletti, G.C. and Bocchini, P. (1996) *J. Chromatogr. A.*, **729**, 393.
[147] Murakami, C., Maruyama, T. and Niiya, I. (1988) *J. Food Hyg. Soc. Japan*, **29**, 235.
[148] Murakami, C., Maruyama, T. and Niiya, I. (1989) *J. Food Hyg. Soc. Japan*, **30**, 306.
[149] Murakami, C., Maruyama, T. and Niiya, I. (1997) *J. Food Hyg. Soc. Japan*, **38**, 105.
[150] Cilliers, J.J.L. and Niekerk, P.J. (1986) *J. Agric. Food Chem.*, **34**, 680.
[151] Matsunaga, A., Yamamoto, A., Mizukami, E., Hayakawa, K. and Miyazaki, M. (1988) *Jpn J. Toxicol. Environ. Health*, **34**, 70.
[152] Nakasone, Y., Morita, Y., Ohsawa. K., Fukuda, F., Tohma, T., Kurihara, O., Shigehara, S., Matsumoto, N. and Iizuka, Y. (1990) *J. Jpn. Vet. Med. Assoc.*, **43**, 896.
[153] Morita, Y., Nakasone, Y., Araki, T., Fukuda, F., Matsumoto, H., Kurihara, O., Matsumoto, N. and Iizuka, Y. (1992) *J. Jpn. Vet. Med. Assoc.*, **45**, 339.
[154] Takeda, N. and Akiyama, Y. (1991) *J. Chromatogr.*, **558**, 175.
[155] Takeda, N. and Akiyama, Y. (1992) *J. Chromatogr.*, **607**, 31.
[156] Tsai, C.-E. and Kondo, F. (1995) *J. AOAC Int.*, **78**, 674.
[157] Pacciarelli, B. (1991) *Mitt. Gebiete. Lebensm. Hyg.*, **82**, 45.
[158] Gehring, T.A., Rushing, L.G. and Thompson, Jr, H.C., (1997) *J. AOAC Int.*, **80**, 751.
[159] Bui, L.V. (1993) *J. AOAC Int.*, **76**, 966.
[160] Boison, J.O.K., Keng, L.J.-Y. and MacNeil, J.D. (1994) *J. AOAC Int.*, **77**, 565.
[161] Boison, J.O., Salisbury, C.D.C., Chan, W. and MacNeil, J.D. (1991) *J. Assoc. Off. Anal. Chem.*, **74**, 497.
[162] Gee, H.-E., Ho, K.-B. and Toothill, J. (1996) *J. AOAC Int.*, **79**, 640.
[163] Ang, C.Y.W., Luo, W., Hansen, Jr, E.B., Freeman, J.P. and Thompson, Jr, H.C. (1996) *J. AOAC Int.*, **79**, 389.
[164] Ang, C.Y.W. and Luo, W. (1997) *J. AOAC Int.*, **80**, 25.
[165] Okayama, A., Kitada, Y., Aoki, Y., Umesako, S., Ono, H., Nishii, Y. and Kubo, H. (1988) *Bunseki Kagaku*, **37**, 221.
[166] Gerhardt, G.C., Salisbury, C.D.C. and MacNeil, J.D. (1994) *J. AOAC Int.*, **77**, 334.

[167] Gerhardt, G.C., Salisbury, C.D.C. and MacNeil, J.D. (1994) *J. AOAC Int.*, **77**, 765.
[168] Hormazabal, V. and Yndestad, M. (1997) *J. Liq. Chrom. & Rel. Technol.*, **20**, 2259.
[169] Nakaya, K., Sugitani, A. and Yamada, F. (1986) *J. Food Hyg. Soc. Japan*, **27**, 258.
[170] Nakaya, K., Sugitani, A. and Kawai, M. (1987) *J. Food Hyg. Soc. Japan*, **28**, 487.
[171] Agarwal, V.K. (1989) *J. Liquid Chromatogr.*, **12**, 613.
[172] Agarwal, V.K. (1989) *J. Liquid Chromatogr.*, **12**, 3265.
[173] Yoneda, Y., Okada, M., Mizoguchi, S., Tokonabe, Y., Shiroishi, M. and Yoshimoto, Z. (1986) *J. Food Hyg. Soc. Japan*, **27**, 369.
[174] Shaikh, B. and Jackson, J. (1989) *J. Liquid Chromatogr.*, **12**, 1497.
[175] Ohguchi, K., Takahara, H., Matsumoto, M., Nakajima, C. and Kido, Y. (1996) *J. Food Hyg. Soc. Japan*, **37**, 319.
[176] Moran, J.W., Turner, J.M. and Coleman, M.R. (1995) *J. AOAC Int.*, **78**, 668.
[177] Moran, J.W., Rodewald, J.M., Donoho, A.L. and Coleman, M.R. (1994) *J. AOAC Int.*, **77**, 885.
[178] Coleman, M.R., Macy, T.D., Moran, J.W. and Rodewald, J.M. (1994) *J. AOAC Int.*, **77**, 1065.
[179] Ericson, J.F., Calcagni, A. and Lynch, M.J. (1994) *J. AOAC Int.*, **77**, 577.
[180] Dimenna, G.P., Creegan, J.A., Turnbull, L.B. and Wright, G.J. (1986) *J. Agric. Food Chem.*, **34**, 805.
[181] Asukabe, H., Murata, H., Harada, K., Suzuki, M., Oka, H. and Ikai, Y. (1993) *J. Chromatogr. A.*, **657**, 349.
[182] Martinez, E.E. and Shimoda, W. (1985) *J. Assoc. Off. Anal. Chem.*, **68**, 1149.
[183] Martinez, E.E. and Shimoda, W. (1986) *J. Assoc. Off. Anal. Chem.*, **69**, 637.
[184] Asukabe, H., Murata, H., Harada, K.Suzuki, M., Oka, H. and Ikai, Y. (1994) *J. Agric. Food Chem.*, **42**, 112.
[185] Miyakawa, H., Horii, S. and Kokubo, Y. (1995) *J. Food Hyg. Soc. Japan*, **36**, 725.
[186] Takatsuki, K., Suzuki, S. and Ushizawa, I. (1986) *J. Assoc. Off. Anal. Chem.*, **69**, 443.
[187] Tolan, J.W., Eskola, P., Fink, D.W., Mzozik, H. and Zimmerman, L.A. (1980) *J. Chromatogr.*, **190**, 367.
[188] Tway, P.C., Wood, Jr, J.S. and Downing, G.V. (1981) *J. Agric. Food Chem.*, **29**, 1059.
[189] Prabhu, S.V., Wehner, T.A. and Tway, P.C. (1991) *J. Agric. Food Chem.*, **39**, 1468.
[190] Reising, K.P. (1992) *J. AOAC Int.*, **75**, 751.
[191] Markus, J. and Sherma, J. (1992) *J. AOAC Int.*, **75**, 757.
[192] Salisbury, C.D.C. (1993) *J. AOAC Int.*, **76**, 1149.
[193] Kijak, P.J. (1992) *J. AOAC Int.*, **75**, 747.
[194] Khunachak, A., Dacunha, R. and Stout, S.J. (1993) *J. AOAC Int.*, **76**, 1230.
[195] Cobin, J.A. and Johnson, N.A. (1996) *J. AOAC Int.*, **79**, 1158.
[196] Cobin, J.A. and Johnson, N.A. (1995) *J. AOAC Int.*, **78**, 419
[197] Allen, J.L. and Meinertz, J.R. (1991) *J. Chromatogr.*, **536**, 217.
[198] Allen. J.L., Gofus, J.E. and Meinertz. J.R. (1994) *J. AOAC Int.*, **77**, 553.
[199] Roybal, J.E., Pfenning, A.P., Munns, R.K Holland, D.C., Hurlbut, J.A. and Long, A.R. (1995) *J. AOAC Int.*, **78**, 453.
[200] Swarbrick, A., Murby, E.J. and Hume, P. (1997) *J. Liq. Chem. and Rel. Technol.*, **20**, 2269.
[201] Degroodt, J.-M., Bukanski, B.W., Groof, J. and Beernaert, H. (1991) *Z. Lebensm. Unters. Forsch.*, **192**, 430.
[202] Degroodt, J.-M., Bukanski, B.W., Beernaert, H. and Courtheyn, D. (1989) *Z. Lebensm. Unters. Forsch.*, **189**, 128.
[203] Hashimoto, T.Miyazaki, T., Kokubo, Y., Suzuki, S. and Nakazawa, H. (1995) *J. Food Hyg. Soc. Japan*, **36**, 595.
[204] Kok, A., Hiemstra, M. and Vreeker, C.P. (1990) *J. Chromatogr.*, **507**, 459.
[205] Kok, A., Hiemstra, M. and Vreeker, C.P. (1992) *J. AOAC Int.*, **75**, 1063.
[206] Page, M.J. and French M. (1992) *J. AOAC Int.*, **75**, 1073.
[207] Patel, B.M., Moye, H.A. and Weinberger, R. (1990) *J. Agric. Food Chem.*, **38**, 127.
[208] Miles, C.J. (1992) *J. Chromatogr.*, **592**, 283.
[209] Takeda, N., Tsuji, M. and Akiyama, Y. (1995) *J. Food Hyg. Soc. Japan*, **36**, 601.
[210] Dorner, J.W. and Cole, R.J. (1988) *J. Assoc. Off. Anal. Chem.*, **71**, 43.
[211] Francis, Jr. O.J., Kirschenheuter, G.P., Ware, G.M., Carman, A.S. Kuan, S.S. (1988) *J. Assoc. Off. Anal. Chem.*, **71**, 725.
[212] Sydenham, E.W., Gelderblom, W.C.A., Thiel, P.G. and Marasas, W.F.O. (1990) *J. Agric. Food Chem.*, **38**, 285.
[213] Sydenham, E.W., Shephard, G.S. Thiel, P.G. (1992) *J. AOAC Int.*, **75**, 313.
[214] Scott, P.M. and Lawrence, G.A. (1992) *J. AOAC Int.*, **75**, 829.
[215] Akiyama, H., Miyahara, M., Toyoda, M. Saito, Y. (1995). *J. Food Hyg. Soc. Japan*, **36**, 77.
[216] Bennet, G.A. Richard, J.L. (1994) *J. AOAC Int.*, **77**, 501.
[217] Holcomb, M., Thompson, Jr. H.C. and Hankins, L.J. (1993) *J. Agric. Food Chem.*, **41**,764.
[218] Velazquez, C., Bloemendal, C., Sanchis, V. and Canela, R. (1995) *J. Agric. Food Chem.*, **43**, 1535.
[219] Miyahara, M., Akiyama, H., Toyoda, M. Saito. Y. (1996) *J. Agric. Food Chem.*, **44**, 842.

[220] Trucksess, M.W., Stack, M.E., Allen, S. and Barrion, N. (1995) *J. AOAC Int.*, **78**, 705.
[221] Bates, H.A. and Rapoport, H. (1975) *J. Agric. Food Chem.*, **23**, 237.
[222] Lawrence, J.F., Menard, C. and Cleroux, C. (1995) *J. AOAC Int.*, **78**, 514.
[223] Lawrence, J.F., Wong, B. and Menard, C. (1996) *J. AOAC Int.*, **79**, 1111.
[224] Oshima, Y. (1995) *J. AOAC Int.*, **78**, 528.
[225] Lawrence, J.F. and Wong, B. (1995) *J. AOAC Int.*, **78**, 698.
[226] Lee, J.S., Yanagi, T., Kenma, R. and Yasumoto, T. (1987) *Agric. Biol. Chem.*, **51**, 877.
[227] Quilliam, M.A. (1995) *J. AOAC Int.*, **78**, 555.
[228] Yoshida, T., Uetake, T., Yamaguchi, H., Nimura, N. and Kinoshita, T. (1988) *Anal. Biochem.*, **173**, 70.
[229] Huumert, C., Shen, J.L. and Luckas, B. (1996) *J. Chromatogr. A.*, **729**, 387.
[230] Carmody, E.P., James, K.J. and Kelly, S.S. (1995) *J. AOAC Int.*, **78**, 1403.
[231] Zonda, F., Stancher, B., Bogoni, P. and Masotti, P. (1992) *J. Chromatogr.*, **594**, 137.
[232] Lawrence, J.F., Roussel, S. and Menard, C. (1996) *J. Chromatogr. A.*, **721**, 359.
[233] Akasaka, K., Ohrui, H., Meguro, H. and Yasumoto, T. (1996) *J. Chromatogr. A.*, **729**, 381.
[234] Akasaka, K., Ohrui, H. and Meguro, H. (1993) *Analyst*, **118**, 765.
[235] Goto, T., Kishi, Y. and Hirata, Y. (1962) *Bull. Chem. Soc. Japan*, **35**, 1045.
[236] Yasumoto, T., Nakamura, M., Ohshima, Y. and Takahata, J. (1982) *Bull. Japan Soc. Sci. Fish.*, **48**, 1481.
[237] Yotsu, M., Endo, A. and Yasumoto, T. (1989) *Agric. Biol. Chem.*, **53**, 893.
[238] Yotsu, M., Hayashi, Y., Khora, S.S., Sato, S. and Yasumoto, T. (1992) *Biosci. Biotech. Biochem.*, **56**, 370.
[239] Murata, M., Legrand, A.-M., Ishibashi, Y. and Yasumoto, T. (1989) *J. Am. Chem. Soc.*, **111**, 8927.
[240] Yasumoto, T., Fukui, M., Sasaki, K. and Sugiyama, K. (1995) *J. AOAC Int.*, **78**, 574.
[241] Sarwar, G. and Botting, G.H. (1994) *J. AOAC Int.*, **77**, 1175.
[242] Yen, G.C. and Hsieh, C.L. (1991) *J. Food Sci.*, **56**, 158.
[243] Sato, M., Nakano, T., Takeuchi, M., Kumagai, T., Kannno, N., Nagahisa, E. Sato, Y. (1995) *Biosci. Biotech. Biochem.*, **59**, 1208.
[244] Malle, P., Valle, M. and Bouquelet, S. (1996) *J. AOAC Int.*, **79**, 43.
[245] Ibe, A., Saito, K., Nakazato, M., Kikuchi, Y., Fujinuma, K. and Nishima, T. (1991) *J. Assoc. Off. Anal. Chem.*, **74**, 695.
[246] Eerola, S., Hinkkanen, R., Lindfors, E. and Hirvi, T. (1993) *J. AOAC Int.*, **76**, 575.
[247] Harduf, Z., Nir, T. and Juven, J. (1988) *J. Chromatogr.*, **437**, 379.
[248] Einarsson, S., Folestad, S., Josefsson, B. and Lagerkvist, S. (1986) *Anal. Chem.*, **58**, 1638.
[249] Velasquez, R., Tsikas, D. and Brunner, G. (1992) *Fresenius J. Anal. Chem.*, **343**, 78.
[250] Achilli, G., Cellerino, G.P. and d'Eril, G.M. (1994) *J. Chromatogr. A.*, **661**, 201.
[251] Suzuki, S., Kobayashi, K., Noda, J., Suzuki, T. and Takama, K. (1990) *J. Chromatogr.*, **508**, 225.
[252] Ohta, H., Takeda, Y., Yoza, K. and Nogata, Y. (1993) *J. Chromatogr.*, **628**, 199.
[253] Veciana-Nogues, M.T., Hernandez-Jover, T., Marine-Font, A. and Vidal-Carou, M.C. (1995) *J. AOAC Int.*, **78**, 1045.
[254] Izquierdo-Pulido, M.L., Vidal-Carou, M.C. and Marine-Font, A.. (1993) *J. AOAC Int.*, **76**, 1027.
[255] Saito, K., Horie, M., Nose, N., Nakagomi, K. and Nakazawa, H. (1992) *J. Chromatogr.*, **595**, 163.
[256] Saito, K., Horie, M., Nose, N., Nakagomi, K. and Nakazawa, H. (1992) *Anal. Sci.*, **8**, 675.
[257] Busto, O., Miracle, M., Guasch, J. and Borrull, F. (1997) *J. Chromatogr. A.*, **757**, 311.
[258] Lehtonen, P., Saarinen, M., Vesanto, M. and Riekkola, M.L. (1992) *Z. Lebensm. Unters. Forsch.*, **194**, 434.
[259] Joosten, H.M.L.J. and Olieman, C. (1986) *J. Chromatogr.*, **356**, 311.
[260] Bartok, T., Borcsok, G. and Sagi, F. (1992) *J. Liquid Chromatogr.*, **15**, 777.
[261] Saito, K., Horie, M. and Nakazawa, H. (1995) *J. Food Hyg. Soc. Japan*, **36**, 639.
[262] Nagami, H., Uno, M., Onji, Y. and Yamazoe, Y. (1988) *J. Food Hyg. Soc. Japan*, **29**, 125.
[263] Pfeffer, M., Gelbe, B., Hampe, P., Steinberg. B., Walenciak-Reddel, E., Woicke, B. and Wykhoff, B. (1992) *Fresenius J. Anal. Chem.*, **342**, 839.
[264] Nakashima, H., Hori, S. and Nakazawa, H. (1990) *Jpn. J. Toxic. Environ. Health*, **36**, 15.
[265] Miyazaki, T., Yasuda, K. and Fujimoto, K. (1988) *Anal. Lett.*, **21**, 1033.
[266] Yamamoto, Y., Brodsky, M.H., Baker, J.C. and Ames, B.N. (1987) *Anal. Biochem.*, **160**, 7.
[267] Yang, G.C. (1992) *Trends Food Sci. Technol.*, **3**, 15.
[268] Akasaka, K., Ijichi, S.Watanabe, K., Ohrui, H. and Meguro, H. (1992) *J. Chromatogr.*, **596**, 197.
[269] Akasaka, K., Suzuki, T., Ohrui, H. and Meguro, H. (1987) *Anal. Lett.*, **20**, 797.
[270] Wu, W.S., Stoyanoff, R.E., Szklar, R.S., Gaind, V.S. and Rakanovic, M. (1990) *Analyst*, **115**, 801.
[271] Damant, A.P., Jickells, S.M. and Castle, L. (1995) *J. AOAC Int.*, **78**, 711.

# 1.3 DERIVATIZATION FOR ENVIRONMENTAL CONTAMINANTS

**Kazuo Iwaki**
Kanagawa, Japan

1.3.1 Introduction 51
1.3.2 Environmental Samples 51
  1.3.2.1 Gas Sample 51
  1.3.2.2 Soil and Various Water Samples 53
1.3.3 Application 55
  1.3.3.1 Aldehydes and Ketones in Air 55
  1.3.3.2 Aliphatic Amines in Air 57
  1.3.3.3 Residual Pesticides and Herbicides in Soil and Water 59
References 62

## 1.3.1 Introduction

In the analysis of environmental contaminants using chromatography, the analysis of GC methods which are superior to LC methods in separation characteristics and sensitivity have, together with the rapid popularization of GC/MS, been applied far more widely than in the analysis of biochemicals due to the applicability of many of the analytes to GC.

Recently, in the field of analysis on residual pesticides and herbicides, HPLC has been increasingly used mainly for pyrolytic compounds and for compounds which require derivatization for GC analysis. However, the 'HPLC with derivatization' methods focused on by this book, are limited in their application, about 80% of these are accounted for partly by the determination of residual pesticides and herbicides and the determination of aldehydes and ketones in the ambient air.

When using chromatography to determine environmental contaminants in environmental samples, the basic procedures for the sampling and pre-treatment methods in GC are naturally the same as those in LC. Here below, we introduce the general structure of the pre-treatment for contaminants analysis in environmental samples, that is the so-called gases (atmosphere, ambient air, etc), various water, and soil, and some recent applications of the HPLC derivatization methods focused on by this book.

## 1.3.2 Environmental Samples

Environmental samples, which can be broadly classified into gases, liquids and solids, vary in detail, and range widely in concentration of contaminants from the ppt level to the percentage level. There is also the possibility of contamination among samples and errors during the measuring sequence. In addition, contamination can occur during the analyzing process since measurement is performed for contaminants originally existing in the environment. Therefore, careful control is required during the analysis operation and regarding the use of tools and reagents. Below a brief outline of the steps in the procedures from sampling through to chromatography is given. For more information on individual operations, refer to the books referenced [1].

### 1.3.2.1 Gas Sample

The sampling methods for contaminants in gas can be generally classified into either of the following: the method using a liquid (the solution trapping method), the method using a solid phase

Edited by Toshimasa Toyo'oka: *Modern Derivatization Methods for Separation Sciences* © 1999 John Wiley & Sons Ltd.

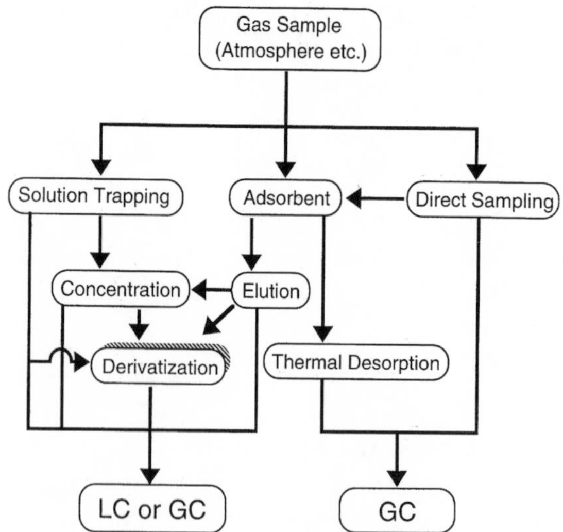


**Fig. 1.3.1.** Scheme of pre-treatment for chromatographic analysis of environmental contaminants in gas sample.

including adsorbents (the adsorbent method) and the method directly sampling the gas (the direct gas sampling method). Fig. 1.3.1 shows the flow from sampling through to chromatography. Refer to the explanation by DeGraff [2].

*Solution Trapping Method*

This method bubbles the sample gas into the trapping solution using an impinger to trap the desired components (Fig. 1.3.2). Although accompanied by complications related to the use of many tools and the problem of processing the exhausted gas when using organic solvent as the trapping solution, the method is suitable for chemicals which do not achieve a satisfactory trapping or recovery rate with the adsorbent method described below.

With regard to the post-treatment, since the trapping solution is often relatively clean, the solution is directly or after being concentrated measured in quantity of trapping solution, and then subjected to GC or LC methods either directly or after derivatization. When the sample with unknown concentration of analytes is to be trapped using the solution trapping method, it is important to connect the multiple impingers in series and to pay attention to any possible break-through of the analytes.

*Adsorbent Method*

This method passes samples through a trapping tube packed with an appropriate adsorbent (carbon molecular sheave, porous polymer, florisil etc.) to adsorb the desired components (Fig 1.3.3), and then measures them using heating desorption or solvent extraction. This method is useful because it allows for very compact sampling. It does, however, require equipment for heating desorption and adsorbent clean-up, which makes the total system expensive. For the selection and combination of appropriate adsorbents for individuals analytes used in this method, it is recommended that you contact the manufactures of adsorbents, who have numerous experience and information in this area.

With regard to post-treatment, the adsorbed analytes are either subjected to heating desorption and then directly introduced into GC with the heating desorption equipment combined with

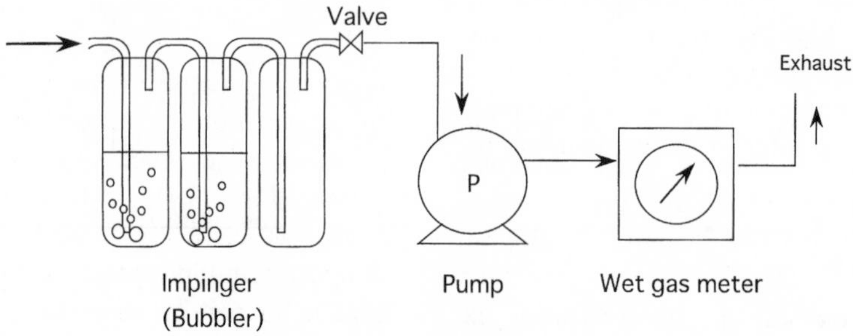


**Fig. 1.3.2.** Scheme of sampling device for solution-trapping method.

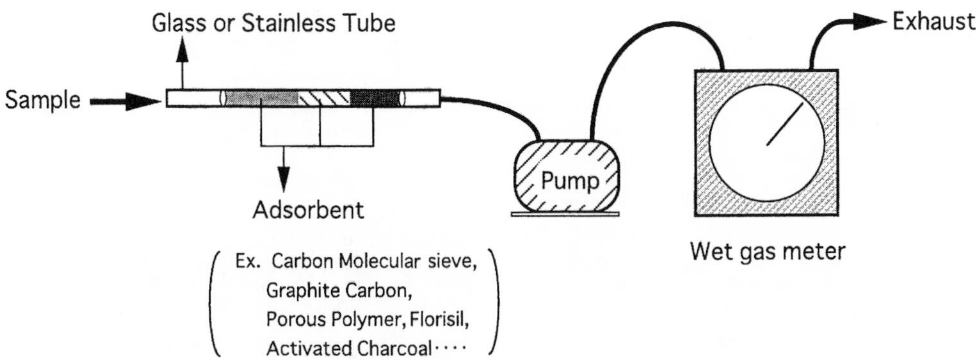


**Fig. 1.3.3.** Scheme of sampling device for adsorbent method.

GC, or are extracted into a solvent and then subjected to GC or LC either directly or following derivatization. The latter is often applied to compounds with a high boiling point. This trapping method also requires attention to the possible break-through of the analytes. When trapping the compounds with an unknown concentration, the trapping tubes should be serially connected, analyzed and checked individually.

*Direct Sampling Method*

This method directly traps the sample gas with TEDLAR bags, vacuum glass bottles or canisters (stainless steel vacuum containers). The method is intended to trap highly volatile compounds for subsequent GC analysis. The trapped compounds are subjected to the determination either directly or after being concentrated with an adsorbent. Among the above mentioned equipment, the canisters are sold on the market as trapping equipment for the analyzing system of EPA TO-14. This system is judged to be a measure which provides accurate results, eliminating human error throughout the process from concentration and measurement through to the determination of collected samples. For more information, refer to the explanation by Rasmussen [3].

*Other Methods*

One of the other methods is an application of the solution trapping method and the adsorbent method. The method is intended for highly volatile compounds which cannot be trapped in either of the two methods. The method reacts analytes in a trapping solution containing reactive agents, or on the solid phase coated with reactive agents, and traps the analytes by derivatizing them to low volatile compounds. When applying this to LC, it is the general rule simultaneously to introduce the chromophore which is advantageous to detection. This method employs the derivatization focused on by this book during trapping and has been applied to the official methods for determination of aldehydes, ketones and isocyanates (Table 1.3.1) [4–7].

Recently, solid phase micro extraction (SPME) has been drawing attention. This method adsorbs any gaseous contaminants in the sample at the tip of a syringe then applies heating desorption in the GC injector and finally determines the adsorbed components [8,9].

### 1.3.2.2 Soil and Various Water Samples

The analysis of residual pesticides and herbicides typifies the environmental analysis of soil and water using chromatography. Described below is an example of the general process flow of pre-treatment. The processing for soil is shown as a flow (Fig 1.3.4) as, following extraction, this extract is generally treated the same as liquid samples. Multiple-point sampling is generally applied to soil because the concentration of the contaminant is naturally expected to differ greatly depending on the location.

**Table 1.3.1.** Environmental contaminants using sampling method with derivatization [7]

| Analyte | Method | Reagent | Solvent or absorbent |
|---|---|---|---|
| Aldehydes and ketones | TO-5 | HCl/2,4-dinitrophenyl hydrazone | iso-octane |
| Ethylene oxide and propylene oxide | NIOSH1614 | HBr | Activated carbon |
| Acroleine | NIOSH2501 | 2-(hydroxymethyl)pyperidine | XAD-2 |
| Diazomethane | NIOSH2515 | Octanoic acid | XAD-2 |
| 2,4-Toluene diisocyanate | NIOSH2535 | *N*-[(4-nitrophenyl)methyl]propylamine | — |
| 2,4-Toluene diamide | NIOSH5516 | 1-(2-methoxyphenyl)pyperazine | Toluene |
| 2,4-Toluene diisocyanate and 4,4′-Methylene diphenyl diisocyanate | NIOSH5521 | 1-(2-methoxyphenyl)pyperazine | Toluene |
| Hydrazine | OSHA20 | $H_2SO_4$ | Gas chrom R |
| Maleic anhydride | OSHA25 | *p*-anisidine | XAD-2 |
| Methyl isocyanate | OSHA54 | 1-(2-Pyridyl)pyperazine | XAD-7 |

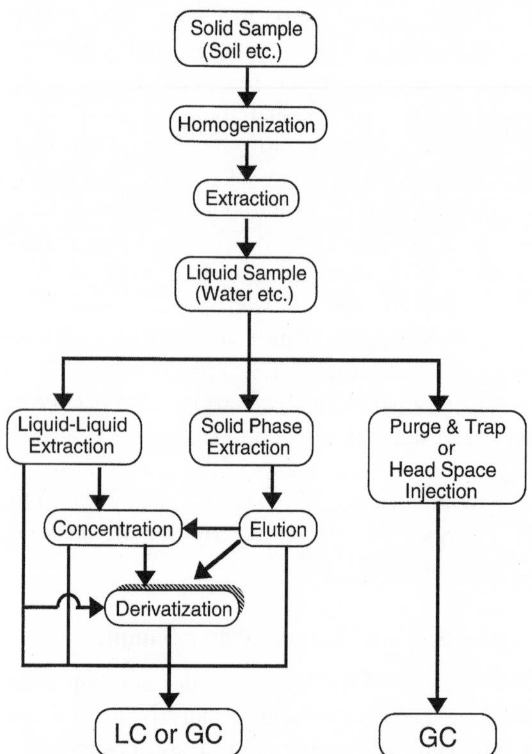


**Fig. 1.3.4.** Scheme of pre-treatment for the chromatographic analysis of environmental contaminants in soil and water samples.

*Extraction from Soil Sample*

Extraction methods from solid samples include the shaking extraction method, the soxhlet extraction method and the supersonic extraction method, all of which use the appropriate solvent for analytes. Recently, the super critical fluid extraction (SFE) method which is applied for chemical compounds having vapor pressure at a specific temperature and pressure changing into the state for super critical fluid for extraction has been used. For more information on this method, refer to the reviews [10,11]. In either of the above methods, it is important to investigate the collected results in detail and select an appropriate solvent.

The extracted solution is often further processed as a liquid sample and is rarely subjected to direct analysis, except in the case of some highly volatile contaminants.

*Liquid–Liquid Extraction*

This method employs shaking extraction mainly on aqueous samples in a non-mixing solvent in order to transfer target compounds into the solvent. It is necessary to select the solvent to be used according to the polarity of the target compounds and to improve recovery by salting-out etc. The extracted solution is subjected to the clean-up operation either directly or following concentration, or is subjected to GC or LC for separation and detection directly or following concentration and derivatization.

Recently, the solid phase extraction (SPE) method, as described below has prevailed due to the following defects of the liquid–liquid extraction method: 1) the disposal of a great amount of solvents used causing related health problems; 2) low operation efficiency; 3) low reproducibility; and 4) low recovery of high-polarity compounds.

*Solid Phase Extraction (SPE)*

This method passes liquid samples through the columns which are packed with silica gel chemically bonded with the octadecyl silyl (ODS) group, with active carbon or other packings used as trapping agents to trap the analytes. Compounds trapped on packing are extracted into solvent or solution with a different polarity from the sample solution, and then concentration and clean-up is carried out. The trapping agents can be classified into adsorbents and chemically bonded reversed-phase silica gel. The former includes active carbon, ion exchange resin, and porous polymer. The latter uses silica gel as a base, chemically bonded with ODS or the octyl group. Nowadays many mini-columns using such packings are sold in the market. It is advisable to contact the manufacturers for the selection of the columns to suit the required use. Regarding post-treatment, the eluent is generally subjected to GC or LC either directly, or following concentration or derivatization, except in the case of samples with complicated matrices.

For the determination of residual pesticides and herbicides in water samples using chromatography, refer to several general remarks referenced in [12–15].

## 1.3.3 Application

In the determination of environmental contaminants, the derivatization HPLC is applied to a far fewer number of compounds than in other fields. Introduced below are the studies on individual compounds which have mainly examined the pre-column derivatization method intended for highly sensitive detection.

### 1.3.3.1 Aldehydes and Ketones in Air

The pre-column derivatization method has been examined to determine aldehydes and ketones in the air, the toxicity of which has recently attracted attention. The main principle is a reaction in which hydrazone compounds are efficiently obtained from the carbonyl group and hydrazine compounds under acidic conditions (Fig. 1.3.5). The basic operation flow is shown in Fig. 1.3.6. 2,4-dinitrophenyl hydrazine (DNPH), which has a strong UV absorption characteristic advantageous to detection, has been the most popular in applications using EPA TO-5 since it was confirmed that various hydrazone

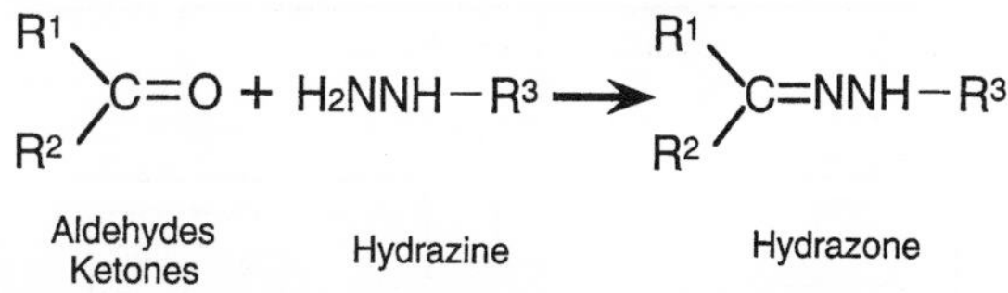


**Fig. 1.3.5.** Reaction of aldehydes and ketones with hydradine compound.

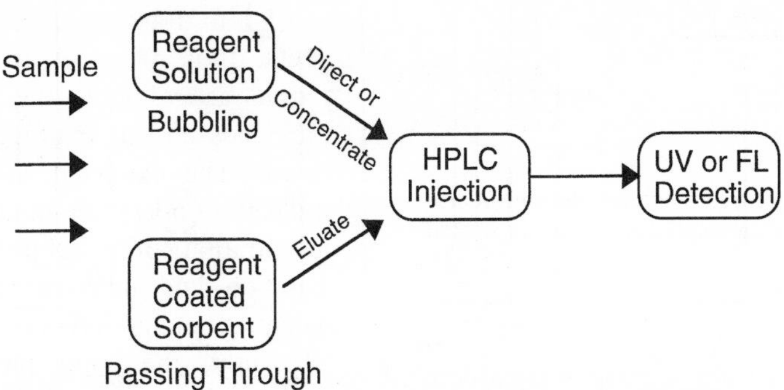


**Fig. 1.3.6.** Scheme of trapping methods with derivatization for the aldehydes and ketones.

derivatives produced from DNPH can be separated by the reverse-phase HPLC method [16]. Recently, Cotrim and others [17] reported the optimum separation conditions for eight types of DNPH derivatives of aldehydes. The applications of the DNPH method are summarized in Table 1.3.2 [18–23]. In addition, Karts and others [24] examined in detail the obstructions to separation and detection by by-products from $NO_2$ using this method. The use of reagents other than DNPH using 2-diphenylacetyl-1,3-indandion-1-hydrazone (DAIH) and dansyl hydrazine (DNSH) has been reported.

Using the apparatus shown in Fig. 1.3.7, Osaki and others [25] bubbled the nitrogen purge gas from environmental water and industrial liquid wastes into a DAIH solution, trapped azine compounds corresponding to acetaldehyde, acrolein, propyonaldehyde and crotonaldehyde, and determined then using HPLC-FL (Ex. 425 nm, Em. 525 nm). The detection limit was at 1.2 μg/l in acetaldehyde. Swarin and others [26] bubbled

**Table 1.3.2.** Various DNPH methods for the aldehydes and ketones in air

| Analyte | Sample | Sampling device | LOD* | Ref |
|---|---|---|---|---|
| Formaldehyde | Air (Industrial location) | Coated silica gel | 0.1 ppm (20L air) | 18 |
| Aldehydes, ketones and carbonyls | Ambient air | Impinger<br>Coated glass beads | 1 ng/Inj | 19 |
| C1 ∼ 4 aldehydes | Atmosphere<br>Industrial and incinerator emissions | Coated sep-PAK C18 | 0.5 ppb (100L air) | 20 |
| Form and acet aldehydes | Ambient air | Impinger | 1 ppb (30L air) | 21 |
| Formaldehyde | Ambient air | Coated florisil | 1 ppb (100L air) | 22 |
| Aldehydes and ketones | Automotive exhaust emission<br>Ambient air | Impinger<br>Coated silica gel | — | 23 |

*LOD: Limit of detection as formaldehyde

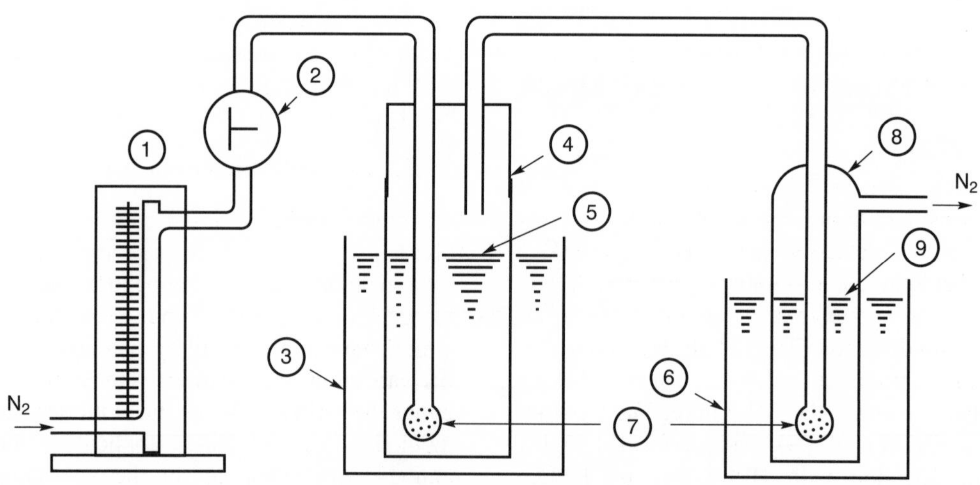


**Fig. 1.3.7.** Equipment for purge and trap of aldehydes: 1, flow meter; 2, 3-way stopcock; 3, 80 °C water bath; 4, gas washing bottle; 5, water sample; 6, 40 °C water bath; 7, bubbler; 8, impinger; 9, trap solution. [From ref. 25, p. 240, Fig. 1.].

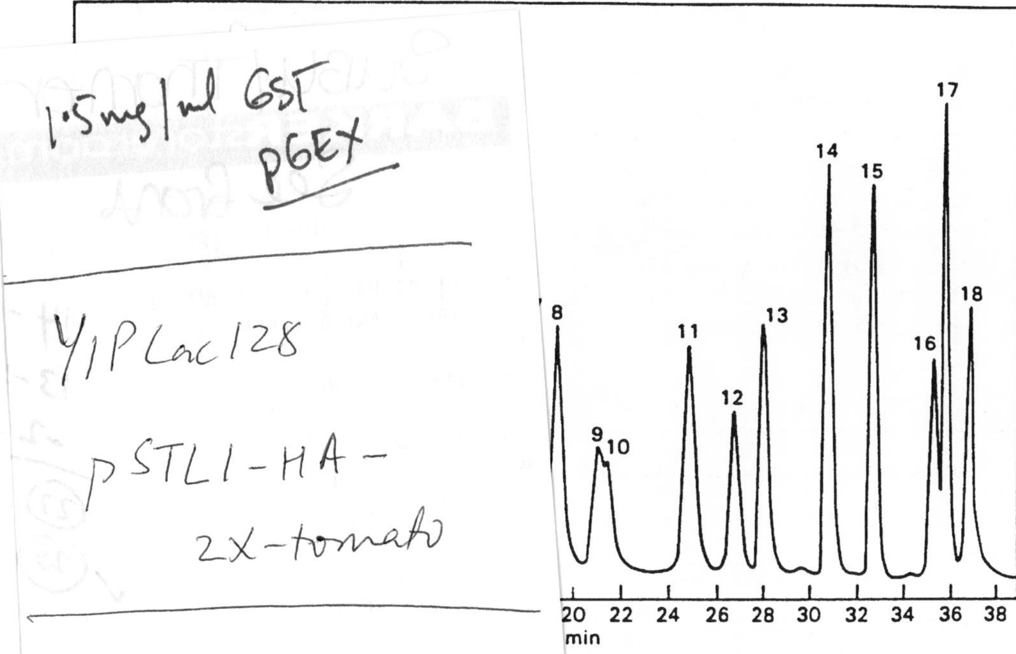


**Fig. 1.** ... atives separated on Zorbax ODS column. Flow rate of 1.0 ... reasing acetonitrile to 100% over 12 min (2%/min), and c... e detection with 20 nm slits and 425 nm excitation a... hyde-azine; 3, Acetaldehyde-azine; 4, Acetone-azine; 5, Acr... ketone-azine; 8, Crotonaldehyde-azine; 9, Butylaldehyde-a... e; 12, Isovaleraldehyde-azine; 13, Valeraldehyde-azine; 14, *p*-... Unknown-from reagent; 17, Heptanaldehyde-azine; 18, Un...

autom... solutio... at a d... (Fig. 1...

Alth... been w... ...ation reagent for highly sensitive fluorescence detection, recently more sensitive detection has been examined in various fields by chemi-luminescence (CL) detection using its peroxyoxalate-chemiluminescence reaction. Noudek and others [27,28] trapped hydrazones, corresponding to aldehydes contained in the atmosphere, using a DNSH impregnated porous glass particle (Fig. 1.3.9), and performed highly sensitive detection with HPLC-FL and HPLC-CL [28].

Recently, Gromping and others [29] developed the HPLC-FL method, using the DAIH method for ...etecting formaldehyde existing in the atmosphere ... the ppb level, and reported a detailed comparison ...ith other derivatization reagents.

### 1.3.3.2 Aliphatic Amines in Air

Although aliphatic amines, which may cause problems to the working environment, do not include any chromophore advantageous to detection, primary and secondary amines, which have high reactivity are advantageous to derivatization, and are used in many applications of either the 'HPLC-UV' or 'FL with derivatization' methods. These classified applications both trap amines with silica gel in the gas sample and derivatize the chemicals in the eluent from silica gel [30–32] and the applications which trap amines as thiourea derivatives on the XAD-2 resin

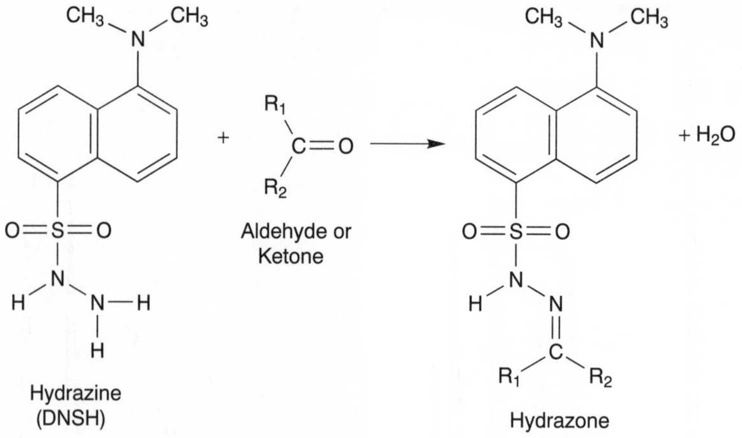

CH3 N CH3 + R1 C=O R2 → + H2O

O=S=O N H N—H H

Hydrazine (DNSH)

Aldehyde or Ketone

Hydrazone

**Fig. 1.3.9.** Reaction of DNSH with aldehydes and ketones. [From ref. 27, p. 1174 scheme.].

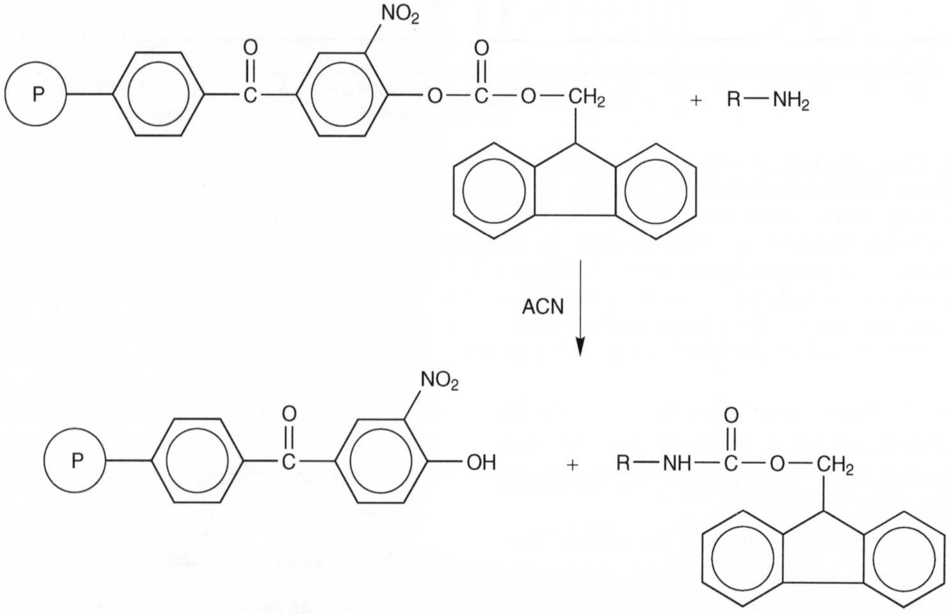

P NO2 O O C O—C—O—CH2 + R—NH2

ACN

P NO2 O C OH + R—NH—C—O—CH2

**Fig. 1.3.10.** Structure of the polymeric FMOC reagent and derivatizations of typical amines in air samples. [From ref. 32, p. 103, Fig. 2.].

which is resin coated with 1-naphtyl isothiocyanate (NITC), separate and detect the eluent from XAD-2 resin with HPLC-UV [33–35].

The methods using silica gel as adsorbent are reported in the HPLC-UV method using *m*-toluoyl chloride as a derivatizing reagent [30], and the HPLC-FL method using 7-chloro-4-nitro-1,2,3-benzoxaziazole [31].

Gao and others [32], by the use of polymeric-activated ester-carbonate fluorenyl reagent (Fig. 1.3.10), developed the HPLC-FL method which allows amine in the eluent to be derivatized on-line. The system according to this method included the column packed with reactive polymer mounted on the HPLC line as shown in Fig. 1.3.11, and enabled on-line operation from the FMOC

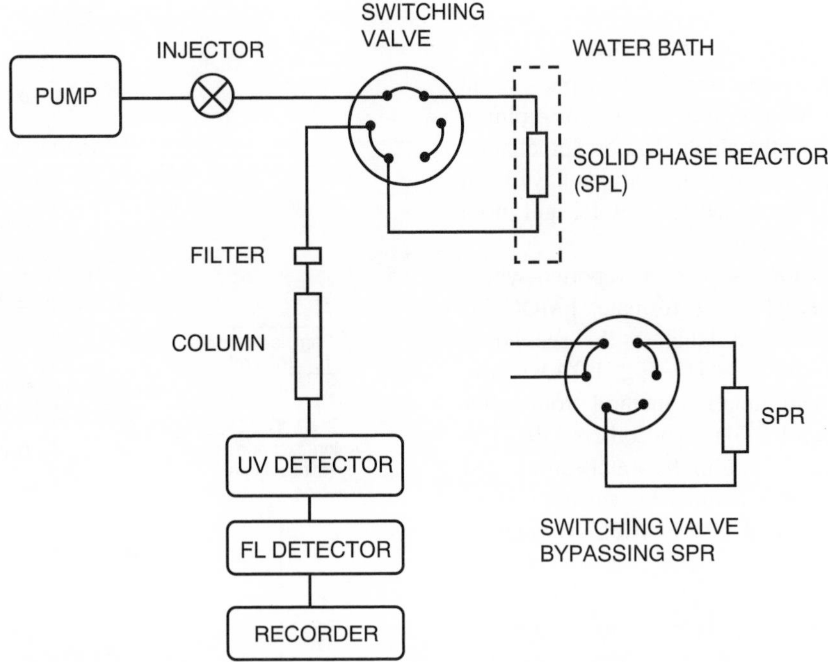


**Fig. 1.3.11.** Schematic diagram of the instrumentation arrangement for performing on-line, pre-column solid-phase derivatization in HPLC-UV/FL. [From ref. 32, p. 103, Fig. 1].

derivatization through to detection by operating valves after the loading of the eluent. The detection limit was 24 ppb in methyl amine.

The HPLC-UV method, which was recently reported, can determine aliphatic amine in the atmosphere down to the detection limit of 50 μg/m$^3$ (8 h sampling) [35,36]. This method employs the diffusive samplers (Fig. 1.3.12) which has eliminated the use of suction pumps during sampling using the NITC coated filter.

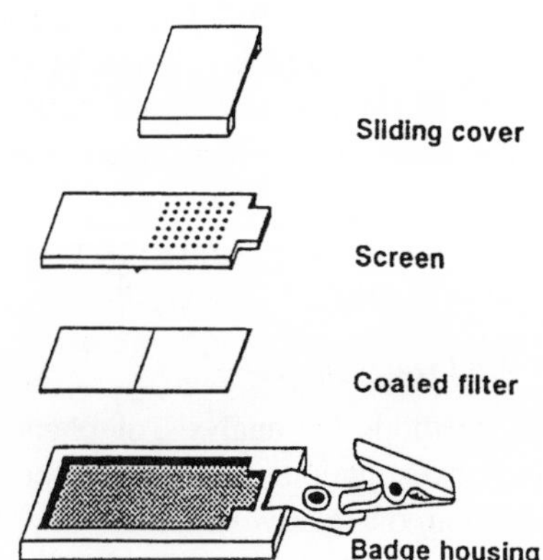


**Fig. 1.3.12.** Diffusive sampler for primary and secondary amines. [From ref. 36, p. 36, Fig. 1.].

### 1.3.3.3 Residual Pesticides and Herbicides in Soil and Water

In the analysis of residual pesticides and herbicides, most samples have complicated matrices. The GC analysis, therefore, is practical for the compounds applicable to the GC methods which have a high level of performance in separation and detection. In fact, GC methods are used to analyze many types of pesticides and herbicides. However, a few applications of the derivatization-HPLC method have been examined for compounds whose properties obstruct the application of GC and those which contain the functional group advantageous to the derivatization intended for highly sensitive detection. Among these, as introduced below, are some applications with the pre-column derivatization aimed at highly sensitive detection.

*Glyphosate*

For determining glyphosate (GP) herbicide and its decomposition products, aminomethyl phosphoric acids (APA) are usually used in the post-column methods with *o*-phthalaldehyde/mercaptoethanol [37,38]. However, pre-column methods have been reported which are intended for simpler and more highly sensitive detection.

Several applications have been reported which use 9-fluorenylmethyl chloro formate (FMOC-Cl) to apply the fluorescent label to the secondary amine of glyphosate [39–41]. Fig. 1.3.13 shows the flow of the determination method from water and soil performed by Glass and others [39]. The detection limit was 10 ppb in the water sample and 5–50 ppm in the soil sample. Gauch and others [41] determined GP and APA in drinking water by HPLC-FL using direct derivatization with FMOC-Cl at the detection limit of 0.02 μg/l.

As an alternative application to FMOC-Cl, recently Sen and others [42] developed the HPLC-CL method which can determine GP in water at the recovery of 67–100% and at the detection limit of 0.005–1 μg/g. This method is the combination of the following: clean-up and concentration of the sample solution by the anion exchange resin mini-columns, nitrosation, separation using amine phase-bonded column-HPLC, production of nitric oxide by the post-column denitrosating reaction, and the CL detection of the liberated nitric oxide.

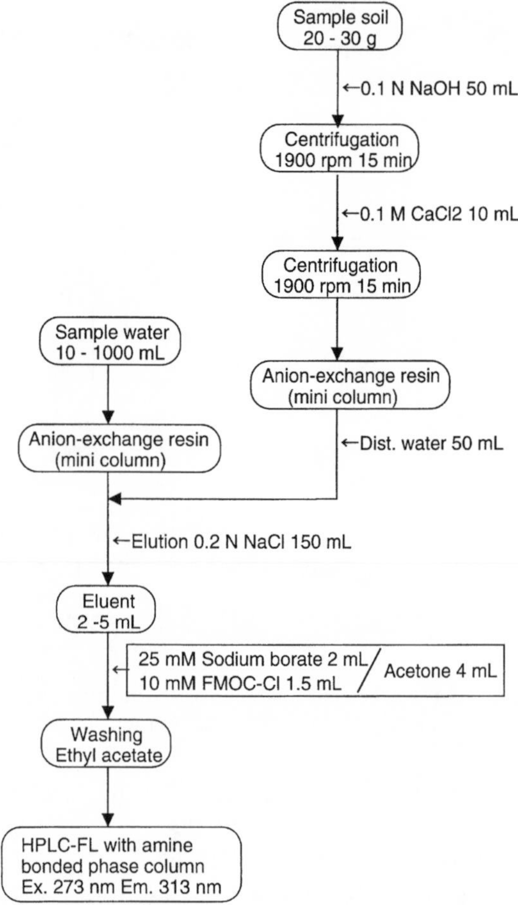


**Fig. 1.3.13.** Procedures for determination of glyphosate and aminomethyl phosphoric acid in soil and water developed by Glass *et al.* [39].

*Phenoxy Acid Herbicides*

In the LC methods for analysis of phenoxy acid herbicides, investigations have been focused on the application of relatively strong UV absorption properties naturally accompanying these compounds. Application of the 'HPLC with derivatization' methods was reported by Suzuki and others [43] which applies 9-anthryldiazonethane (ADAM) to carboxyl residues to obtain anthrylmethyl ester derivatives (Fig. 1.3.14). They adjusted water samples to be acidic, extracted the components in hexane-ethyl acetate mixture, concentrated the solution, derivatized the concentrate with ADAM, and conducted separation and detection by HPLC-FL (Fig. 1.3.15). The recovery was 93% or higher and the detection limit was 500 pg/inj.

*Other Compounds*

Sanchez and others [44] enabled the simultaneous analysis of asulam and amitrole, which are agricultural chemicals belonging to different chemical categories, by applying HPLC-FL with the direct derivatization with fluorescamine. The detection limit was 0.04 ng for asulam and 7.5 ng for amitrole.

Lantos and others [45] reported the pre-column derivatization HPLC-FL method of methoxaron and its metabolite, 3-chloro-4-methoxyaniline,

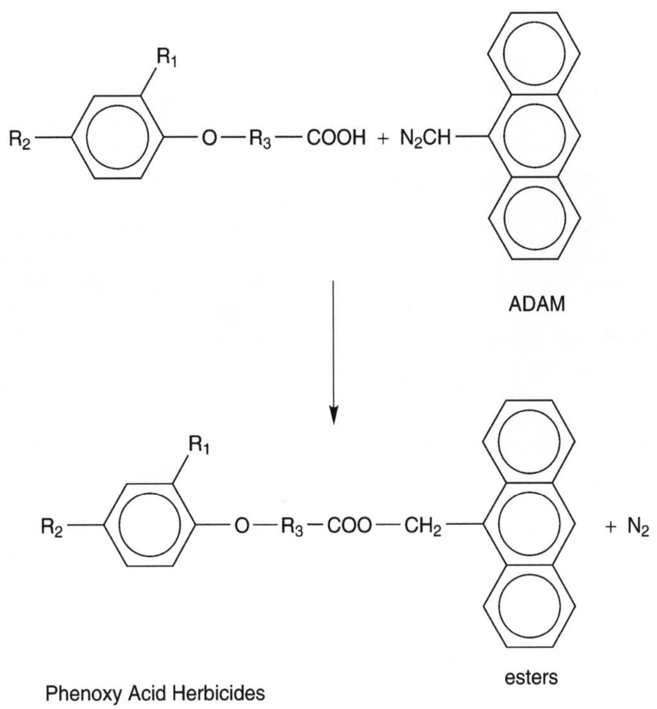

$R_2$—$C_6H_3(R_1)$—O—$R_3$—COOH + $N_2CH$—(anthracene)

ADAM

↓

$R_2$—$C_6H_3(R_1)$—O—$R_3$—COO—$CH_2$—(anthracene) + $N_2$

esters

Phenoxy Acid Herbicides

| | $R_1$ | $R_2$ | $R_3$ |
|---|---|---|---|
| 2,4-D | Cl | Cl | $CH_2$ |
| MCPA | $CH_3$ | Cl | $CH_2$ |
| MCPP | $CH_3$ | Cl | $CH(CH_3)$ |
| MCPB | $CH_3$ | Cl | $CH_2$-$CH_2$-$CH_2$ |

**Fig. 1.3.14.** Reaction course of phenoxy acid herbicides with ADAM. [From ref. 43, p. 361, Fig. 1.].

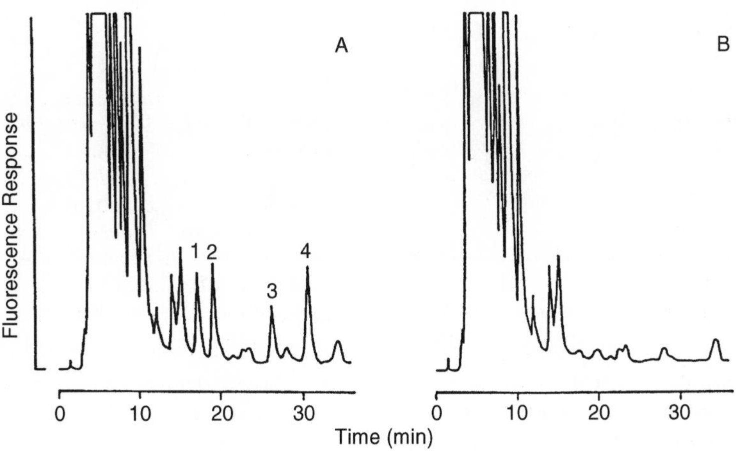


**Fig. 1.3.15.** Chromatograms of ground water extracts treated with ADAM: (A) Ground water sample spiked with 0.5 mg/l concentrations of phenoxy acid herbicides. Peaks: 1, 2,4-D; 2, MCPA; 3, MCPP; 4, MCPB as shown in Fig. 1.3.14. (B) Blank ground water sample. [From ref. 43 p. 363, Fig. 3.].

with Dns-Cl which uses thin-layer chromatography for pre-treatment.

## REFERENCES


[1] Barcelo, D. (editor) (1993) *Environmental Analysis: Techniques, Applications and Quality Assurance*, Elsevier, Amsterdam.
[2] DeGraff, I.D., (1995) *Air Monitoring Devices: Selection and Application*, Supelco, Inc., Bellefonte.
[3] Rasmussen, R.A. (1995) *Air Toxies Analysis by U.S. EPA*, Oregon Graduate Institute of Science & Technology, March 5.
[4] *Compendium of Methods for the Determination of Toxic Organic Compounds in Air*, PB90-127374, EPA/600/4-89/017, USEPA. Research Triangle Park, NC, June 1988.
[5] *NIOSH Manual of Analytical Methods*, 3 rd Edition (1984) Vol. 1–2, U.S. Department of Health and Human Services, U.S. Government Printing Office, Washington, D.C., February.
[6] *OSHA Analytical Methods Manual*, Second Edition, Part, I and II (1990) U.S. Department of Labor, Occupational Safety and Health Administration, Salt Lake City, Utah, January.
[7] Suzuki, S., (1995) *Bunseki*, **6** 444.
[8] Boyd-Boland, A.A., Chai, M., Luo, Y.Z., Zhang, Z., Yang, M.J., Pawliszyn, J.B. and Gorecki, T., (1994) *Environ. Sci. Technol.*, **28** 569A.
[9] Chai, M. and Pawliszyn, J. (1995) *Environ. Sci. Technol.*, **29** 693.
[10] Chester, T.L., Pinkston, J.D. and Raynie, D.E. (1994), *Anal. Chem.*, **66** 106R.
[11] Barnabas, I.J., Dean, J.R. and Owen, S.P. (1994), *Analyst (London)* **119** 2381.
[12] Font, G., Manes, J., Molto, J.C. and Pico, Y. (1993), *J. Chromatogr.*, **642** 135.
[13] Tekel, J. and Kovacicova, J. *J. Chromatogr.*, (1993) **643** 291.
[14] Hatrik, S. and Tekel, J., *J. Chromatogr.*, (1996) **733** 217.
[15] Liska, I. and Slobodnik, J., *J. Chromatogr.*, (1996) **733** 235.
[16] Selim, S., *J. Chromatogr.*, (1977) **136** 271.
[17] Countrim, M.X., Nakamura, L.A. and Collins, C.H., (1993) *Chromatographia*, **37** 185.
[18] Beasley, R.K., Hoffmann, C.E., Ruppel, M.L. and Worley, J.W. (1980) *Anal. Chem.*, **52** 1110.
[19] Fung, K. and Grosjean, D. (1981) *Anal. Chem.*, **53** 168.
[20] Kuwata, K., Uebori, M., Yamasaki, H., Kuge, Y. and Kiso, Y. (1983) *Anal. Chem.*, **55** 2013.
[21] Tanner, R.L. and Meng, Z. (1984) *Environ. Sci. Technol.*, **18** 723.
[22] Lipari, F. and Swarin, S.J. (1985) *Environ. Sci. Technol.* **19** 70.
[23] Tejada, S.B. (1986) *Int. J. Environ. Anal. Chem.*, **26** 167.
[24] Karst, U., Binding, N., Camman, K. and Witting, U. (1993) *Fresenius J. Anal. Chem.*, **345** 48.
[25] Osaki, Y., Nagase, M. and Matsueda, T. (1989) *Bunseki Kagaku*, **38** 239.
[26] Swarin, S.J. and Lipari, F. (1983) *J. Liq. Chromatogr.*, **6** 425.
[27] Nondek, L., Rodler, D.R. and Birks, J.W. (1992) *Environ. Sci. Technol.*, **26** 1174.
[28] Nondek, L., Milofsky, R.E. and Birks, J.W. (1991) *Chromatographia*, **32** 33.
[29] Gromping, A.H.J. and Cammann, K. (1993) *Chromatographia*, **35** 142.
[30] Simon, P. and Lemacon, C. (1987) *Anal. Chem.*, **59** 480.
[31] Nishikawa, Y. and Kuwata, K. (1984) *Anal. Chem.*, **56** 1790.
[32] Gao, C.X., Krull, I.S. and Trainor, T. (1990) *J. Chromatogr. Sci.*, **28** 102.
[33] Andersson, K., Hallgren, C., Levin, J.O. and Nilsson, C.-A. (1985) *Am. Ind. Hyg. Assoc. J.*, **46** 225.
[34] Levin, J.O., Andersson, K., Fangmark, I. and Hallgren, C. (1989) *Appl. Ind. Hyg.*, **4** 98.
[35] Levin, J.O., Lindahl, R., Andersson, K. and Hallgren, C. (1989) *Chemosphere*, **18** 2121.
[36] Lindahl, R., Levin, J.O. and Andersson, K. (1993) *J. Chromatogr.*, **643** 35.
[37] Bardalaye, P.C., Wheeler, W.B. and Moye, H.A. (1985) in *The Herbicide Glyphosate*, E. Grossbard and D. Atkinson (editors), Butterworths, London, UK, pp. 263.
[38] Oppenhuizen, M.E. and Cowell, J.E. (1991) *J. Assoc. Off. Anal. Chem.*, **74** 317.
[39] Glass, R.L. (1983) *J. Agric. Food. Chem.*, **31** 280.
[40] Miles C. J., Wallance L. R. and Moye H. A. (1986) *J. Assoc. Off. Anal. Chem.*, **3** 458.
[41] Gauch, R., Leuenberger, U. and Muller, U. (1989) *Z. Lebensm. Unters. Forsch.*, **188** 36.
[42] Sen, N.P. and Baddoo, P.A. (1996) *Int. J. Environ. Anal. Chem.*, **63** 107.
[43] Suzuki, T. and Watanabe, S. (1991) *J. Chromatogr.*, **541** 359.
[44] Sanchez, F.G., Diaz, A.N., Pareja, A.G. and Bracho, V. (1997) *J. Liq. Chromatogr. & Rel. Technol.*, **20** 603.
[45] Lantos, J., Brinkman, U.A.Th. and Frei, R.W. (1984) *J. Chromatogr.*, **292** 117.

# 2

# Reagent for UV-VIS Detection

**Kiyoshi Zaitsu*, Masaaki Kai and Kenji Hamase**
Faculty of Parmaceutical Science, Kyushu University, Japan

2.1 Introduction 64
2.2 Label of Amines and Amino Acids ($-NH_2$, NH) 65
2.2.1 Nitrobenzenes 65
2.2.1.1 1-Fluoro-2,4-dinitrobenzene (FDNB, 2,4-Dinitro-1-fluorobenzene, Sanger reagent) 65
2.2.1.2 4-Fluoro-3-nitrotrifluoromethylbenzene (4-Fluoro-3-nitrobenzotrifluoride) (FNBT) 66
2.2.1.3 2,4,6-Trinitrobenzene-1-sulfonic Acid (TNBS) 67
2.2.1.4 Na-(2,4-Dinitro-5-fluorophenyl)-L-alanine amide (FDAA, 1-Fluoro-2,4-dinitrophenyl-5-L-alanine Amide, Marfey's Reagent) 67
2.2.2 Acetic Anhydride 67
2.2.3 Isocyanates and Isothiocyanates 67
2.2.3.1 Isocyanates 67
2.2.3.2 Phenyl Isothiocyanate (PITC, Edman Reagent) 68
2.2.3.3 Butylisothiocyanate (BITC) 70
2.2.3.4 4-*N*,*N*-Dimethylaminoazobenzene-4′-isothiocyanate (DABITC) 70
2.2.4 1,2-Naphthoquinone-4-sulfonate (NQS) 70
2.2.5 Acyl Chlorides 71
2.2.5.1 Benzoyl Chloride 71
2.2.5.2 Dansyl Chloride (DNS-Cl) 72
2.2.5.3 4-*N*,*N*-Dimethylaminoazobenzene-4′-sulfonyl Chloride (DABS-Cl, Dabsyl-Cl) 73

Edited by Toshimasa Toyo'oka: *Modern Derivatization Methods for Separation Sciences* 

2.2.6 9-Fluorenylmethyl Chloroformate (9-Fluorenylmethoxycarbonyl Chloride, FMOC-Cl, Fluorescence and UV Detection) 73
2.2.7 Ninhydrin 74
2.2.8 Diethylethoxymethylenemalonate (DEMM) 75
2.2.9 6-Aminoquinolyl-*N*-hydroxyuccinimidyl Carbamate (AQC) 76
2.2.10 Disuccinimido Carbonate (DSC) 76
2.2.11 Solid-phase Reagent with UV or VIS Light Absorbing Moiety 76
2.2.11.1 Polymeric 3,5-dinitrobenzoyl Tagged Derivatization Reagent 77
2.2.11.2 Polymeric 6-Aminoquinoline (6-AQ) Tagged Derivatization Reagent 78
2.2.11.3 Polymeric Benzotriazole Activated Reagent Containing FMOC Group 78
2.3 Label of Carboxyl (−COOH) 78
2.3.1 Alkyl Halides 78
2.3.2 Aromatic Amines 79
2.3.3 Hydrazines 81
2.3.4 Hydroxylamine 82
2.4 Label of Hydroxyl (−OH) 84
2.4.1 Acyl Halides 84
2.4.2 Acid Anhydrides 86
2.4.3 Isocyanates 87
2.4.4 Other Reagents 88
2.5 Label of Reducing Carbohydrate 89
2.5.1 Reductive Amination 89
2.5.2 1-Phenyl-3-methyl-5-pyrazolone 91
2.5.3 Post-column Derivatizing Reagent 92
2.6 Label of Thiol (−SH) 92
2.6.1 2-Halopyridinium Salt 92
2.6.2 Disulfhide Reagent 93
2.6.3 Other Reagents 94
2.7 Labelling of Other Compounds 94
2.7.1 1-(2-Pyridyl)-piperazine (PYP) 94
2.7.2 Diethyldithiocarbamate (DDTC) 94
2.7.3 9-Methylamino-methylanthracene (MAMA) 95
References 95

## 2.1 INTRODUCTION

Whenever a new or a more sensitive and selective method has been developed for the detection and determination of biologically important compounds, drugs or their metabolites, biopolymers, and also environmental pollutants, important scientific knowledge has evolved to a greater or lesser extent. In the 1970s and 1980s, various derivatization methods for ultraviolet and visible light (UV-VIS) absorption detection have been developed [1–5] for high-performance liquid chromatography (HPLC), and many of the fundamental UV-VIS derivatizing reagents of the 1990s, in practical use are derived from the findings of the 70s and 80s [6]. In the 1990's, fluorescence derivatization has been a major detection technique when highly sensitive detection is required in HPLC and capillary electrophoresis (CE) [7,8]

In this chapter, the derivatization method for HPLC and CE connecting with UV-VIS detection will be described. The term derivatization in analysis means basically the reactions caused by chemical reagents or physical methods to convert a poor detector-responding analyte into a highly detectable product which has enhanced chromatographic or electrophoretic properties. The properties are: (1) detector response; (2) linear response range; and (3) ability of separation. In other words, derivatization aims to increase the detectability of the analyte and, sometimes to achieve the addition of a chromatographic or electrophoretic handle onto the analyte molecule.

Derivatization procedures both in the pre and post-column or capillary cause problems for HPLC or CE. On-line and off-line derivatization also has inherent problems which need to be considered [8].

We do not discuss derivatization with an enzyme, chelating reagent for the determination of metals and photochemical derivatizations, rather we have tried to select those papers which help to construct a view of UV-VIS derivatization of organic compounds involved in a creature's life.

## 2.2 LABEL OF AMINES AND AMINO ACIDS ($-NH_2$, NH)

The detection and determination of amines are concerned not only with biochemical but also industrial analysis. Thus various derivatization reagents for the chromatographic determination of primary and secondary amines as well as tertiary amines, are available. In many biologically important compounds, an amino groups exists as amino acids and polyamines. When HPLC or CE analysis of these amino compounds with direct UV-VIS detection is difficult because of the weak UV-VIS absorption of the compounds, derivatization of amines with a reagent having a strong UV-VIS absorbing structure is one of the considerable methods to choose from.

### 2.2.1 Nitrobenzenes

#### 2.2.1.1 1-Fluoro-2,4-dinitrobenzene (FDNB, 2,4-Dinitro-1-fluorobenzene, Sanger Reagent)

Pre-column derivatization of aminoglycosides with FDNB prior to HPLC used an important analytical technique for the determination of various antibiotics. Aminoglycosides, such as amikacin [9,10], tobramycin [11,12], gentamicin [13], sisomicine [13], neamine, neomycin B, neomycin C [14] in microliter quantities of biological fluids were measured by HPLC using FDNB derivatization. The 2,4-dinitrophenyl (DNP) derivative produced is stable. A drawback of FDNB is its toxicity; it must be handled with protective gloves. FDNB reacts with primary and secondary amino groups [14]. The chemical structure of FDNB derivative of amikasin is shown in Fig. 2.1 [10]. Also the major FDNB derivatives of tobramycin (Tb) were considered as a Tb derivative with all five amino groups 2,4-dinitrophenylated $TB(NDNB)_5$ and one hydroxyl group 2,4-dinitrophenylated $TB(NDNB)_5$ (ODNB) [11].

The amino acid $N^6, N^6, N^6$-trimethyllysine was also derivatized by FDNB. Trimethyllysine and $N^6, N^6, N^6$-triethyllysine (internal standard, IS) were isolated from urine specimens

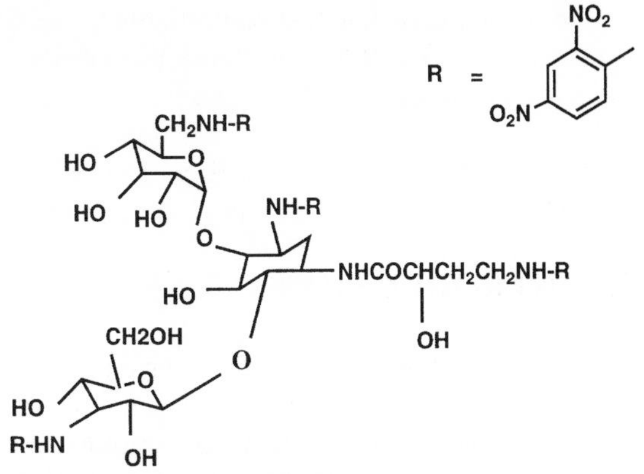

**Fig. 2.1.** 2,4-Dinitrophenyl derivative of amikacin.

by ion-exclusion chromatography, pre-column derivatization by reaction with FDNB, and determined by RP-paired-ion HPLC with UV detection at 405 nm. The detection limit corresponded to 0.1 nmol trimethyllisine per injection [15].

FDNB can modify the side chain of lysine, tyrosine, histidine, and cysteine residue in proteins, and the *N*-terminal amino group. Thus, FDNB is suitable for chromophoric labeling for selective mapping [16]. In addition, the DNP group can be removed from tyrosine, histidine, and cysteine residues by thiolysis with 2-mercaptoethanol [16].

*FDNB Application: Determination of Amikacin in Serum by HPLC [10]*

Remove most of the stem of a Pasteur pipette and plug the remainder of the stem with glass wool. Incubate CM-Sephadex C-25 cation exchanger in a 0.2 M sodium phosphate solution in water at room temperature for at least 24 h. Fill the Pasteur pipette with sufficient of the Sephadex slurry to obtain a column height of 1.5 cm (5.4 mm I.D.). Pipette 200 μl of the serum sample into a centrifuge tube and add 20 μl of a solution containing 250 mg/l kanamycin sulfate in water (IS), Vortex and dispense the contents of the centrifuge tube on top of the column. Elute the column with 2 ml of a solution containing 1 mM HCl and 0.2 M sodium sulfate (initial eluent). Discard the eluate (dead volume of the column). Elute the column with 1 ml of 50 mM NaOH and collect this eluate in a 5-ml ampoule. Add 2.5 ml of a solution of FDNB in methanol (30 g/l). The formation of a precipitate is observed, which redissolves upon mixing. Heat-seal the ampoule and place in boiling water for 5 min. After cooling, a 150-μl portion of the final reaction mixture is applied onto RP-HPLC with UV detection at 365 nm. The sensitivity is 1 mg/l for amikacin with samples of 200 μl. Precision, expressed R. S. D., is about 3%.

*FDNB Application: Derivatization Used in Peptide Mapping [16]*

For peptide mapping, each protein (<5 mg) was first modified with an equal weight of maleic anhydride at pH 8.5, using a pH stat. Urea (1 g/ml of sample volume) and $NaHCO_3$ (0.1 g/ml) were dissolved in the sample, and 50% FDNB in $CH_3CN$ (50 μl) was added. The mixtures were stirred in the dark at room temperature overnight, then dialyzed extensively against water, then 0.2 M $NH_4CO_3$, using dialysis tubing with a 3500 molecular weight cut-off (Spectra-Por3). The modified protein was digested with thermolysin (1:25 enzyme: substrate ratio) at 37 °C for 8 h, and freeze-dried. The modified peptides were separated by RP-HPLC (monitored at 270 and 320 nm). The collected peptide was hydrolyzed in 6 N HCl at 110 °C for 22 h for amino acid analysis.

#### 2.2.1.2 4-Fluoro-3-nitrotrifluoromethylbenzene [17] (4-Fluoro-3-nitrobenzotrifluoride) (FNBT)

The reaction of FNBT with primary amines is shown in Fig. 2.2. FNBT reacts with polyamine and produces the *N*-2′-nitro-4-trifluoromethylphenyl polyamine (NTP-polyamine) (Fig. 2.2). NTP-polyamines could be extracted by organic solvents. 2-methylbutane was selected because it extracted less by-products of the reaction and could easily be removed by evaporation. FNBT does not react with secondary amines. The reaction products of polar compounds such as amino acids with FNBT were not extracted by 2-methylbutane. Thus histidine was used to scavenge excess FNBT. NTP-polyamines have absorption maxima at 242 and 410 nm.

*FNBT Application: Derivatization of Polyamines [17]*

The dry residue obtained from a sample solution containing putrescine, spermidine and spermine together with diaminooctane (IS) was redissolved in 0.1 ml of 1 M $Na_2CO_3$. The solution was mixed with 0.3 ml of FNBT reagent (10 μl FNBT/ml DMSO) and kept at 60 °C for 20 min. To the resulting mixture, 40 μl of 1 M histidine in 1 M

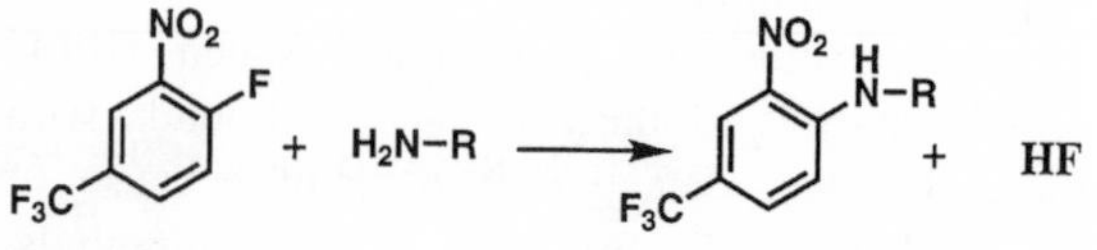

**Fig. 2.2.** The reaction of FNBT with primary amines.

$Na_2CO_3$ was added and incubated for a further 5 min. After cooling the mixture, NTP-polyamines were extracted twice with 2 ml of 2-methylbutane. The organic phase was evaporated to dryness. The residue was reconstitute with 50 µl of MeOH and 10-µl aliquots were subjected to PR-HPLC with the detection at 242 nm. The detection limits were 5 pmol for putrescine, 10 pmol for spermidine and 25 pmol for spermine.

#### 2.2.1.3 2,4,6-Trinitrobenzene-1-sulfonic Acid (TNBS) [18]

TNBS reacts with primary amino groups of amino acids and peptides in aqueous solution at pH 8 and at room temperature without any undesirable side reactions. The *N*-trinitrophenyl derivatives produced (Fig. 2.3) have a high molar absorptivity at 340 nm. The optimum pH for the reaction of TNBS with amikacin is between 9.5 and 10.0. At temperatures below 70 °C and reaction times shorter than 30 min, multiple derivatives of amikacin formed, due to the incomplete reaction.

#### 2.2.1.4 Na-(2,4-Dinitro-5-fluorophenyl)-L-alanine amide (FDAA, 1-Fluoro-2,4-dinitro-phenyl-5-L-alanine amide, Marfey's Reagent)

An HPLC-UV procedure for the separation and quantification of enantiomeric amino acid mixtures [19] and D- and L-phosphoserine [20] were employed using pre-column derivatization with a chiral reagent, FDAA (Fig. 2.4). The absolute sensitivity of the detector to the phosphoserine was found to be linear to 2 ng (11 pmol) on column.

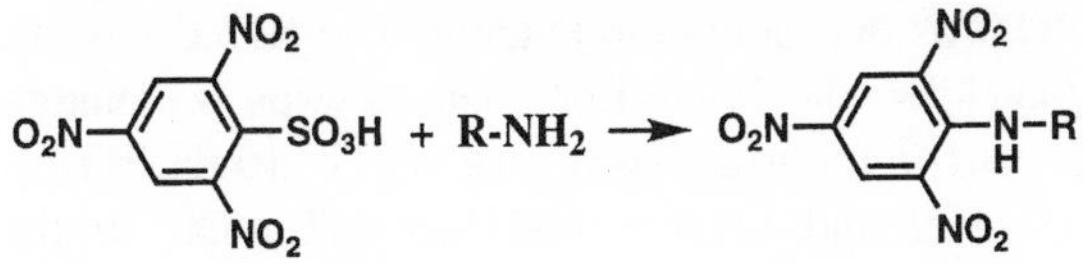

**Fig. 2.3.** The reaction of TNBS with primary amines.

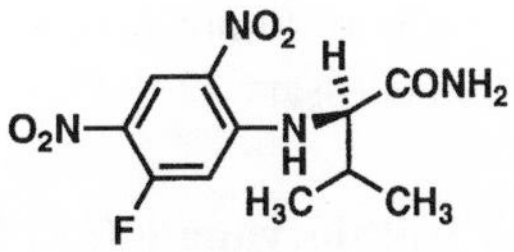

**Fig. 2.4.** Na-(2,4-Dinitro-5-fluorophenyl)-L-alaninea-mide (FDAA).

### 2.2.2 Acetic Anhydride [21]

Tertiary aliphatic amines can be separated on a HPLC column packed with Nucleosil $5N(CH_3)_2$ resin, and were detected by a post-column derivatization based on a reaction with a color reagent consisting of an acetic anhydride solution of citric acid. The reaction temperature was 120 °C and the wavelength of detection of wine red color was 550 nm. Primary and secondary amines do not react or interfere. The lower limit of determination is 0.01% for aliphatic mono, di- and trialkylamines.

### 2.2.3 Isocyanates and Isothiocyanates

#### 2.2.3.1 Isocyanates

Phenyl isocyanate (PIC) reacts with primary and secondary amines to form *N,N′*-disubstituted ureas (Fig. 2.5). [22]

1-Naphthyl isocyanate (NIC) can be used for an HPLC pre-column derivatization reagent of amino acids [23]. Derivatization is carried out by adding the isocyanate dissolved in dry acetone to a buffered amino acid solution followed by extraction of the excess reagent with cyclohexane. The resulting naphthylcarbamoyl amino acids are stable and highly fluorescent (excitation maxima at 385 and 305 nm, and an emission maximum at 385 nm). UV detection near 222 nm, the absorption maximum, can also be employed. UV detection is considerably less sensitive than fluorometry. However, for cystine and tryptophane

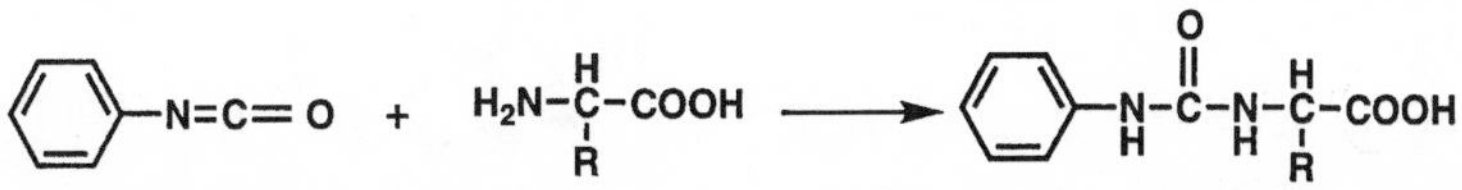

**Fig. 2.5.** The reaction of PIC with primary amines.

determination, where fluorescence is quenched, UV detection is necessary [23].

### 2.2.3.2 Phenyl Isothiocyanate (PITC, Edman Reagent)

In 1949, PITC was first introduced by Edman for peptide sequencing [24]. Practically, the chemistry of this reaction has remained unchanged until now. The derivatives are usually cyclized and rearrange under acid condition to form 3-phenyl-2-thiohydantoin (PTH) derivatives (Fig. 2.6). Automation of the method was originated by Edman and Begg in 1967 [25], and Zimmerman *et al.* developed the RP-HPLC method for the identification of PTH-amino acids in 1977 [26]. This contributed to the increase in the sensitivity of the microsequence analysis. After some improvement [27,28], the detection limit of 5 pmol for PTH-amino acid was attained, thus permitting sequencing at the low-nanomole level of peptide. Identification at the femtomole level of PTH-amino acid was achieved by using microbore columns of 2 mm I.D. and RP-HPLC column with isocratic elution and UV detection at 265 nm [29].

PITH reacts with a wide variety of primary and secondary amines including free amino acids. Heinrikson and Meredith [30] developed pre-column derivatization using PITC in an alkaline medium to form phenylthiocarbamyl (PTC) derivatives of amino acids (Fig. 2.7) including proline and hydroxyproline in protein hydrolyzates. The derivatives were separated by RP-HPLC and detected at 254 nm. This method permits 1–10 pmol levels of amino acids. Sensitive analysis of asparagine and glutamine in physiological fluid and cells was also developed by Lavi and Holcenberg [31].

The applicability in biological research of four currently used pre-column derivatization methods for HPLC were compared with OPA, FMOC-Cl, PITC and DNS-Cl [32] (Table 2.1).

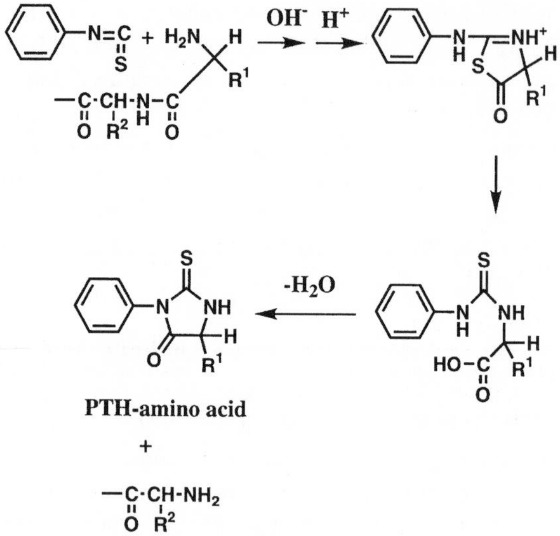

$OH^-$ $H^+$ -$H_2O$ PTH-amino acid

**Fig. 2.6.** Analysis of sequential *N*-terminal amino acid analysis using phenyl isothiocyanate (Edman degradation).

The comparison concluded that: 1) application of the PITC method might be beneficial in the area of clinical and/or protein chemistry, if sufficient sample material is available; 2) the OPA method is suitable for routine analyses of primary free amino acids (except cystine) in a biological fluid and 3) if determination of a secondary amino acid is desirable, the use of the FMOC-Cl method or a combination of the OPA and FMOC-Cl methods is recommended.

S-Alk(en)yl-L-cysteine sulfoxides (Fig. 2.8) in various *Allium* can be determined for the allien content of various samples of garlic acid by use of the Waters Pico-Tag method, which employs PITC as derivatization reagent. The HPLC system used for the Pico-Tag method was a Waters chromatograph consisting of a UV detector with a wavelength fixed at 254 nm [33]. The origin of this method is developed by Bidlingmeyer *et al.* in Waters Associates. In 1984 they had

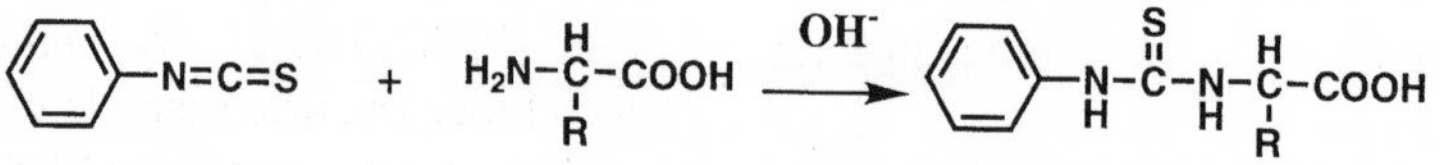

**Fig. 2.7.** The formation of phenylthiocarbamoyl derivatives by reaction of phenylisothiocyanate with amino acids.

**Table 2.1.** HPLC analyses of free amino acids: comparison of four derivatization methods

| Parameter | PITC | OPA | FMOC-Cl | DNS-Cl |
|---|---|---|---|---|
| Limit of sensitivity, pmol $S/N = 2.5$ | 5.0 | 0.8 | 1.0 | 1.5 |
| Reproducibility(R.S.D.) | 2.6–5.5 | 0.4–2.2 | 1.9–4.6 | 1.5–4.1 |
| Stable adducts | yes | no | yes | yes |
| Detection of secondary amines/cystine | yes/yes | no/no | yes/no | yes/yes |
| Problematic amino acids | Orn, Trp His, (Cys)2 | Asp, Trp | His, Trp | His, Asn |
| Effort required | ++ | − | +++ | + |

PITC: phenyl isothiocyanate, OPA: *o*-phthalaldehyde, FNOC-Cl: 9-fluorenylmethyl chloroformate (9-fluorenylmethoxycarbonyl chloride, DNS-Cl: dansyl chloride. Reproduced with permission from P. Fürst, L. Pollack, T.A. Graser, H. Godel and P. Stehle (1990) *J. Chromatogr.*, **499**, 557 [32]

$$R-\overset{\overset{O}{\uparrow}}{S}-\underset{H_2}{C}-\overset{NH_3^+}{\underset{H}{C}}-COO^- \qquad R{=}S-\underset{H_2}{C}-\overset{NH_3^+}{\underset{H}{C}}-COO^-$$

**Fig. 2.8.** *S*-Alk(en)yl-L-cystine sulfoxides.

already described the liquid chromatographic system which allows for the separation with UV detection of the common amino acid with 12 min analysis time. The detection limit had an S/N ratio of 5, the detectable limit for most of the amino acid derivatives can be considered to be 1 pmol level. However, starting with greater than 500 ng of protein before hydrolysis is practically, best suited for the amino acid analysis of proteins when using a UV detector at 0.005 a.u.f.s [34].

Picomoles of phosphoamino acids were determined by a RP-HPLC using C18 column under conditions both with and without the presence of a large excess of non-phosphorylated amino acids [35]. A UV absorption at 254 nm is employed for detection of the PITC-derivatized amino acids. Phosphoserine, phosphothreonine and phosphotyrosine are resolved within the first 8 min. The sensitivity is in the picomple range, and the separation time, injection to injection, is 36 min. In practical analysis of PVDF membrane-bound phosphoproteins, the destruction of the phosphoester bonds of phosphoamino acids during acid hydrolysis of phosphoproteins is the remaining difficult problem to solve.

PTH derivatives of amino acids, electrically neutral compounds, were separated by micellar electrokinetic chromatography (MEKC). Various conditions for the separation of PTH amino acids were tested [36–38], Terabe *et al.* successfully separated 23 PTH amino acids and detected them with UV absorption at 210–220 nm [39]. Later, 24 amino acids were separated by MEKC with another derivatization reagent, Dansyl-Cl as described in §2.2.5.2 [40]. Various other amino acid derivatives had been used in MEKC and were summarized by Matsubara and Terabe [41]. Enantiomeric separation of PTH-DL-amino acids by MEKC with b-escin was completed; the maximum number (8) of enantiomeric pairs of PTH amino acids can be separated and detected with UV absorbance at 220 nm [42].

### *PITC Application: Derivatization of Amino Acids [43]*

Standards or sample preparations which had been evaporated to dryness were treated with 10 µl of a mixture of methanol-1 M sodium acetate -triethylamine (TEA) (2:2:1, v/v) and the drying process was repeated under vacuum until a pressure of 70 mTorr was reached. This redrying step ensured that any residual acid was neutralized and that the alkaline pH which favored derivatization was attained. The derivatization reagent consisted of MeOH-TEA-water-PITC (7:1:1:1, v/v) and was freshly prepared each time. The dried sample was dissolved in 20 µl of derivatizing reagent and left to react for 20 min at ambient temperature. Excess reagent, TEA and other volatile products were then removed by evaporation under vacuum. A derivatized sample could be stored dry at 4 °C for 24 h or frozen at −20 °C for a week or more. The derivatives were dissolved either in 50 μl (cerebrospinal fluid, blood spots) or 100 μl (plasma, amniotic fluid, urine) of a diluent composed of 5 mM Na phosphate buffer (pH 7.4) containing 50 ml/l $CH_3CN$. Usually 20 μl of sample solution were injected onto the column provided by the Waters Pico-Tag amino acid analysis system.

#### 2.2.3.3 Butylisothiocyanate (BITC) [44]

BITC reacts with amino acids including hydroxyproline and proline and produces butylthiocarbamyl (BTC) derivatives. All derivatives of 22 protein amino acid were quantitatively resolved in 27 min by C18 RP-HPLC, except asparagine and serine, and detected at UV 250 nm (absorption maximum wavelength is ca. 234 nm, but the wavelength is the most efficient wavelength to avoid the absorption spectra of the impurities and the electrolyte, ammonium acetate, in the solvent). With the PTC derivatives, the derivatization reaction was completed in 10–20 min at room temperature, but with the BTC derivatives 120 min were required. The calibration graphs showed good linearity in the range 0.5–2.5 nmol. The BTC derivatives were stable at room temperature for ca. 8 h.

#### 2.2.3.4 4-*N,N*-Dimethylaminoazobenzene-4′-isothiocyanate (DABITC) [45–50]

This modified Edman reagent can be used to derivatize peptides (Fig. 2.9) after a tryptic or chymotryptic treatment and offers distinct advantages as described in detail by Chang [49]. The introduction of the DABITC reagent to the derivatization of peptides and to the microsequencing analysis of proteins and peptides with the DABITC/PITC double coupling method [51,52] is thought to be a very efficient tool for microsequencing studies at picomole levels [53]. The thiohydantoin amino acid derivatives (DABTHs) of DABITC are red and molar absorptivity at 436 nm = 34 000. Quantitative coupling with DABITC requires a temperature of 75 °C. Thus, the double coupling method is performed instead at 52 °C. The second coupling is with PITC to drive the reaction to completion [54].

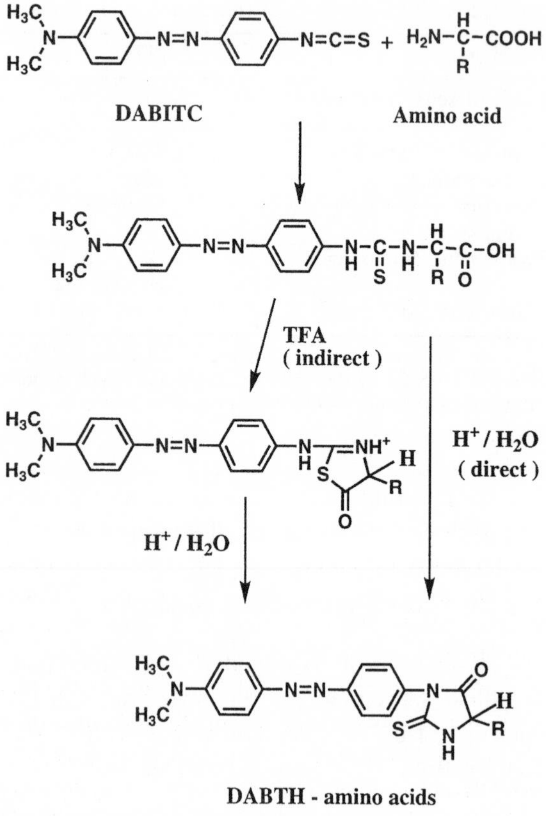

DABITC + Amino acid → TFA (indirect) → $H^+/H_2O$ → DABTH - amino acids ($H^+/H_2O$ (direct))

**Fig. 2.9.** Formation of DABTH-amino acids by reaction of DABITC with amino acids.

### 2.2.4 1,2-Naphthoquinone-4-sulfonate (NQS)

NQS reacts with amino acids in an alkaline medium giving a derivative which is detected at 305 and 480 nm [55–57].

NQS reacts with primary and secondary amino groups under relatively mild conditions. The reaction of amino acid with NQS is thought to be as in Fig. 2.10.

A new chromatographic method for the determination of amino acids has been developed

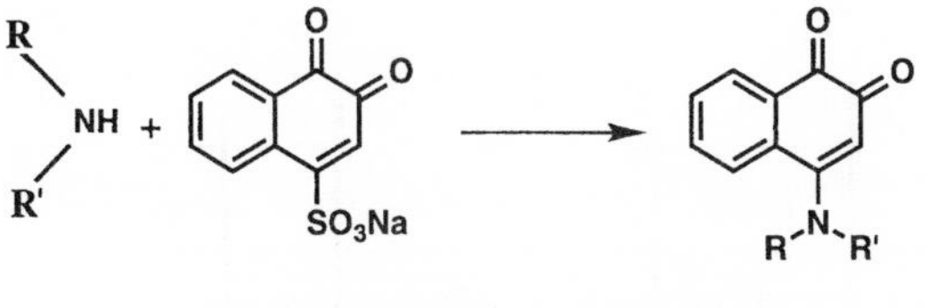

(R= alkyl, aryl, R'=H, alkyl or aryl)

**Fig. 2.10.** The reaction of sodium NQS with amines.

[55]. The method is based on the separation of amino acids by ion-pair LC and post-column derivatization using NQS. The derivatization reaction took place at 65 °C in a reaction coil. The spectrophotometric detection was performed at 305 nm. The detection limit for lysine is 90 pmol. The linear range for lysine is up to 32 nmol. The method was applied to the determination of amino acids in animal feed and powdered milk. The method involves the one-line post-column derivatization, thus automation should be easy, sample preparation is minimized and problems in dealing with derivative instability are overcome. OPA and fluorescamine fail in the labeling of a secondary amino acid, while NQS reacts with both primary and secondary amino acids. The separation of derivatives is performed on an RP column which is cheaper than the cation-exchange columns used in a prevailing ninhydrin method.

A general spectrophotometric method for the determination of amines and amino acids based on their reaction with NQS in a basic medium was developed by flow injection analysis [56,57]. NQS is water soluble, reacts with primary and secondary amino groups under mild conditions and is inexpensive. The main problem with this reagent is its instability in an alkaline medium (pH 10). The flow injection technique can overcome the problem. Reproducibility was 2.5% and the limit of detection was $2 \times 10^{-6}$ M for phenylalanine at 470 nm [56].

## 2.2.5 Acyl Chlorides

### 2.2.5.1 Benzoyl Chloride [58]

As described in another chapter, polyamine derivatization with the fluorogenic reagent, OPA-2-mercaptoethanol and fluorescamine, gives specific and sensitive determination of polyamines (putrescine, cadaverine, spermidine and spermine). However, the reaction products have a short life and the method requires pre-separation of di- and polyamines before derivatization because of insufficient resolution of fluorescamine derivatives [59]. Use of tosyl-, dansyl- or benzoyl chloride is preferred as they derivatize most of the naturally occurring di and polyamines and the reaction products are more stable. In these chlorides, the use of benzoyl chloride is advantageous due to the lengthy derivatization procedure with tosyl chloride, and the long elution time in HPLC with dansyl derivatives. A rapid and simple procedure using benzoyl chloride was originally described by Redmond and Tseng [60]. The method was improved by dissolving benzoyl chloride in MeOH thereby enhancing the reaction with polyamines. The benzoylated polyamines are eluted by RP-HPLC using MeOH-$H_2O$ (60:40) as the mobile phase. The sensitivity of this method is 100 pmol. The benzoylated polyamines can be stored up to three weeks at −20 °C.

A one hundred times more sensitive method [61] employs extraction of the derivatives by $CHCl_3$, separation by RP-HPLC under isocratic elution and detection by UV detection at 229 nm. The detection limit was ca. 1 pmol. The derivatization procedure [61] is given below.

On the other hand, Mei [62] found that the absorbance of benzoylated polyamines increases ca. 50 times from 254 to 198 nm in $CH_3CN$ as solvent. However, the use of MeOH/water as the mobile phase disturbs detection at wavelengths shorter than 205 nm, the UV cut-off point of MeOH. Benzoylated polyamines were well separated by the use of 42% $CH_3CN/H_2O$ as the mobile phase of RP-HPLC using C8 column. This allowed detection at 198 nm without significant background noise. The detection limits in the HPLC are: 0.8 pmol for putrescine, 1 pmol for spermidine and 1.3 pmol for spermine.

*Benzoyl Chloride Application: HPLC Determination of Polyamines [61]*

In glass tubes 2 ml of 2 M NaOH and 10 µl of benzoyl chloride were added to 100–400 µl of the 0.3 M perchloric acid extract containing a known amount of 1,6-hexanediamine (IS). Tubes were closed with caps with PTFE-coated rubber closure, and the contents were mixed vigorously and incubated for 30 min at room temperature under gentle rotation. The resulting solution was extracted with 2 ml of $CHCl_3$. After centrifugation the $CHCl_3$ layer was removed, washed with 2 ml of water, centrifuged and transferred to another tube. The solvent was evaporated under a stream of air and the residue was dissolved in 100–500 μl of methanol. Typical chromatograms were given for benzoylated polyamines from tumor tissue grown in nude mice (Fig. 2.11).

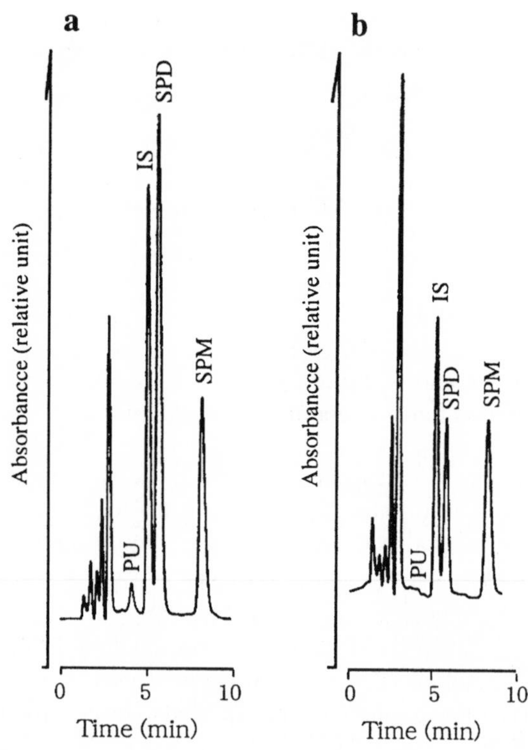


**Fig. 2.11.** Chromatograms of benzoylated polyamines from PC-93 tumour tissue grown in nude mice, monitored by their absorbance at 229 nm. Elution with methanol-water (65:35, v/v): (a) Untreated control; (b) tumour tissue from a nude mouse treated with DL-α-difluoromethylornithine (DFMO) for seven days by oral administration (2% DMFO in drinking water) plus intraperitoneal injection (500 mg/Kg). Benzoylated putrescine (PU), 1,6-hexanediamine (internal standard, IS), spermidine (SPD), spermine (SPM) were monitored by their absorbance at 229 nm. [Reproduced from ref. 61, p. 50, Fig. 4].

### 2.2.5.2 Dansyl Chloride (DNS-Cl)

Dansyl chloride is used in the derivatization of primary and secondary amines in which fluorimetric detection of dansyl derivatives is usually performed [63]. UV-detection of dansylated compounds is not a common procedure except on some occasions. The fluorescence of the dansyl amino acids is quenched in aqueous solutions and the quantum yield obtained is about 1/10th that in organic solvents. In an RP-HPLC of dansyl amino acids [64], the necessity of an aqueous buffer for elution, which quenches fluorescence, the UV absorbance of the dansyl amino acids at 250 nm, is used for detection giving a sensitivity of about 100 pmol for a single dansyl amino acid. A rapid HPLC method for the determination of the dansyl derivative of pentaazapentacosane (PAPC, Fig. 2.12) pentahydrochloride, an anticancer agent, has been developed [65].

Like spermine and spermidine, PAPC does not absorb in the UV region and therefore, it must be derivatized. In the chromatographic system, an RP-C8 column, a mobile phase of 40 mM $CH_3COOH/NH_4OH$ (pH 5)-$CH_3CN$ (10:90) and UV detection at 254 nm is conducted.

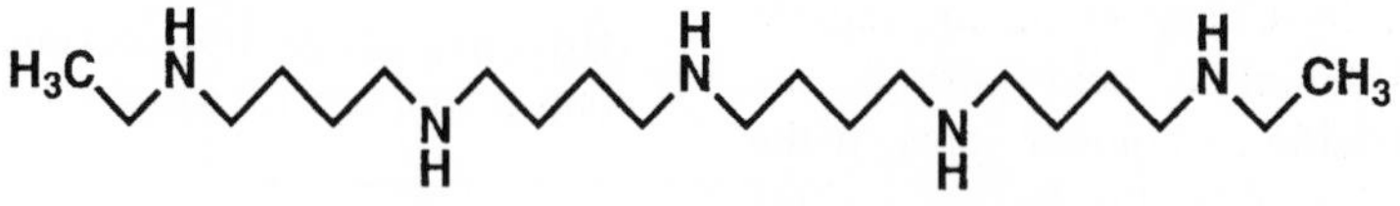

**Pentaazapentacosane (PAPC)**

**Fig. 2.12.** Pentaazapentacosane (PAPC).

Dansyl derivatives of 24 amino acids can be separated by EMKC with UV absorption at 214 nm [40].

Recent interest in the field of amino acid analysis is placed on chiral separation of D and L-amino acids by HPLC and CE in which high sensitivity and high resolution are essential. A maximum number of six pairs of DL-dansylamino acids was separated by cyclodextrin-modified MEKC in a single run [66].

*DNS-Cl Application: Optimized Derivatization Procedure for PAPC [64]*

Dansylation was carried out by adding 2.0 ml of TEA buffer (at pH 11), 2.0 ml PAPC solution and 3.0 ml DNS-Cl solution to a 10-ml volumetric. The sample was then diluted to volume with $CH_3CN$ and water to give a final ratio of 60:40 organic to aqueous. The final concentrations in the derivatization solutions were 14 nmol/ml PAPC, 17 μmol/ml TEA and 1.3 μmol/ml dansyl chloride. The sample solutions were allowed to react at room temperature in the dark for 4, 8 and 24 h. The reaction was stopped by adding a 2.0 ml aliquot of the reaction solution to 2 ml of glycine (2 mg/ml) in a 10 ml volumetric. After 10 min the neutralized solutions were diluted to volume with 0.04 M pH 5 acetic acid-CH3CN (30:70). HPLC was performed under the above conditions, detection limits for DNS-PAPC were found by injecting 0.7 pmol in column at an S/N ratio of 2.1.

#### 2.2.5.3 4-*N,N*-Dimethylaminoazobenzene-4′-sulfonyl Chloride (DABS-Cl, Dabsyl-Cl) [53]

The sulfonyl group of DABS-Cl reacts with primary and secondary amino groups, thiol, imidazole, phenols and aliphatic hydroxyl groups. In an original study by Chang *et al.*, the intense chromophoric dabsyl amino acid permits the detection of amino acids as colored spots in the range 0.1–0.01 nmol visible directly on a silica gel TLC plate [67] and on a polyamide layer sheets [68]. The dabsyl derivatives of various amino acids show absorption maxima ranging from 448 to 468 nm. Actually absorbance detection wavelength at 436 nm [69,70], 456 nm [69] and 425 nm [71] was also utilized in HPLC. The stability of the dabsyl amino acid derivaties is as good as that of dansyl amino acid.

*Dabsyl-Cl Application: Derivatization of Amino Acids [69,70]*

The derivatization reagent consisted of a solution of 4 nmol/l DABS-Cl in $CH_3CN$. This solution was prepared daily. $CH_3CN$ (40 μl) and DABS-Cl reagent (40 μl) were added to 20 μl of the standard solution. For each sample, 100 μl of sample were added to 100 μl of DABS-Cl reagent. The mixture was heated for 10 min at 70 °C in an oven (the mixture will become completely soluble after heating). A 900 μl volume of a solution of sodium hydrogen phosphate (0.1 mol/l, pH 7.0)-CH3CN-MeOH (70:20:10, v/v/v) was added to the reaction mixture. A 20 μl volume of this mixture was injected on to a C18 HPLC column.

### 2.2.6 9-Fluorenylmethyl Chloroformate (9-Fluorenylmethoxycarbonyl Chloride, FMOC-CL, Fluorescence and UV Detection)

An automatic pre-column derivatization procedure is developed for the determination of amino acids. Carbonyl chlorides react with primary, secondary and also tertiary amines and alcohols. The reactivity of the chloride is thought to be higher than that of sulfonyl chlorides. Carratù *et al.* [72] reported that the results of determination for amino acids obtained with FMOC-Cl by UV detection (265 nm) were similar to those obtained by fluorescence detection (excitation at 265 nm and emission at 340 nm) except for tryptophan, which was not detectable by fluorimetry because the fluorescence of the derivative is quenched. The derivative obtained in the reaction of amino acid with FMOC-Cl is shown in Fig. 2.13.

*FMOC-Cl Application: Derivatization Procedure [72]*

A 50-μl volume of test solution of amino acids and 50 μl of norvaline solution were added to

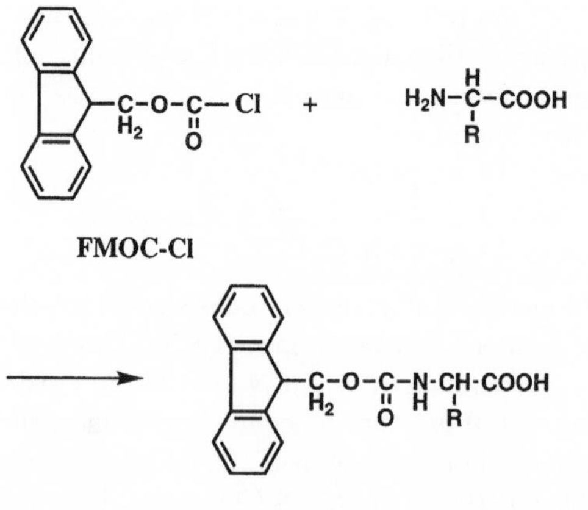

FMOC-Cl

Fig. 2.13. The reaction of FMOC-Cl with amino acids.

100 μl of borate buffer in glass sample vials. Subsequent steps followed the autosampler injector program: 200 μl of FMOC-Cl reagent (30 mM in dry acetone) were added, the mixture was quickly shaken and the reaction was allowed to proceed for 10 min at room temperature. Excess fluorescent and UV-absorbing FMOC-Cl was destroyed by the addition of 200 μl 1-aminoamantadine solution (25 mM in methyl alcohol) and after 2 min the solution was subjected to HPLC.

## 2.2.7 Ninhydrin

Traditionally, the two types of liquid chromatographic system used for the determination of amino acids in protein are the post-column and pre-column derivatization methods. The former is characterized by an ion-exchange separation and post-column derivatization with ninhydrin. Ninhydrin reacts with only primary amines (especially for $\alpha$-amino acids except for cysteine), giving Ruhemann purple, blue-violet colored indanic compounds (diketohydrindylidene-diketohydrindamine). The derivatization reaction of amino acids with ninhydrin is shown in Fig. 2.14. The ninhydrin method originated by Moore and Spackman *et al.* [73,74] has been used for amino acid determinations in the wide field of natural science and has become a classical method because of its suitability for automation, reproducibility and accuracy in spite of the low-sensitivity and high-cost instrumentation and long run times. In the automatic technique, ninhydrin reagent is introduced into the effluent flowing from the ion-exchange column. Color development ensues as the mixture is pumped through a reaction coil immersed in a boiling water bath, after which the intensity of the blue-violet color produced when amino acids are present is determined by continuous photometry at 570 nm. The yellow colors from proline and hydroxyproline can be monitored at 440 nm. The peaks on the recorded curves can be integrated with a precision of 100 ± 3% for loads from 0.1 to 3.0 μmol of each amino acid. A hydrolyzate of a protein or peptide may be analyzed in less than 24 h [73].

Ninhydrin/collidin reagent was used for the qualitative and quantitative determination of $\varepsilon$-aminocaproic acid, which is a potential contaminant of polyamide 6, by instrumental TLC [75]. Post-chromatographic derivatization was performed with ninhydrin/collidin reagent, by the ‘over pressured derivatization technique’. Detection was carried out using a photodensitometer scanner at an absorption maximum wavelength of 558 nm for aminocaproic acid. The detection limit for $\varepsilon$-aminocaproic acid was 0.02 mg.

Glyphosphate (*N*-(phosphonomethyl)glycine), a widely used herbicide under the trade name of ‘Roundup,’ and aminomethylphosphonic acid (AMPA) (Fig. 2.15) in biological fluids are determined by the use of a routine amino acid analyzer [76]. Glyphosphate is a secondary amine not expected to give Ruhemann purple. Despite

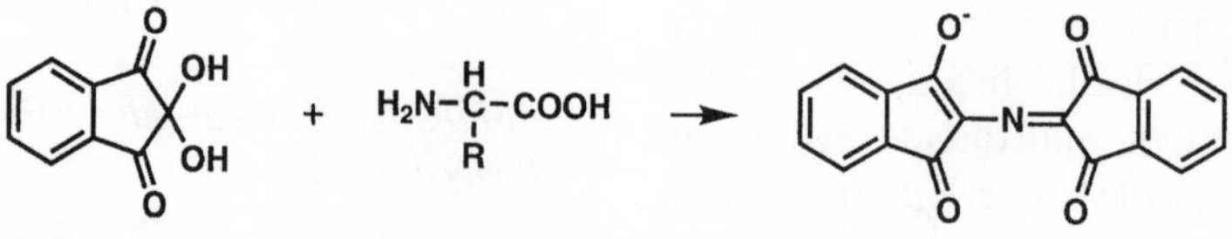

Fig. 2.14. The reaction of ninhydrin with amino acids.

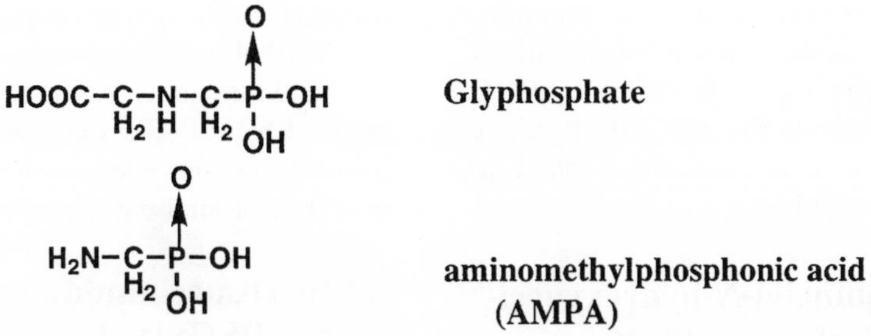

$HOOC-CH_2-NH-CH_2-P(O)(OH)-OH$ **Glyphosphate**

$H_2N-CH_2-P(O)(OH)-OH$ **aminomethylphosphonic acid (AMPA)**

**Fig. 2.15.** Glyphosphate and its metabolite, AMPA.

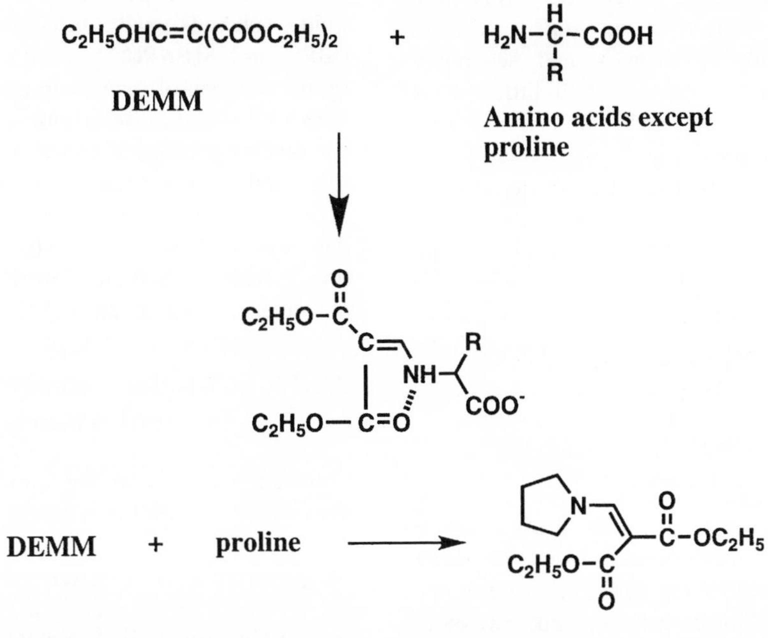

$C_2H_5OHC{=}C(COOC_2H_5)_2$ + $H_2N-CH(R)-COOH$

**DEMM** **Amino acids except proline**

**DEMM** + **proline** ⟶

**Fig. 2.16.** Derivatives obtained in the reaction of DEMM with amino acids.

its chemical structure, it reacts with ninhydrin as much as AMPA. It can be assumed that during the ninhydrin reaction, glyphosphate is converted into AMPA. For biological fluids, the minimum detectable amount is 80 pmol.

## 2.2.8 Diethylethoxymethylenemalonate (DEMM) [77]

Amino acids were determined by pre-column derivatization with DEMM and RP-HPLC with UV detection at 280 nm. The limit of detection was 3 pmol with a signal-to-noise ratio of 5. This derivatization of amino acids with DEMM and their subsequent RP-HPLC is recommended as an alternative to the dedicated amino acid analyzer when subpicomole sensitivity is not required. Formations of *N*-[2,2-bis(ethoxycarbonyl)vinyl] derivatives (Fig. 2.16) are stable for several weeks at room temperature.

*DEMM Application: Derivatization of Amino Acids [77]*

To a dried sample of standard amino acid mixture, or a protein hydrolysate (2–200 μg), in 1 M sodium borate buffer (pH 9.0) (1 ml) containing 0.02% of sodium azide was added 0.8 μl of DEMM. The reaction was carried out at 50 °C for 50 min with vigorous shaking. The resulting mixture was cooled to room temperature and 15 μl were subjected to an HPLC.

### 2.2.9 6-Aminoquinolyl-*N*-hydroxyuccinimidyl Carbamate (AQC)

Cohen and Michaud [78] originated a pre-column derivatization method in which AQC was used for amino acid derivatization. AQC can react quantitatively with all primary and secondary amino acids in a few seconds with little matrix interference and single and stable fluorescent derivatives are formed (Fig. 2.17). Recently, the method was modified by Liu [79]. UV detection at 248 nm was used for the assay of AQC derivatives of amino acids. All nineteen amino acids were separated in 35 min with resolutions larger than 1.6. The detection limits for all common amino acids including cystine and tryptophan were in the range 0.07–0.3 pmol.

*AQC Application: RP-HPLC Analysis of Amino Acids in Feed Sample [78,79]*

A 59 mg amount of feed powder was added to 10 ml of 6 M HCl and hydrolyzed for 24 h. After filtration through 0.45 μm Millex-HV filter, the hydrolysate was diluted with ultra pure water to a concentration of ca. 1.5 mg/ml of amino acids and 10 μl of the diluted solution were pipetted into a 6 × 50 mm tube. After drying under vacuum, the feed hydrolysate was reacted with AQC. The recovery of the method for feed hydrolysates should be around 98%.

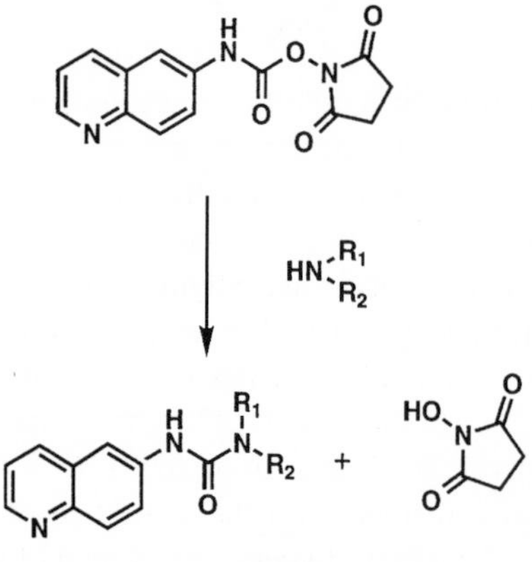

**Fig. 2.17.** The reaction of AQC with amines.

### 2.2.10 Disuccinimido Carbonate (DSC) [80]

Activated carbamate reagents can be used for HPLC determination of amino compounds. Succinimido phenyl- and *p*-bromophenylcarbamates (SIPC and SIBr-PC, respectively) are produced by the reaction of disuccinimido carbonate (DSC) with aniline and p-bromoaniline, respectively. The carbamates produce a urea derivative by reaction with both primary and secondary amines such as alkyl amines or amino acids (Fig. 2.18). Urea derivatives of SIPC and SIBr-PC have absorption maxima at 240 and 250 nm, and their molar absorptivities are 26 000 and 28 000, respectively.

### 2.2.11 Solid-phase Reagent with UV or VIS Light Absorbing Moiety

Solid supports containing immobilized reagents, also known as solid-phase or polymeric reagents. The polymeric reagents used today are based on polystyrene cross-linked with small amounts of divinylbenzene. The use of solid-phase derivatizing agents in place of the solution-phase reaction has certain advantages, including in some cases the ability to filter and reuse the derivatizing agents. Krull's group contribute greatly to this kind of approach [81,82]. Their work was thought to be one of the superior work's of the 1990's in the field of analytical chemistry, especially in HPLC. In the early stages of this kind of work, Chou *et al.* used a polymeric *o*-acetylsalicyl activated anhydride for improved UV detection at 196 nm or electrochemical detection of primary and secondary amines [83]. After that, a polymeric benzotriazole *o*-acetylsalicyl activated reagent [84] was developed and used for the derivatization of primary and secondary amines, as well as polyamines such as cadaverine and putrescine.

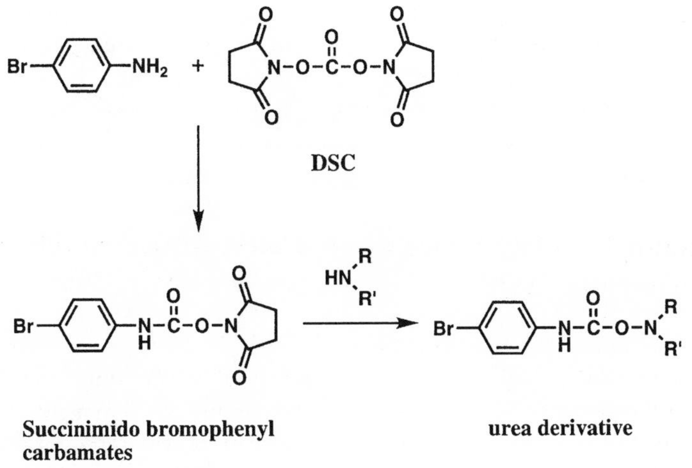

DSC

Succinimido bromophenyl carbamates

urea derivative

**Fig. 2.18.** Preparation of activated carbamate reagent and formation of urea derivative from activated carbamate reagent, succinimido bromophenyl carbamates with amines.

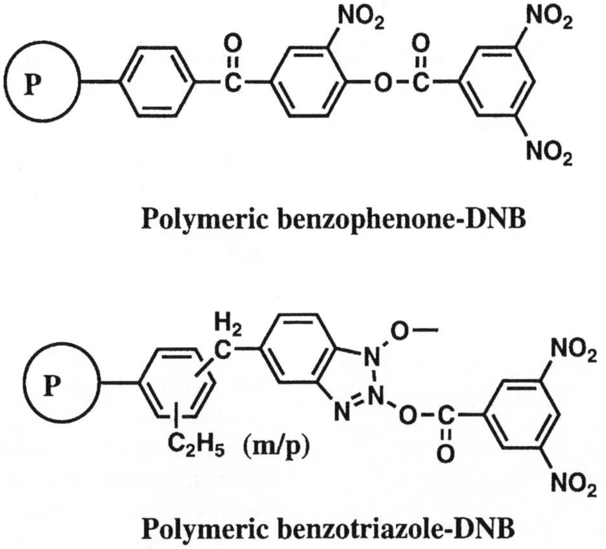

**Polymeric benzophenone-DNB**

**Polymeric benzotriazole-DNB**

**Fig. 2.19.** Polymeric benzophenone-DNB.

#### 2.2.11.1 Polymeric 3,5-Dinitrobenzoyl Tagged Derivatization Reagent

Bourque and Krull described a solid-phase technique for improved derivatization of chiral and achilal aliphatic amines, amino alcohols and amino acids in HPLC with UV detection [85]. Amphetamine in urine can be determined using 3,5-dinitrobenzoyl (DNB) moiety in polymeric benzophenone-DNB and polymeric benzotriazole-DNB [86] (Fig. 2.19). For amphetamine derivatives, the limit of detection is 14 ng/ml, and the limit of quantification is 47 ng/ml. The calibration curves are linear from 0.001 to 4.0 μg/ml. The above method used the same line of reagent having 3,5-DNB moiety developed by Zhou *et al.* [81].

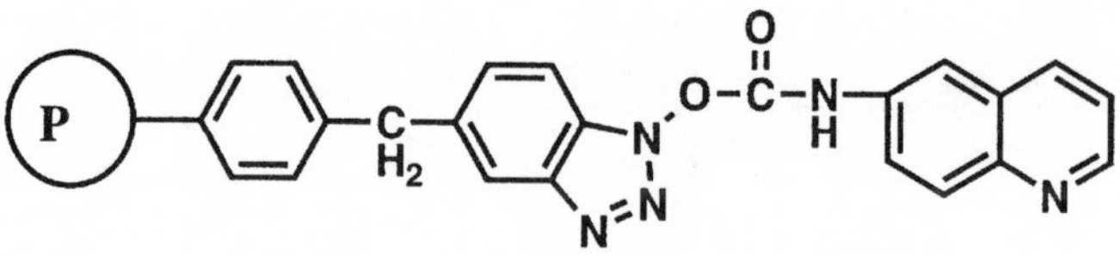

**Polymeric benzotriazole-6-AQ**

**Fig. 2.20.** Polymeric Benzotriazole-6-AQ.

*Polymeric 3,5-Dinitrophenyl Activated Reagent Application: Solid-phase Off-line Derivatizations [85]*

A volume of analyte (30–100 μl) was added to the polymer (30–150 mg) such that the polymer was wetted without the analyte reaching the tissue plug. The solid-phase reaction was allowed to proceed and was washed with 500 μl $CH_3CN$. The eluant was then diluted with water to match the mobile phase. For normal sensitivity work (ppm and above), 20 μl of the final solution was subjected to the HPLC.

#### 2.2.11.2 Polymeric 6-Aminoquinoline (6-AQ) Tagged Derivatization Reagent [87]

6-AQ, an activated carbamate reagent, was developed and applied for the off-line derivatization of amines and amino acids in HPLC. Derivatized amines and amino acids were separated under conventional RP-HPLC with UV and fluorescence detection (Fig. 2.20).

#### 2.2.11.3 Polymeric Benzotriazole Activated Reagent Containing FMOC Group

A polymeric benzotriazole reagent containing the FMOC group (Fig. 2.21) for UV and fluorescence detection, was applied in the determination of cadaverine and putrescine, normally occurring polyamines in human urine [88]. Polymeric FMOC reagent was also used for the determination of volatile amines in air by on-line solid phase derivatization and HPLC with UV and fluorescence detection [89]. The polymeric benzotriazole described above are too sensitive to moisture and elevated temperature. An improved polymeric reagent, a fluorenyl-attached polymeric *o*-nitrobenzophenone reagent (Fig. 2.22) was presented by Gao *et al.* [90].

## 2.3 LABEL OF CARBOXYL (−COOH)

Absorbance of the carboxyl group is generally very low. Thus, several derivatization reagents that have UV or visible absorbing moieties in their molecules have been used in order to enhance the sensitivity of detection for biologic carboxylic compounds such as fatty acids (FAs) and prostaglandins (PGs).

As the carboxyl moiety shows low reactivity, high activity is required derivatizing reagents for carboxylic acids (Fig. 2.23). The reagents involving; 1) alkyl halides; 2) aromatic amines; 3) hydrazines; and 4) hydroxylamine as the reactive site, are mainly used for pre-column derivatization reaction and the produced derivatives are separated on a reversed phase or silica gel column in HPLC with UV or visible detection.

### 2.3.1 Alkyl Halides

Alkyl halide reagents such as phenacyl bromide (PHB) [91–93], *p*-bromophenacyl bromide (BPB) [94–99], α-bromo-2′-acetonaphtone (BAN) [100–102] and panacyl bromide (PB) [103] react with carboxylic acids in acetonitrile under mild conditions, and the reactions are catalyzed by a crown ether and potassium ion, or organic bases such as triethylamine and ethylamine (Fig. 2.23a).

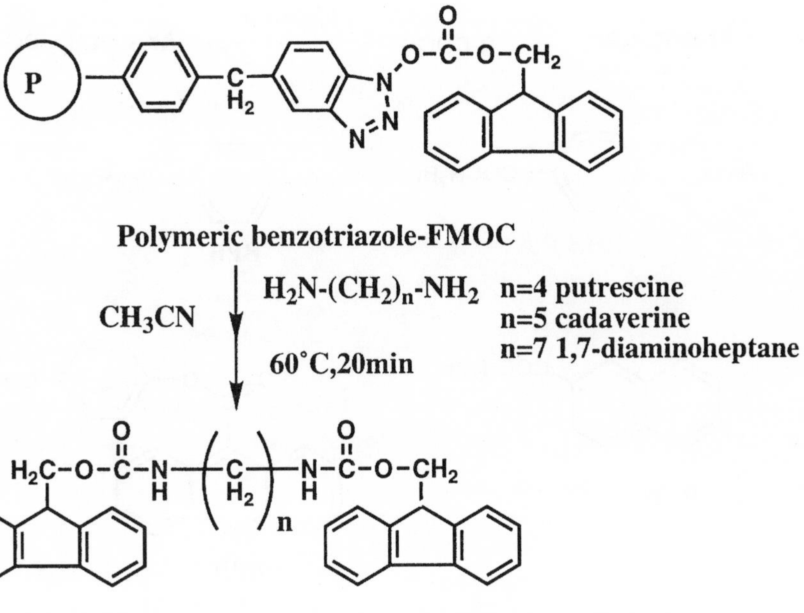

Polymeric benzotriazole-FMOC

$CH_3CN$ $H_2N$-$(CH_2)_n$-$NH_2$ n=4 putrescine n=5 cadaverine n=7 1,7-diaminoheptane 60°C,20min

**Fig. 2.21.** The derivatization of polyamines with the polymeric benzotriazole-FMOC reagent.

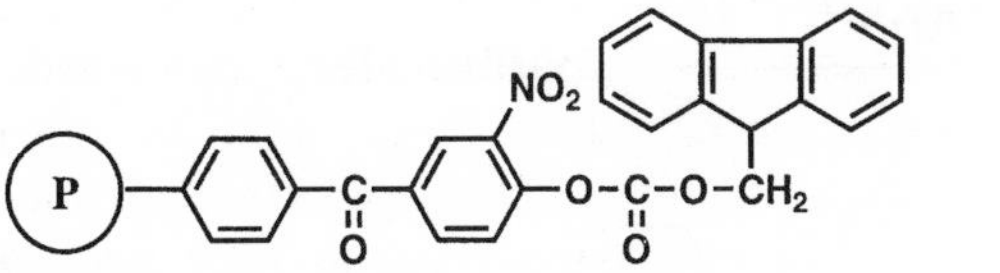

**Fig. 2.22.** Polymeric 3-nitro-4-[[9-fluorenylmethoxy)-carbonyl]oxy]benzophenone.

*BAN Application: HPLC Determination of Prostaglandins $A_1$ and $B_1$ in Alprostadil (PGE1) [102]*

*Sample preparation*: Accurately weigh ca. 2 mg of alprostadil, transfer to a suitable container, dissolve in ca. 2 ml of absolute ethanol, and gently evaporate to dryness under a stream of nitrogen. *Derivatization procedure*: Add ca. 200 μl of BAN (20 mg/ml in acetonitrile) to each sample vial and swirl to wash the sides. Add 100 μl of di-isopropylethylamine (10 μl/ml in acetonitrile) and swirl again. Allow to stand at room temperature for at least 90 min, then evaporate to dryness under a stream of nitrogen. Re-dissolve the sample preparation in 2.0 ml of internal standard solution (20 μg/ml methyltestosterone in methylene chloride). Chromatograph the solutions.

*HPLC conditions and separation (Fig. 2.24)*: UV detection at 254 nm is performed. The mobile phase consists of 1000 ml of methylene chloride, 7.5 ml of tert.-amyl alcohol and 1.0 ml of deionized water. The mobile phase is filtered and degassed before use. The flow-rate is 2.0 ml/min and the injection volume is 25 μl.

## 2.3.2 Aromatic Amines

Aromatic amine-type reagents such as *p*-methoxyaniline [104], *p*-chloroaniline [105] and 1-naphthylamine [106] readily react with acid chlorides of the carboxyl group. Therefore, the amide derivatives can be formed with the aromatic amine by means of conversion of carboxylic acids into the acid chlorides (Fig. 2.23b). Triethylphosphin [104] or oxalyl chloride and thionyl chloride [106] were used to form the acid chlorides.

*p-Methoxyaniline Application: HPLC Determination of Fatty Acids as p-Methoxyanilides [104]*

*Derivatization procedure*: A mixture of fatty acid, 5–10 mg of each (about 0.7 mmol total acids),

a: Ar-$COCH_2Br$ + R-COOH —Catalyst→ Ar-$COCH_2OCO$-R

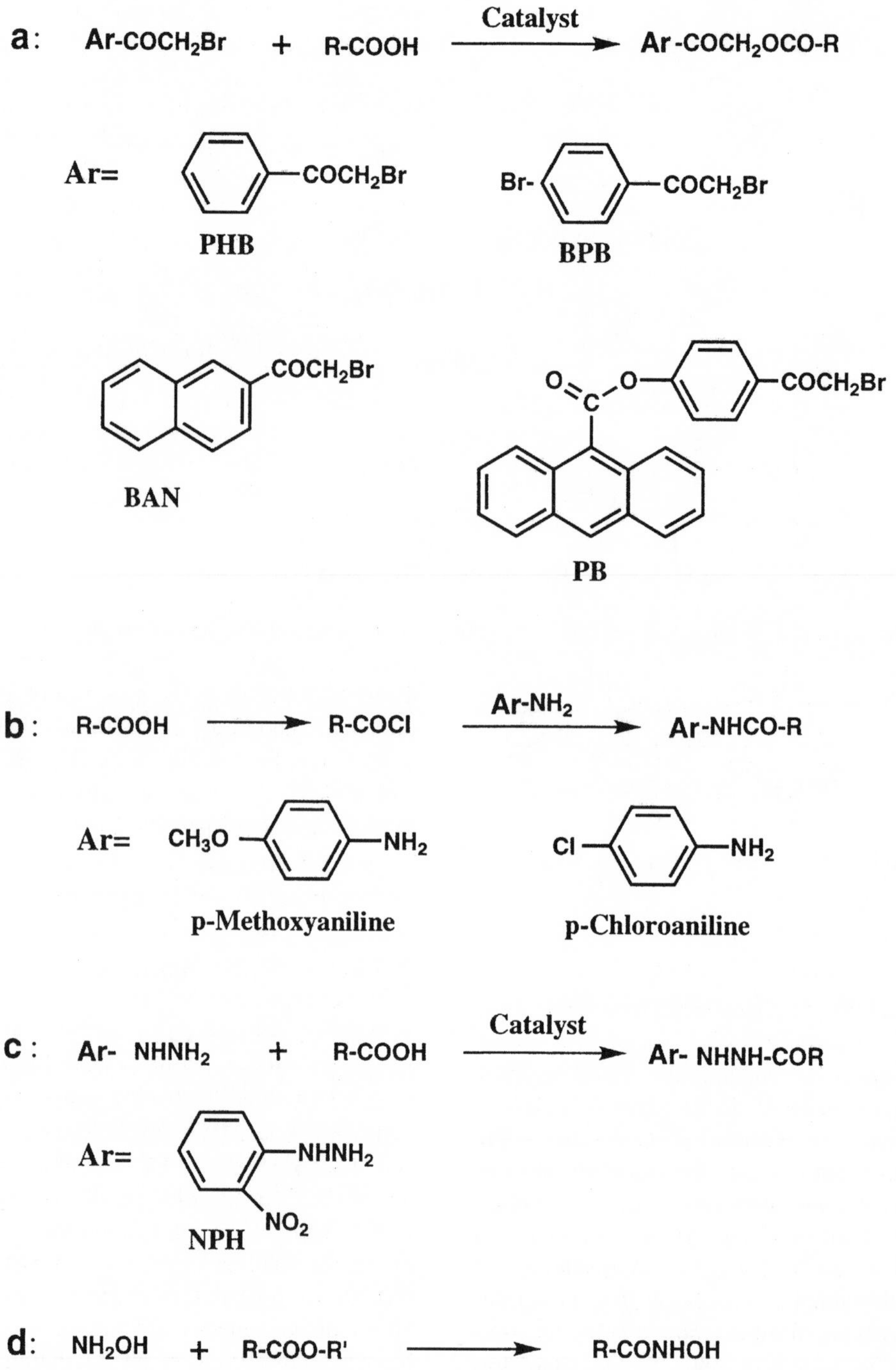

Ar= PHB, BPB, BAN, PB

b: R-COOH → R-COCl —Ar-$NH_2$→ Ar-NHCO-R

Ar= p-Methoxyaniline, p-Chloroaniline

c: Ar- $NHNH_2$ + R-COOH —Catalyst→ Ar- NHNH-COR

Ar= NPH

d: $NH_2OH$ + R-COO-R' → R-CONHOH

**Fig. 2.23.** Four reaction schemes between (a) an alkyl halide, (b) an aromatic amine, (c) a hydrazine or (d) hydroxylaime, and a carboxylic acid.

**Fig. 2.24.** Chromatogram of $PGA_1$ and $PGB_2$ in alprostadil. Peaks: A = excess derivatizing reagent; B = methyl testosterone (I.S.); C = $PGA_1$; D = $PGB_1$. [Reproduced from ref. 102, p 284, Fig. 2.].

0.5 g (1.9 mmol) triphenylphosphine and 2 ml dry carbon tetrachloride were placed in a 15 ml vial. The vial was sealed with a Teflon-lined screwcap and kept at 80 °C for 5 min. Cloudiness appears when the reaction is complete. After cooling to room temperature, the vial contents were mixed with 0.5 g (4.1 mmol) of *p*-methoxyaniline dissolved in 8 ml of dry ethyl acetate. The vial was heated at 80 °C for 1 h. An insoluble oil appears at this point but, because it is insoluble, it has no effect on chromatography. If the reaction temperature is raised to 140 °C, the first step can be completed in 2 min and the second step in 20 min. *HPLC conditions and separation (Fig. 2.25)*: The chromatograph has a UV detector that was operated at 254 nm. The flow rate of the chromatographic solvent was 1.0 ml/min. A 1/4-in. o.d. by 30 cm $\mu$-Bondapak C18 column (Waters Associates) was used. The column's plate number was 3000 measured with *p*-methoxylauranilide and a solvent of 83% acetonitrile and 17% water. In all runs, 2 µl of the solution containing the anilides was injected with a 10-µl syringe. The gradient elution program was begun immediately upon sample injection. However, the used injector has a 2 ml solvent loop between the pumped solvent and the point of injection. Therefore, 2 ml of programmed solvent must be pumped before it reaches the point of injection (approximately the head of the column). The gradient curve in Fig. 2.25 plots the solvent composition at the head of the column and show this 2 ml lag.

### 2.3.3 Hydrazines

NPH [107–110] reacts with short and long-chain FAs, and also straight and branched-chain dicarboxylic acids in the presence of 1-ethyl-3-(3-dimethylaminopropyl)carbodiimide hydrochloride (EDC), as shown in Fig. 2.23c. The HPLC method with UV detection (230 nm) provides the detection limit of 2–5 pmol per injection.

*NPH Application: HPLC Determination of Straight and Branched-chain Dicarboxylic Acids in Urine as their 2-Nitrophenylhydrazides [109]*

*Reagent solutions*: an NPH solution (0.02 M) was prepared by dissolving the reagent in ethanol. Solutions of pyridine (3% v/v) in ethanol and EDC hydrochloric acid (0.25 M) in water were prepared, and then a working EDC solution was prepared by mixing equal volumes of the pyridine and EDC hydrochloric acid solutions. A potassium hydroxide solution (15%, w/v) in methanol-water (80:20) was prepared. *Derivatization procedure*: To 50 ml of an aqueous mixture of standard dicarboxylic acids, 100 µl of NPH solution and 200 µl of working EDC solution were added, and the mixture was heated at 60 °C for 20 min. After the addition of 50 µl of potassium hydroxide solution, the mixture was further heated at 60 °C for 15 min

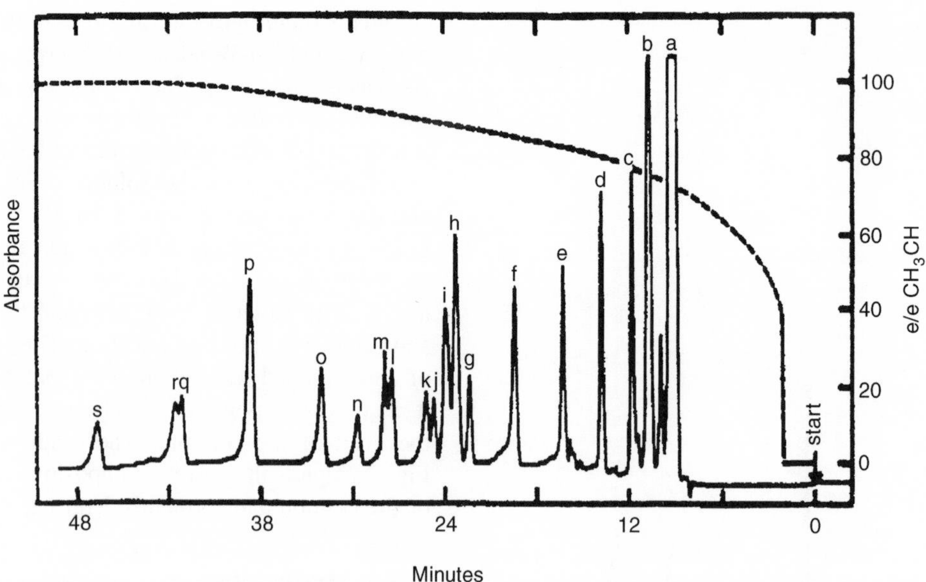


**Fig. 2.25.** Chromatogram of fatty acid *p*-methoxyanilides by water-acetonitrile gradient elution. Peaks: a = *p*-methoxyanilline; b = unknown; c = C6; d = C8; e = C10; f = C12; g = C18:3; h = C14 + C22:6; i = C20:4 + C16:1; j = C18:2; k = C15; l = C16; m = C18:1; n = C17; o = C18; p = C20 + C22; q = C24:1; r = C22; s = C24. [Reproduced from ref. 104, p. 1105, Fig. 1.].

and then cooled in running water. *Sample preparation:* To 500 µl of human urine, 100 µl of water containing 100 nmol of 3,3-dimethylglutaric acid as the internal standard was added. The sample was acidified to pH 1-2 by adding 6 M hydrochloric acid and saturated with sodium chloride. The urinary acids were extracted twice with 2 ml of ethyl acetate. The combined organic extracts were washed with 2 ml of 0.1 M hydrochloric acid solution saturated with sodium chloride. After the organic extracts were dried over anhydrous sodium sulfate, the solvents were evaporated. The residue was dissolved in 50 µl of water and treated according to the derivatization procedure. The resulting hydrazides mixture was neutralized by adding 2 ml of 1/30 M phosphate buffer (pH 6.4)-0.5 M hydrochloric acid (3.8:0.4, v/v) and washed twice with 2.5 ml of ethyl acetate to remove interfering substances and all acid hydrazides other than and dicarboxylic acid hydrazides. A 1.5 ml portion of the aqueous layer was taken and acidified by adding 2 ml of 0.5 M hydrochloric acid. The dicarboxylic acid hydrazides were extracted with 2.5 ml of ethyl acetate. The organic extract was washed twice with 2 ml of 0.1 M hydrochloric acid and evaporated with a stream of nitrogen at room temperature. The residue was redissolved in 50 µl of methanol and an aliquot (2–10 µl) was injected into the chromatograph. *HPLC conditions and separation (Fig. 2.26):* A visible detector was set at 400 nm. The separation was performed with a YMC-AQ (ODS) column (particle size 5 µm, 250 × 4.6 mm i.d.). The column temperature was kept constant at 40 °C. All analyses were carried out isocratically using an acetonitrile-phosphate buffer as the eluent at a flow rate of 1.2 ml/min. The pH was adjusted to the desired value by mixing acetonitrile-5 mM $KH_2PO_4$ with acetonitrile-5 mM $Na_2HPO_4$ and then dissolving counter ions at a concentration of 5 mM. The counter ions studied were TMA, TEA, and TPA (as bromides). The solvents were filtered through a filter (pore size 2 µm) and degassed before use.

## 2.3.4 Hydroxylamine

Hydroxylamine is usable for simple and rapid derivatization of fatty acid esters to their

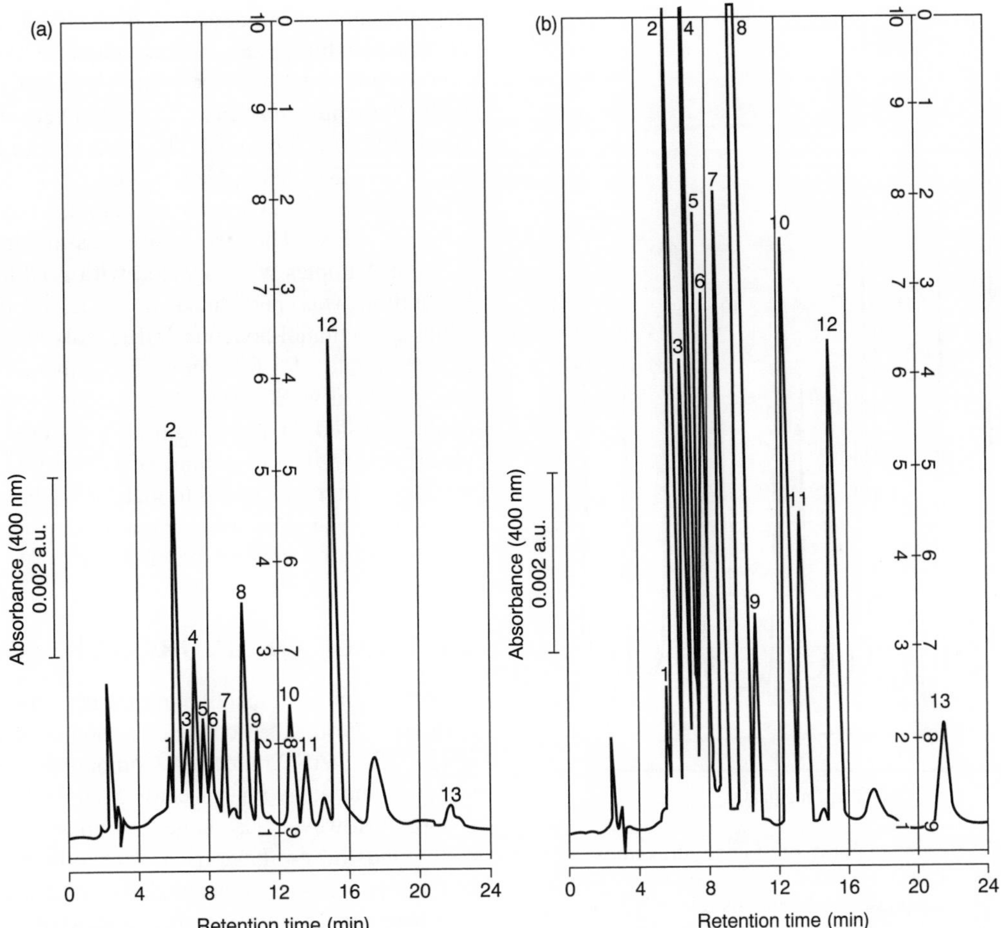


**Fig. 2.26.** Chromatograms of the derivatized dicarboxylic acids in (a) urine and in (b) urine supplemented with the dicarboxylic acids (25 nmol in each). Peaks: 1 = malonic; 2 = succinic; 3 = methylmalonic; 4 = glutaric; 5 = methylsuccinic; 6 = methylsuccinic; 7 = adipic; 8 = 3-methylglutaric; 9 = ethylmalonic; 10 = pimelic and 3-methyladipic; 11 = 3-methyladipic; 12 = 3,3-dimethylglutaric (I.S.); 13 = suberic acid hydrazide. The urinary levels of methylmalonic and succinic acids were in the range of 4.5–30.6 nmol/ml and in the range of 45.2–378.8 nmol/ml, respectively. [Reproduced from ref. 109, p. 306, Fig. 2.].

hydroxamic acids [111], which absorb strongly in the low wavelength UV region (206 or 213 nm). The derivatives are separable by a reversed phase HPLC. The detection limit (2 × baseline noise) was 150 pmol at 213 nm.

*Hydroxylamine Application: HPLC Determination of Fatty Acids as Hydroxamic Acids [111]*

*Reagent solution:* The hydroxamation reagent, 1.5 M hydroxylammonium perchlorate was prepared by adding 18.6 g of sodium perchlorate to 100 ml of tert.-butanol and stirring to comminute and partially dissolve the solid. Finely ground hydroxylamine hydrochloride (10.5 g) was dried for 1 h at 110 °C and, after cooling, added, along with one drop of sodium methoxide solution, to the stirring sodium perchlorate suspension. After stirring the mixture for at least 4 h at room temperature, the suspension was centrifuged and the reagent solution decanted from the resulting sodium chloride solid. *Derivatization procedure:*

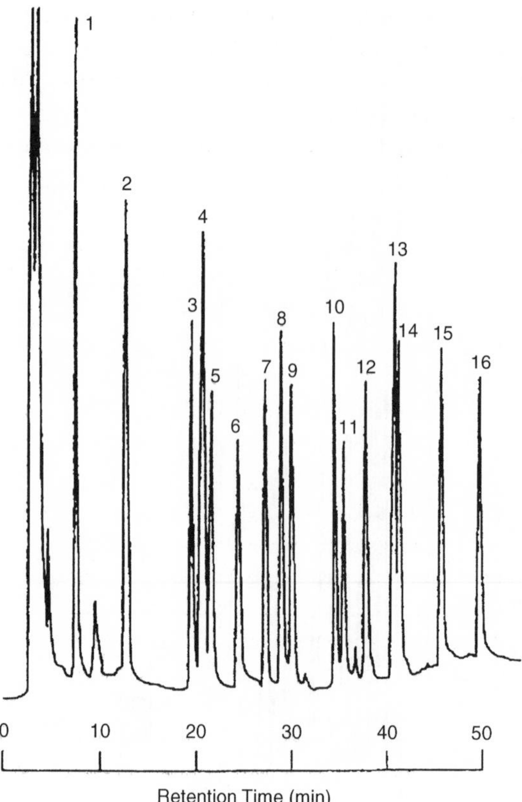


**Fig. 2.27.** Chromatogram of the hydroxamic acid derivatives of a fatty acid methyl esters separated on a reversed-phase column (Nucleosil) with a methanol-phosphate buffer gradient (70–95% methanol). Peaks: 1 = capric; 2 = lauric; 3 = myristic; 4 = linolenic; 5 = palmitoleic; 6 = linoleic; 7 = palmitic; 8 = oleic (cis-18:1); 9 = elaidic (trans-18:1); 10 = stearic; 11 = eicosenoic; 12 = nonadecanoic; 13 = arachidic; 14 = erucic; 15 = behenic; 16 = lignoceric. Each peak corresponds to 10–30 nmol fatty acid. [Reproduced from ref. 111, p. 294, Fig. 1.].

A sample of fat or oil (5–10 mg) was placed in a PTFE-lined screw-capped vial (45 × 10 mm) and dissolved in 1 ml of tert.-butyl methyl ether. A 6 μmol amount of methyl nonadecanoate was added as internal standard. A 150 μl volume of the hydroxamation reagent was pipetted in, followed by 150 μl of 25% sodium methoxide solution, and the mixture was shaken for a few seconds to ensure mixing. After 1 min, 2 ml of quench solution (5%, v/v, glacial acetic acid in methanol) was added and a 10 μl aliquot of this mixture was injected directly into the liquid chromatograph for analysis. *HPLC conditions and separation (Fig. 2.27):* A reversed-phase column (250 × 4.6 mm I. D.) packed with Nucleosil C18 (Macherey-Nagel, Duren, Germany), 5 μm particle size. Column temperature was maintained at 40 °C. The absorbance was monitored at 213 nm. Samples were injected with a 10 μl loop. Separation was performed by gradient elution, using a methanol-aqueous buffer solvent system at a flow-rate of 1.0 ml/min. The aqueous buffer was prepared as 20 mM sodium dihydrogenphosphate adjusted to pH 3 with concentrated phosphoric acid. A 125 μl volume of 1% (w/w) sodium nitrate per liter was added to match the absorbance of methanol at 213 nm. Solvents were degassed continuously by helium sparging.

## 2.4 LABEL OF HYDROXYL –OH

There are many bioactive molecules containing hydroxyl groups. Alcohols, carbohydrates and lipids are very important compounds in life systems, and many pharmaceuticals have hydroxyl groups. However, a large number of these compounds do not have the UV absorbing groups of fluorophores, they must be derivatized with suitable reagents for sensitive ultraviolet-visible detection. In addition to this the hydrophilic property of hydroxyl groups is a major disadvantage for chromatographic separation. For the derivatization of hydroxyl groups, acyl halides, acid anhydrides, isocyanates and other compounds having a light absorbing group are often used.

### 2.4.1 Acyl Halides

Benzoyl chloride is widely used for derivatizing reagents for ultraviolet detection. It reacts with the hydroxyl group in alkaline conditions to form esters and the derivatives are detectable at about 230 nm. The reaction is shown in Fig. 2.28. Vollmer *et al.* [112] reported the rapid analytical method of ethylene glycol and propylene glycol in human serum using RP-HPLC. The method is free from interference by other common volatile

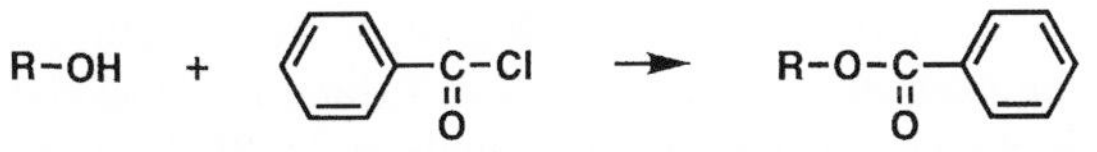

R–OH + C–Cl → R–O–C
O O

Fig. 2.28. The reaction of hydroxyl group and benzoyl chloride.

alcohols, and the derivatized and purified serum sample was analyzed within 8 min. A typical chromatogram is shown in Fig. 2.29. Detection of ethylene glycol and propylene glycol was carried out at 237 nm, and the detection limit was 10 mg/l serum. Kasama *et al.* [113] show the simultaneous quantification of serum 5$\alpha$-cholestan-3$\beta$-ol and cholesterol. Reversed-phase HPLC was used for the separation and the detection wavelength was 228 nm. This method needed only 0.1 ml serum to give a reproducible result, and it has been used for the biochemical diagnosis of cerebrotendinous xanthomatosis, which is a hereditary disorder of cholesterol metabolism. Benzoyl chloride is also used to derivatize acyl glycerols. Ioneda *et al.* [114] reported the classification and fractionation of 1-monomycoloyl glycerols using RP-HPLC and structural investigation by mass spectrometry. This method was applied to the 1-monomycoloyl glycerol fraction from *Rhodococcus lentifragmentus*, and the *o*-benzoyl derivatives were amenable to fractionation into homologous components. Uzawa *et al.* [115] developed a general method to evaluate stereoselectivities of lipase catalyzed hydrolysis of tri-*o*-acyl glycerols. Enzymatic hydrolysis products 1,2(or 2,3)-di-*o*-acylglycerols were first derivatized with *tert*-butyldimethylsilyl chloride to form silylated compounds. The products were deacylated using ethylmagnesium bromide (Grignard reagent) followed by the benzoylation with benzoyl chloride. The key compound, 1,2-di-*o*-benzoyl-3-*o*-*tert*-butylmethylsilyl-*sn*-glycerol and its enantiomer 2,3-di-*o*-benzoyl-1-*o*-*tert*-buthylmethylsilyl-*sn*-glycerol were separated by the HPLC with a chiral stationary phase. The derivatives were detected at 245 nm and the detection limit was

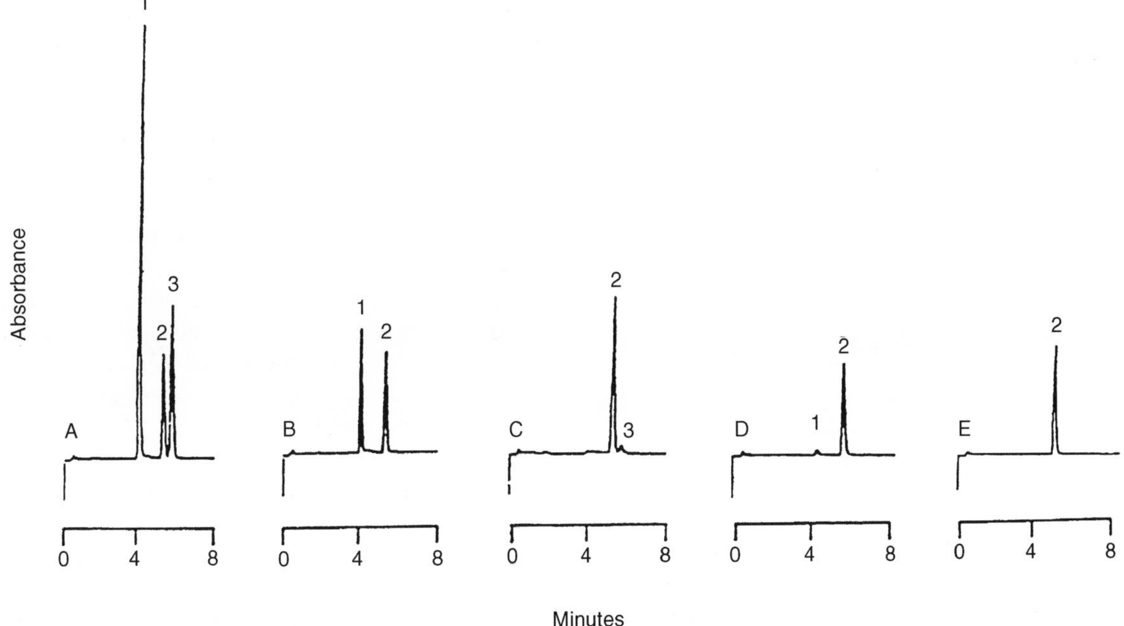


Fig. 2.29. Chromatograms of ethylene glycol (1), 1,3-propanediol (I.S., 2) and propylene glycol (3). (A) Standard mixture in bovine serum. (B) Patient sample positive for ethylene glycol. (C) Patient specimen positive for propylene glycol. (D) Patient sample positive for ethylene glycol. (E) Patient sample negative for ethylene glycol and propylene glycol. [Reproduced from ref. 112, p. 373, Fig. 2.].

20 ng. Davey *et al.* [116] showed a rapid and quantitative analytical method for cyanobacterial heterocyst type glycolipids. The type of glycolipids includes hydroxyalcohol or hydroxycarboxylic acid, which were derivatized with benzoyl chloride and analyzed NP-HPLC. Hawkins *et al.* [117] investigated the steric configuration of hydroxyeicosatetraenoates (HETEs) and other hydroxy fatty acids. The HETEs were derivatized with benzoyl or naphthoyl chloride following the methylation by diazomethane. The derivatives were purified by RP-HPLC and enantiomers were separated using a Pirkle type chiral stationary phase.

*Benzoyl Chloride Application: Determination of Alcohols in Human Serum [112]*

Serum sample (100 μl) was mixed with 100 μl of 1,3-propanediol (500 mg/l, internal standard) and 500 μl of 8 M NaOH. Benzoyl chloride (50 μl) was added and the solution was incubated at ambient temperature for 5 min. The derivatizing reaction was terminated by adding 30 μl of 1% glycine. After purification by solid phase extraction, the derivatives were analyzed by C18 RP-HPLC within 8 min. The eluent was acetonitrile/water = 1/1 with a flow rate of 2.5 ml/min. Detection was carried out at 237 nm and the detection limit was 10 mg/l serum, the calibration range was 20–2000 mg/l serum. Recovery of ethylene glycol addition in the serum was 96%. Within-run and between-run imprecision were less than 5% and 9.5% R.S.D. respectively.

## 2.4.2 Acid Anhydrides

Acid anhydride reacts with hydroxyl groups to form esters. For the ultraviolet detection of biological molecules, benzoic anhydride was widely used (Fig. 2.30). Formato *et al.*

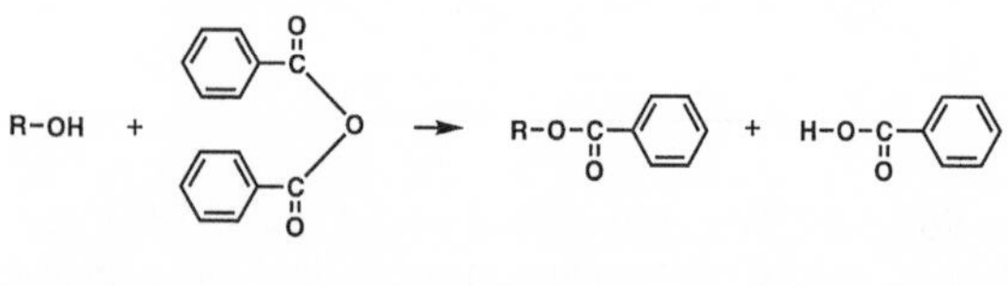

R–OH + (benzoic anhydride) → R–O–C(=O)–Ph + H–O–C(=O)–Ph

**Fig. 2.30.** The reaction of hydroxyl group and benzoic anhydride.

[118, application] reported the quantitation of monosaccharide (D-glucose, 1,6-anhydro-L-idose and D-galactosamine) in galactosaminoglycan from human arterial proteoglycan. The method includes hydrolysis of proteoglycan, benzoylation of resultant monosaccharide using benzoic anhydride and RP-HPLC separation and determination of each derivatives. Benzoyl derivatives of monosaccharide were monitored at 230 nm and the detection limit was 10 ng (S/N = 2). Species analysis of cardiolipin (CL, 1,3-bisphosphatidyl-*sn*-glycerol) was shown by Schlame *et al.* [119] using benzoylation and RP-HPLC. CL from bovine heart was derivatized by methylation of free phosphate groups with diazomethane followed by the benzoylation using benzoic anhydride with 4-(dimethylamino)pyridine as a catalyst. The resultant 1,3-bisphosphatidyl-2-benzoyl-*sn*-glycerol dimethyl ester was analyzed by C18 RP-HPLC within 150 min and monitored at 228 nm. CL from bovine heart was resolved into 11 molecular species by this method. Cantafora *et al.* [120] reported the determination of individual molecular species of phosphatidylcholine (PC) in human bile, liver and plasma. PC was converted into 1,2-diradylglycerobenzoate by hydrolysis with phospholipase C and reaction with benzoic anhydride. Distearoylphosphatidylcholine was used as an internal standard. The benzoates were separated by RP-HPLC within 45 min and the detection was carried out at 230 nm. Oshima *et al.* [121] reported the analysis of glycosphingolipids in human urine sediment. Glycosphingolipids were purified by extraction and silica-gel open column and derivatized with benzoic anhydride. The derivatives were analyzed within 15 min by silica-gel NP-HPLC monitored at 230 nm. By this method, glucosyl ceramide (GlcCer), lactosyl ceramide (LacCer) and globotriaosyl ceramide (GbOse3Cer) were clearly separated and the detection limit was 50 pmol. This method was applicable to the rapid urinary diagnosis of Fabry's disease, which is a inborn error of glycoshpingolipid metabolism caused by lysosomal *a*-galactosidase A deficiency. The typical chromatograms obtained from a normal and a patient sample were shown in Fig. 2.31.

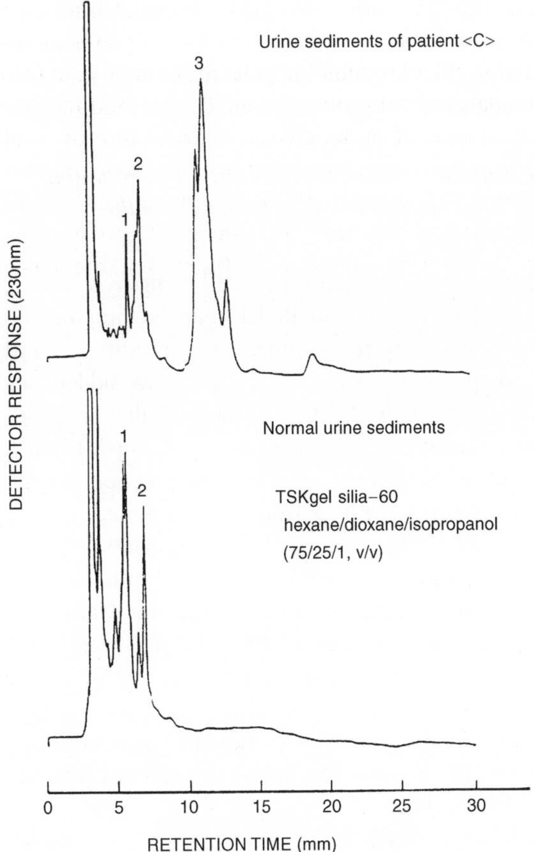


**Fig. 2.31.** Chromatogram of per-*o*-benzoylated sphingoglycolipids of urinary sediments from a Fabry patient and that of a normal control. 1, GlcCer; 2, LacCer and Gal-GalCer; 3, GbOse3Cer. [Reproduced from ref. 121, p. 159, Fig. 3.].

Ghosh *et al.* [122] reported the analysis of saccharides using isatoic anhydride as a derivatizing reagent. The reagent reacts with the primary hydroxyl group in monosaccharides to form mono-anthranilate derivatives (Fig. 2.32). The derivatives were separated using RP-HPLC and monitored at 254 nm.

*Benzoic Anhydride Application: Determination of Monosaccharide in Galactosaminoglycan [118]*

Galactosaminoglycan obtained from human aorta was dissolved in 2 M TFA and heated at 100 °C for

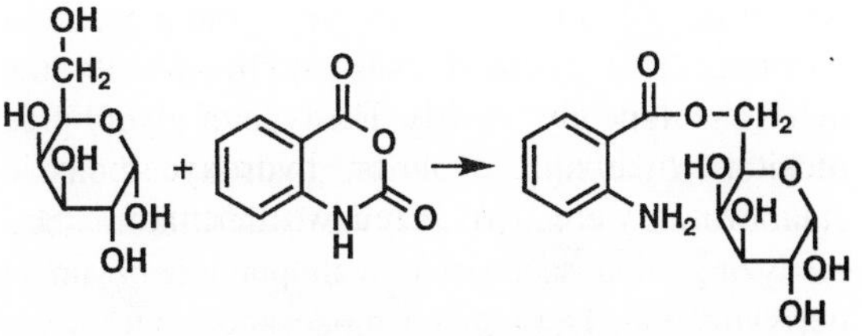

**galactose isatoic anhydride galactose anthranilate**

**Fig. 2.32.** The reaction of galactose and isatoic anhydride.

9 h. The hydrolyzed mixture was lyophilized and 0.5 ml of pyridine containing 10% (w/v) benzoic anhydride and 5% (w/v) 4-dimethylaminopyridine were added. The solution was heated at 37 °C for 90 min, then quenched with 4.5 ml water. After the reaction, the mixture was purified by C18 cartridge and centrifugation at 9000 g, the benzoyl derivatives were subject to RP-HPLC. Gradient elution using acetonitrile and water was performed and the separation of derivatives was monitored at 230 nm. Using this method, D-glucose, 1,6-anhydro-L-idose and D-galactosamine were separated within 60 min. The detection limit was 10 ng (S/N = 2), and the calibration range was 0.05–4.0 μg.

### 2.4.3 Isocyanates

Isocyanates are also used for the derivatization of hydroxyl group. They react by nucleophilic addition to form carbamate derivatives. For ultraviolet detection, phenylisocyanate is widely used and the resultant phenylcarbamate is a stable, UV absorbing compound (230–240 nm). The reaction is shown in Fig. 2.33. Indyk *et al.* [123] reported a simple method to determine myo-inositol in infant formulation, which is based on cow's milk. Inositol was derivatized with phenylisocyanate and separated by C18

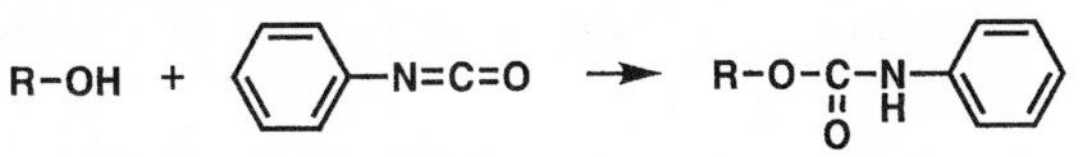

**Fig. 2.33.** The reaction of hydroxyl group and phenylisocyanate.

RP-HPLC within 50 min and monitored at 240 nm. This method was applicable to the determination of endogenous levels of free inositol in the milk of cows, goats and humans. Rakotomanga *et al.* [124] investigated the method analyzing monosaccharide with phenylisocyanate derivatization. Derivatized monosaccharides were separated using RP-HPLC within 20 min. The detection limit of each monosaccharide at 240 nm was 0.2 ng. Determination of deoxysugars and methylglycosides were also performed by this method. Dethy *et al.* [125] showed a sensitive method of measuring polyols in biological samples by RP-HPLC. Sorbitol and galactitol in rat and human tissues were derivatized with phenylisocyanate and separated within 20 min. The detection limit at 240 nm was 3 ng. This method is applicable to estimate aldose reductase activity in various tissues.

3,5-Dinitrophenyl isocyanate was also used to derivatize the hydroxyl group. Nakagawa *et al.* [126] reported chiral separation of hydroxy fatty acid using 3,5-dinitrophenyl isocyanate as a derivatizing reagent. Hydroxy fatty acids with various chain lengths and different positions of the hydroxyl group, were separated within 50 min by a HPLC with a chiral stationary phase. The derivatives were detected at 245 nm and the detection limit was 2.5 μg/ml. This method was applied to determine the optical configuration of 2 and 3 hydroxy fatty acids in bacterial lipids. Severini *et al.* [127] showed a stereospecific assay of methocarbamol, (R,S)-3-(2-methoxyphenoxy)-1,2-propanediol-1-carbamate, in human plasma and urine. It is a skeletal muscle relaxant and was derivatized with (S)-(+)-1-(1-naphthyl)ethyl isocyanate to form diastereomers (Fig. 2.34). The derivatives were separated using silica gel NP-HPLC within 50 min. Detection was carried out at 280 nm, and the detection limit was 10 ng/ml (S/N = 3). The method is suitable for stereoselective pharmacokinetic studies in human models.

*Phenylisocyanate Application: Determination of Myo-inositol in Milk (Infant Formula) [123]*

Milk sample was deproteinized using hydrochloric acid and lyophilized under reduced pressure. To each lyophilized sample, dry pyridine (70 μl) and phenylisocyanate (20 μl) were added and incubated at 55 °C for 70 min. Methanol (20 μl) was added and incubated at 55 °C for 10 min to destroy the excess reagent. The reaction mixture was diluted with pyridine and applied to C18 RP-HPLC. Gradient elution using acetonitrile and water was performed within 50 min and the derivatives were monitored at 240 nm. Calibration curves for myo-inositol (0.1–4.0 μg) were linear with the correlation coefficient 0.9998. The recovery of myo-inositol in the milk sample was almost 100%, instrument precision was 1.8% (RSD) and the analytical repeatability was 4.8% (RSD).

## 2.4.4 Other Reagents

Phenyldimethylsilyl chloride is also used for the derivatization reagent for the hydroxyl group. White *et al.* [128] described a method to analyze monosaccharide and disaccharide as phenyldimethylsilyl derivatives. The derivatives were separated using silica gel NP-HPLC within 20 min and monitored at 254 nm. Doehl *et al.* [129] determined prostaglandins of the E type (PGE) in human seminal fluid. PGEs were derivatized with pyridinium dichromate (oxidation) to the

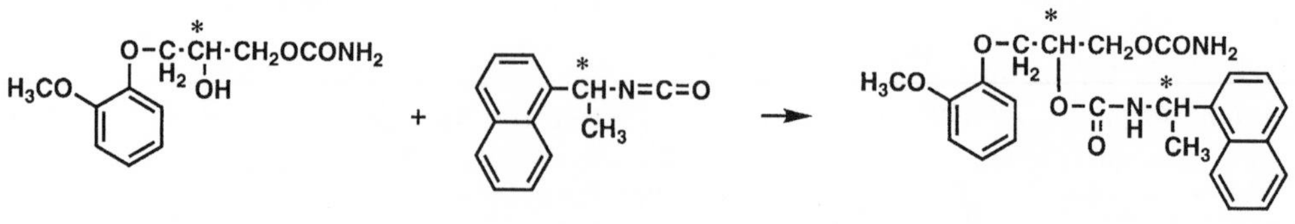

**Fig. 2.34.** The reaction of methocarbamol and 1-naphthylethylisocyanate.

corresponding 15-oxoprostaglandins. The derivatives were separated by C18 RP-HPLC within 10 min and monitored at 230 nm.

## 2.5 LABEL OF REDUCING CARBOHYDRATE

Carbohydrates have a reducing end in addition to the primary and secondary hydroxyl groups. To derivatize the reducing end of carbohydrates, reductive amination is widely used with various UV-absorbing amines, 1-phenyl-3-methyl-5-pyrazolone as pre-column derivatizing reagents. Some post-column derivatizing reagents are also used.

### 2.5.1 Reductive Amination

Reductive amination is widely used for the analysis of reducing carbohydrates, and numbers of fluorophores and chromophores were reported. Amino compounds react with reducing ends to form corresponding aminoalditol (Fig. 2.35). Kwon *et al.* [130] showed a simple method to determine monosaccharides in glycoproteins using *p*-aminobenzoic ethyl ester. Monosaccharide released from glycoproteins were derivatized and separated using C18 RP-HPLC within 20 min. The chromatograms obtained from some glycoproteins are shown in Fig. 2.36. The derivatives were detected at 254 nm, and the calibration range was 0.5–10 nmol. Pauly *et al.* [131] reported the rapid determination of neutral and sialylated oligosaccharide. Oligosaccharide was derivatized with *p*-nitrobenzylhydroxylamine (PNB) to form a Schiff base and reduced using cyanoborohydride. The derivatives were analyzed by anion-exchange HPLC and hydrophilic interaction HPLC. The elution was monitored at 275 nm. By this method, sialylated oligosaccharide can be detected without detectable desialylation. Zhang *et al.* [132] reported *N*,*N*-(dinitrophenyl)octylamine derivatization suitable for purification and characterization of oligosaccharide using UV detection HPLC and FABMS. Oligosaccharides were derivatized with octylamine and cyanoborohydride, then the chromophore 2,4-dinitrofluorobenzene was added to the reaction mixture. The derivatives were separated using C18 RP-HPLC within 15 min and the detection wavelength was 392 nm. With this method, maltodextrins with a degree of polymerization 1–16 were clearly separated within 10 min. HPCE (high-performance capillary electrophoresis) was also used to analyze carbohydrates. Plocek *et al.* [133] reported a simple and rapid procedure to analyze oligosaccharide cleaved from glycoconjugates. Carbohydrate samples were derivatized with *N*-(4-aminobenzoyl)-L-glutamic acid (ABG) and cyanoborohydride. The derivatives

$$R{-}C(=O){-}H + H_2N{-}R' \rightarrow R{-}CH_2{-}NH{-}R'$$

**p-aminobenzoic ethyl ester** **N,N'-(2,4-dinitrophenyl)octylamine** **6-aminoquinoline**

**p-nitrobenzylhydroxylamine** **N-(4-aminobenzoyl)-L-glutamic acid**

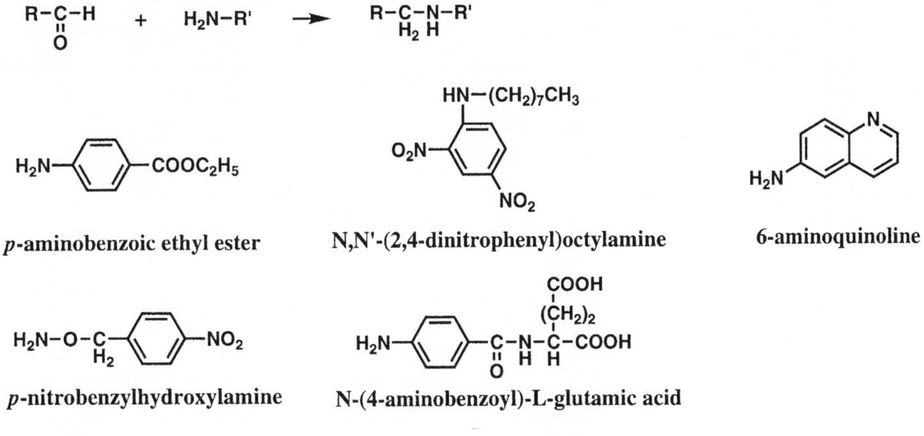

**Fig. 2.35.** The reaction of reducing carbohydrate and amino compounds.

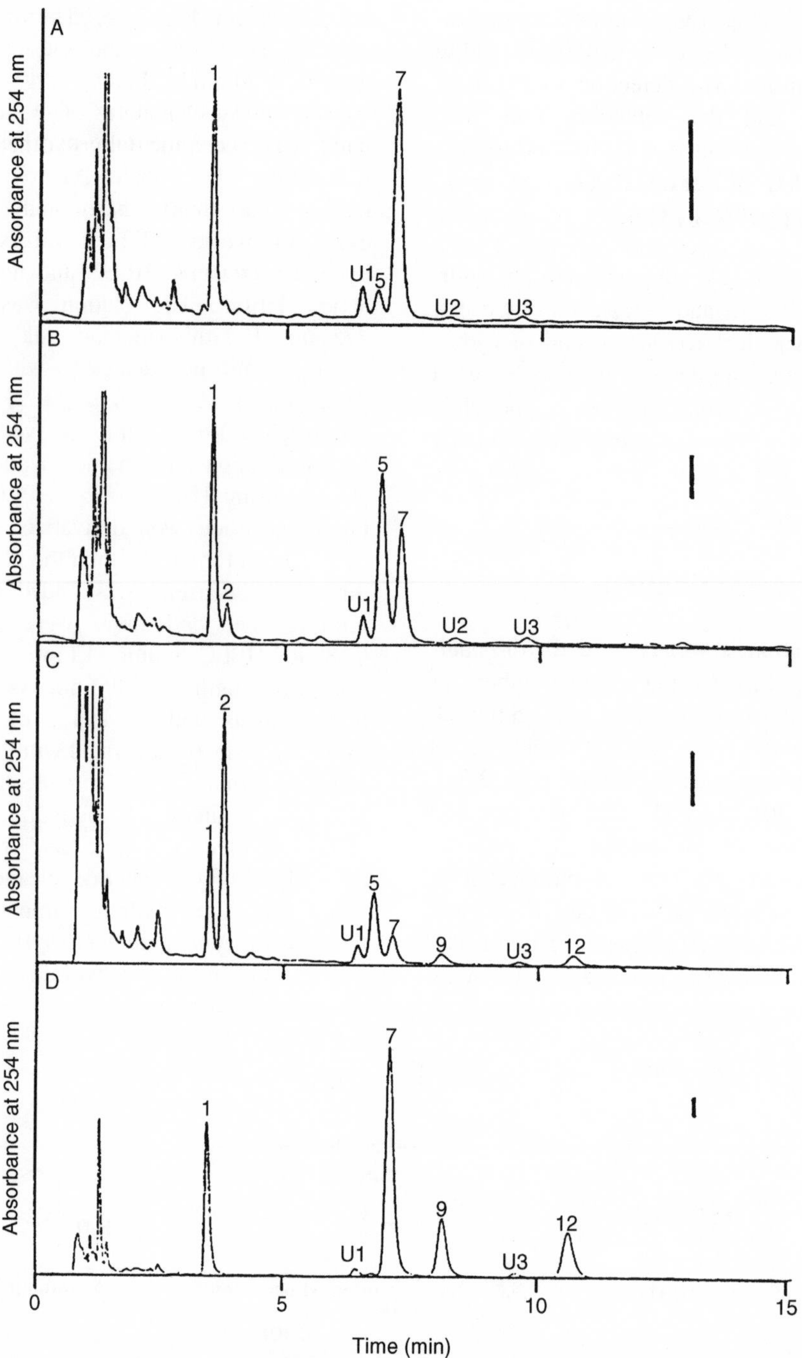


**Fig. 2.36.** Analysis of monosaccharide composition and contents of glycoproteins. (A) Ovalbumin (chicken egg), (B) fetuin (fetal calf serum), (C) mucin (bovine submaxillary glands), (D) peroxidase (horseradish). (1) GlcN, (2) GalN, (5) Gal, (7) Man, (9) Xyl, (12) Fuc, (U) unknown. [Reproduced from ref. 130, p. 250, Fig. 7.].

were separated with CE-CZE (capillary zone electrophoresis) using borate buffer (pH 10.0) as a electrophoresis buffer. The detection was carried out at 291 nm and the detection limit was 170 fmol. Calibration range was 10–500 μg/ml. Rydlund *et al.* [134] showed an efficient CZE separation of wood-derived saccharides. Carbohydrates released from hemicellulose were derivatized with 6-aminoquinoline (6-AQ) and cyanoborohydride and separated by CE-CZE as borate complexes within 15 min. Derivatives were monitored at 245 nm and the detection limit was in the few fmol range.

*Reductive Amination Application: Determination of Oligosaccharide Derivatized with N-(4-Aminobenzoyl)-L-glutamic Acid Using CE-CZE [133]*

An oligosaccharide sample (up to 2 mg/ml) was mixed with the derivatizing reagent solution (4% *N*-(4-aminobenzoyl)-L-glutamic acid 2% cyanoborohydride in 50% acetic acid). The mixture was heated at 90 °C for 2–3 h. After cooling, the sample was diluted appropriately prior to electrophoresis. Separation of derivatized oligosaccharide was performed using CE-CZE with a fused silica capillary (75 μm i.d. × 70 cm). The electrophoresis buffer was a 200 mM borate buffer (pH 10.0) and detection was carried out at 291 nm. The detection limit of maltopentaose was 170 fmol and the calibration range was 10–500 μg/ml.

## 2.5.2 1-Phenyl-3-methyl-5-pyrazolone

Honda *et al.* [135] reported a pre-column derivatization of mono and oligosaccharide using 1-phenyl-3-methyl-5-pyrazolone (PMP) as a reagent. The reagent reacts with reducing carbohydrates almost quantitatively to yield 2:1 compounds (Fig. 2.37). The derivatives were separated by C18 RP-HPLC and monitored at 245 nm. The detection limit of aldoses and *N*-acetylhexosamines was 1 pmol, and the calibration range was 5–1000 pmol. Kakehi *et al.* [136] improved the reagent and reported that 1-(*p*-methoxy)phenyl-3-methyl-5-pyrazolone (PMPMP) was more active and sensitive than PMP. The derivatizing reaction is mild and applicable to neutral and sialic acid containing oligosaccharide. The derivatives were separated by C18 RP-HPLC and the detection wavelength was 249 nm. The detection limit was 500 fmol (S/N = 5) and the calibration range using Glc-PMPMP as an internal standard was 2–2000 pmol. Oligosaccharide in glycoproteins is successfully determined using both of the two methods.

*PMPMP Application: Oligosaccharide Analysis in Glycoproteins [136]*

An oligosaccharide sample obtained by either hydrazinolysis or glycopeptidase A digestion was mixed with 0.3 M NaOH and 0.5 M PMPMP in methanol. The mixture was heated at 70 °C for 20 min, then after cooling the mixture was neutralized with 0.3 M HCl. The derivatives were

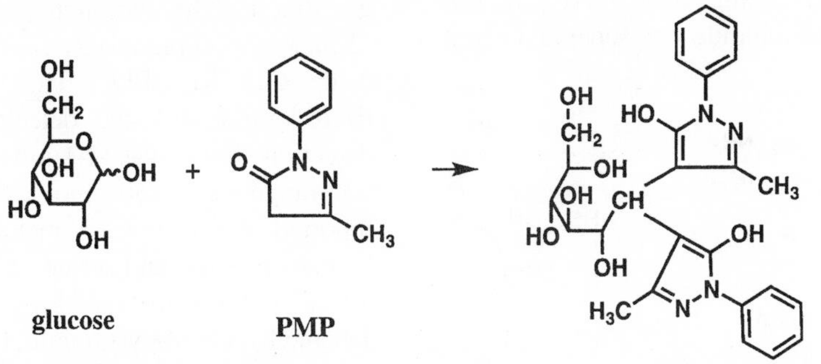

glucose + PMP →

**Fig. 2.37.** The reaction of glucose and PMP.

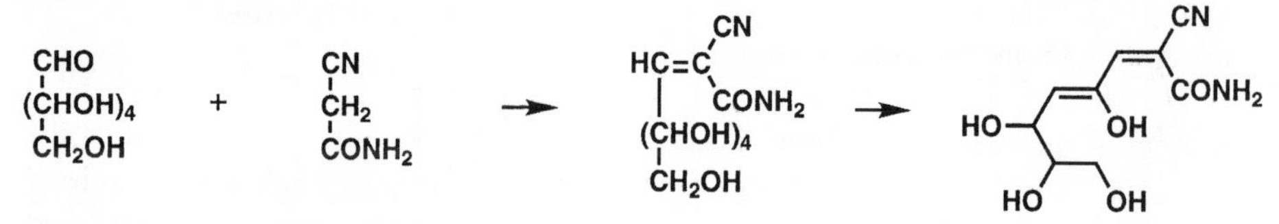

glucose 2-cyanoacetamide

Fig. 2.38. The reaction of glucose and 2-cyanoacetamide.

extracted with ethyl acetate/water, and the aqueous layer was evaporated to dryness. The residue was dissolved in acetonitrile and water, and applied to HPLC analysis. The derivatives were separated by C18 RP-HPLC using 100 mM phosphate buffer (pH 7.0) containing 15% acetonitrile as a mobile phase. Detection was carried out at 249 nm, and the detection limit was 500 fmol. The repeatability of the method using NeuAc-Lac was 2% (R.S.D).

### 2.5.3 Post-column Derivatizing Reagent

2-Cyanoacetamide (CA) was used for the post-column derivatization of reducing carbohydrate (Fig. 2.38). Cramer *et al.* [137] reported the separation of hyaluronic acid oligomers using an RP-ion pair-HPLC. The separated oligomers were mixed with a derivatizing reagent (1% 2-CA in 0.2 M borate buffer (pH9.0)) and detected at 276 nm. The method is linear and reproducible, and the detection limit was 25 pmol. Nozal *et al.* [138] investigated the analytical method of neutral monosaccharide using Purpald(4-amino-3-hydrazino-5-mercapto-1,2,4-triazole; AHMT) as a post-column derivatizing reagent (Fig. 2.39). The reducing monosaccharide was separated using

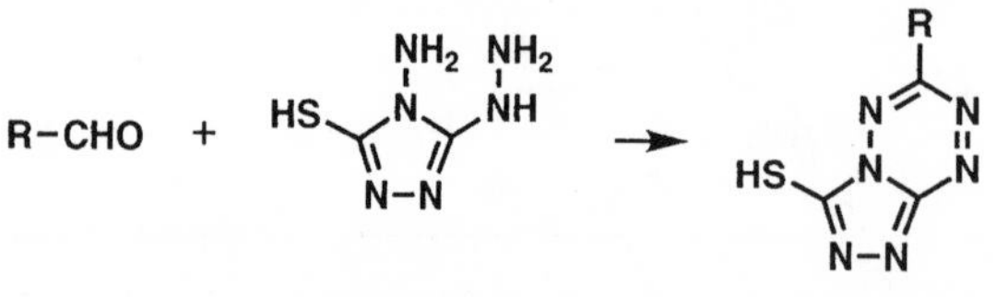

Purpald

Fig. 2.39. The reaction of carbonyl compound and Purpald.

cation exchange HPLC and derivatized with AHMT in the basic medium in the presence of hydrogen peroxide. The derivatives were detected at 550 nm and the detection limit of ribose was 15 ng, the calibration range was 20–500 ng.

## 2.6 LABEL OF THIOL −SH

Cysteine (Cys), glutathione (GSH) and many pharmaceuticals contain the sulfhydryl group in the molecule. Derivatization of these thiols was performed by many reagents. Bimanes, aziridines, maleimides and benzofurazans are used for fluorescence derivatization. For UV-visible detection, 2-halopyridinium salt, disulfhide compound and other reagents are used.

### 2.6.1 2-Halopyridinium Salt

2-Iodo-1-methylpyridinium chloride (IMPC) [139] and 2-chloro-1-methylpyridinium iodide (CMPI) [140] is used for the pre-column ultraviolet derivatization reagent of thiols. The reaction was shown in Fig. 2.40. Sypniewski *et al.* [140] reported the determination of cysteine, glutathione, homocysteine, acetylcysteine, *N*-(2-mercaptopropionyl)glycine and its metabolite using C18 RP-HPLC. The thiols were derivatized with CMPI and detected at 314 nm. The detection limit was 2.1 pmol and the calibration range was 1–50 μmol/l. Bald *et al.* [141,142] reported the analytical method of captopril in human plasma and urine. Captopril, an orally active antihypertensive drug, is derivatized with 1-benzyl-2-chloropyridinium bromide (BCPB) to form *S*-pyridinium derivative. The derivative was separated by C18 RP-HPLC and detected at

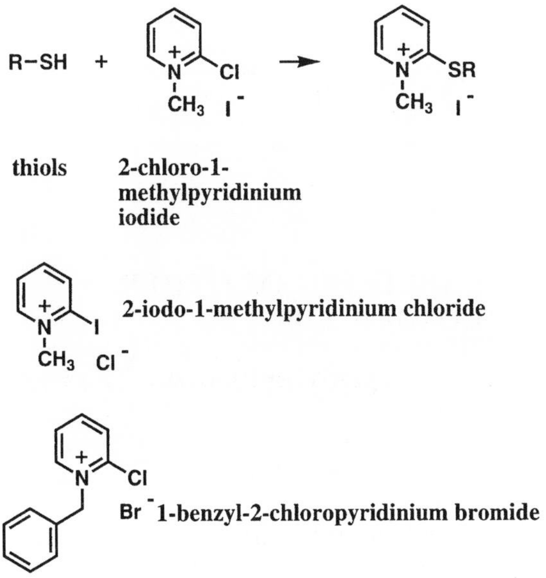

R–SH + [structure] → [structure]

**thiols** **2-chloro-1-methylpyridinium iodide**

**2-iodo-1-methylpyridinium chloride**

**1-benzyl-2-chloropyridinium bromide**

**Fig. 2.40.** The reaction of 2-halopyridinium compounds and thiols.

314 nm. Oxidized and protein-bound captopril was converted to reduced form using triphenylphosphine and derivatized and quantified in the same manner. The detection limit was 0.3 ng/ml and the calibration range was 10–500 ng/ml.

*BCPB Application: Determination of Captopril in Human Plasma [141]*

Human plasma was deproteinized using 3 M perchloric acid and neutralized with 1 M NaOH. To the solution, 1 M Tris buffer (pH 8.2) and 20 mg/ml BCPB solution were added and derivatized for 15 min. After acidifying by adding 4 M phosphoric acid, the reaction mixture was centrifuged and the derivative was extracted with C18 disposable cartridge. The extract was evaporated, and the residue was dissolved with the mobile phase and subject to HPLC. BCPB derivative of captopril and the internal standard (captopril derivatized with 1-benzyl-2-chloro-4-methylpyridinium bromide) were separated using RP-HPLC within 20 min. The chromatograms are shown in Fig. 2.41. The quantitation limit at 314 nm was 10 ng, and the calibration range was 10–500 ng/ml plasma. This method

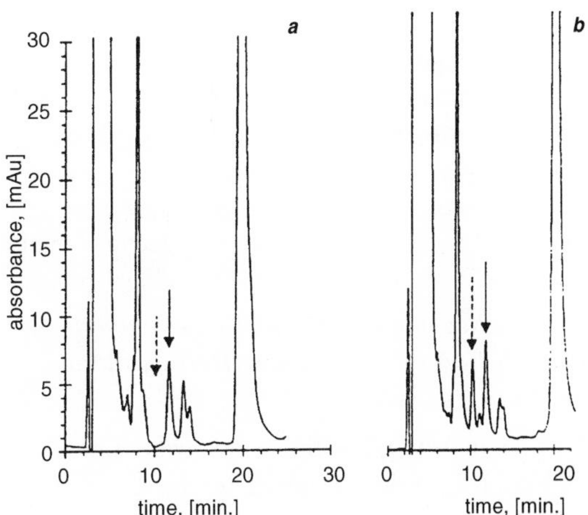


**Fig. 2.41.** Typical chromatograms of captopril (50 mg oral dose) in human plasma as BCPB derivative. (a) Blank, (b) patient sample. Broken arrow represents captopril and solid arrow represents the internal standard. [Reproduced from ref. 141, p. 288, Fig. 3.].

enabled sensitive determination of captopril and its disulfides in human plasma after oral administration to patients.

## 2.6.2 Disulfide Reagent

The sulfhydryl-disulfide exchange reaction was used for the post column derivatization of thiols (Fig. 2.42). Ellman's reagent (5,5′-dithiobis(2-nitrobenzoic acid), DTNB) and related compounds are widely used. Yamato *et al.* [143, see Ellman's reagent application] reported an analytical method specific to acetyl-coenzyme A. Acetyl-CoA separated by RP-ion pair HPLC was enzymatically converted to thio-CoA with an immobilized enzyme reactor, and determined spectrophotometrically after post-column derivatization with Ellman's reagent. Detection was carried out at 412 nm and the detection limit was 0.05 nmol. *n*-Octyl-5-dithio-2-nitrobenzoic acid (ODNB) was also used to derivatize thiols. Faulstich *et al.* [144] investigated the derivatization of protein thiols by ODNB and reported the reaction of ODNB to be faster than Ellman's reagent. The reagent reacts with thiol groups, which are not reactive with Ellman's reagent.

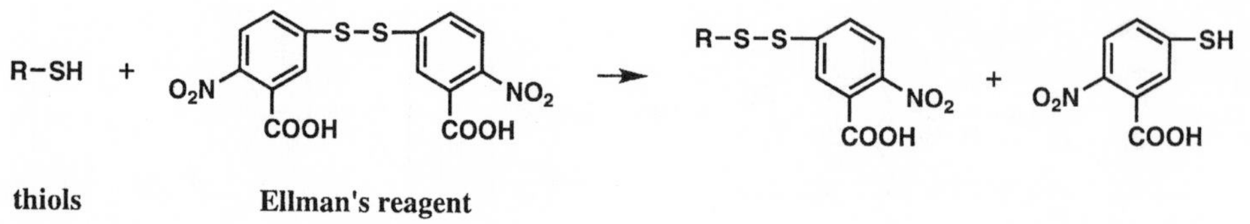

Fig. 2.42. The reaction of Ellman's reagent and thiols.

*Ellman's Reagent Application: Determination of Acetyl-CoA by Post-column Derivatization [143]*

Short-chain acyl-CoA thioesters were first separated by C8 RP-HPLC using tetra-*n*-butylammonium phosphate as an ion-pairing reagent. The separated thioesters were mixed with a reagent solution (30 mM $Na_2HPO_4$ as the substrate, 15 mM $(NH_4)_2SO_4$ as the enzyme activator and 0.1 mM DTNB as the color-producing reagent) and introduced to the immobilized enzyme reactor. Acetyl-CoA in the elution mixture was converted to thio-CoA by the immobilized phosphotransacetylase and reacted with Ellman's reagent (DTNB) in the reagent stream. The calibration curve at 412 nm was linear between 0.2 and 10 nmol, with a detection limit of 0.05 nmol.

### 2.6.3 Other Reagents

Shen *et al.* [145] reported a simple method of determining captopril in biological fluids using *p*-bromophenacyl bromide as a pre-column derivatizing reagent. The derivative was separated by C18 RP-HPLC and detected at 260 nm. The calibration range was 20–1000 ng/ml plasma and 10–200 μg/ml urine. Pietra *et al.* [146] determined glutathione and L-cysteine in pharmaceuticals after derivatization with ethacrynic acid. Derivatives were separated by C18 RP-HPLC and monitored at 270 nm. Gotti *et al.* [147] reported the determination of glutathione in pharmaceuticals and cosmetics. Glutathione was derivatized with 4-(6-methylnaphthalene-2-yl)-4-oxo-2-buteneoic acid, and the thiol adduct was chromatographed on C8 RP-HPLC. Derivatives were detected using absorbance at 254 nm or fluorescence (Ex. 300 nm, Em. 460 nm).

## 2.7 LABELLING OF OTHER COMPOUNDS

### 2.7.1 1-(2-Pyridyl)-piperazine (PYP) [148]

Chromosorb coated with PYP was evaluated for collection/derivatization of Phosgene ($COCl_2$) gas. Phosgene reacts with PYP to form a substituted urea derivative (Fig. 2.43) which is desorbed with $CH_3CN$ and determined by PR-HPLC with UV absorbance detection at either 254 or 270 nm.

### 2.7.2 Diethyldithiocarbamate (DDTC)

The CE method is developed for the quantitative determination of the anti-cancer drug prospidin in human tissue after its derivatization with sodium diethyldithiocarbamate (DDTC, Fig. 2.44) [149]. Detection is by UV absorption at 254 nm. The

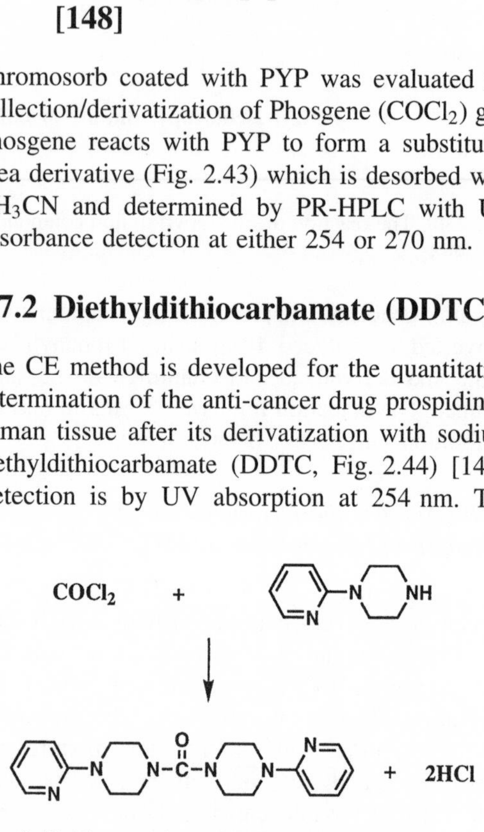

Fig. 2.43. The reaction of PYP with phosgene.

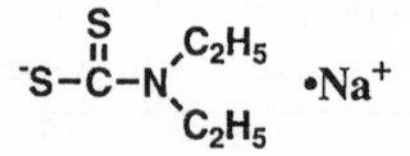

Fig. 2.44. Sodium diethyldithiocarbamate (DDTC).

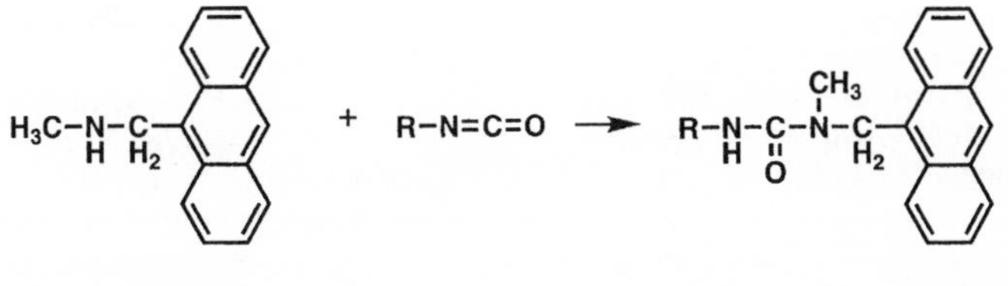

$$H_3C-\underset{H}{N}-\underset{H_2}{C}-(\text{9-anthryl}) + R-N=C=O \longrightarrow R-\underset{H}{N}-\underset{O}{\overset{\|}{C}}-\overset{CH_3}{\overset{|}{N}}-\underset{H_2}{C}-(\text{9-anthryl})$$

**MAMA**

**Fig. 2.45.** Reaction of reagent MAMA with airborne csocyanates.

studied range of concentrations of prospidin is 1–50 μg/ml.

DDTC was also used for the HPLC determination of busulfan, OH compound, in human serum [150].

### 2.7.3 9-methylamino-methylanthracene (MAMA) [151]

MAMA is a secondary amino compound commonly used for derivatization and quantitation of specific commercial isocyanate compounds by HPLC with fluorescence and UV light absorbance. The determination of airborne total reactive isocyanate group compounds is of interest in industrial hygiene. Detection is by UV absorption at 370 nm. MAMA produces substituted urea in the reaction with airborne isocyanate monomers (Fig. 2.45).

## References


[1] Lawrence, J.F. and Fri, R.W. (1976) *Chemical Derivatization in Liquid Chromatography*, Elsevier, Amsterdam, Netherlands.

[2] Frei, R.W., in Fri, R.W. and Lawrence, J.F. (Eds) (1981) *Chemical Derivatization in Analytical Chemistry*, Vol. 1, *Chromatography*, Plenum Press, New York.

[3] Knapp, D.R. (1979) *Handbook of Analytical Derivatization Reactions*, J. Wiley and Sons, New York.

[4] Blau, K. and King, G.S. (Eds) (1977) *Handbook of Derivatives for Chromatography*, Heyden, London.

[5] Meulendijk, J.A.P. and Underberg, W.J.M. (1990) in *Detection Oriented Derivatization Techniques in Liquid Chromatography*, H. Lingeman and W.J.M. Underberg (Eds), pp 247–281, Marcel Dekker, New York.

[6] Li, F. and Lim, C.K. (1993) in *Handbook of Derivatives for Chromatography*, 2nd ed, K. Blau and J. Halket (Eds), pp 158–174, John Wiley & Sons, Chicheste, New York, Brisbane, Toronto and Singapore.

[7] Szule, M.E. and Krull, I.S. (1994) *J. Chromatogr. A*, **659**, 231.

[8] Krull, I.S., Deyl, Z. and Lingeman, H. (1994) *J. Chromatogr. B.*, **659**, 1.

[9] Wong, L.T., Beaubien, A.R. and Pakuts, A.P. (1982) *J. Chromatogr. Biomedical Applications*, **231**, 145.

[10] Barends, D.M., Blauw, J.S., Smits, M.H. and Hulshoff, A. (1983) *J. Chromatogr. Biomedical Applications*, **276**, 385.

[11] Barends, D.M., Blauw, J.S., Mijnsbergen, C.W., Govers, C.J.L.R. and Hulshof, A. (1985) *J. Chromatogr.*, **322**, 321.

[12] Barends, D.M., Zwaan, C.L. and Hulshoff, A. (1981) *J. Chromatogr. Biomedical Applications*, **225**, 417.

[13] Barends, D.M., Zwaan, C.L. and Hulshoff, A. (1981) *J. Chromatogr. Biomedical Applications*, **222**, 316.

[14] Helboe, P. and Kryger, S. (1982) *J. Chromatogr.*, **235**, 215.

[15] Hoppel, C.L., Weir, D.E., Gibbons, A.P., Ingalls, S.T., Brittain, A.T. and Brown, F.M. (1983) *J. Chromatogr. Biomedical Applications*, **272**, 43.

[16] Jackson, G.E.D. and Young, N.M. (1987) *Anal. Biochem.*, **162**, 251.

[17] Spragg, B.P. and Hutchings, A.D. (1983) *J. Chromatogr.*, **258**, 289.

[18] Kabra, P.M. Bhatnager, P.K. and Nelso, M.A. (1984) *J. Chromatogr. Biomedical Applications*, **307**, 224.

[19] Scaloni, A. Simmaco, M and Bossa, F. (1995) *Amino Acids*, **8**, 305.

[20] Goodnough, D.B. Lutz M.P. and Wood, P.L. (1995) *J. Chromatogr. B*, **672**, 290.

[21] Kudoh, M., Matoh I. and Fudano, S. (1983) *J. Chromatogr.*, **261**, 293.

[22] Bjorkqvist, B. (1981) *J. Chromatogr.*, **204**, 109.
[23] Neidle, A. Banay-Schwartz, M. Sacks S. and Dunlop, D.S. (1989) *Anal. Biochem.*, **180**, 291.
[24] Edmann, P. (1949) *Acta Chem. Scand.*, **4**, 283.
[25] Edman, P. and Begg, G. (1967) *European J. Biochem.*, **1**, 80.
[26] Zimmerman, C.L. Appella, E. and Pisano, J.J. (1977) *Anal. Biochem.*, **77**, 569.
[27] Pucci, P., Sannia G. and Marin, G. (1983) *J. Chromatogr.*, **270**, 371.
[28] Hawke, D., Yuan P.-M. and Shively, J.E. (1982) *Anal. Biochem.*, **120**, 302.
[29] Lottspeich, F. (1985) *J. Chromatogr.*, **326**, 321.
[30] Heinrikson, R.L. and Meredith, S.C. (1984) *Anal. Biochem.*, **136**, 65.
[31] Lavi, L.E., Holcenberg, J.S., Cole D. and Jolivet, J. (1986) *J. Chromatogr. Biomedical Applications*, **377**, 155.
[32] Fürst, P., Pollack, L., Graser, T.A., Godel H. and Stehle P. (1990) *J. Chromatogr.*, **499**, 557.
[33] Auger, J., Mellouki, F., Vannereau, A., Boscher, J., Cosson, L. and Mandon, N. (1993) *Chromatographia*, **36**, 347.
[34] Bidlingmeyer, B.A. Cohen, S.A. and Tarvin, T.L. (1984) *J. Chromatogr., Biomedical Applications*, **336**, 93.
[35] Murthy, L.R. and Iqbal, K. (1991) *Anal. Biochem.*, **193**, 299.
[36] Otsuka, K., Terabe, S. and Ando, T. (1985) *J. Chromatogr.*, **332**, 219.
[37] Castagnola, M., Rossetti, D.V., Cassiano, L., Rabino, R., Nocca, G. and Giardina B. (1993) *J. Chromatogr.*, **638**, 327.
[38] Tsai, P. Patel, B. and Lee, C.S. (1993) *Anal. Chem.*, **65**, 1439.
[39] Terabe, S. Ishihama, Y. Nishi, H. Fukuyama, T. and Otsuka, K. (1991) *J. Chromatogr.*, **545**, 359.
[40] Matsubara, N. and Terabe, S. (1994) *J. Chromatogr. A*, **680**, 311.
[41] Matsubara, N. and Terabe, S. (1996), in *Capillary Electrophoresis in Analytical Biotechnology*, P. G. Righetti (Ed), p 155, CRC Press, Boca, Raton, New York, London and Tokyo.
[42] Ishihama, Y. and Terabe, S. (1993) *J. Liq. Chromatogr.*, **16**, 933.
[43] Davey, J.F. and Ersser, R.S. (1990), *J. Chromatogr. Biomedical Applications*, **528**, 9.
[44] Woo, K.-L. and Lee, S.-H. (1994) *J. Chromatogr. A*, **667**, 105.
[45] Chang, J.-Y. Creaser, E.H. and Bentley, K.W. (1976), *Biochem. J.*, **153**, 607.
[46] Chang, J.-Y. and Creaser, E.H. (1976) *Biochem. J.* **157**, 77.
[47] Chang, J.-Y. (1979), *Biochim. Biophys. Acta*, **578**, 175.
[48] Chang, J.-Y. (1979), *Biochim. Biophys. Acta*, **578**, 188.
[49] Chang, J.-Y. (1980) *Anal. Biochem.*, **102**, 384.
[50] Chang, J.-Y. (1981) *Biochem. J.*, **199**, 557.
[51] Chang, J.-Y., Brauer, D. and Wittmann-Liebold, B. (1978) *FEBS Lett.*, **93**, 205.
[52] Chang, J.-Y. (1983), in *Methods in Enzymology*, (Eds) Hirs, C.H.W. and Timashff, S.N. Vol. 91, pp 78–84, Academic Press, New York.
[53] Stocchi, V., Piccoli, G., Magnani, M., Palma, F., Biagiarelli B. and Cucchiarini, L. (1989) *Anal. Biochem.*, **178**, 107.
[54] Lundblad, R.L. (1995) *Techniques in Protein Modification.*, pp 35–50, CRC Press, Boca Raton, Ann Arbor, London and Tokyo.
[55] Saurina, J. and Hernández-Cassou, S. (1994) *J. Chromatogr. A*, **676**, 311.
[56] Saurina, J. and Hernández-Cassou, S. (1993) *Anal. Chim. Acta*, **283**, 414.
[57] Saurina, J. and Hernández-Cassou, S. (1993) *Anal. Chim. Acta*, **281**, 593.
[58] Asotra, S. Mladenov, P.V. and Burke, R.D. (1987) *J. Chromatogr.*, **408**, 227.
[59] Samejima, K. Kawase, M. Sakamoto, S. Okada, M. and Endo, Y. (1976) *Anal. Biochem.*, **76**, 392.
[60] Redmond, J.W. and Tseng, A. (1979) *J. Chromatogr.*, **170**, 479.
[61] Verkoelen, C.F. Romijn, J.C. Schroeder, F.H. van Schalkwijk, W.P. and Splinter, T.A.W. (1988) *J. Chromatogr. Biomedical Applications*, **426**, 41.
[62] Mei, Y.-H. (1994) *J. Liq. Chromatogr.*, **17**, 2413.
[63] Chen, P.F. (1967) *Arch. Biochem. Biophys.*, **120**, 609.
[64] Michael Wilkinson, J. (1978) *J. Chromatogr. Sci.*, **16**, 547.
[65] Heimbecher, S., Lee, Y.-C., Tabibi, S.E. and Yalkowsky, S.H. (1997) *J. Chromatogr. B*, **691**, 173.
[66] Terabe, S., Miyashita, Y., Ishihama, Y. and Shibata, O. (1993), *J. Chromatogr.*, **636**, 47.
[67] Lin, J. and Chang, J.-Y. (1975) *Anal. Chem.*, **47**, 1634.
[68] Chang, J.Y. and Creaser, E.H. (1976) *J. Chromatogr.*, **116**, 215.
[69] Jansen, E.H.J.M., Vandenberg, R.H., Both-miedema, R. and Doorn, L. (1991) *J. Chromatogr.*, **553**, 123.
[70] Knecht, R. and Chang, J.-Y. (1986) *Anal. Chem.*, **58**, 2375.
[71] Lin J.-K. and Wang, C.-H. (1980) *Clin. Chem.*, **26**, 579.
[72] Carratù, B., Boniglia, C. and Bellomonte, G. (1995) *J. Chromatogr. A*, **703**, 203.
[73] Spackman, D.H., Stein, W.H. and Moore, S. (1958) *Anal. Chem.*, **30**, 1190.
[74] Moore, S., Spackman D.H. and Stein, W.H. (1958) *Anal. Chem.*, **30**, 1185.

[75] Sarbach, C., Postaire, E. and Sauzieres, J. (1994) *J. Liq. Chromatogr.*, **17**, 2737.
[76] Parrot, F., Bedry, R., Favvarel.-Garrigues, J.-C. (1995) *J. Toxicol. Clin. Toxicol.* **33**, 695.
[77] Alaiz, M., Navarro, J.L., Girón, J. and Vioque, E. (1992) *J. Chromatogr.*, **591**, 181.
[78] Cohen, S.A. and Michaud, D.P. (1993) *Anal. Biochem.*, **211**, 279.
[79] Liu, H.J. (1994) *J. Chromatogr. A*, **670**, 59.
[80] Nimura, N., Iwaki, K., Kinoshita, T., Takeda, K. and Ogura, H. (1986) *Anal. Chem.*, **58**, 2372.
[81] Zhou, F.-X. Wahlberg, J. and Krull, I.S. (1991) *J. Liq. Chromatogr.*, **14**, 1325.
[82] Krull, I.S., Szulc, M.E., Bourque, A.J., Zhou, F.-X. and Yu, J. Strong, R. (1994) *J. Chromatogr. B*, **659**, 19.
[83] Chou, T.-Y., Colgan, S.T., Kao, D.M., Krull, I.S., Dorschel, C. and Bidlingmeyer, B. (1986) *J. Chromatogr.*, **367**, 335.
[84] Gao, C.-X. Chou, T.-Y. Colgan, S.T. Krull, I.S. Dorschel, C. and Bidlingmeyer, B. (1988) *J. Chromatogr. Sci.*, **26**, 449.
[85] Bourque, A.J. and Krull, I.S. (1991) *J. Chromatogr.* **537**, 123.
[86] Fisher, D.H. and Bourque, A.J. (1993) *J. Chromatogr. Biomedical Applications*, **614**, 142.
[87] Yu, J.H., Li, G.D., Krull I.S. and Cohen, S. (1994) *J. Chromatogr. B*, **658**, 249.
[88] Chou, T.-Y. Gao, C.-X. Colgan, S.T. and Krull, I.S. (1988) *J. Chromatogr.*, **454**, 169.
[89] Gao, C.-X. Krull, I.S. and Trainor, T.M. (1989) *J. Chromatogr.*, **463**, 192.
[90] Gao, C.-X. Chou, T.-Y. and Krull, I.S. (1989) *Anal. Chem.*, **61**, 1538.
[91] Borch, R.F. (1975) *Anal. Chem.*, **47**, 2437.
[92] Hanis, T., Smrz, M., Klir, P. and Macek, K. (1988) *J. Chromatogr.*, **452**, 443.
[93] Damyanova, B.N. Herslof, B.G. and Christie, W.W. (1992) *J. Chromatogr.*, **609**, 133.
[94] Durst, H.D., Milano, M., Kikta, E.J., Connelly, S.A. and Grushka, E. (1975) *Anal. Chem.*, **47**, 1797.
[95] Fitzpatrick, F.A. (1976) *Anal. Chem.*, **48**, 499.
[96] Patience, R.L. and Thomas, J.D. (1982) *J. Chromatogr.*, **234**, 225.
[97] Halgunset, J. Lund, E.W. and Sunde, A. (1982) *J. Chromatogr.*, **237**, 496.
[98] Desbene, P.-L. Coustal, S. and Frappier, F. (1983) *Anal. Biochem.*, **128**, 359.
[99] Korte, K. Chien, K.R. and Casey, M.L. (1986) *J. Chromatogr.*, **375**, 225.
[100] Alric, R., Cociglio, M., Blayac, J.P. and Puech, R. (1981) *J. Chromatogr.*, **224**, 289.
[101] Zoutendam, P.H. Bowman, P.B. Ryan, T.M. and Rumph, J.L. (1984) *J. Chromatogr.*, **283**, 273.
[102] Zoutendam, P.H. Bowman, P.B. Rumph, J.L. and Ryan, T.M. (1984) *J. Chromatogr.*, **283**, 281.
[103] Watkins, W.D. and Peterson, M.B. (1982) *Anal. Biochem.*, **125**, 30.
[104] Hoffman, N.E. and Liao, J.C. (1976) *Anal. Chem.*, **48**, 1104.
[105] Bissinger, J.M., Rullo, K.T., Stoklosa, J.T., Shearer, C.M. and De Angelis, N.J. (1983) *J. Chromatogr.*, **268**, 102.
[106] Ikeda, M., Shimada, K. and Sakaguchi, T. (1982) *Chem. Pharm. Bull.*, **30**, 2258.
[107] Miwa, H. and Yamamoto, M. (1986) *J. Chromatogr.*, **351**, 275.
[108] Miwa, H., Yamamoto, M., Nishida, T., Nunoi, K. and Kikuchi, M. (1987) *J. Chromatogr.*, **416**, 237.
[109] Miwa, H. and Yamamoto, M. (1988) *Anal. Biochem.*, **170**, 301.
[110] Kondoh, Y. Yamada, A. and Takano, S. (1991) *J. Chromatogri.*, **541**, 431.
[111] Gutnikov, G. and Streng, J.R. (1991) *J. Chromatogr.*, **587**, 292.
[112] Vollmer, P.A. Harty, D.C. Erickson, N.B. Balhon, A.C. and Dean, R.A. (1996) *J. Chromatogr. B*, **685**, 370.
[113] Kasama, T. Byun, D.-S. and Seyama, Y. (1987) *J. Chromatogr.*, **400**, 241.
[114] Ioneda, T. and Ono, S.S. (1996) *Chem. Phys. Lipids*, **81**, 11.
[115] Uzawa, H., Ohrui, H., Meguro, H., Mase, T. and Ichida, A. (1993) *Biochim. Biophys. Acta*, **1169**, 165.
[116] Davey, M.W. and Lambein, F. (1992) *Anal. Biochem.*, **206**, 323.
[117] Hawkins, D.J., Kühn, H., Petty, E.H. and Brash, A.R. (1988) *Anal. Biochem.*, **173**, 456.
[118] Formato, M., Senes, A., Soccolini, F., Coinu, R. and Cherchi, G.M. (1994) *Carbohydr. Res.*, **255**, 27.
[119] Schlame, M. and Otten, D. (1991) *Anal. Biochem.*, **195**, 290.
[120] Cantafora, A. and Masella, R. (1992) *J. Chromatogr.*, **593**, 139.
[121] Oshima, M., Asano, K., Shibata, S., Suzuki, Y. and Masuzawa, M. (1990) *Biochim. Biophys. Acta*, **1043**, 157.
[122] Ghosh, D., Mathur, N.K. and Narang, C.K. (1993) *Chromatographia*, **37**, 543.
[123] Indyk, H.E. and Woollard, D.C. (1994) *Analyst*, **119**, 397.
[124] Rakotomanga, S., Baillet, A., Pellerin, F. and Ferrier, D.B. (1992) *J. Pharm. Biomed. Anal.*, **10**, 587.
[125] Dethy, J.-M., Deveen, B.C., Janssens, M. and Lenaers, A. (1984) *Anal. Biochem.*, **143**, 119.
[126] Nakagawa, Y., Kishida, K., Kodani, Y. and Matsuyama, T. (1997) *Microbiol. Immunol.*, **41(1)**, 27.
[127] Severini, S.A., Coutts, R.T., Jamali, F. and Pasutto, F.M. (1992) *J. Chromatogr.* **582**, 173.

[128] White, C.A., Vass, S.W., Kennedy, J.F. and Large, D.G. (1983) *J. Chromatogr.*, **264**, 99.
[129] Doehl, J. and Greibrokk, T. (1990) *J. Chromatogr.*, **529**, 21.
[130] Kwon, H. and Kim, J. (1993) *Anal. Biochem.*, **215**, 243.
[131] Pauly, M., York, W.S., Guillen, R., Albersheim, P. and Darvill, A.G. (1996) *Carbohydr. Res.*, **282**, 1.
[132] Zhang, Y., Cedergren, R.A., Nieuwenhuis, T.J. and Hollingsworth, R.I. (1993) *Anal. Biochem.*, **208**, 363.
[133] Plocek, J. and Novotny, M.V. (1997) *J. Chromatogr. A*, **757**, 215.
[134] Rydlund, A. and Dahlman, O. (1996) *J. Chromatogr. A*, **738**, 129.
[135] Honda, S., Akao, E., Suzuki, S., Okuda, M., Kakehi, K. and Nakamura, J. (1989) *Anal. Biochem.*, **180**, 351.
[136] Kakehi, K., Suzuki, S., Honda, S. and Lee, Y.C. (1991) *Anal. Biochem.*, **199**, 256.
[137] Cramer, J.A. and Bailey, L.C. (1991) *Anal. Biochem.*, **196**, 183.
[138] Nozal, M.J.D., Bernal, J.L., Hernandez, V., Toribio L. and Mendez, R. (1993) *J. Liq. Chromatogr.*, **16**, 1105.
[139] Bald, E. and Sypniewski, S. (1993) *J. Chromatogr.*, **641**, 184.
[140] Sypniewski, S. and Bald, E. (1994) *J. Chromatogr. A*, **676**, 321.
[141] Bald, E., Sypniewski, S., Drzewoski, J. and Stepien, M. (1996) *J. Chromatogr. B*, **681**, 283.
[142] Sypniewski, S. and Bald, E. (1996) *J. Chromatogr. A*, **729**, 335.
[143] Yamato, S., Nakajima, M., Wakabayashi, H. and Shimada, K. (1992) *J. Chromatogr.*, **590**, 241.
[144] Faulstich, H., Tews, P. and Heintz, D. (1993) *Anal. Biochem.*, **208**, 357.
[145] Shen, G., Weirong, T. and Shixiang, W. (1992) *J. Chromatogr.*, **582**, 258.
[146] Pietra, A.M.D., Gotti, R., Bonazzi, D., Andrisano V. and Cavrini, V. (1994) *J. Pharm. Biomed. Anal.*, **12**, 91.
[147] Gotti, R., Andrisano, V., Cavrini V. and Bongini, A. (1994) *Chromatographia*, **39**, 23.
[148] Rando, R.J., Poovey, H.G. and Chang, S.-H. (1993) *J. Liq. Chromatogr.*, **16**, 3291.
[149] Okun, V.M., Aak, O.V. and Kozlov, V.Y. (1996) *J. Chromatogr. B*, **675**, 313.
[150] Funakoshi, K., Yamashita, K., Chao, W., Yamaguchi, M. and Yashiki, T. (1994) *J. Chromatogr. B*, **660**, 200.
[151] Rando, R.J., Poovey, H.G. and Lefante, J.J. and Esmundo, F.R. (1993) *J. Liq. Chromatogr.*, **16**, 3977.

# 3

# Reagent for FL Detection

**Masatoshi Yamaguchi, and Junichi Ishida**
Fukuoka University, Fukuoka, Japan

---


Abbreviations 100
3.1 Introduction 102
3.2 Reagents for Amines and Amino Acids 103
3.2.1 General Amino Compounds 103
3.2.1.1 Primary Amines and Amino Acids 103
3.2.1.2 Primary and Secondary Amines and Amino Acids 106
3.2.2 Particular Amines and Amino Acids 116
3.2.2.1 Catecholamines 116
3.2.2.2 Tryptophan and Indoleamines 117
3.2.2.3 5-hydroxyindolamines (Serotonin Related Compounds) 117
3.2.2.4 Guanidino Compounds 119
3.2.3 Peptides 121
3.2.3.1 General Peptides 121
3.2.3.2 Arginine-containing Peptides 121
3.2.4 Fluorescence Derivatization for CE 121
3.3 Reagents for Organic Acids 121
3.3.1 Carboxylic Acids 121
3.3.1.1 Fatty Acids 129
3.3.1.2 Prostaglandins 130
3.3.1.3 Steroids 131
3.3.1.4 Glucuronic Acid Conjugates 132
3.3.2 $\alpha$-Keto Acids 133
3.3.3 Sialic Acids and Dehydroascorbic Acid 133
3.4 Reagents for Alcohols 135
3.5 Reagents for Phenols 139


---

Edited by Toshimasa Toyo'oka: *Modern Derivatization Methods for Separation Sciences* 

3.6 Reagents for Thiols 141
3.7 Reagents for Aldehydes and Ketones 146
3.8 Reagents for Carbohydrates 150
3.9 Reagents for Dienes 152
3.10 Reagents for Nucleic Acid Related Compounds 153
References 158

## ABBREVIATIONS

OPA: *o*-phthalaldehyde
NDA: 2,3-naphthalenedialdehyde
FQCA: 3-(2-furoyl)quinoline-2-carbaldehyde
BQCA: 3-benzoyl-2-quinoline carbaldehyde
CBQCA: 3-(4-carboxybenzoyl)-2-quinoline carbaldehyde
MDF: 2-methoxy-2,4-diphenyl-3[2*H*]-furanone
DNS-Cl: 1-dimethyaminonaphthalene-5-sulfonyl chloride
Dabsyl-Cl: 1-di-*n*-butylaminonaphthalene-5-sulfonyl chloride
Mansyl-Cl: 2-methylanilinonaphthalene-6-sulfonyl chloride
Phisyl-Cl: 4-(*N*-phthalimidyl)benzenesulfonyl chloride
BHBT-SOCl: 2-(5′,6′-dimethoxybenzothiazolyl)-benzenesulfonyl chloride
Fmoc-Cl: 9-fluorenylmethyl chloroformate
NT-COCl: 2-naphthyl chloroformate
PE-COCl: 2-(1-pyrenyl)ethyl chloroformate
DMEQ-COCl: 3,4-dihydro-6,7-dimethoxy-4-methyl-3-oxoquinoxaline-2-carbonyl chloride
MMSQ-COCl: 6-methoxy-2-methylsulfonylquinoline-4-carbonyl chloride
DAM-F: 7-dimethylaminocoumarin-3-carbonyl fluoride
BHBT-COF: 4-(5′,6′-dimethoxybenzothiazolyl)benzoyl fluoride
FITC: fluorescein isothiocyanate
DANITC: 4-(*N*,*N*-dimethylamino)-1-naphthylisothiocyanate
DNSITC: 4-(*N*,*N*-dimethylaminonaphthalene-5-sulfonylamino)phenylisothiocyanate
3- and 4-POPITCs: 3- and 4-(2-phenanthraoxazolyl)phenyl isothiocyanates
BHBT-NCS: 4-(5′,6′-dimethoxybenzothiazolyl) phenyl isothiocyanate
CIPITC: 4-(2-cyanoisoindolyl)phenyl isothiocyanate
DTDITC: 5-isothiocyanato-1,3-dioxo-2-*p*-tolyl-2,-3-dihydro-1*H*-benzo[*de*]isoquinoline
SINC: *N*-succinimidyl-1-naphthylcarbamate
SIPC: *N*-succinimidyl-1-fluorenylcarbamate
AQC: 6-aminoquinolyl-*N*-hydroxysuccinimidylcarbamate
NBD-Cl: 4-chloro-7-nitro-2,1,3-benzoxadiazole
NBD-F: 4-fluoro-7-nitro-2,1,3-benzoxadiazole
DBD-F: 4-(*N*,*N*-dimethylaminosulfonyl)-7-fluoro-2,1,3-benzoxadiazole
DBD-COCl: 4-(*N*-chloroformylmethyl-*N*-methyl)-amino-7-*N*,*N*-dimethylaminosulfonyl benzofurazan
DIFOX: 2-fluoro-4,5-diphenyloxazole
SAOX-Cl: 2-chloro-4,5-bis(*p*-*N*,*N*-dimethylaminosulfonyl)oxazole
ED: ethylenediamine
DPE: 1,2-diphenylethylenediamine
PGO: phenylglyoxal
PQ: 9,10-phenanthraquinone
Br-MMC: 4-bromomethyl-7-methoxycoumarin
Br-DMC: 4-bromomethyl-6,7-dimethoxycoumarin
Br-MDC: 4-bromomethyl-6,7-methylenedioxycoumarin
Br-MAC: 4-bromomethyl-7-acetoxycoumarin
Br-DMEQ: 4-bromomethyl-6,7-dimethoxy-1-methyl-2(1*H*)quinoxalinone
Br-MMEQ: 4-bromomethyl-6,7-methylenedioxy-1-methyl-2(1*H*)quinoxalinone
Br-MB: 3-bromomethyl-7-methoxy-1,4-benzoxazin-2-one
Br-MA: 9-bromomethylacridine
Br-AMC: 3-bromoacetyl-7-methoxycoumarin

Br-ADMC: 3-bromoacetyl-6,7-methylenedioxycoumarin
Br-AA: (*N*-9-acridinyl)bromoacetamide
Br-AP: 1-bromoacetylpyrene
DNS-BAP: *N*-bromoacetyl-*N*′-(dansyl)piperazine
9-APB: *p*-(9-anthroyloxy)phenacylbromide
Br-AMN: 2-bromoacetyl-6-methoxynaphthalene
ADAM: 9-anthryldiazomethane
PDAM: 1-pyrenyldiazomethane
DAM-MC: 4-diazomethyl-7-methoxycoumarin
9-AP: 9-aminophenanthrene
NEDA: *N*-(1-naphthyl)ethylenediamine
MBPA: 2-[*p*-(5,6-methylenedioxy-2*H*-benzotriazol-2-yl)]phenethylamine
DNS-CD: dansyl cadaverine
DNS-PZ: dansyl semipiperazine
DMEQ-PAH: 6,7-dimethoxy-1-methyl-2(1*H*)-quinoxalinone-3-propionylcarboxylic acid hydrazide
DMBI-BH: 4-(5,6-dimethoxy-2-benzimidazol)-benzohydrazide
BHBT-BH: 5,6-dimethoxy-2-(4′-hydrazinocarbonylphenyl)benzothiazole
HCPI: 2-(hydrazinocarbonylphenyl)-4,5-diphenylimidazole
TM-BH: 2-(5-hydrazinocarbonyl-2-thienyl)-5,6-methylenedioxybenzofuran
FM-BH: 2-(5-hydrazinocarbonyl-2-furyl)-5,6-methylenedioxybenzofuran
OM-BH: 2-(5-hydrazinocarbonyl-2-oxazolyl)-5,6-methylenedioxybenzofuran
MPBI-BH: 4-(1-methylphenanthro[9,10-*d*]imidazol-2-yl)benzohydrazide
PTM: 5-(4-pyridyl)-2-thiophenemethanol
HMA: 9-hydroxymethylanthracene
DNS-AE: 2-dansylaminoethanol
HMC: 4-hydroxymethyl-7-methoxycoumarin
1-PM: 1-pyrenemethanol
NE-OTf: 2-(2,3-naphthalimido)ethyl trifluoromethanesulphonate
AE-OTf: 2-(2,3-antrathenedicarboxyimide)-ethyltrifluoromethanesulfonate
NBD-PZ: 4-nitro-7-*N*-piperazino-2,1,3-benzoxadiazole
DBD-PZ: 4-(*N*,*N*-dimethylaminosulfonyl)-7-*N*-piperazino-2,1,3-benzoxadiazole
ABD-PZ: 4-(aminosulfonyl)-7-(1-piperazinyl)-2,1,3-benzoxadiazole
ABD-AE: 4-(aminosulfonyl)-7-*N*-(2-aminoethylamino)-2,1,3-benzoxadiazole
DBD-CD: 4-(*N*,*N*-dimethylaminosulfonyl)-7-cadaverino-2,1,3-benzoxadiazole
NBD-OH: 4-(*N*-methyl-*N*-(2-hydroxyethyl)-amine-7-nitrobenz-3-oxa-1,3-diazole
DBD-COHz: 4-(*N*-hydrazinoformyl-*N*-methyl)-amino-7-*N*,*N*-dimethylaminosulfonyl)-2,1,3-benzoxadiazole
DBD-ProCZ: 4-(*N*,*N*-dimethylaminosulfonyl)-7-(2-carbazoylpyrrolidin-1-yl)-2,1,3-benzoxadiazole
NBD-ProCZ: 4-nitro-7-(2-carbazoylpyrrolidin-1-yl)-2,1,3-benzoxadiazole
DDB: 1,2-diamino-4,5-dimethoxybenzene
DMB: 1,2-diamino-4,5-methylenedioxybenzene
CPCI: *N*-cyclohexyl-*N*′-(1-pyrenyl)carbodiimide
ACCBI: *N*-(anthracenyl)-*N*′-cyclohexylcarbodiimide
NE-PS: poly[2-(1-naphthyl)ethyl *p*-styrenesulfonate]
OMB-COCl: 2-(5-chlorocarbonyl-2-oxazolyl)-5,6-methylenedioxybenzofuran
7MC-COCl: 7-methoxycoumarin-3-carbonyl chloride
3MC-COCl: 3-methylcoumarin-7-carbonyl chloride
NBI-SOCl: 1,2-naphthoylenebenzimidazole-6-sulfonyl chloride
*m*-Phibyl-$N_3$: 3-(2-phthalimidyl)benzoyl azide
*p*-Phibyl-$N_3$: 4-(2-phthalimidyl)benzoyl azide
7MC-4-$CON_3$: 7-methoxycoumarin-4-carbonyl azide
DMEQ-$CON_3$: 3,4-dihydro-6,7-dimethoxy-4-methyl-3-oxoquinoxaline-2-carbonyl azide
9-AN: 9-anthroylnitrile
PCN: pyrene-1-carbonylnitrile
1-AN: 1-anthroylnitrile
DNN: 4-dimethylamino-1-naphthoylnitrile
MBC: (*aS*)-2′-methoxy-1,1′-binaphthalene-2-carbonylnitrile
DNS-ECF: 2-dansyl ethylchloroformate
FBIBT: 1,2-(difluoro-1,3,5-triazinyl)benz[*f*]isoindolo-[*1,2-b*][*1,3*]benzothioazolidine
PSCL: pyrene sulfonyl chloride

DPM: *N*-[4-(5,6-dimethoxy-2-phthalimidyl)-phenyl]maleimide
DBPM: *N*-[4-(6-dimethylamino-2-benzofuranyl)-phenyl]maleimide
NAM: *N*-(9-acridinyl)maleimide
BPM: *N*-[*p*-(2-benzoxazolyl)phenyl]maleimide
SBD-Cl: 7-chloro-2,1,3-benzoxadiazole-4-sulfonate
SBD-F: 7-fluoro-2,1,3-benzoxadiazole-4-sulfonate
ABD-F: 4-(aminosulfonyl)-7-fluoro-2,1,3-benzoxadiazole-4-sulfonate
1-, 2- or 9-AMH: *O*-(1-, 2- or 9-anthrylmethyl)hydroxylamine
Luminarin 3: 1*H*,5*H*,11*H*-[*1*] benzopyrano [*6,7, 8-ij*] quinolozine-9-acetic acid 2,3,6,7-tetrahydro-11-oxohydrazide
PBH: 4-(2-phthalimidyl)benzohydrazide
DBD-H: 4-(*N*,*N*-dimethylaminosulfonyl)-7-hydrazino-2,1,3-benzoxadiazole
MPIB: 4-(1-methyl-2-phenanthro[*9,10-d*]imidazol-2-yl)-benzohydrazide
DEB: 1,2-diamino-4,5-ethylenedioxybenzene
DTAD: 2,2′-dithiobis(1-amino-4,5-dimethoxybenzene)
BA: benzamidine
*p*-MBA: 4-methoxybenzamidine
*p*-MOED: *meso*-1,2-bis(4-methoxyphenyl)ethylenediamine
2-AP: 2-aminopyridine
P-TAD: 4-(1-pyrenyl)-1,2,4-triazoline-3,5-dione
DMEQ-TAD: 4-[2-(6,7-dimethoxy-4-methyl-3-oxo-3,4-dihydroquinoxalinyl)ethyl]-1,2,4-triazoline-3,5-dione
A-TAD: 4-(1-anthryl)-1,2,4-triazoline-3,5-dione
DNS-EDA: 5-dimethylaminonaphthalene-1-[*N*-(2-aminoethyl)] sulfonamide
DMPG: 3,4-dimethoxyphenylglyoxal

## 3.1 INTRODUCTION

High-performance liquid chromatography (HPLC) has been one of the most widely used of the effective separation methods. Capillary electrophoresis (CE) is rapidly becoming an accepted routine analytical technique, characterized by short run times and high efficiencies. Of the various detection methods used in HPLC and CE, fluorescence detection has been frequently utilized for the determination of trace levels of bioactive compounds in complicated matrices such as biological and environmental samples, owing to its high selectivity and sensitivity. However, most biologically and environmentally important substances are weakly fluorescent or non-fluorescent. Thus, various reagents have been developed for the fluorescence detection of their substances in HPLC and CE.

The reagents utilized for fluorescence detection are divided into two groups: ‘fluorogenic reagent’ and ‘fluorescence labeling reagent’. The fluorogenic reagents are generally non-fluorescent and react with target compounds to form conjugated-ring molecules, resulting in the generation of fluorescence. The fluorescence labeling reagents are composed of a highly fluorescent aromatic moiety (fluorophore) and a reactive group, and the reactive group attaches to an analyte to give a fluorescent derivative.

There are two approaches to derivatization for fluorescence detection in HPLC and CE: pre-column (pre-capillary) derivatization and post-column (post-capillary) derivatization. The former and latter are based on the fluorescence derivatization of analytes before and after separation by HPLC or CE, respectively. The pre-column derivatization should produce, individually, the different derivatives from the analytes. However, the optimum reaction conditions are not limited by the HPLC separation conditions. On the other hand, post-column derivatization should proceed rapidly because a prolonged reaction time causes peak broadening in chromatography. The reagents for post-column fluorescence derivatization need to be non-fluorescent or markedly different from the derivatives in their fluorescence excitation and/or emission spectra in the mobile phase; they are allowed to give multiple fluorescent derivatives, provided that their reactions are reproducible. In general, the fluorogenic reagents have the possibility of being used for both pre and post-column derivatization methods, although fluorescence labeling reagents are used only for pre-column derivatization.

Determination of sample components at extremely low concentrations (ppb—ppt level) are found to be important to many areas of science. Because of its extremely high sensitivity, laser-induced fluorescence (LIF) detection has been successfully introduced to HPLC and CE analyses, and the LIF instrument equipped with helium-cadmium (He-Cd) and argon (Ar) ion lasers is now available commercially. The successful design of derivatization reagents for He-Cd and Ar ion LIF detections, relies on the development of excellent fluorophore with highly intense fluorescence and an excitation maximum nearing the 325 nm and 488 nm output wavelengths of the He-Cd and Ar ion lasers, respectively. Numerous reagents for LIF detection, using these lasers, are proposed. Recently, (visible) diode LIF has been focused as an ultrasensitive and selective on-line detection method in both HPLC and CE, because of the reduction of background fluorescence and Raman scattering light. The diode lasers have several advantages; they are stable, compact, easy to perform, highly efficient and inexpensive. Most naturally occurring substances do not show fluorescence in the far-red and near-infrared wavelength regions in which the diode lasers emit. Therefore, derivatization reagents are required for the diode LIF detection of the substances. In order to utilize the potential diode LIF detection, it is essential to develop new labeling reagents with special fluorophores that absorb, and fluorescence in the far-red and near-infrared regions.

In this section, analytical features of fluorescence derivatization of biologically and environmentally important compounds and their applications to HPLC and CE with fluorescence detection are surveyed.

## 3.2 REAGENTS FOR AMINES AND AMINO ACIDS

Amines and amino acids are present in most biological and environmental samples. Hence, numerous fluorescence derivatization reagents have been proposed for the determination of the amino compounds. The amino moieties of the compounds are so reactive that the fluorescence derivatization generally proceeds under mild conditions.

### 3.2.1 General Amino Compounds

#### 3.2.1.1 Primary Amines and Amino Acids

OPA has been widely used for the derivatization of primary amines and amino acids. OPA reacts only with primary amino groups in alkaline pH in the presence of an alkylthiol such as 2-mercaptoethanol to give highly fluorescent isoindole derivatives (Fig. 3.1A) [1]. The derivatization is complete within about 2 min in a mixture of borate buffer (pH 6–8 for amines, pH 9.7–10.0 for amino acids) and methanol, even at room temperature. The OPA reaction can be applied to post-column derivatization because OPA itself is non-fluorescent [2]. As OPA derivatives, however, are not sufficiently stable to afford reproducible results, the post-column OPA reaction is introduced for an automatic amino acid analyzer including cation-exchange chromatography. An automated system, based on the OPA pre-column derivatization is also developed. Using the system, the highly sensitive and reproducible analysis for amino compounds are attained [3–6]. 2-Ethanethiol [7], 3-mercaptopropionic acid [8] and *N*-acetyl-L-cysteine [9,10] afford more stable fluorescent derivatives than 2-mercaptoethanol. Sample preparation and OPA reagent stability are still improving. The addition of nitriloacetic acid to reaction mixture results in a four-fold improvement in stability [11]. The detection limits [signal-to-noise ratio (S/N) = 3] of amino acids in pre-column derivatization HPLC are around 1 pmol on-column.

*General Derivatization Procedure of Primary Amines with OPA*

OPA reagent (27 mg) is dissolved in 0.5 ml of ethanol, and 5 ml of 0.1 M sodium tetraborate (or 0.4 M boric acid adjusted to pH 9.5 with NaOH) are added, followed by 20 μl of mercaptoethanol (or the equivalent among of another thiol). The reagent is kept overnight before use, and 10 μl of mercaptoethanol are added each week to maintain maximum yield.

An amine sample solution (50 μl) is mixed with four volumes of the OPA reagent. At a timed

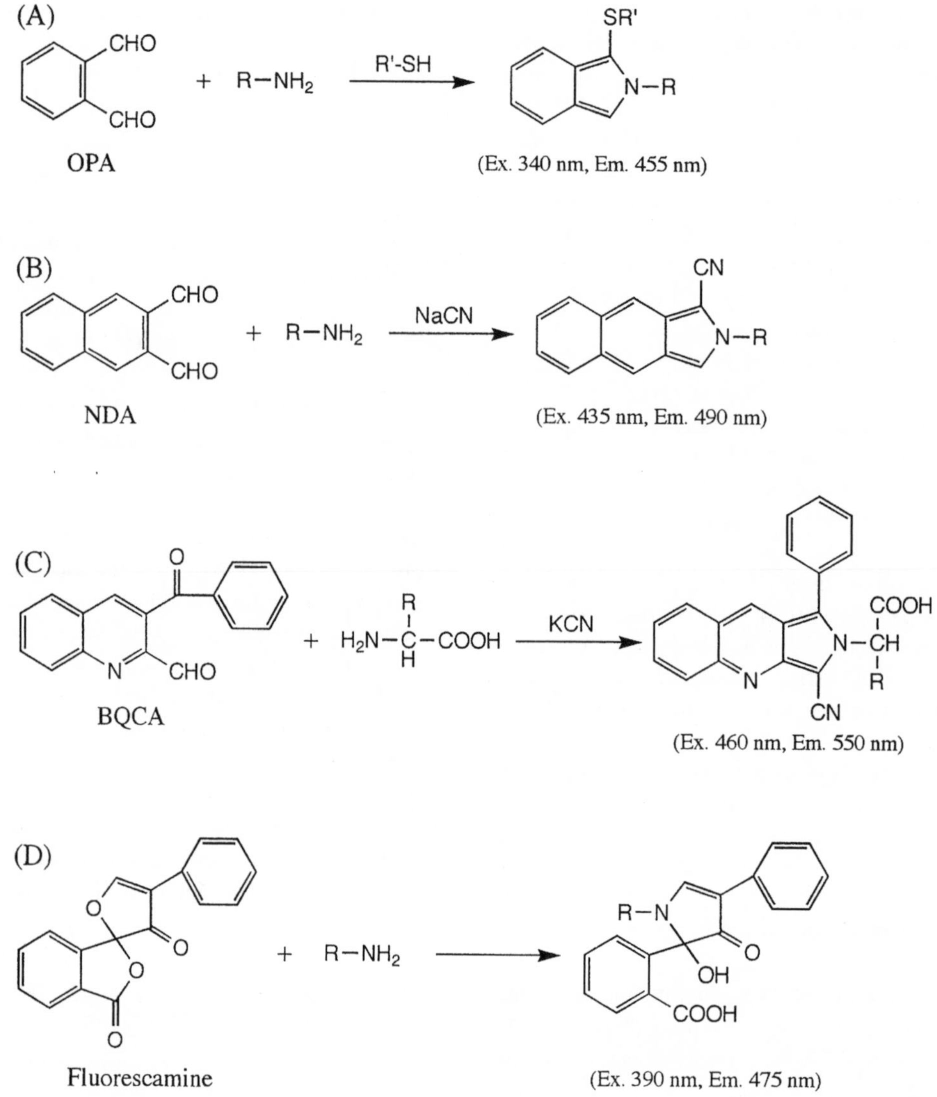

(A) OPA + R–$NH_2$ —R'-SH→ (Ex. 340 nm, Em. 455 nm)

(B) NDA + R–$NH_2$ —NaCN→ (Ex. 435 nm, Em. 490 nm)

(C) BQCA + $H_2N$–C(R)(H)–COOH —KCN→ (Ex. 460 nm, Em. 550 nm)

(D) Fluorescamine + R–$NH_2$ → (Ex. 390 nm, Em. 475 nm)

**Fig. 3.1.** Fluorogenic reactions of primary amines with (A) OPA, (B) NDA, (C) BQCA and (D) fluorescamine.

2 min intervals, a portion of the resulting mixture is injected into an RP-HPLC system. Extraction with chloroform or ethyl acetate (100 μl) removes excessive reagent and may be advantageous.

*OPA Application: HPLC Assay for 1-aminocyclopropanecarboxylic Acid (ACPC) from Plasma and Brain [12]*

Tissue samples (90 μl) from ACPC-treated mice or samples from untreated animals are thawed on the day of the assay. All samples are spiked with 10 μl of 3.8 mM cycloleucine (internal standard; IS). Samples are denatured by the addition of 300 μl of 6.2% perchloric acid, followed by 2 min of centrifugation at 14 000 g. The supernatants are transferred to clean microfuge tubes, placed on ice and alkalinized by adding 1 ml of 0.4 M KOH. The samples are centrifuged for 2 min at 14 000 g to remove precipitated potassium perchlorate. Aliquots of the supernatants (800 μl)

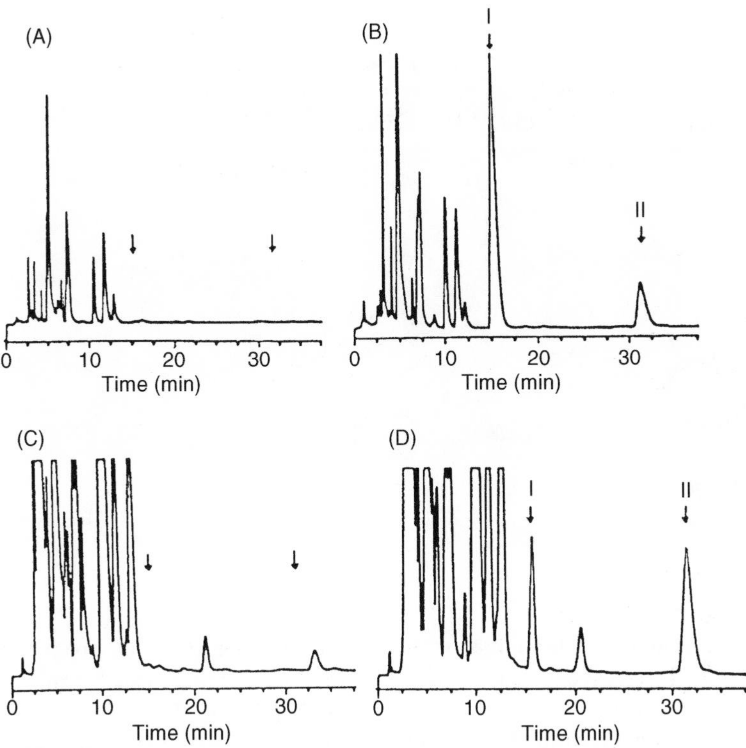


**Fig. 3.2.** Representative chromatograms of: (A) blank mouse plasma; (B) plasma collected 1 h after ACPC administration to a mouse; (C) blank mouse brain; and (D) brain extract collected 1 h after administration to a mouse. The concentrations of ACPC in (B) and (D) are 0.14 mg/mg and 0.89 mg/mg of protein, respectively. Arrows identify the peak retention times of ACPC (peak I, 15 min) and cycloluecine (peak II, 31 min). Column: Alltech C18 (3 μm; 100 × 4.6 mm I.D.). Mobile phase: methanol-acetonitrile-0.1 M sodium phosphate buffer (pH 6.0) (28:5:67, v/v). Flow-rate: 1 ml/min. [Reproduced from ref. 12, p. 105, Fig. 1.].

are derivatized with 400 μ1 of the OPA solution for 5 min. A 1-ml aliquot of the derivatized solution is injected for RP-HPLC analysis (Fig. 3.2).

An OPA solution is prepared by dissolving 54 mg of OPA in 0.5 ml of methanol, 10 ml of 2-mercaptoethanol and 4.5 ml of 0.1 M borate buffer (pH 9.0). This solution is prepared daily and protected from ambient light.

OPA-analogous reagents, NDA [13–15], FQCA [16], BQCA [17] and CBQCA [18], are reported (Figs. 3.1 and 3.3). The reagents react with primary amino compounds in a similar manner except that cyanide ion (KCN and NaCN) in place of the thiol compound is used (Figs. 3.1B and C). Although these reactions require longer reaction times (15–60 min at room temperature) than the OPA reaction, the resulting derivatives are so stable that these reagents are suitable for pre-column derivatization. The detection limits (S/N = 3) of the NDA method for amino acids are 10–50 fmol on-column, and the application to an Ar-ion LIF detection system can increase the sensitivity around 100-fold [19].

BQCA gives femtogram detection limits, with more stability than those of OPA. CBQCA (Fig. 3.3) allows quick, ambient-temperature chemistry, freedom from excess reagent problems, and the detection of amino acids and peptides in low amol range using LIF detection (Fig. 3.3) [18]. The reagent is used only with CE [18].

Fluorescamine is a reagent with no inherent fluorescence properties, but offers enhanced

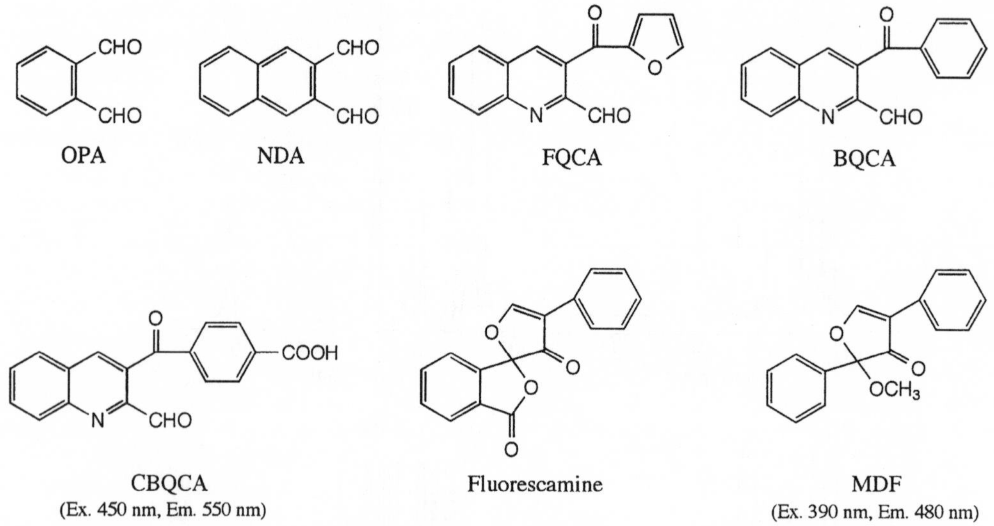

OPA NDA FQCA BQCA

CBQCA (Ex. 450 nm, Em. 550 nm) Fluorescamine MDF (Ex. 390 nm, Em. 480 nm)

**Fig. 3.3.** Fluorogenic reagents for primary amines.

detectability of primary amino compounds with fluorescence detection. The reagent reacts with primary amines and amino acids in borate buffer (pH 9.5–10) at room temperature (Fig. 3.1D), and this reaction is complete in a few minutes [20,21]. On the other hand, the reaction of secondary amines with fluorescamine gives non-fluorescent derivatives. Fluorescamine is, therefore, a specific reagent for compounds with primary amino groups. Since the reagent is non-fluorescent, the reaction is used for pre [22] and post-column [23,24] derivatization of amino acids and small peptides. The fluorescent derivatives are not very stable in aqueous eluents, and the sensitivity is 2–5 times lower than in the OPA method. However, this method is useful for the determination of labile substances.

An analogous reagent, MDF, reacts with primary amines in a similar manner [25]. Although the reagent requires a long reaction time (about 30 min at room temperature), the derivatives are very stable.

*Fluorescamine Application: Simple, Rapid and Sensitive Determination of Plasma Taurine by HPLC [26]*

Heparinized plasma is collected from four yellowtail fish, *S. quinqueradiata*. Samples for analysis are prepared from 0.2 ml of plasma by deproteinization with 2.0 ml of 5% trichloroacetic acid (TCA). TCA is extracted with 2 ml of diethyl ether (three times) from the sample solution. The solution is evaporated to dryness *in vacuo*, and the residue is dissolved in 1.0 ml of water. A 20 μl volume of the solution is treated with 20 μl of phosphate buffer (200 μM, pH 7.8) and 100 μl of fluorescamine reagent solution (25 mg/100 ml acetone) for 15 min at room temperature, and centrifuged for 1 min at 5000 g. A 10 μl volume of the supernatant is injected into an RP column (Fig. 3.4). The column eluent is monitored at 400 nm (excitation) and 480 nm (emission).

### 3.2.1.2 Primary and Secondary Amines and Amino Acids

Fluorescence labeling reagents (Fig. 3.5) having a reactive group such as (A) sulfonyl chloride, (B) carbonyl chloride and fluoride, (C) isothiocyanate and (D) succinimidyl moiety, have been developed for the HPLC or CE determination of primary and secondary amines. The reagents react with the amines to give the corresponding fluorescent derivatives.

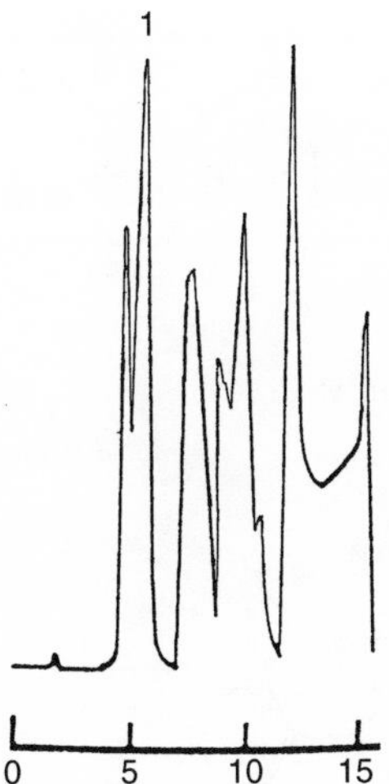


**Fig. 3.4.** Chromatogram obtained with a plasma sample of yellowtail fish. Peak: 1 = taurine. Column: LiChrospher 100 RP-8 (5 μm; 250 × 4 mm I.D.). Mobile phase: 23% acetonitrile in 15 mM phosphate buffer (pH 1.9). Flow-rate: 1 ml/min. [Reproduced from ref. 26, p. 156, Fig. 1D.].

*(1) Sulfonyl Chlorides*

DNS-Cl (Fig. 3.6) works on secondary and primary amino compounds under weakly alkaline conditions (Fig. 3.5A); the optimized reaction times (30–120 min) vary depending on the type of amino compound [27,28]. DNS derivatives are fairly stable and have long Stokes shifts. The reagent is hydrolyzed in the derivatization procedure to produce highly fluorescent 1-dimethylaminonaphthalene-5-sulfonic acid. Hence this reaction is used mainly for pre-column derivatization. The derivatives of amino acids can be separated on an RP column [28]. The sensitivity of this method is comparable to that of the OPA method. Bovine serum albumin (BSA) is found to be a fairly selective fluorescence enhancement reagent for DNS amino acids, formed by reaction between amino acids and DNS-Cl. Based on this finding, an HPLC method with fluorescence enhancement detection using BSA as a post-column modification reagent has been developed for the determination of DNS amino acids. Dansyl amino acids can be measured 8–109 times more sensitively with the conventional fluorescence detection [29]. As DNS-Cl also acts on a phenolic hydroxyl moiety, amino compounds containing this moiety such as catecholamines, tyramine and tyrosine, give multiple fluorescent derivatives [30].

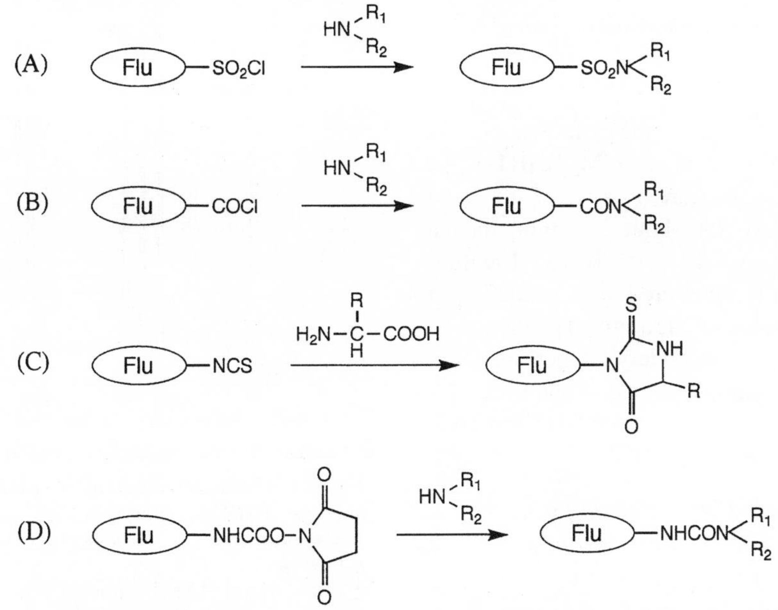

(A) Flu—$SO_2Cl$ + $HN(R_1)(R_2)$ → Flu—$SO_2N(R_1)(R_2)$

(B) Flu—COCl + $HN(R_1)(R_2)$ → Flu—$CON(R_1)(R_2)$

(C) Flu—NCS + $H_2N-CH(R)-COOH$ → Flu—N (thiohydantoin: S, NH, R, O)

(D) Flu—NHCOO—N (succinimide, O, O) + $HN(R_1)(R_2)$ → Flu—$NHCON(R_1)(R_2)$

**Fig. 3.5.** Fluorescence labeling reagents for amines.

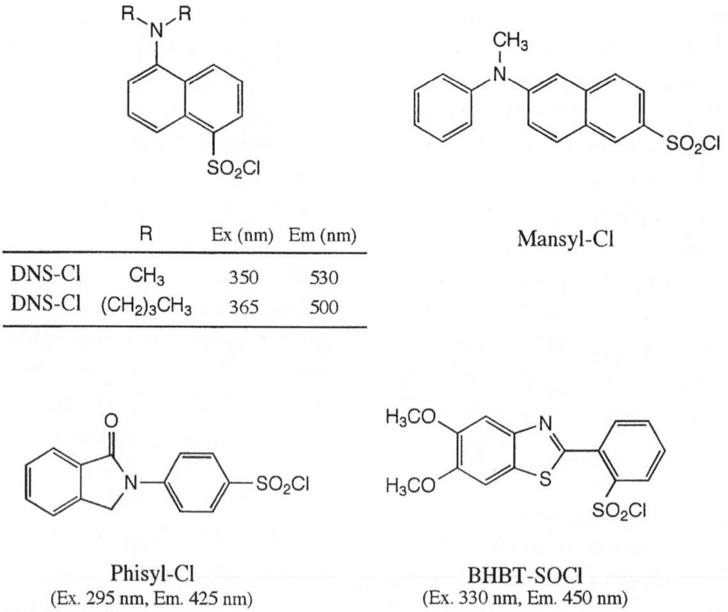

| | R | Ex (nm) | Em (nm) |
|---|---|---|---|
| DNS-Cl | $CH_3$ | 350 | 530 |
| DNS-Cl | $(CH_2)_3CH_3$ | 365 | 500 |

Mansyl-Cl

Phisyl-Cl
(Ex. 295 nm, Em. 425 nm)

BHBT-SOCl
(Ex. 330 nm, Em. 450 nm)

**Fig. 3.6.** Sulfonyl chloride reagents.

Analogous sulfonyl chlorides (Fig. 3.6), dabsyl-Cl [31], mansyl-Cl [32], phisyl-Cl [33], BHBT-SOCl [34], also act on primary and secondary amino compounds in a similar manner.

*General Derivatization Procedure of Amines with BHBT-SOCl*

To 100 μl of a test solution of amines in acetonitrile are added 100 μl of 10 mM triethylamine in acetonitrile and 100 μl of 10 mM BHBT-SOCl in acetonitrile. The mixture is allowed to stand at room temperature for about 2–3 min, then a portion (20 μl) is injected into RP-HPLC. The fluorescence intensity is monitored with excitation at 330 nm and emission at 450 nm. The detection limits for primary and secondary amines are about 3 and 300 fmol, respectively, on column. Fig. 3.7A shows a typical chromatogram of BHBT derivatives of some amines.

*DNS-Cl Application: HPLC Method for the Quantification of Iodothyronines in Body Tissues and Fluids [35]*

Rats are anesthetized with diethyl ether and, following a mid-line incision, blood is withdrawn

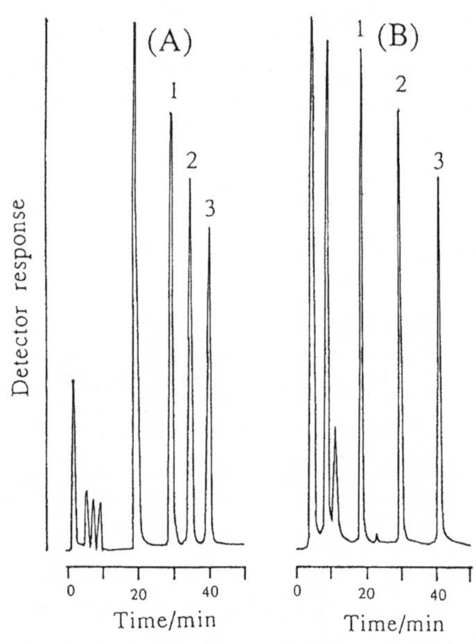


**Fig. 3.7.** Chromatograms of (A) BHBT-sulfonyl amide and (B) BHBT-carbonyl amide derivatives. Peaks: 1 = *n*-propylamine; 2 = *n*-heptylamine; 3 = *N*-methyl-*n*-hexylamine. Column: TSK gel ODS-120T(5 μm; 250 × 4.6 mm I.D.). Mobile phase: methanol-water (70:30, v/v). Flow-rate: 1 ml/min. [Reproduced from ref. 34, p. 477, Fig. 3.].

from the inferior vena cava. The animals are then infused rapidly via the left ventricle with three times their estimated blood volume with a solution containing 0.9% NaCl and 100 μM phloretin [(3′),4′-4,6-(tetra)trihydroxyaurone] at pH 7.4 to clear the tissues of blood. Phloretin as well as other aurone flavonoids can displace thyroid hormones from their binding proteins and block deiodination, therefore, immediately inhibiting the peripheral metabolism of iodothyronines. The tissues to be utilized, usually brain and liver, are removed rapidly and generally 1.0 g of tissue is used for analysis. The tissues are homogenized immediately in 6.0 ml of cold 80% ethanol, 0.02 M NaOH and 100 μM phloretin at pH 11.5. The samples are centrifuged and the supernatant is poured into a small beaker. This procedure is repeated two more times, and the supernatants are combined. The supernatants are dried in a vacuum oven at 40 °C. The residues are then resuspended in 4 ml of water containing 100 μM phloretin at pH 6.0 in order to separate free amino acids from the iodo compounds. The samples are then centrifuged at 105 000 g for 30 min. The supernatants are discarded and the residues are solubilized in 0.02 M NaOH and 100 μM phloretin at pH 11.5 and frozen until analysis.

Small (2.0 ml volume) vials are used for the derivatization procedure. A 100 μl volume of 0.5 M $NaHCO_3$ at pH 9.5 is added to the vials. This is followed with 25–100 μl of tissue or serum extract being added to the buffer. The amount of sample depended upon the tissue being analyzed, the age of the animal and prior treatment of the animal. Either 25 or 50 μl (0.5 or 1.0 pmol) of the mixed standards are added to each vial and finally 100 μl of DNS-Cl solution (6.0 mg/ml of acetone) are added to each vial. Several vials containing only standards (0.5 or 1.0 pmol) as well as a blank with buffer only are prepared for each run. The samples are vortex-mixed and placed in a refrigerator, and the reaction is allowed to proceed overnight. The following morning the volume of each vial is brought to 1.0 ml with water. Volumes of 50 μl are injected onto the RP column (Fig. 3.8).

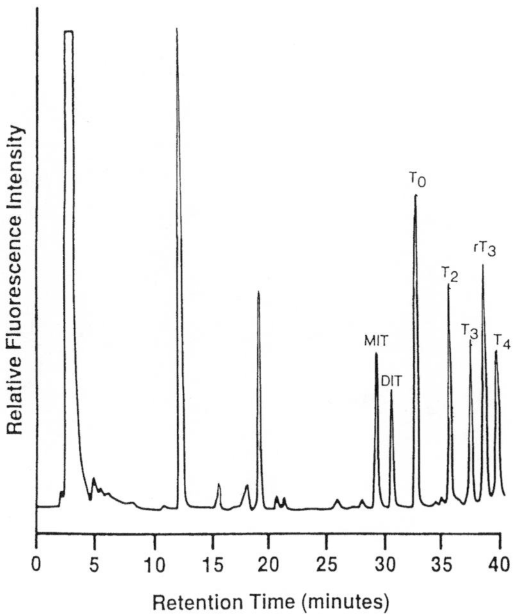


**Fig. 3.8.** Chromatogram obtained with rat brain extract. Peaks: MIT = 3-monoiodo-L-tyrosine; T0 = L-thyronine; T2 = 3,5-diiodo-L-thyronine; T3 = 3,5,3′-triiodo-L-thyronine; rT3 = reverse 3,3′,5′-triiodo-L-thyronine; T4 = 3,3′,5,5′-tetraiodo-L-thyronine (thyroxine). Column: Spherisorb ODS-2 12% (5 μm; 250 × 4 mm I.D.). Mobile phase: gradient elution. Flow-rate: 1 ml/min. [Reproduced from ref. 35, p. 21, Fig. 1.].

*(2) Carbonyl Chlorides and Fluorides*

The derivatization of these type reagents are based on the amidation of amines with carbonyl chloride and fluoride groups (Fig. 3.5B). 9-Fluorenylmethyl chloroformate (Fmoc-Cl) is used as a pre-column derivatization reagent for primary and secondary amino compounds (Fig. 3.9) [36]. The reaction proceeds in borate buffer (pH 8) within 2 min, and the derivatives are stable. However, Fmoc-Cl and its hydrolyzed product are highly fluorescent, hence they have to be removed by extraction with an organic solvent, pentane. As with DNS-Cl, Fmoc-Cl is also reactive towards a phenolic hydroxyl moiety. Further, the reagent combines with the imidazole ring of amino compounds. In this reaction, tyrosine and histidine give the corresponding mono and disubstituted derivatives.

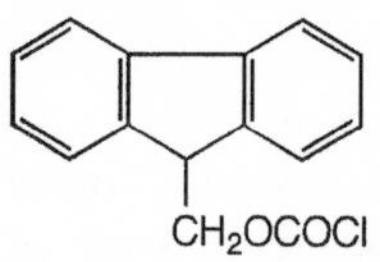

Fmoc-Cl
(Ex. 260 nm, Em. 315 nm)

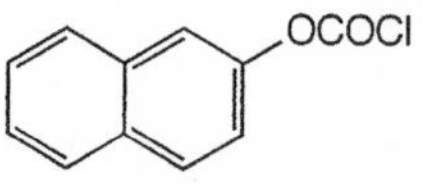

NT-COCl
(Ex. 275 nm, Em. 335 nm)

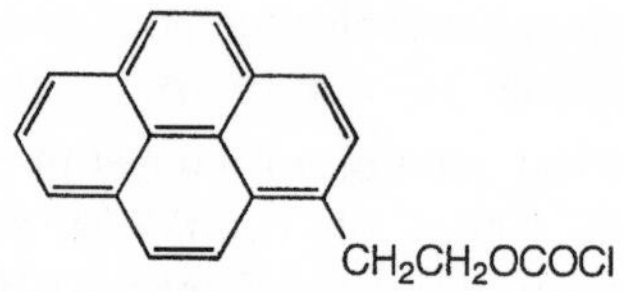

PE-COCl
(Ex. 275 nm, Em. 335 nm)

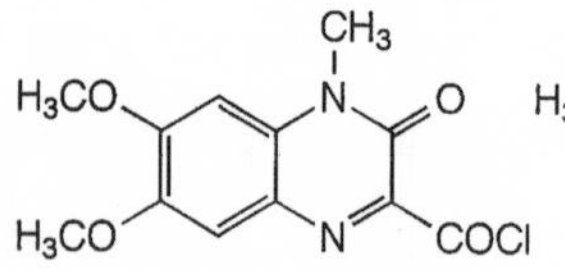

DMEQ-COCl
(Ex. 400 nm, Em. 500 nm)

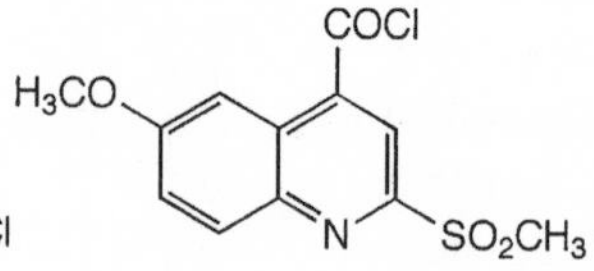

MMSQ-COCl
(Ex. 342 nm, Em. 448 nm)

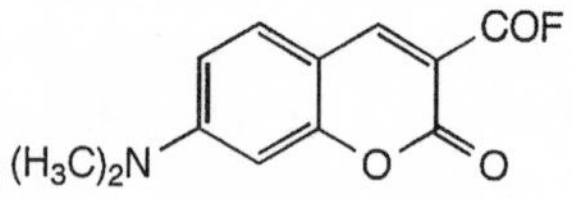

DAM-COF
(Ex. 350 nm, Em. 405 nm)

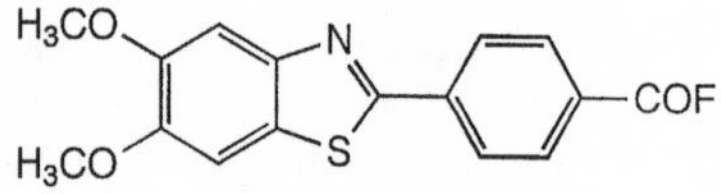

BHBT-COF
(Ex. 350 nm, Em. 450 nm)

**Fig. 3.9.** Carbonyl chloride and fluoride reagents.

Despite these disadvantages, this reaction has become popular through simplification [37,38] and/or automation [37,39] of the derivatization procedure. The sensitivity is comparable to that of the OPA method. The limitation of OPA to primary amines has lead to the development of a mixed system of OPA and Fmoc-Cl [40,41]. The system has been used with an automatic handling device for fully automated analyses.

Analogous reagents (Fig. 3.9), NT-COCl [42], PE-COCl [43], DMEQ-COCl [44], MMSQ-COCl [45], DAM-COF [46] and BHBT-COF [34], have been developed. In general, these reactions proceed in an aprotic solvent, benzene or acetonitrile, and are complete within 5 min at room temperature, except for NT-COCl (100 °C, 1 h). The reactions may be useful for the pre-column derivatization of hydrophobic amines. The detection limits (S/N = 3) of DMEQ-COCl and BHBT-COF are 2–5 fmol on column. DMEQ-COCl has been applied to the determination of $\beta$-phenylethylamine, a neuromodulator in the central nervous system, in human plasma [47], 1,2,3,4-terahydroisoquinoline, a parkinsonian-induced-substance, in rat brain [48] and amantadine, a brain metabolic stimulant, in human plasma [49].

### *General Derivatization Procedure of Amino Acids with Fmoc-Cl*

To 0.5 ml of sample solution is added 0.5 ml of 15 mM fmoc-Cl in acetone. After about 40 s, the mixture is extracted with three 2 ml portions of *n*-pentane. The aqueous solution with the amino acid derivatives is ready for chromatographic separation. If the reaction volume is reduced to 10–50 µl and the amount of the reagent is minimized, it is not necessary to extract the surplus reagent, even though it is fluorescent.

*General Derivatization Procedure of Amines with BHBT-COF*

To 100 µl of a test solution of amines in acetonitrile are added 100 µl of 10 mM quinuclidine in acetonitrile and 100 µl of 3 mM BHBT-COF in acetonitrile. The mixture is allowed to stand at 37 °C for 20 min, then a portion (20 µl) is subjected to RP-HPLC. The fluorescence intensity is monitored with excitation at 350 nm and emission at 450 nm. The detection limits for primary and secondary amines are about 3 and 30 fmol, respectively, on-column. Fig. 3.7B shows a typical chromatogram of BHBT derivatives of some amines.

*DMEQ-COCl Application: HPLC Determination of β-phenylethylamine in Human Plasma [47]*

A 1.0 ml aliquot of plasma is mixed with 50 µl of *p*-methylbenzylamine (MBA, IS) and 0.5 ml of 70 mM HCl and the mixture is poured into a Toyopak SP cartridge. The cartridge is washed successively with 5 ml of water (twice) and 1.8 ml of aqueous 40% acetonitrile (twice). The adsorbed amines are eluted with 3 ml of a mixture of acetonitrile and 1.0 M NaCl (2:3, v/v). To the eluate, 0.6 ml of 0.5 M NaOH and 6 ml of ethyl acetate are added. Phenylethylamine and MBA are extracted by shaking for 10 min. The organic layer (about 6 ml) is evaporated to dryness *in vacuo* at 50 °C and the residue is dissolved in 200 µl of acetonitrile containing 2.0% Triton X-405. To the solution, 100 µl of 2 mM DMEQ-COCl in acetonitrile and about 3 mg of $K_2CO_3$ are added. The mixture is allowed to stand at room temperature for about 1 min. The supernatant (20 µl) of the final reaction mixture is applied onto RP-HPLC. Figs. 3.10A and B are typical chromatograms obtained with a standard and plasma, respectively.

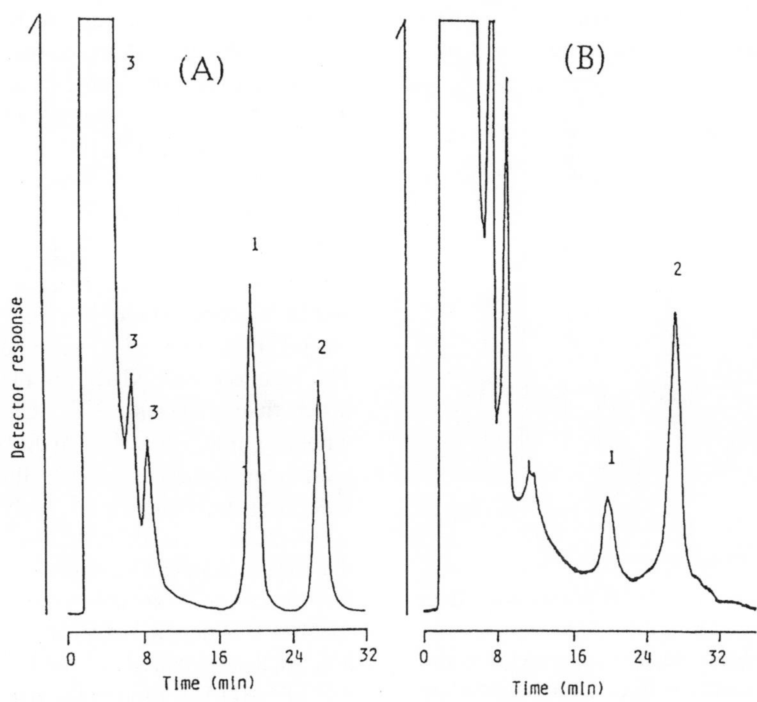


**Fig. 3.10.** Chromatograms obtained with: (A) a standard; and (B) a human plasma sample. Peaks: 1 = phenylethylamine; 2 = *p*-methylbenzylamine (IS); 3 = reagent blank. Column: TSK gel ODS-120T(5 µm; 250 × 4.6 mm I.D.). Mobile phase: acetonitrile-50 mM ammonium acetate (33:67, v/v). Flow-rate: 1 ml/min. [Reproduced from ref. 47, p. 87, Figs. 1 and 2.].

*DMEQ-COCl Application: Determination of Amantadine in Human Plasma by HPLC*

A 50-μl aliquot of plasma is pipetted into a 10-ml screw-capped test tube together with 10 μl of 1-(1-adamantyl)ethylamine (IS: 1.0 μg/ml) solution, 50 μl of 1 M NaOH and 2 ml of toluene. After vortex-mixing for about 2 min, the mixture is centrifuged at 1000 g for 10 min. The supernatant (1.5 ml) is transferred to another tube and is evaporated to dryness under a nitrogen stream. To the residue are added 1.0 mM DMEQ-COCl solution (100 μl) and 5.0 mM triethylamine solution (50 μl). The resulting mixture is allowed to stand at room temperature (15–40 °C) for 5 min. A 20 μl portion of the final reaction mixture is applied onto RP-HPLC. Figs. 3.11A and B are typical chromatograms obtained with drug-free plasma and plasma spiked with amantadine, respectively [49].

*(3) Isothiocyanates (Edman's Reagent)*

Isothiocyanates and isocyanates react with primary amines to give thiourea (thiohydantoin) and urea

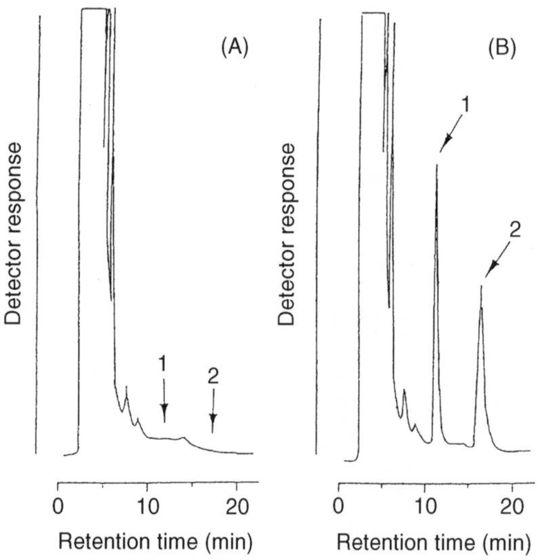


**Fig. 3.11.** Chromatograms obtained from: (A) drug-free plasma; and (B) plasma spiked with amantadine. Peaks: 1 = amantadine; 2 = (1-adamantyl)ethylamine (IS). Column: YMC Pack C8 (5 μm; 250 × 4.6 mm I.D.). Mobile phase: methanol-acetonitrile-water (35:35:30, v/v). Flow-rate: 1 ml/min. [Reproduced from ref. 49, p. 470, Fig. 4.].

derivatives, respectively (Fig. 3.5C). However, isocyanates react quite readily with water and alchohols to give urethanes. For this reason they are generally replaced by the less reactive isothiocyanates. Hence bolt the former and latter groups are used as reactive groups for the derivatization of amines and alcohols, respectively.

Edman's reagents are used not only in the pre-column determination of amino acids, but also in the microanalysis of peptide sequences. The following reagents have been developed: FITC [50], DANITC [51], DNSITC [52], 3- and 4-POPITCs [53], BHBT-NCS [54], CIPITC [55] and DTDITC [56] (Fig. 3.12). These reagents do not show great differences in sensitivity and reactivity.

*(4) Succinimidyls*

SINC reacts with amino acids in 0.5 M borate buffer (pH 9.5) within 1 min at room temperature to form naphthylcarbamoyl derivatives [57] (Figs. 3.5D and 3.13). The rapid reaction and the removal of excessive reagent by hydrolysis make SINC suitable for automated pre-column derivative formation [58]. The intense fluorescence allows the determination of subpicomole quantities.

SIPC [59] and AQC [60] also are developed.

*(5) Benzofurazans*

Halogenobenzofurazan reagents, NBD-Cl [61] and NBD-F [62–64], react with both primary and secondary amino compounds (Fig. 3.14) at 50–60 °C under alkaline conditions (pH 8–9). The reaction with NBD-F is more than ten times faster than that with NBD-Cl, and is complete within 1 min. As the hydrolyzed product of the reagents (NBD-OH) also fluoresces intensely, both reagents are used only for precolumn derivatization. The derivatives of amino acids other than tryptophan can be separated by RP-HPLC and their detection limits are in the subpicomole range.

DBD-F is developed a a fluorogenic reagent for amines. The sensitivity of the reagent is higher than that of NBD-F, although the reactivity is less [65].

A noble electrophilic reagent for amines, DBD-COCl, is synthesized [66]. The reagent reacts with amines (primary and aromatic amines) in benzene solution at room temperature to give

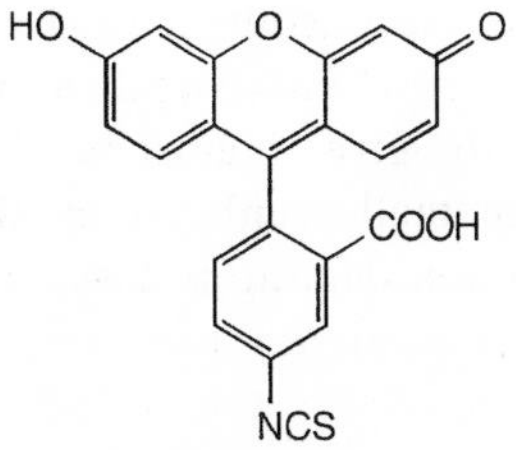

FITC
(Ex. 490 nm, Em. 520 nm)

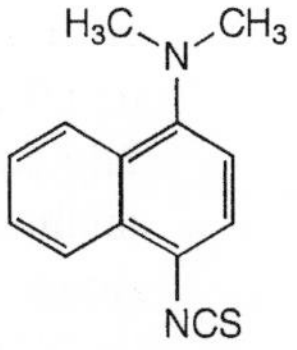

DANITC

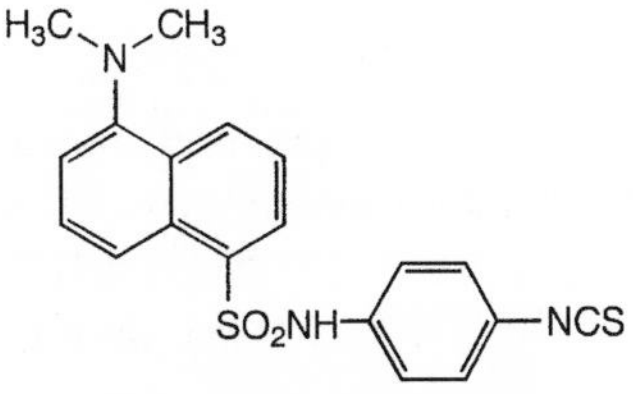

DNSITC

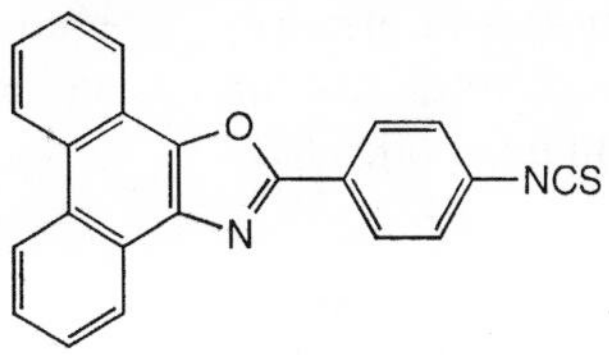

4-POPITC
(Ex. 330 nm, Em. 400 nm)

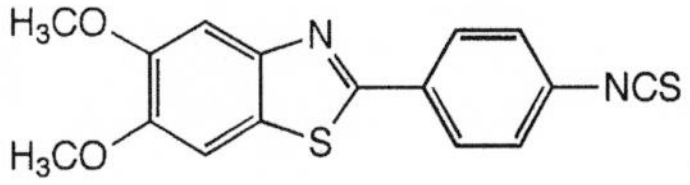

BHBT-NCS
(Ex. 355 nm, Em. 420 nm)

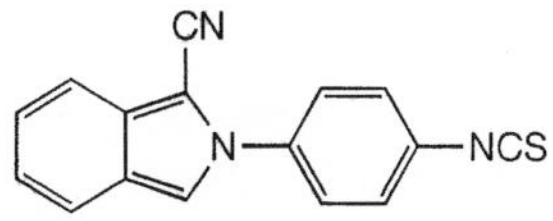

CIPITC
(Ex. 260 nm, Em. 410 nm)

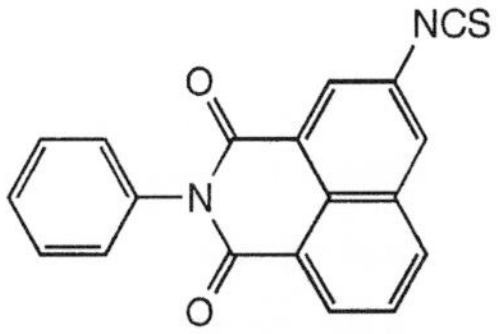

DIDITC
(Ex. 300 nm, Em. 530 nm)

**Fig. 3.12.** Edman's reagents.

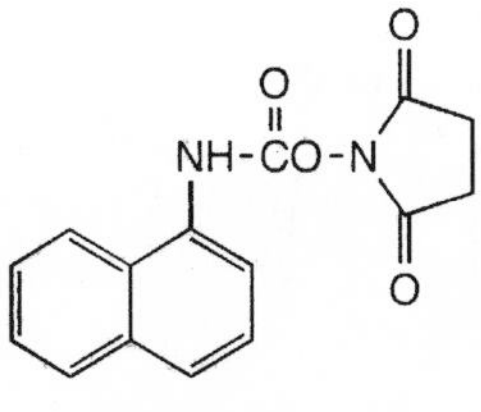

SINC
(Ex. 305 nm, Em. 378 nm)

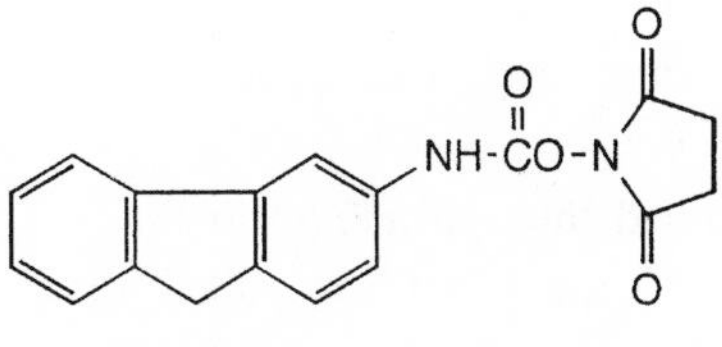

SIPC
(Ex. 286 nm, Em. 343 nm)

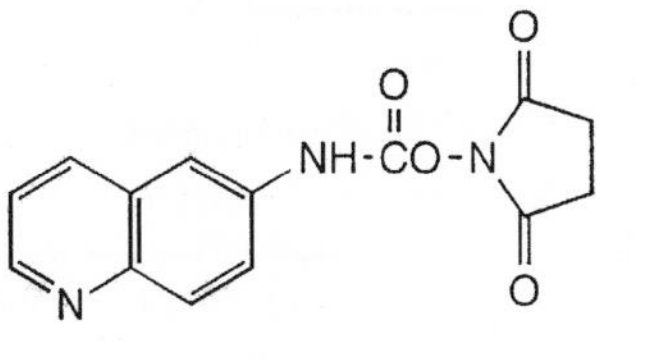

AQC
(Ex. 250 nm, Em. 395 nm)

**Fig. 3.13.** Succinimidyl reagents.

*Non-fluorescent* *Highly-fluorescent*

| | R | X |
|---|---|---|
| NBD-Cl | $NO_2$ | Cl |
| NBD-F | $NO_2$ | F |
| DBD-F | $SO_2N(CH_3)_2$ | F |

DBD-COCl

**Fig. 3.14.** Benzofurazan reagents.

fluorescent products (Ex around 440 nm, Em around 550 nm). The reagent reacts not only with amines, but also alcohols, thiols and phenols.

Some benzofurazan derivatives, which have an excitation wavelength near 480 nm, can be determined by HPLC with Ar ion LIF detection.

*NBD-F Application: Determination of D-amino Acids, Derivatized with NBD-F, in Wine Samples by HPLC [67]*

A wine sample is passed through a 0.5 μm filter and a 10 μl-aliquot of the filtrate is added to 90 μl of 0.2 M borate buffer (pH 8.0) containing 4 mM EDTA 2Na. Then 50 μl of 50 mM NBD-F in acetonitrile is added and heated at 60 °C for 5 min. After the addition of 850 μl of 1% acetic acid in methanol, a 50-μl aliquot of the resultant mixture is applied onto RP-HPLC.

*DBD-F Application: HPLC with Fluorescence Detection of Ebiratide [68]*

A portion (100 μl) of rat plasma is extracted with 300 μl of methanol, dried under the nitrogen stream, and reconstituted to 100 μl with 0.1 M borate buffer (pH 9.0), and then mixed with 100 μl of 30 mM DBD-F in acetonitrile. The mixture is allowed to stand at 50 °C for 30 min. A portion of the reaction mixture is applied onto RP-HPLC. (Fig. 3.15).

*(6) Oxazoles*

Oxazole reagents, DIFOX and SAOX-Cl (Fig. 3.16), have been used to derivatize amines, thiols and alcohols to produce the corresponding derivatives, making this an interesting prospect for some peptides [69].

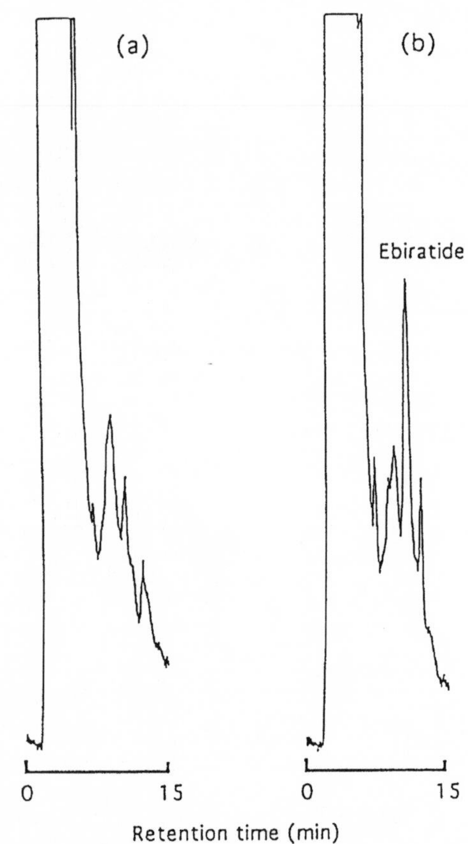


**Fig. 3.15.** Chromatograms obtained with: (a) rat plasma; and (b) ebiratide-spiked rat plasma (1 nmol/ml). Column: Vydac protein and peptide column 5C18 (5 μm; 150 × 4.6 mm I.D.). Mobile phase: 50 mM $Na_2HPO_4$-acetonitrile (3:2, v/v). Flow-rate: 1 ml/min. [Reproduced from ref. 68, p. 219, Fig. 7.].

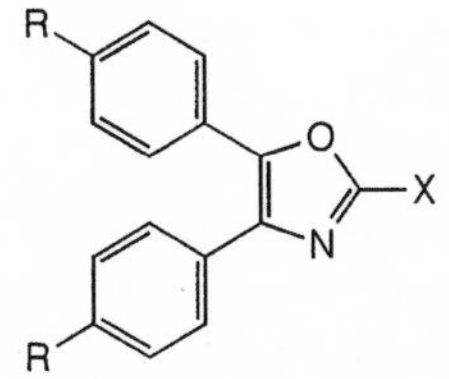

| | R | X | Ex (nm) | Em (nm) |
|---|---|---|---|---|
| DIFOX | H | F | 320 | 420 |
| SAOX-Cl | $SO_2N(CH_3)_2$ | Cl | 361 | 475 |

**Fig. 3.16.** Oxazole reagents.

*(7) Solid Phase Reagents*

Compared with liquid-phase derivatizations, the solid-phase derivatization approach has become increasingly attractive due to some unique features, such as good selectivity and ease of operation. Some polymeric reagents for the solidphase derivatization have been proposed (Fig. 3.17) [70–72].

*(8) Labeling Reagents for LIF Detection*

Various reagents having pyronin, thionine and cyanine fluorophores (Fig. 3.18) are utilized for

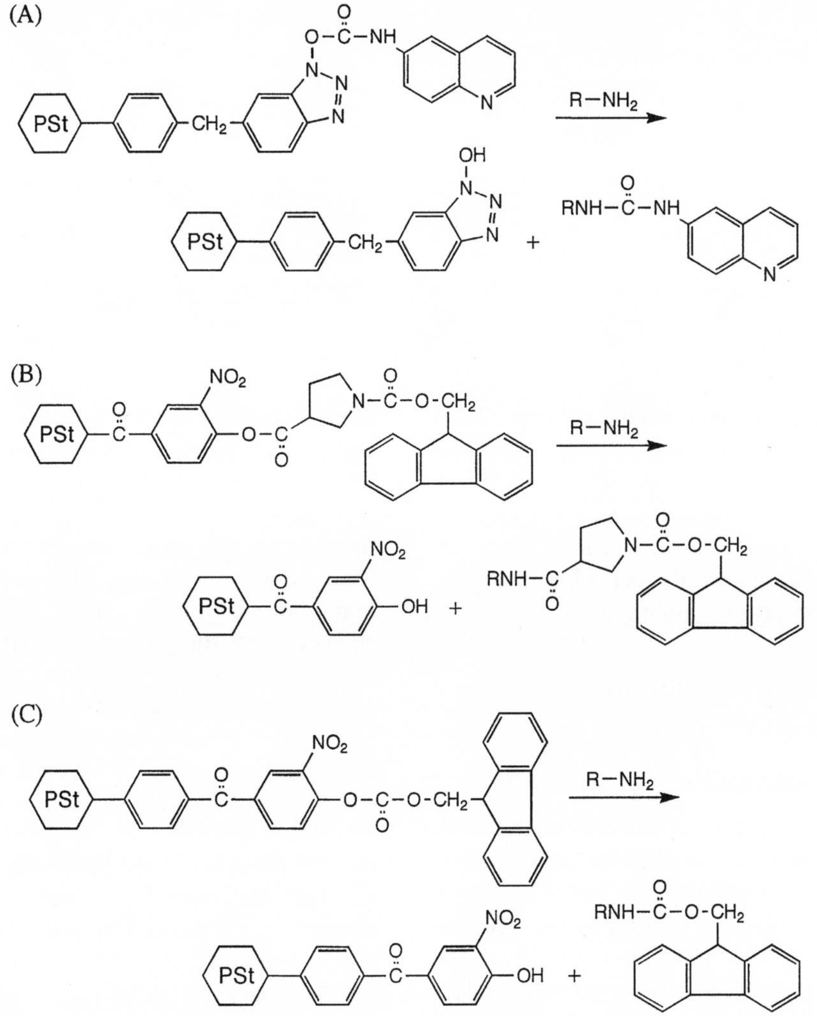

**Fig. 3.17.** Polymeric reagents for amines.

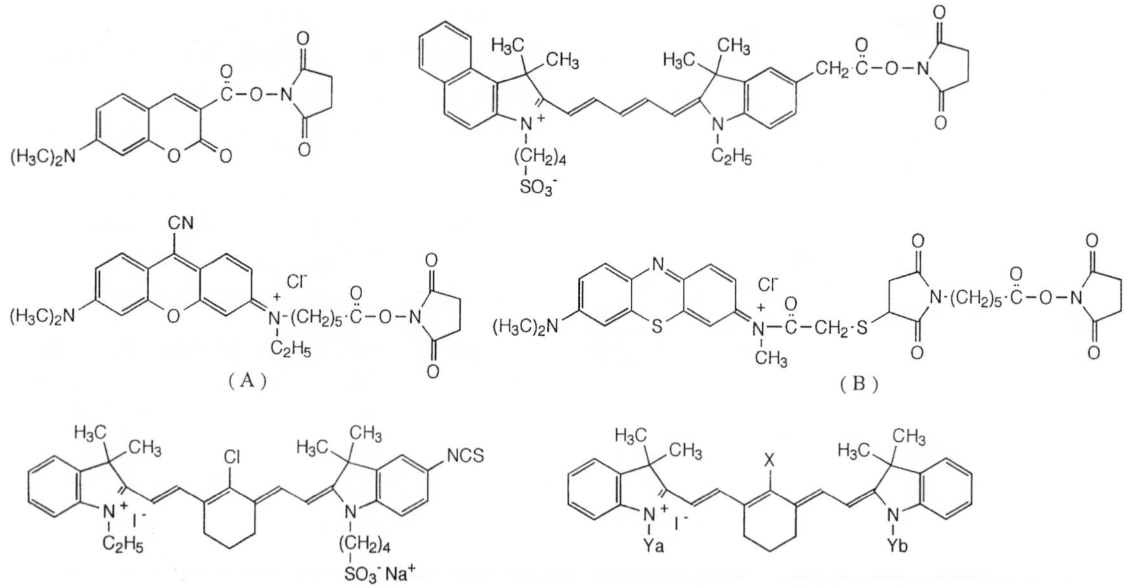

| | X | Ya | Yb |
|---|---|---|---|
| a | $p$-O-$C_6H_4$-NCS | $C_2H_5$ | $C_2H_5$ |
| b | $p$-O-$C_6H_4$-NCS | $(CH_2)_4SO_3^-$ | $(CH_2)_4SO_3^-\,Na^+$ |
| c | $p$-S-$C_6H_4$-NCS | $C_2H_5$ | $C_2H_5$ |
| d | $p$-S-$C_6H_4$-NCS | $(CH_2)_4SO_3^-$ | $(CH_2)_4SO_3^-\,Na^+$ |

**Fig. 3.18.** LIF-labeling reagents for amines.

the determination of amino compounds by HPLC (or CE) with various types of LIF detections [73–76]. In general, LIF detection is more useful in sensitivity for CE than for HPLC. The CE-LIF detection method with some reagents such as reagents A and B, allows the detection of amino compounds at less than 1 amol [73,74].

## 3.2.2 Particular Amines and Amino Acids

### 3.2.2.1 Catecholamines

Catecholamines are present at extremely small amounts in biological samples, in which many precursors and metabolites co-exist. Therefore, a highly selective and sensitive reagent is required for the determination of catecholamines. The reagents for general amino compounds is not sufficient in selectivity for this purpose. Trihydroxyindole (THI) [77,78] and ethylenediamine (ED) [79,80] methods have been reported as highly selective reactions for catecholamines (Fig. 3.19). Of these methods, the THI method is the most selective, but it cannot afford fluorescence with dopamine. Further, the methods have limited sensitivity. The condensation reaction between catecholamines and ED has been adapted only for post-column HPLC of urinary catecholamines.

1,2-Diphenylethylenediamine (DPE) [81] has been reported as a highly selective and sensitive fluorogenic reagent for catecholamines. DPE reacts selectively with catecholamines under mild conditions (pH 6.5–6.8, 37–50 °C) in the presence of potassium hexacyanoferrate (III) to yield a single fluorescent derivative from each catecholamine; it is therefore applicable to both pre and post-column derivatization in HPLC. DPE derivatives of catecholamines can be separated on an RP column. The DPE method is very sensitive and selective for catecholamines, and requires only

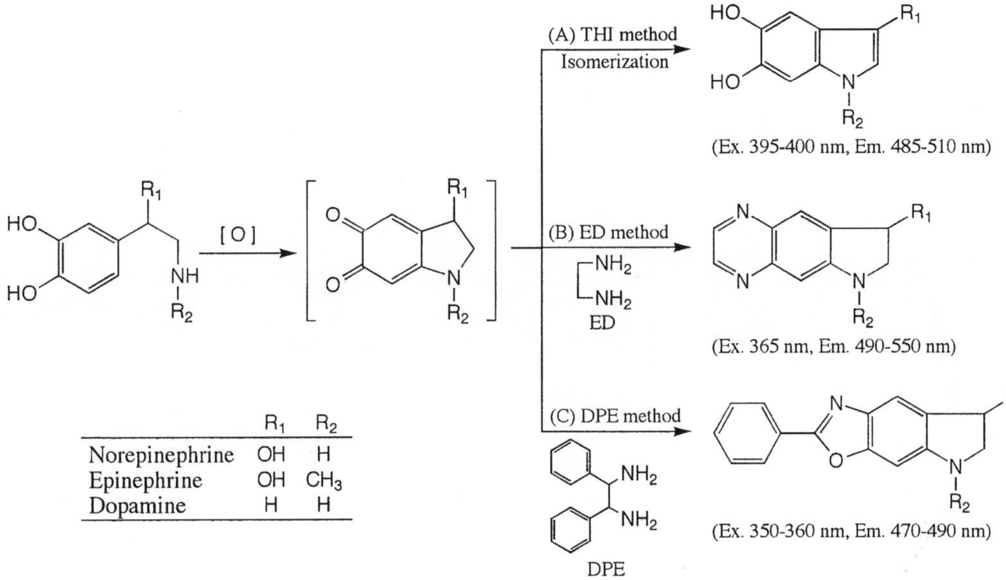

| | $R_1$ | $R_2$ |
|---|---|---|
| Norepinephrine | OH | H |
| Epinephrine | OH | $CH_3$ |
| Dopamine | H | H |

**Fig. 3.19.** Fluorogenic reactions of catecholamines by: (A) THI, (B) ED and (C) DPE method.

simple clean-up of plasma by solid-phase extraction using a cation-exchange cartridge. Clean-up by liquid-liquid extraction is also effective [82,83]. An on-line automated pre-column derivatization has been devised for reproducible results [84]. An automated catecholamine analyzer system has been constructed based on the DPE reaction applied to post-column HPLC. For the quantification of catecholamines and their metabolites (normetanephrine *et al.*), an ion-pair RP-HPLC system with post-column derivatization involving coulometric oxidation followed by fluorescence reaction with DPE has been established.

*DPE Application: HPLC Determination of Plasma Catecholamines [81]*

To 0.5 ml of plasma, 25 μl of 10 nM isoproterenol (IS) solution and 0.5 ml of 0.2 M lithium phosphate buffer (pH 5.8) are added. The mixture is poured into a Toyopak SP cartridge. The cartridge is washed successively with 5 ml of water (twice) and 1 ml of aqueous 50% acetonitrile. The adsorbed amines are eluted with 300 μl of 0.6 M KC1-acetonitrile (1:1, v/v) containing 0.6 mM potassium hexacyanoferrate(III). To the resulting solution, 50 μl of 0.1 M DPE in 0.1 M HCI are added and the mixture is allowed to stand at 37 °C for 40 min. The reaction is stopped by cooling the mixture in ice-water. A 100-μl aliquot of the mixture is applied onto RP chromatography (Fig. 3.20).

### 3.2.2.2 Tryptophan and Indolamines

Tryptophan and indolamines react with formaldehyde [86], chloroacetaldehyde [86] and methoxyaldehyde [87] to form fluorescent derivatives. The reaction conditions are fairly drastic (acidic medium, 80–100 °C, 15–60 min) in the presence of an oxidizing agent, and they are used for precolumn derivatization.

PGO reacts selectively with tryptophan under relatively mild conditions [88]. This reaction is applied to the determination of free and total tryptophan in human serum [89] (Fig. 3.21).

### 3.2.2.3 5-hydroxyindoleamines (Serotonin Related Compounds)

5-Hydroxyindoles are metabolites of tryptophan and play physiologically important roles in the

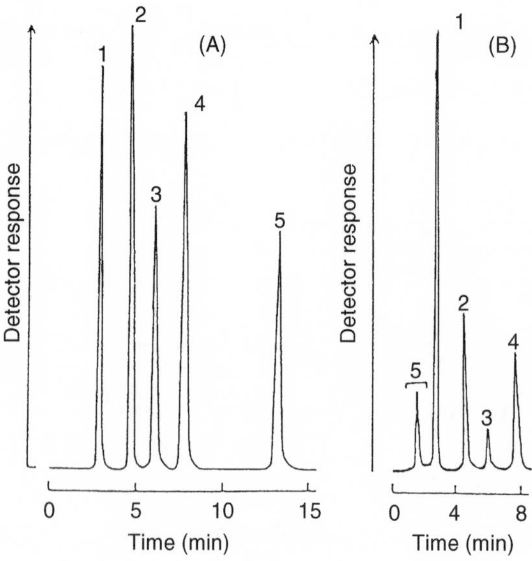


**Fig. 3.20.** Chromatograms obtained with: (A) a standard; and (B) a plasma. Peaks: 1 = norepinephrine; 2 = epinephrine; 3 = dopamine; 4 = isoproterenol; 5 = *N*-methyldopamine. Column: TSK-gel ODS120T (5 μm; 150 × 4.6 mm I.D.). Mobile phase: methanol-acetonitrile-50 mM Tris-HCl buffer (pH 7.0) (1:5:4, v/v). Flow-rate: 1.0 ml/min. [Reproduced from ref. 81, p. 54, Fig. 1 and p. 67, Fig. 4a.].

human body. Therefore, the determination of the amines in biological samples is useful for the elucidation of tryptophan metabolism and for the diagnosis of certain mental disorders such as schizophrenia and migraine.

Benzylamine reacts with 5-hydroxyindolamines such as serotonin and 5-hydroxyindol-3-ylacetic acid under fairly mild conditions (pH 9.0, 37 °C, 20 min) (Fig. 3.22A) in the presence of potassium hexacyanoferrate(III) to give highly fluorescent oxazole compounds [90]. This reaction can be applied to pre and post-column derivatization HPLC [91]. 5-Hydroxyindoles in urine, plasma and rat brain are quantified by the post-column RP-HPLC [92]. Tryptophan hydroxylase activity in rat brain is also assayed by post-column RP-HPLC [93]. The pre-column derivatization is applied to the determination of the indoles in microdialysate from rat brain [94].

Benzylamine also reacts with catechoamines under slightly different conditions from those for the indoles [95] (Fig. 3.22B). Thus, a pre-column HPLC method for the simultaneous determination of catecholamines and 5-hydroxyindoles is developed, and is applied to their quantification in urine [96].

*Benzylamine Application: HPLC Method for the Determination of Serotonin and 5-hydroxyindole-3-acetic Acid in Human Plasma [92]*

To a 200-μl aliquot of platelet-poor plasma are 40 μl of 0.50 mM 5-hydroxyindole-3-acetamide (IS) and 50 μl of 1.5 M $HClO_4$. After centrifugation, the supernatant (100 μl) is applied onto RP-HPLC. The derivatives are monitored at an excitation wavelength of 345 nm and an emission wavelength of 481 nm. Figs. 3.23A and B are of a typical chromatogram obtained with a standard and platelet-poor plasma from a healthy person, respectively.

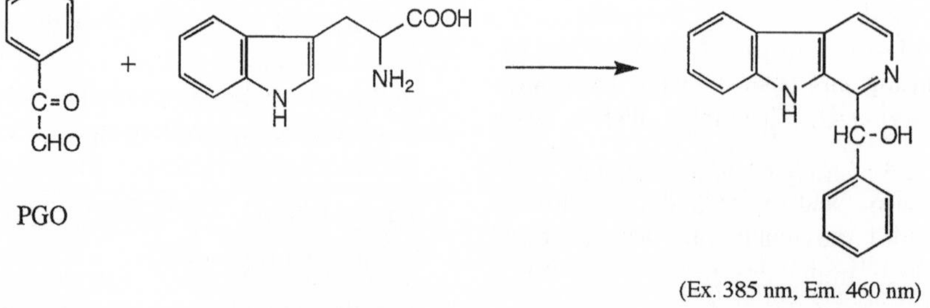

C=O
CHO
PGO
+
COOH
NH₂
N
H
N
H
N
HC-OH
(Ex. 385 nm, Em. 460 nm)

**Fig. 3.21.** Fluorogenic reaction of tryptophan with PGO.

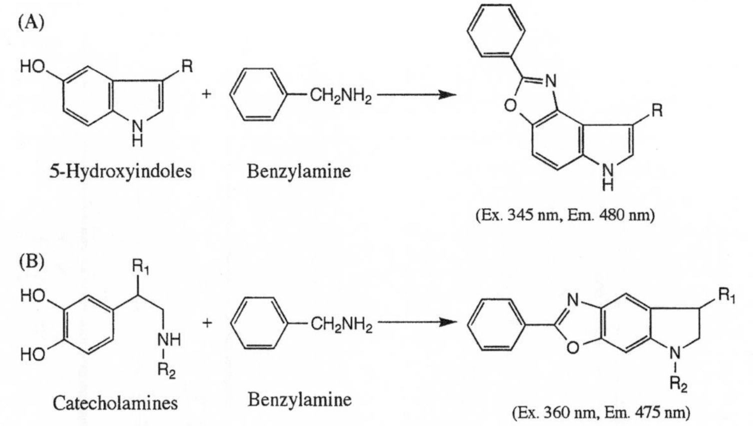

(A)

5-Hydroxyindoles + Benzylamine → (Ex. 345 nm, Em. 480 nm)

(B)

Catecholamines + Benzylamine → (Ex. 360 nm, Em. 475 nm)

**Fig. 3.22.** Fluorogenic reactions of: (A) 5-hydroxyindoles and (B) catecholamines with benzylamine.

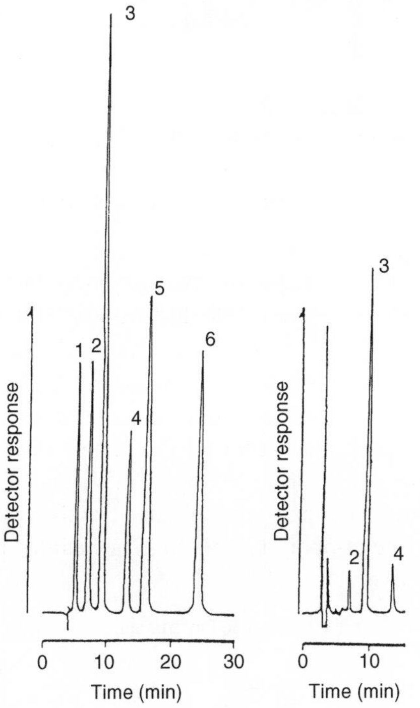


**Fig. 3.23.** Chromatograms obtained with: (A) a standard solution; and (B) platelet-poor plasma from a healthy person. Peaks: 1 = 5-hydroxytryptophan; 2 = serotonin; 3 = 5-hydroxyindole-3-acetamide; 4 = 5-hydroxyindole-3-acetic acid; 5 = 5-hydroxytryptophol; 6 = *N*-acetyl-5-hydroxytryptamine. Column: TSK-gel ODS-120T (5 μm; 150 × 4.6 mm I.D.). Mobile phase: acetonitrile-10 mM acetate buffer (pH 4.7) (5:95, v/v). Flow-rate: 1.0 ml/min. [Reproduced from ref. 92, p. 2355, Fig. 1.].

*Benzylamine Application: Simultaneous Determination of Catecholamines and 5-hydroxyindoleamines by HPLC [96]*

Urine is diluted 20 times with water and passed through a disposable filter (0.45 μm). To a 200-μl portion of the diluted urine sample, 200 μl of 0.15 M benzylamine in aqueous 90% methanol, 100 μl of a mixture of 0.1 M CAPS buffer (pH 11.0) and methanol (3:7, v/v) and 100 μl of 50 mM potassium hexacyanoferrate(III) in aqueous 50% methanol are added successively. The mixture is allowed to stand at 50 °C for 20 min. A 50-μl portion of the final reaction mixture is applied onto RP-HPLC. Fig. 3.24 shows a typical chromatogram obtained with a standard mixture and a human urine sample.

### 3.2.2.4 Guanidino Compounds

Guanidino compounds and arginine can be selectively converted into fluorescent derivatives by reaction with PQ [97] or benzoin [98,99] (Fig. 3.25).

PQ reacts with guanidino compounds under alkaline and subsequent acidic conditions to give fluorescent derivatives. The reagent has been employed only for the post-column HPLC.

On the other hand, benzoin yields a single fluorescent derivative for guanidino compound. Thus, the reaction with benzoin is used for both pre and post-column derivatization in HPLC.

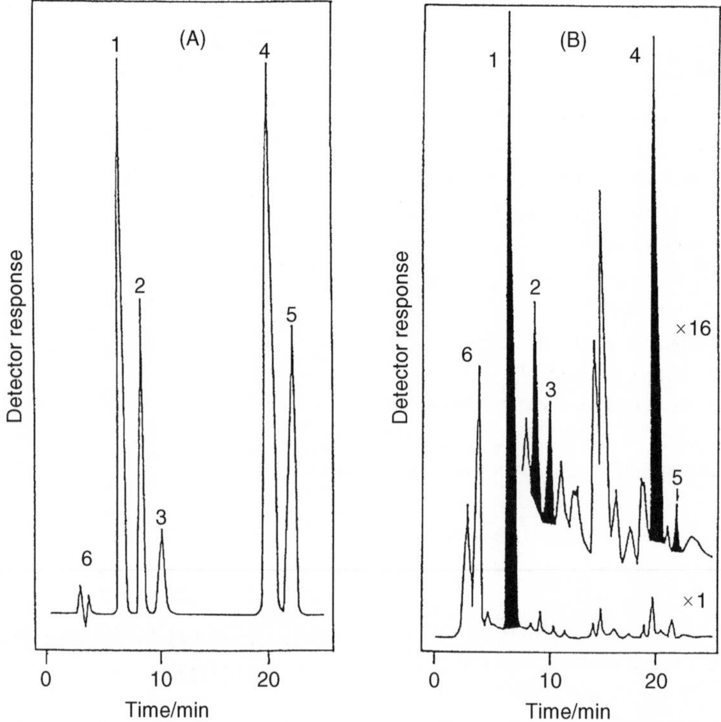


**Fig. 3.24.** Chromatograms obtained with: (A) a standard mixture; and (B) a human urine sample. Peaks: 1 = 5-hydroxyindole-3-acetic acid; 2 = norepinephrine; 3 = serotonin; 4 = 5-hydroxyindole-3-acetamide; 5 = epinephrine; 6 = reagent blank. Column: Cosmosil 5C18 (5 μm; 250 × 4.6 mm I.D.). Mobile phase: acetonitrile-10 mM acetate buffer (pH 6.0) (35:65, v/v). Flow-rate: 1.0 ml/min.

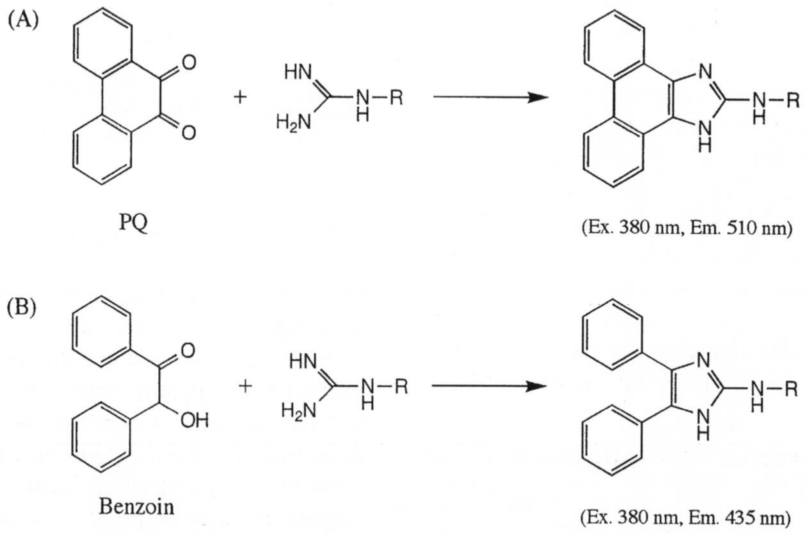

(A)

PQ

(Ex. 380 nm, Em. 510 nm)

(B)

Benzoin

(Ex. 380 nm, Em. 435 nm)

**Fig. 3.25.** Fluorogenic reactions of guanidino compounds with (A) PQ and (B) benzoin.

## 3.2.3 Peptides

### 3.2.3.1 General Peptides

Fluorogenic reagents such as OPA (Fig. 3.1A), NDA (Fig. 3.1B) and fluorescamine (Fig. 3.1D) are used for the sensitive HPLC determination of general peptides. These reagents react with primary amines, so that peptides containing free $\alpha$- and $\varepsilon$- amino moieties can be detected by pre or post-column derivatization methods [100–103]. Using chromatographic methods, the peptides can be determined at the 0.05–100 pmol levels on-column. However, these reagents are not selective for peptides because nearly all peptides have a primary amino group. Therefore, the separation conditions with a high resolution for peptides is required, especially when endogenous peptides in biological matrices are determined by HPLC.

A pre-column HPLC method with NDA reagent is developed for the determination of enkephalin peptides [102]. In the method, the column-switching technique is necessary for the separation of the biogenic enkephalins.

### 3.2.3.2 Arginine-containing Peptides

The post-column derivatization HPLC method has been established for the determination of arginine-containing peptides [104,105]. The method is based on the conversion of the guanidino moiety of an arginyl residue in an alkaline solution (100 °C, 90 s) with benzoin into their respective fluorescent derivatives (Fig. 3.25). The fluorescence detection is specific for the arginine-containing peptides, and this method can detect them down to about 10 pmol on-column (S/N = 2).

The pre-column derivatization HPLC method using benzoin is more sensitive for arginine-containing peptides and permits possible detection at levels as low as 100 fmol on-column [106].

## 3.2.4 Fluorescence Derivatization for CE

Fluorescamine and OPA are reagents for pre and post-capillary derivatization of primary amino acids and small peptides. Fmoc-Cl, FITC, NDA and CBQA can only be used for the pre-capillary derivatization of both primary and secondary amino compounds. For the simultaneous derivatization of primary and secondary amino compounds, parallel evaluation of more than one amino-selective reagent have been reported [107]. The fluorescent derivatives from these reagents are usually determined by various types of LIF detection such as He-Cd, Ar-ion, far-red and near-infrared diode lasers, depending on the reagents used.

Benzoin and 4-methoxy-1,2-phenylenediamine are specific reagents for the derivatization of arginine and tyrosine containing peptides, respectively. LIF detection employing a He-Cd laser is well suited to detect the derivatives from both reagents.

# 3.3 REAGENTS FOR ORGANIC ACIDS

Various organic acids (e.g. fatty acids, bile acids, prostaglandins, $\alpha$-keto acids, sialic acids and ascorbic acids) play important physiological roles at extremely low concentrations in biological, food and environmental samples. In general, the acids have no strong absorption and fluorescence, and show low reactivity. Therefore, fluorescence pre-column derivatization is widely applied due to its sensitivity and selectivity. A number of highly sensitive and selective derivatization reagents have been proposed for quantification of the acids.

## 3.3.1 Carboxylic Acids

Most of the derivatization reagents for carboxylic acids are of the fluorescence labeling type; the reagents, generally, have a fluorophore (coumarin, phenanthrene, pyrene, quinoxaline, benzofurazan and so on) and a reactive site (bromomethyl, diazomethyl, amine, hydrazine moieties etc.) for a carboxylic group (Fig. 3.26). Therefore, these reagents are used for pre-column derivatization and the derivatives are generally separated on an RP column.

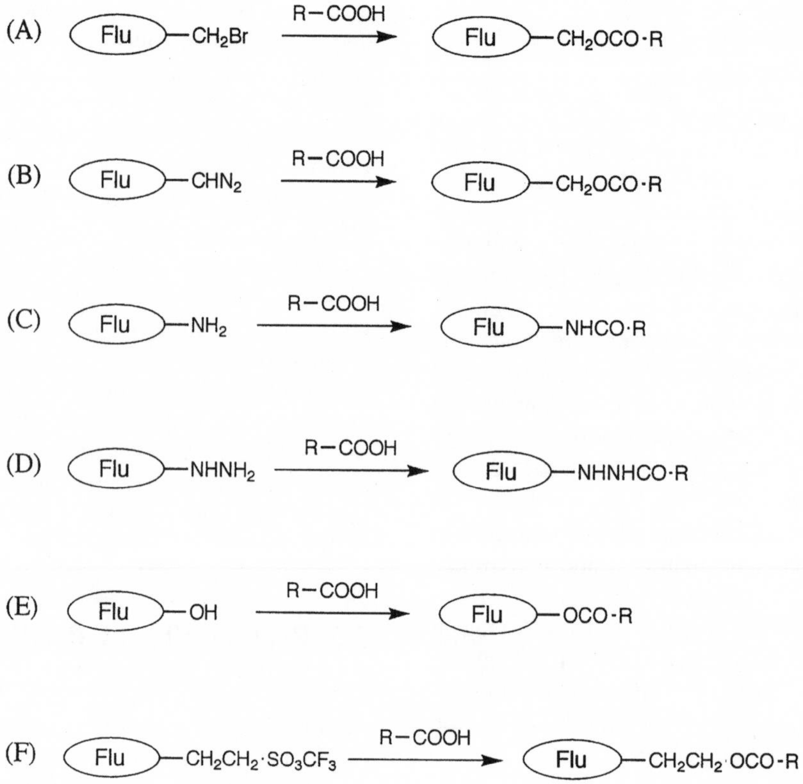

(A) Flu—$CH_2Br$ $\xrightarrow{R-COOH}$ Flu—$CH_2OCO \cdot R$

(B) Flu—$CHN_2$ $\xrightarrow{R-COOH}$ Flu—$CH_2OCO \cdot R$

(C) Flu—$NH_2$ $\xrightarrow{R-COOH}$ Flu—$NHCO \cdot R$

(D) Flu—$NHNH_2$ $\xrightarrow{R-COOH}$ Flu—$NHNHCO \cdot R$

(E) Flu—OH $\xrightarrow{R-COOH}$ Flu—$OCO \cdot R$

(F) Flu—$CH_2CH_2 \cdot SO_3CF_3$ $\xrightarrow{R-COOH}$ Flu—$CH_2CH_2 \cdot OCO \cdot R$

**Fig. 3.26.** Fluorescence labeling reagents for carboxylic acids.

*(1) Bromomethyl Reagents (Fig. 3.27)*

Br-MMC was first used as a fluorescence labeling reagent for carboxylic acids (Fig. 3.27A) [108]. The application of this reagent is reviewed in detail [109]. The reagent reacts with carboxylic acids in acetonitrile or acetone under fairly drastic conditions and the reactions are catalyzed by a crown ether (18-crown-6, dibenzo-18-crown-6) and potassium ion ($K_2CO_3$, $KHCO_3$) [110–113]. In general, the reaction of the acids with Br-MMC is carried out by heating at 60 °C for 30–60 min for monocarboxylic acids and 80 °C for about 60 min for dicarboxylic acids. The MMC esters of fatty acids are determined at 10 pmol levels. Br-MMC have the following disadvantages; the reagent has a limited sensitivity and the fluorescence intensity of the MMC esters of fatty acids are strongly affected by their chain length and the eluent composition [114]. In order to overcome these problems, various bromomethyl reagents are developed. MMC esters are detected by He-Cd LIF laser, and permit the detection of the acids at the fmol levels [115].

Br-DMC [116] and Br-MDC [117] (Fig. 3.27B) are reported as higher sensitive reagents for carboxylic acids than Br-MMC. Br-DMC is applied successfully to the analysis of 5-fluorouracil, an antineoplastic agent, in human serum [118]. Br-MDC is used in a pharmocokinetic study of loxoprofen in human plasma and urine [119].

Br-MAC is reported as a highly sensitive reagent [120]. This high sensitivity is based on the strong fluorescence intensity of 4-methyl-7-hydroxycoumarin in alkaline media. After derivatization of carboxylic acids with Br-MAC and HPLC separation, the eluent is mixed with a 0.1 M borate buffer (pH 11.0). The MAC esters are hydrolyzed to 4-methyl-7-hydroxycoumarin, and measured fluorimetrically.

(A)



(B) Coumarins



| Reagents | $R_1$ | $R_2$ | $R_3$ | $R_4$ | Ex (nm) | Em (nm) |
|---|---|---|---|---|---|---|
| Br-MMC | H | $OCH_3$ | $CH_2Br$ | H | 330 | 390 |
| Br-DMC | $OCH_3$ | $OCH_3$ | $CH_2Br$ | H | 345 | 425 |
| Br-MDC | O—$CH_2$—O | | $CH_2Br$ | H | 355 | 435 |
| Br-MAC | H | $OCOCH_3$ | $CH_2Br$ | H | 370 | 470 |
| Br-AMC | H | $OCH_3$ | H | $COCH_2Br$ | 365 | 415 |
| Br-ADMC | O—$CH_2$—O | | H | $COCH_2Br$ | 388 | 475 |

**Fig. 3.27.** Bromomethyl reagents with a coumarin fluorophore.

Some reagents having a fluorophore other than coumarin are synthesized (Fig. 3.28). 6,7-Dimethoxy-1-methyl-2(1*H*)quinoxalinone (DMEQ) structure is found to be an extremely highly fluorescent compound. Thus, Br-DMEQ is synthesized as a fluorescence labeling reagent for carboxylic acids [121]. The sensitivity of the method with Br-DMEQ is at least 100 times and about 10 times higher than those with Br-MMC and Br-MAC, respectively [the detection limits for fatty acids (S/N = 3): 0.3–1 fmol]. Br-DMEQ is applied to the determination of benzoylecgonine (main matabolite of cocaine) in urine [122], cystein protease inhibitor in mouse serum [123], phenylacetic acid in plasma and urine [124,125], 5-fluorouracil in plasma [126] and gangliosides in plasma [127].

Br-MMEQ gives around a 1.6 times higher sensitivity than Br-DMEQ [128]. Aqueous methanol is found to be suitable as an eluent in RP-HPLC of DMEQ (or MMEQ) esters of fatty acids with gradient elution, because the fluorescence intensity only slightly decreased with an increase in the amount of water in the eluent. The derivatization conditions for Br-DMEQ and Br-MMEQ are almost the same as those for Br-MMC and Br-MDC.

Br-MB having benzoxazin structure is reported to be a more reactive reagent than Br-MMC and Br-MDC [129].

Br-MA derivatizes carboxylic acids in DMF or DMSO in the presence of tetraethylammonium carbonate as a base within 10 min at room temperature [130].

A bromoacetyl group shows a higher reactivity with carboxylic acids than a bromomethyl group. Thus, various bromoacetyl group bearing reagents are developed as highly reactive reagents, e.g. Br-AMC [131], Br-ADMC [132] (Fig. 3.27B), Br-AA [133], Br-AP [134,135], DNS-BAP [136], 9-APB [137] and Br-AMN [138] (Fig. 3.28).

Br-AMC readily reacts with carboxylic acids at room temperature in the presence of $KHCO_3$ and 18-crown-6 or a weak basic anion-exchange

| | $R_1$ | $R_2$ | Ex (nm) | Em (nm) |
|---|---|---|---|---|
| Br-DMEQ | $CH_3O$ | $CH_3O$ | 375 | 450 |
| Br-MMEQ | $O-CH_2-O$ | | 365 | 445 |

Br-MB
(Ex. 345 nm, Em. 400 nm)

| | R | Ex (nm) | Em (nm) |
|---|---|---|---|
| Br-MA | $CH_2Br$ | 362 | 420 |
| Br-AA | $NHCOCH_2Br$ | 358 | 482 |

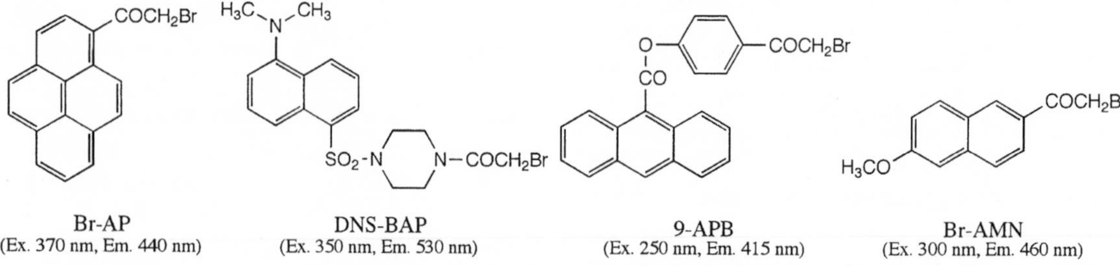


**Fig. 3.28.** Bromomethyl reagents with a fluorophore other than coumarin fluorophore.

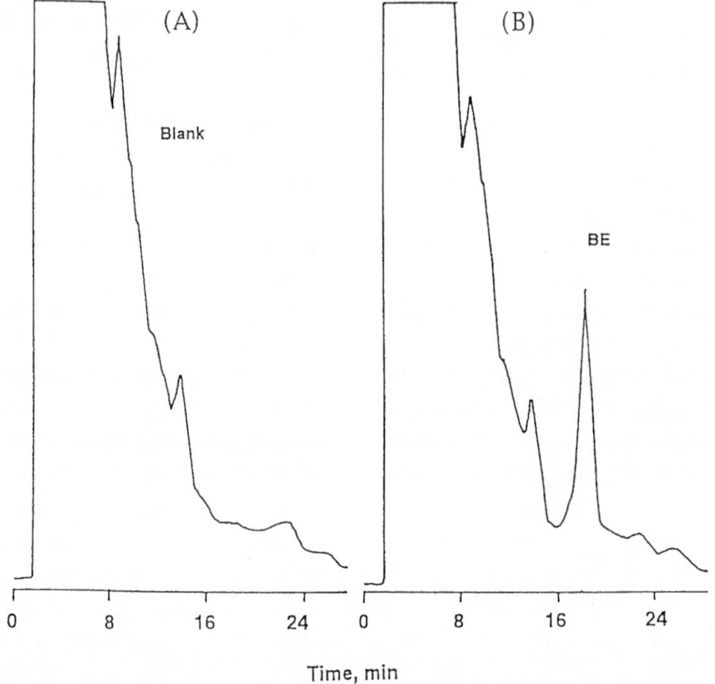


**Fig. 3.29.** Chromatograms of: (A) blank; and (B) benzoylecgonine-spiked urine. Column: TSK-gel ODS-80TM (5 μm; 150 × 4.6 mm I.D.). Mobile phase: methanol-phosphate-borate buffer (0.1 M $KH_2PO_4$-0.05 M $Na_2B_4O_7$ $10H_2O$, pH 7.5) (60:40, v/v) containing 5 mM tetra-*n*-butylammonium bromide. Flow-rate: 0.8 ml/min. [Reproduced from ref. 122, p. 278, Fig. 5].

resin (Duolite A-375) as catalysts. The He-Cd LIF detection method with 9-APB allows the detection of alkylphosphonic acid at fmol levels [139].

9-APB and Br-AA are powerful reagents for phase-transfer-catalyzed fluorescence derivatization. Most bromomethyl and bromoacetyl reagents reacted in organic solvents such as acetonitrile and acetone with carboxylic acids. On the other hand, Br-MA, Br-AA and 9-APB react with carboxylic acids in an aqueous matrix. Br-MA completely derivatized fatty acids within 10 min at room temperature even in aqueous DMF and DMSO. Br-MA and Br-MMC are used in micellar-mediated labeling reactions [111,140]. Fatty acids and valproic acid can be determined in plasma with these methods. Further, 9-APB, a phase transfer reagent, is also advantageously used [137].

Most of this bromomethyl type reagents are easily decomposed on exposure to moisture or light to form fluorescent by-products.

*Br-DMEQ Application: Preparation of a Fluorescent Derivative of Benzoylecgonine, and Preliminary Studies of its Application to the Analysis of Urine [122]*

The pH of a urine sample (1 ml) is adjusted to about 9 with ammonia water. After 5 ml of chloroform-isopropyl alcohol (9:1, v/v) has been added, the mixture is mechanically shaken for 10 min and centrifuged at 1000 g for 10 min. The organic layer (4 ml) is evaporated, the residue is dissolved in 500 μl of acetonitrile. To the resulting solution is added 250 μl of 10 mM 18-crown-6 in acetonitrile and about 5 mg of $K_2CO_3$ and about 2 mg of $Na_2SO_4$, and sonicated for 1 min. After adding 250 μl of 2 mM Br-DMEQ in acetonitrile, the mixture is heated at 80 °C for 20 min. After cooling to room temperature, 20 μl aliquots of the mixture are applied in the HPLC system (Fig. 3.29).

*(2) Diazomethyl Reagents (Figs. 3.26 and 3.30)*

ADAM [141], PDAM [142,143] and DAM-MC [144] have been reported. These diazomethyl reagents offer a number of advantages as derivatization reagents. The reagents are relatively easy to synthesize from readily available compounds and work under mild conditions without a catalysis even in the presence of water.

In particular, ADAM has been widely used as a fluorescence labeling reagent for biologically important carboxylic acids, and permits picomole level determination of the acids by RP-HPLC. ADAM is used for the quantification of fatty acids including 5,8,11,14,17-eicosapentaenoic acid and arachidonic acid, prostaglandins, oxalic acid, amino acids, carnitine, biotin, herbicides, okadaic acid, phosphoinositides, phosphonic acid and polyether antibiotics, including monesin and laslocid [145,146]. The disadvantage of the reagent is the fact that the reagent is very unstable even at −10 °C. In order to overcome this drawback, a simple method for the preparation of ADAM *in situ* is established [147].

PDAM is more stable than ADAM, and the fluorescence intensity of PDAM esters is about five times higher than that of ADAM derivatives (detection limits: 15–30 fmol). However, PDAM requires a longer reaction time than ADAM to reach a constant derivatization yield. PDAM

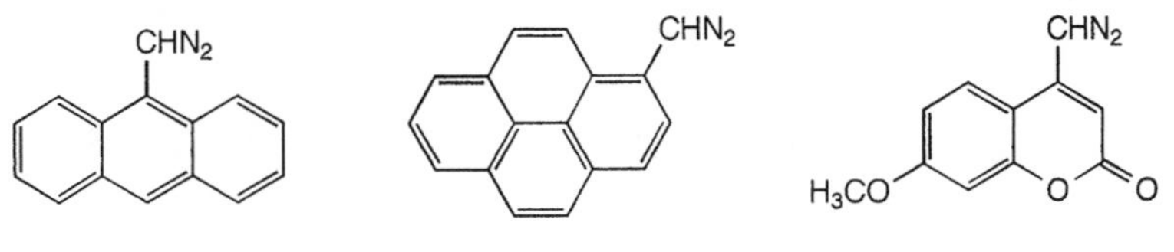

$CHN_2$ $CHN_2$ $CHN_2$ $H_3CO$ O O

ADAM (Ex. 255 nm, Em. 415 nm)

PDAM (Ex. 340 nm, Em. 390 nm)

DAM-MC (Ex. 325 nm, Em. 385 nm)

**Fig. 3.30.** Diazomethyl reagents.

is used for neurotransmitter carboxylic acid quantification [142].

DAM-MC may be applied to He-Cd LIF detection and allows a highly sensitive determination of carboxylic acids, although the reactivity of DAM-MC is less than those of ADAM and PDAM.

*General Derivatization Procedure of Fatty Acids with ADAM*

Fatty acids are dissolved in methanol in the concentration range 0.05–100 μg. To 50 μl of a fatty acid solution are added 50 μl of 0.1% (w/v) methanolic ADAM solution. The resulting mixture is allowed to stand at room temperature for 60 min and a 10 μl aliquot of the mixture is injected directly on the HPLC column.

*(3) Amine Reagents (Figs. 3.26C and 3.31)*

Some amine-type reagents have been reported as labeling reagent for carboxylic acids: 9-AP [148], NEDA [149,150], MBPA [151], DNS-CD [152,153] and DNS-PZ [154].

In general, activation of carboxylic groups is required prior to reaction with the amine reagents. In the 9-AP and NEDA methods, carboxylic acids are chlorinated with thionyl chloride (or oxalyl chloride) in the presence of triethylamine, and the resulting acid chlorides are derivatized with the amine reagents to form the fluorescent amides. The 9-AP method is successfully applied for the determination of serum valproic acid which is widely used in epileptic therapy. The derivatives of 9-AP and NEDA are detected by xenon-chloride excimer laser [148] and Ar ion laser [150], respectively.

MBPA reacts with carboxylic acids in the presence of 2-bromo-1-ethylpyridinium tetrafluoroborate to give the corresponding amides at room temperature. The derivatization reaction of carboxylic acids with MDC proceeds very rapidly in the presence of diethylphosphorocyanide as an effective activation reagent in DMF at room temperature.

*(4) Hydrazine Reagents (Figs. 3.26D and 3.32)*

The development of highly sensitive and selective HPLC reagents to derivatize carboxylic acids in aqueous solution under mild reaction conditions, is important and challenging. A hydrazino group is generally reacted with carboxylic acids in aqueous solution in the presence of a coupling reagent to give the corresponding acid hydrazide. Various coupling reagents [*N*-ethyl-*N*′-(3-dimethylaminopropyl)carbodiimide (EDC), *N*,*N*′-carbonyldiimidazole] make the reaction

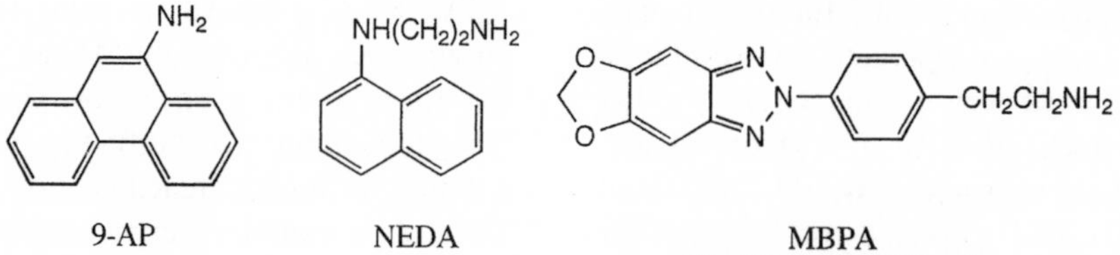

$NH_2$ 9-AP

$NH(CH_2)_2NH_2$ NEDA

$CH_2CH_2NH_2$ MBPA (Ex. 333 nm, Em. 372 nm)

$H_3C$ N $CH_3$ $SO_2$-R

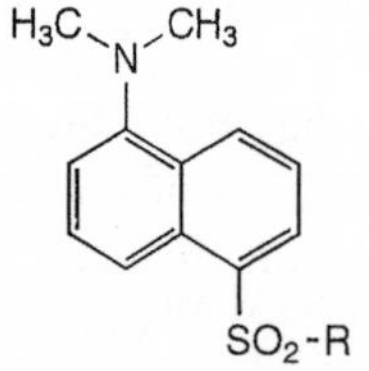

| | R | Ex (nm) | Em (nm) |
|---|---|---|---|
| DNS-CD | $-NH(CH_2)_5NH_2$ | 340 | 520 |
| DNS-PZ | -N NH | 340 | 518 |

**Fig. 3.31.** Amine reagents.

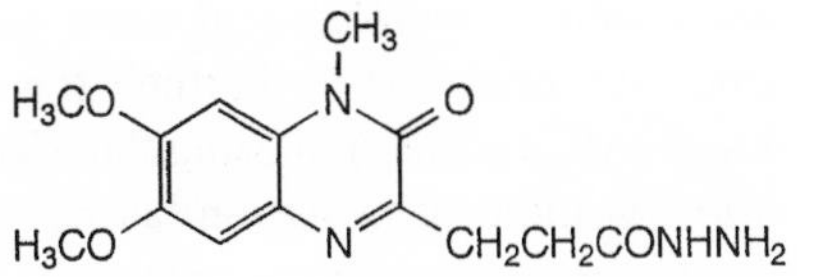

DMEQ-PAH
(Ex. 360 nm, Em. 440 nm)

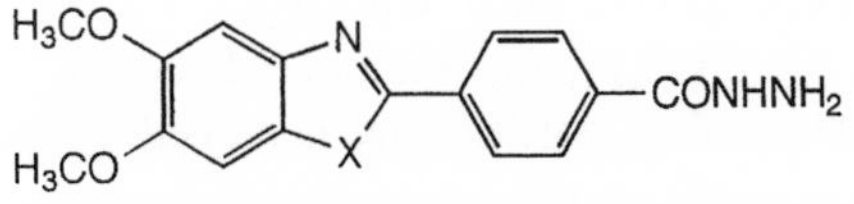

| | X | Ex (nm) | Ex (nm) |
|---|---|---|---|
| DMBI-BH | NH | 360 | 460 |
| BHBT-BH | S | 365 | 447 |

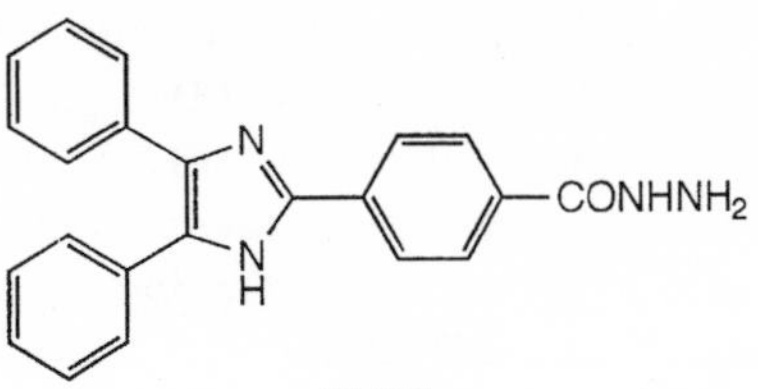

HCPI
(Ex. 360 nm, Em. 440 nm)

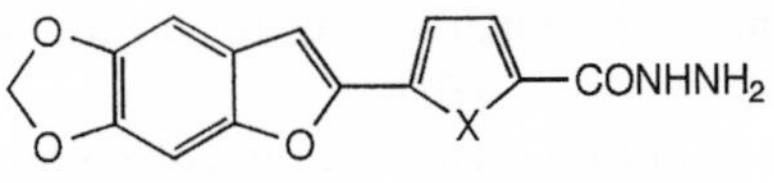

| | X | Ex (nm) | Ex (nm) |
|---|---|---|---|
| FM-BH | NH | 363 | 439 |
| TM-BH | S | 376 | 468 |

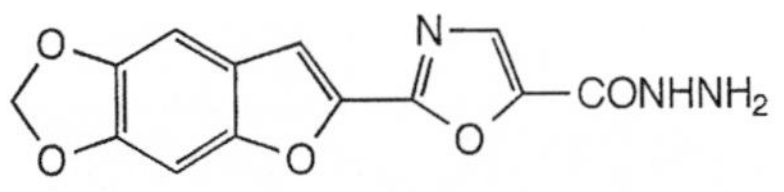

OM-BH
(Ex. 350 nm, Em. 450 nm)

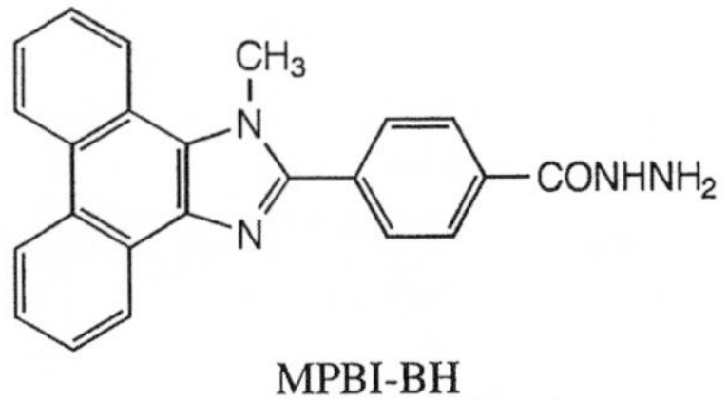

MPBI-BH
(Ex. 325 nm, Em. 460 nm)

**Fig. 3.32.** Hydrazine reagents.

proceed under mild conditions (room temperature, 10–30 min). Some reagents have been particularly developed as this type of reagent: e.g. DMEQ-PAH [156,157], DMBI-BH [157], BHBT-BH [158], HCPI [159], FM-BH [160], TM-BH [160], OM-BH [161] and MPBI-BH [162].

DMEQ-PAH reacts with carboxylic acids even in water in the presence of water soluble carbodiimide (EDC) at room temperature to give highly fluorescent hydrazides. The reagent is very stable and ideal for the derivatization of thermally labile polyunsaturated fatty acids. The detection limits are in the range of 2.5–5 fmol. The reagent is used for the direct determination of phenylacetic acids in plasma and urine [162].

DMBI-BH is developed as a more sensitive reagent for the acid (detection limit: 1–3 fmol).

As described above, some labeling reagents (e.g. Br-MMC and Br-AMN) have been applied to HPLC determination of carboxylic acids with He-Cd LIF detection. However, these reagents are not always ideal for He-Cd LIF detection because of their excitation maxima and intensities. The successful design of derivatization reagents for He-Cd LIF detection relies on the development of an excellent fluorophore with highly intense fluorescence and excitation maximum nearing the 325 nm output wavelength of the He-Cd laser. MPBI-BH has been developed as a highly sensitive labeling reagent for carboxylic acids in HPLC. The resulting fluorescent derivatives are separated by RP-HPLC with aqueous methanol. The fluorescent derivatives display an excitation maximum at 325 nm, which coincides closely with the light emission of the He-Cd laser. Hence HPLC with the reagent is combined with He-Cd LIF detection. Using this system, attomole detection limits (70–200 amol on-column) are achieved.

*(5) Alcohol Reagents (Figs. 3.26E and 3.33)*

Some alcohol reagents have been reported: PTM [161], HMA [162], DNS-AE [165], HMC [166] and 1-PM [167,168]. Activation of the carboxyl group is also required prior to their reaction with alcohol reagents, in a similar way as for the amine reagents.

PTM reacts with carboxylic acids in chloroform in the presence of 1-isopropyl-3-(3-dimethylaminopropyl)carbodiimide at 60 °C for 30 min.

HMA is reacted with carboxylic acids in the presence of 2-bromo-1-ethylpyridinium tetrafluoroborate and diethylphosphorocyanide at 50 °C for 30 min. HMA is also used for complete conversion of benzoic acid to its ester (detection limit: 100 fmol) [168]. Further, the ester is determined by LIF detection using the 350–360 nm UV emission lines of an Ar ion laser [169].

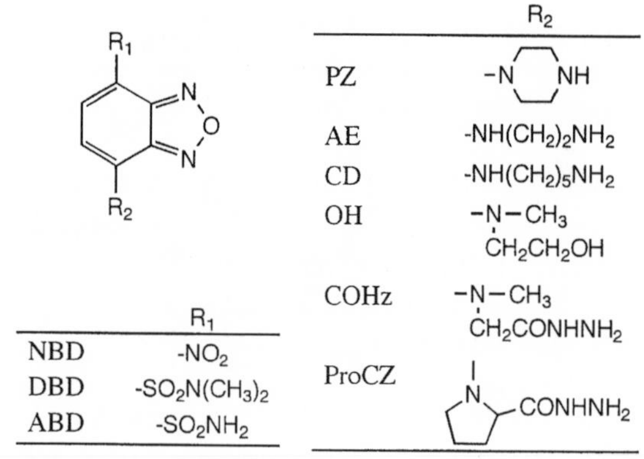
R1
N
O
N
R2
R1
NBD -NO2
DBD -SO2N(CH3)2
ABD -SO2NH2
R2
PZ
-N NH
AE -NH(CH2)2NH2
CD -NH(CH2)5NH2
OH -N−CH3
CH2CH2OH
COHz -N−CH3
CH2CONHNH2
ProCZ
N
CONHNH2

**Fig. 3.35.** Benzofrazan reagents.

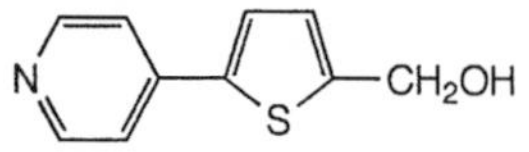
N
S
CH2OH

PTM
(Ex. 300 nm, Em. 360 nm)

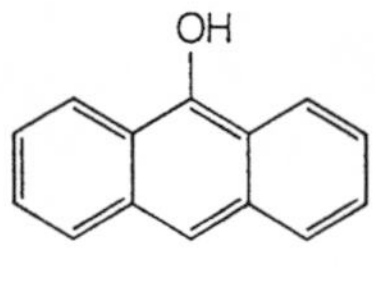
OH

HMA
(Ex. 360 nm, Em. 420 nm)

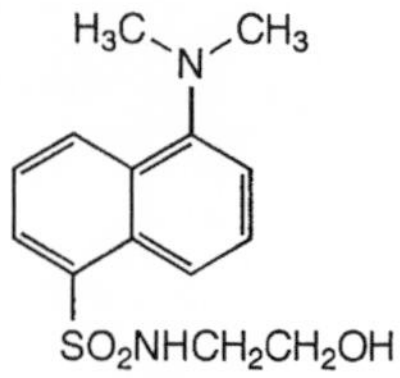
H3C
N
CH3
SO2NHCH2CH2OH

DNS-AE
(Ex. 330 nm, Em. 500 nm)

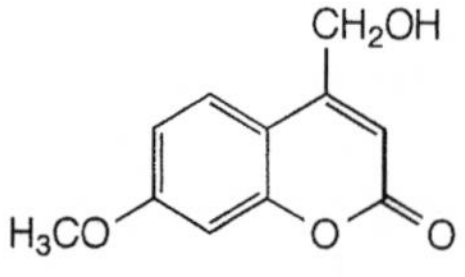
CH2OH
H3CO
O
O

HMC

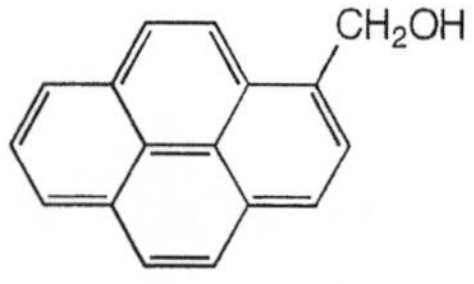
CH2OH

1-PM

**Fig. 3.33.** Alcohol reagents.

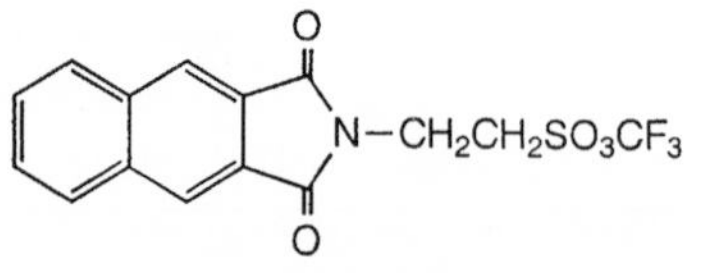
O
N−CH2CH2SO3CF3
O

NE-OTf
(Ex. 259 nm, Em. 394 nm)

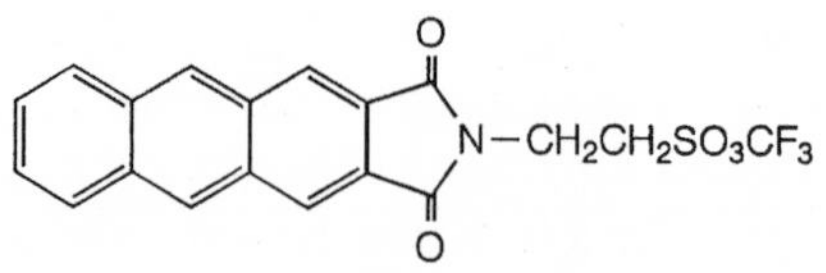
O
N−CH2CH2SO3CF3
O

AE-OTf
(Ex. 298 nm, Em. 456 nm)

**Fig. 3.34.** Sulfonate reagents.

*(6) Sulfonate Reagents (Figs. 3.26F and 3.34)*

NE-OTf and AE-OTf are synthesized as fluorescence derivatization reagents for carboxylic acids [170,171]. Thermolabile carboxylic acids (e.g. $\alpha$-ketocarboxylic acids) can be labeled with these reagents without isomerization or decomposition. NE-OTf is successfully applied to the determination of carboxylic acids in mouse brain.

*(7) Benzofurazans (Fig. 3.35)*

Various reagents having NBD, DBD and ABD structures such as fluorophores and PZ, AE, OH, CD, COHz and ProCZ as active sites against carboxylic acids, are reported [172–184].

*(8) LIF-labeling Reagents for Carboxylic Acids*

Some LIF-labeling reagents for carboxylic acids are proposed (Fig. 3.36) [185,186].

### 3.3.1.1 Fatty Acids

Serum fatty acids can be measured by using the following reagents: Br-MMC [187–189], Br-MAC [190], Br-DMEQ [191,192], Br-MMEQ [193], ADAM [194,195], DNS-PZ [196], 9-APB [197], HCPI [198] and DMEQ-PAH [199]. The high sensitivity of Br-MAC, Br-DMEQ and ADAM are successfully applied for the determination of arachidonic acid metabolites in biological fluids.

*DMEQ-PAH Application: Simple and Highly Sensitive Determination of Free Fatty Acids in Human Serum [199]*

A 5 µl-aliquot of serum is mixed with 50 µl of 4% pyridine in 20 mM HCl ethanolic solution. The mixture is vortex mixed for about 10 s and then 25 µl of 50 mM DMEQ-PAH in DMF and 15 µl of 2 M EDC in water are added. The resulting mixture is warmed at 37 °C for 10 min

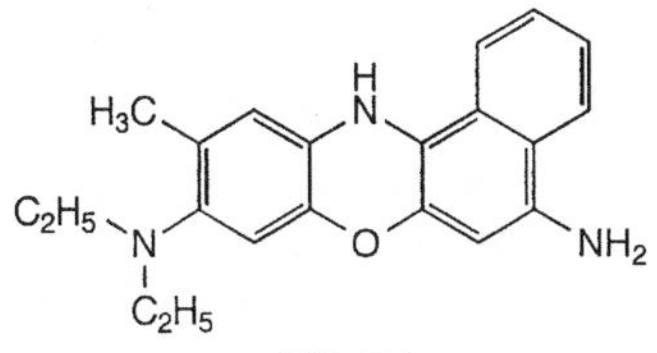

Nile Blue

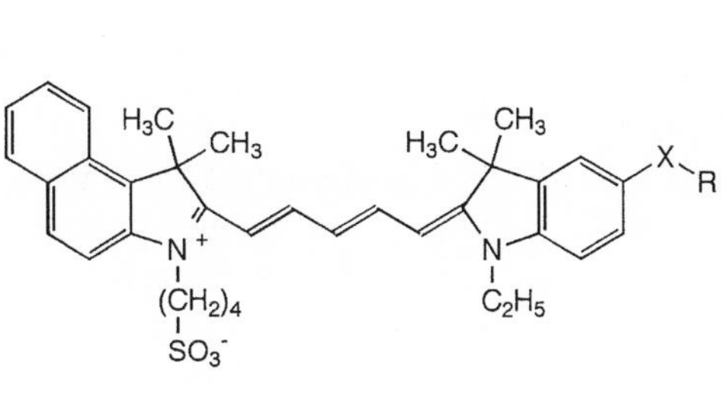

| | X | R |
|---|---|---|
| CY-5.3a-H | NH | $R'CONHNH_2$ |
| CY-5.3a-BrA | NH | $COCH_2Br$ |
| CY-5.3a-IA | NH | $COCH_2I$ |
| CY-5.4a-H | $CH_2NH$ | $R'CONHNH_2$ |
| CY-5.4a-BrA | $CH_2NH$ | $COCH_2Br$ |
| CY-5.4a-IA | $CH_2NH$ | $COCH_2I$ |

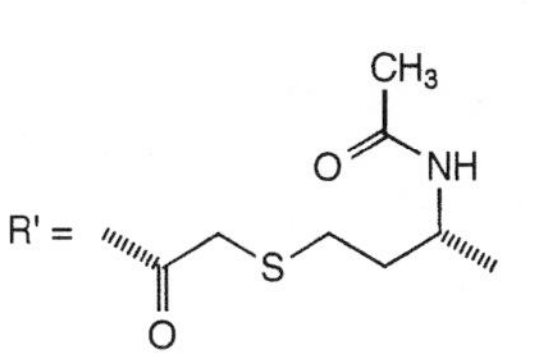

**Fig. 3.36.** LIF-labeling reagents for carboxylic acids.

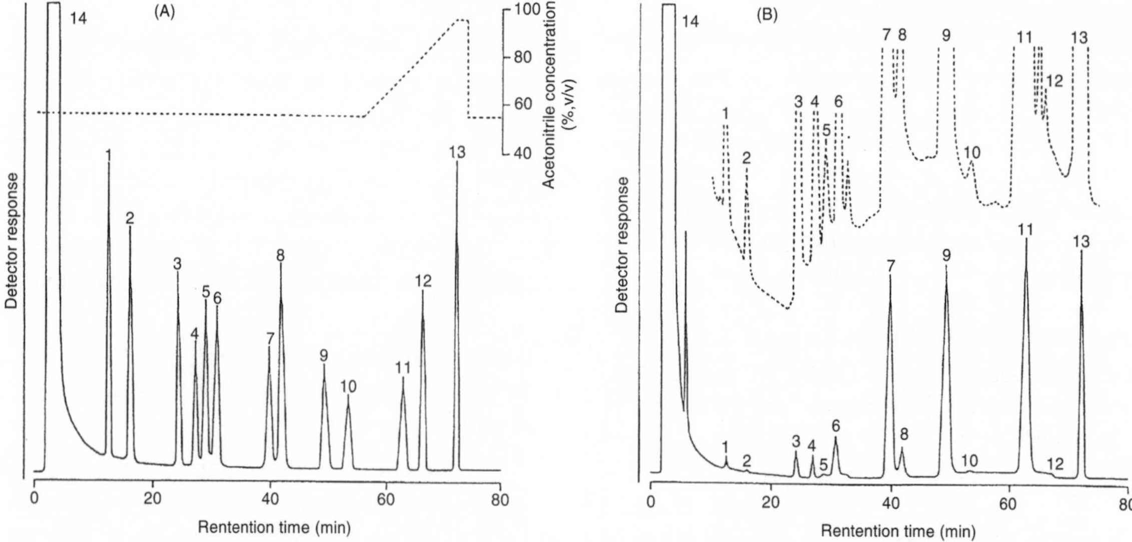


**Fig. 3.37.** Chromatograms obtained with: (A) a standard and (B) a normal human serum sample. Peaks: 1 = C12:0; 2 = C14:1; 3 = C14:0; 4 = C18:3; 5 = C20:5; 6 = C16:1; 7 = C18:2; 8 = C22:6 and C20:4; 9 = C16:0; 10 = C20:3; 11 = C18:1; 12 = C17:0; 13 = C18:0; 14 = DMEQ-PAH. Column: YMC-Pack C8 (10 μm; 250 × 4.6 mm I.D.). Mobile phase: aqueous 55–95% acetonitrile (Fig. 3.37A) (gradient elution). Flow-rate: 1.0 ml/min. [Reproduced from ref. 199, p. 121, Fig. 1 and p. 122, Fig. 2.].

and is centrifuged at 1000 g for about 5 min. The supernatant (10 μl) is subjected to RP-HPLC (Fig. 3.37).

### 3.3.1.2 Prostaglandins

As prostaglandins (Pgs) are generally labile, the derivatizations under mild conditions are preferable. Measurement of PGs in biological samples requires higher sensitivity than that of fatty acids. Several methods have been developed using the following reagents: Br-MAC [200,201], Br-DMEQ [202], ADAM [203], DNS-PZ [204] and DBD-PZ [205].

The reagents, Br-MAC and Br-DMEQ, are usable for the determination of endogenous PGs in human seminal fluids [200,202]. Fig. 3.38 shows chromatograms of the DMEQ derivatives of synthetic and biogenic PGs. The detection limits in the method are at the femtomole level; even higher sensitivity is required for the measurement of serum PGs. 9-APB is used for the HPLC assay of PGs generated in the gastric mucosa of rats and swine and PGs released from human lung tissues [206]. The reagent is also applied for simultaneous quantification of PGs and thromboxane B2 in blood [207].

*Br-DMEQ Application: Determination of Prostaglandins in Human Seminal Fluid [202]*

A 5 μl of seminal fluid sample is diluted with 50 μl of 2.5 mM 16-methyl-PGF1 $\alpha$ (IS) in methanol and allowed to stand for about 2 min to coagulate serum protein. The resulting solution is mixed with 1 ml of diluted HCl (pH 3.0–3.5) and 2.5 ml of ethyl acetate. The mixture is vortexed for about 2 min and centrifuged at 1000 g for 5 min. The organic layer (about 1 ml) is evaporated to dryness in vacuo and the residue is dissolved in 200 μl of acetonitrile. A 100 μl of the solution is placed in a screw-capped 10 ml vial, to which are added 2.5–5.0 mg of $KHCO_3$ and 50 μl each of 10 mM Br-DMEQ and 5.7 mM 18-crown-6 in acetonitrile. The vial is tightly closed and warmed at 50 °C for

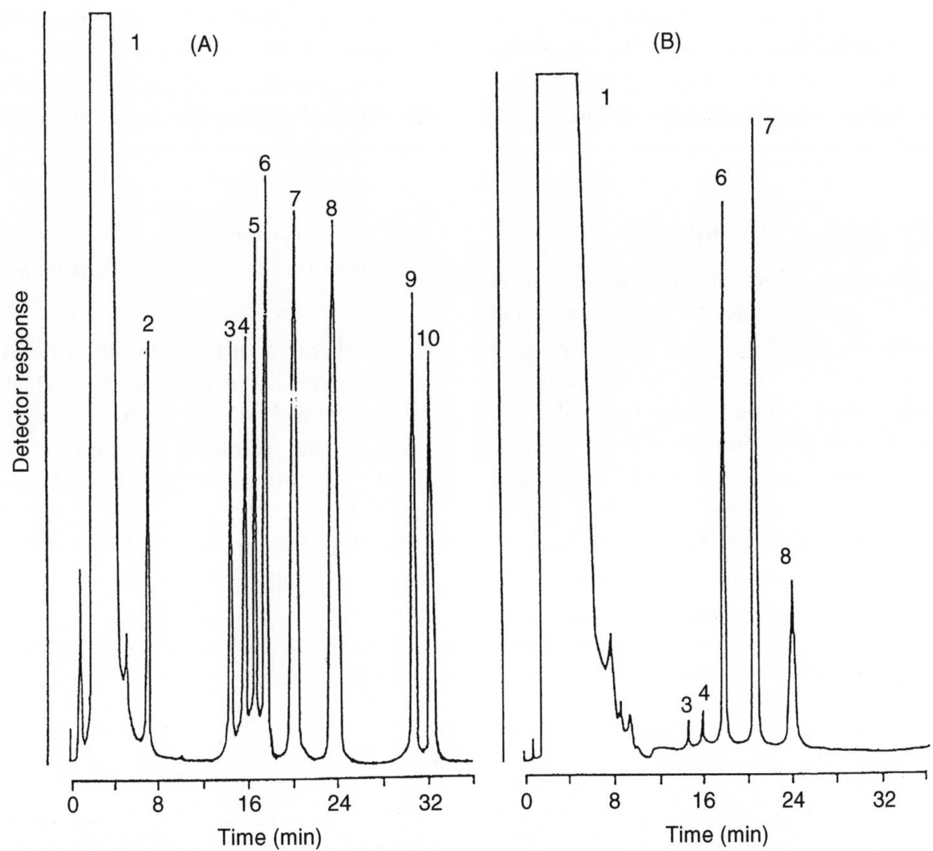


**Fig. 3.38.** Chromatograms obtained with: (A) a standard mixture and (B) human seminal fluid. Peaks: 1 = Br-DMEQ; 2 = 6-keto-PGF1a; 3 = PGF2a; 4 = PGF1a; 5 = PGD2; 6 = PGE2; 7 = PGE1; 8 = 16-methyl-PGF1a; 9 = PGA2 and PGB2; 10 = PGA1 and PGB1. Column: YMC-Pack C8 (10 μm; 250 × 6 mm I.D.). Mobile phase: acetonitrile-methanol-water (0–26 min = 35:10:15; 26–36 min = 35:30:35, v/v). Flow-rate: 2.0 ml/min. [Reproduced from ref. 202, p. 260, Fig. 1A and p. 263, Fig. 4A].

15 min in the dark. After cooling, 10 μl of the resulting mixture are applied onto the RP-HPLC column (Fig. 3.38).

### 3.3.1.3 Steroids

As biological steroids have hydroxyl, carbonyl and/or carboxyl groups, fluorogenic reactions selective for these groups are used in their determinations. In this section, a method using fluorescence labeling reagents for carboxylic acids are described.

Fluorophores reactive toward the carboxylic function, such as Br-MMC [207], Br-AP [208, 134], Br-AA [209], Br-MA [138], Br-AP [210] and Br-AMN [211], have been successfully applied to the fluorimetric HPLC determination of bile acids in a variety of matrices.

A method for the quantitative analysis of unconjugated and conjugated bile acids in serum is reported [212]. After separation of the free, glycine and taurine conjugated fractions by solid-phase extraction, the isolated taurine conjugates are hydrolysed enzymatically using cholyglycine hydrolase. The bile acid fractions are derivatized using Br-AMN and detected fluorimetrically (Ex. 300 nm, Em. 460 nm). The derivatization reaction is performed under mild conditions (10 min at

40 °C) in an aqueous medium in the presence of tetrakis(decyl)ammonium bromide. The separation is achieved using RP-HPLC. The method allows a sensitive detection of each bile acid in serum with a detection limit of about 1–2 pmol.

### 3.3.1.4 Glucuronic Acid Conjugates

A HPLC method based on the derivatization of carboxylic acid moiety in glucuronic acid conjugates with Br-MMC has been reported [213]. The method is not very sensitive and requires dried aprotic solvents and long heating at a high temperature for the derivatization. A highly sensitive RP-HPLC method with DMEQ-PAH has been developed and applied to the measurement of etiocholanorone-3-glucuronide and androsterone-3-glucuronide [214] and estriol 3 and 16-glucuronides in urine [215]. The method is based on the derivatization of the conjugates with DMEQ-PAH in the presence of pyridine and 1-ethyl-3-(3-dimethylaminopropyl)carbodiimide (EDC) in aqueous solution at 0–37 °C.

*DMEQ-PAH Application: HPLC Determination of Urinary Glucuronides [214]*

To 5 μl of a urine sample are added 50 μl of 4% pyridine in ethanol, 50 mM DMEQ-PAH in DMF and 4.0 M EDC in water, successively. The resulting mixture is warmed at 37 °C for 20 min. The derivatives are separated by RP-HPLC and detected fluorimetrically (Ex. 367 nm, Em. 445 nm) (Fig. 3.39). The detection limits (S/N = 3) for the glucuronides are 150–200 fmol in 5 μl of urine.

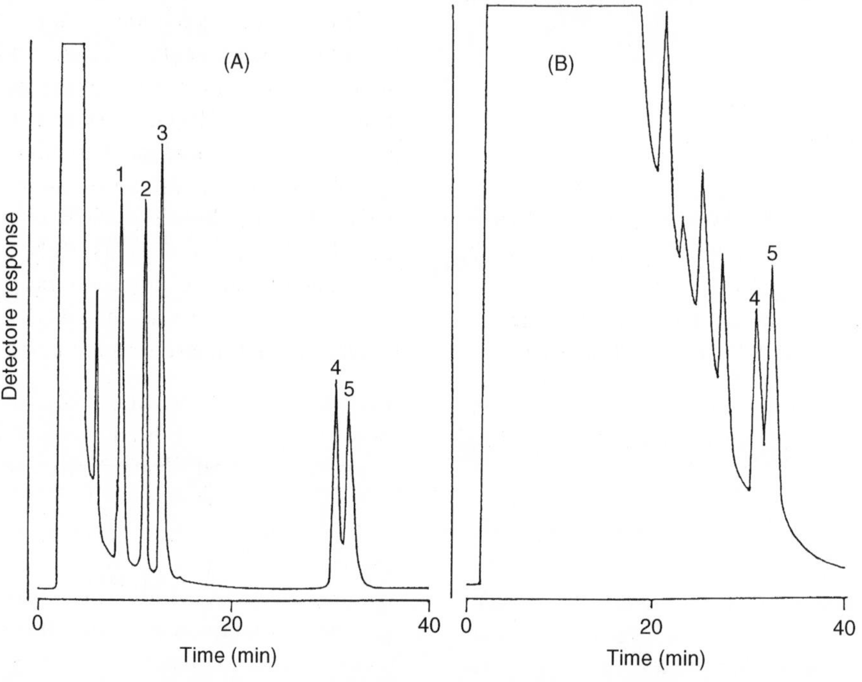


**Fig. 3.39.** Chromatograms obtained with: (A) a standard; and (B) normal human urine. Peaks: 1 = estrone-3-glucuronide; 2 = testosterone-3-glucuronide; 3 = dehydroisoandrosterone-3-glucuronide; 4 = etiocholanorone-3-glucuronide; 5 = androsterone-3-glucuronide. Column: L-column ODS (5 nm; 250 × 4.6 mm I.D.). Mobile phase: methanol-acetonitrile-0.5% triethylamine in water (1:1:2, v/v). Flow-rate: 1.0 ml/min. [Reproduced from ref. 214, p. 174, Fig. 2 and p. 1751, Fig. 4].

### 3.3.2 $\alpha$-Keto Acids

$\alpha$-Keto acids are important intermediates in the biosyntheses of amino acids, carboxylic acids and sugars. The $\alpha$-keto acids can be derivatized in dilute hydrochloric acid with *o*-phenylenediamine, DDB [216–218] and DMB [219] to give fluorescent 2-quinoxalinone compounds (Fig. 3.40) (reaction temperatures and times: 80 °C, 2 h, or 100 °C, 30 min for *o*-phenylenediamine; 100 °C, 2 h for DDB; 100 °C, 45–50 min for DMB). A pre-column RP-HPLC method using DDB have been successfully applied to the simultaneous quantification of these acids in human serum and urine. The method with DDB is highly sensitive; $\alpha$-keto acids in human serum and urine can be determined with the detection limits (S/N = 3) of 10–300 fmol on-column [218]. A typical chromatogram obtained with normal serum and urine is shown in Fig. 3.41. The HPLC method with DMB is more sensitive than that with DDB [219].

*DDB Application: Determination of $\alpha$-keto Acids in Serum and Urine [218]*

Serum sample solution: A 5 µl portion of serum is mixed with 5 µl of 40 µM ketovaleric acid (KV, IS) and 80 µl of 0.8 M $HClO_4$. The mixture is centrifuged at 1000 g for 10 min.

Urine sample solution: To 5 µl of urine are added 5 µl of 40 µM KV and 90 µl of water.

To 50 µl of the serum or urine sample solution placed in a screw-capped 1.5-ml vial, 50 µl of 50 mM DDB in 0.5 M HCl containing 0.7 M $\beta$-mercaptoethanol are added, and the vial is tightly closed and heated at 100 °C for 2.5 h in the dark. The reaction mixture is cooled in ice-water to stop the reaction. A 10 µl aliquot of the resulting solution is subjected to RP-HPLC (Fig. 3.41).

### 3.3.3 Sialic Acids and Dehydroascorbic Acid

Sialic acids [i.e. *N*-acetyl and *N*-glycolylneuramic acids (NANA and NGNA, respectively)] often occur as components of sialoglycoproteins and sialoglycolipids in biological materials. The concentration of NANA has been found to increase in the sera of cancer patients. These acids behave as $\alpha$-keto acids and can be derivatized with DDB or DMB under conditions similar to those for biogenic $\alpha$-keto acids. NANA and NGNA in human serum and urine and animal sera can be measured by an RP (ODS column) HPLC method involving sulfuric acid- or neuraminidase-mediated hydrolysis of samples, followed by fluorescence derivatization [220–223]. DMB is more sensitive than DDB [detection limits (S/N = 3), 25 fmol each on-column]. Chromatograms of the DMB derivatives of NANA and NGNA in normal human, rat and mouse sera are given in Fig. 3.42.

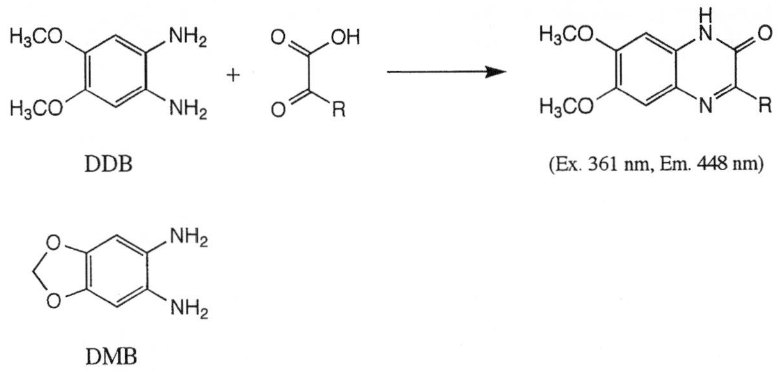

**Fig. 3.40.** Fluorogenic reaction of $\alpha$-keto acids with DDB (or DMB).

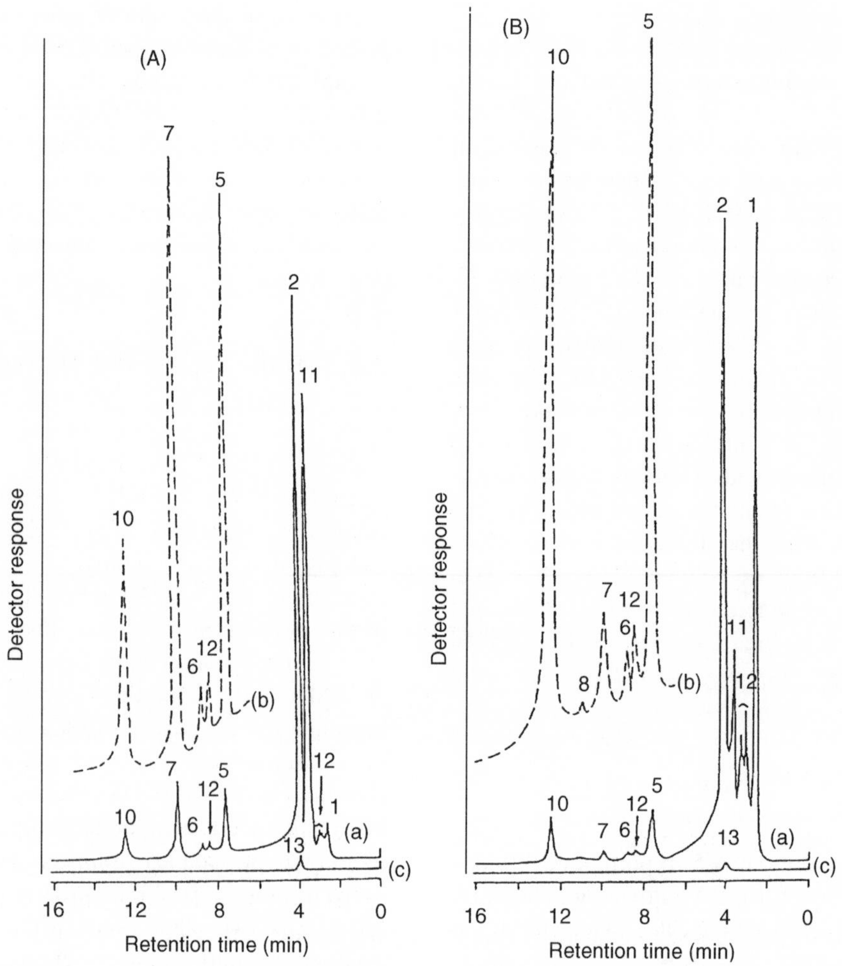


**Fig. 3.41.** Chromatograms of the DDB derivatives in normal human: (A) serum and (B) urine. Peaks: 1 = $\alpha$-ketoglutaric acid; 2 = pyruvic acid; 3 = $p$-hydroxyphenylpyruvic acid; 4 = $\alpha$-ketobutyric acid; 5 = $\alpha$-ketovaleric acid; 6 = $\alpha$-ketoisovaleric acid; 7 = $\alpha$-ketoisocaproic acid; 8 = phenylpyruvic acid; 9 = $\alpha$-ketocaproic acid; 10 = $\alpha$-keto-$\beta$-methylvaleric acid; 11 = glyoxylic acid; 12 = unidentified; 13 = DDB. Column: Radial-Pak C18 (5 μm; 100 × 8 mm I.D.). Mobile phase: methanol-acetonitrile-40 mM phosphate buffer (pH 7.0) (60:40:95, v/v). Flow-rate: 1.2 ml/min. [Reproduced from ref. 218, p. 37, Figs. 3 and 4.].

*Application: Determination of N-acetyl- (NANA) and N-glycolylneuraminic Acids (NGNA) in Human and Animal Sera [221]*

An (5 μl) aliquot of human (or animal) serum placed in a screw-capped 1.5 ml vial is mixed with 200 μl of 25 mM $H_2SO_4$. The vial is tightly closed and heated at 80 °C for 1 h to hydrolyze the serum sample. After cooling, 200 μl of 7.0 mM DMB in water 1.0 M $\beta$-mercaptoethanol 18 mM sodium hydrosulfite is added and the mixture is heated at 60 °C for 2.5 h in the dark to derivatize the resulting NANA and NGNA. The reaction mixture is cooled in ice-water to stop the reaction. A 10 μl-aliquot of the resulting solution is applied to an RP-HPLC (Fig. 3.42).

Dehydroascorbic acid fluoresces when treated with DDB under mild conditions (37 °C, 30 min) in an acetate buffer (pH 4.5) [224]. A highly sensitive quantification of the total ascorbic acid

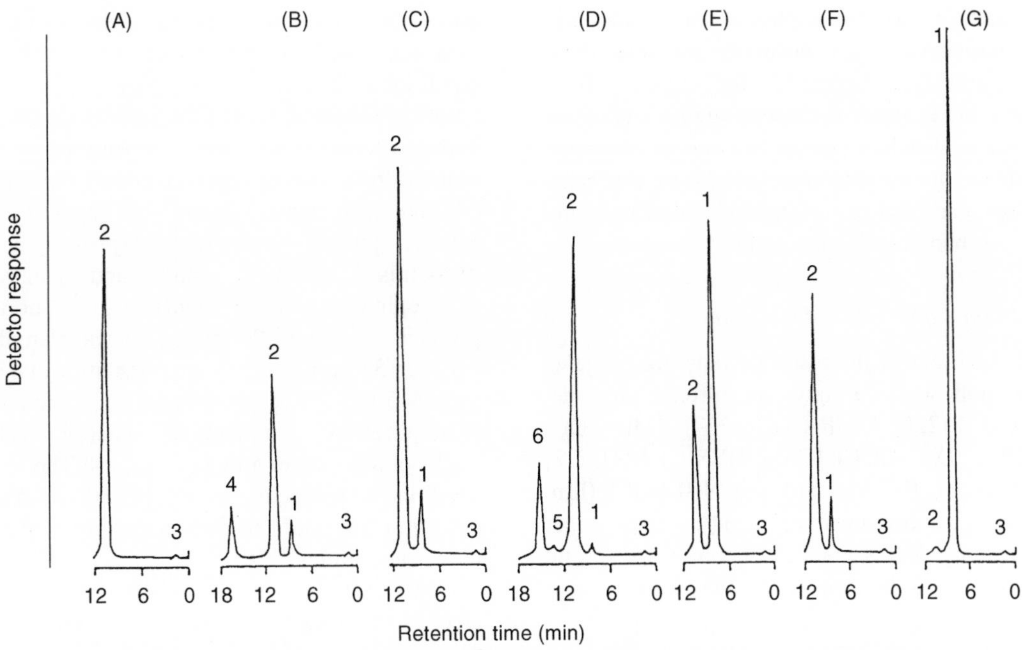


**Fig. 3.42.** Chromatograms obtained with sera from: (A) human; (B) horse; (C) swine; (D) rat; (E) bovine; (F) sheep; and (G) mouse. Peaks: 1 = NGNA; 2 = NANA; 3 = DMB; 4–6 = unknown. Column: Radial-Pak C18 (5 μm; 100 × 8 mm I.D.). Mobile phase: methanol-acetonitrile-water (25:4:91, v/v). Flowrate: 1.2 ml/min. [Reproduced from ref. 221, p. 143, Fig. 4.].

(the sum of dehydroascorbic acid and ascorbic acid) has thus been achieved by means of pre-column HPLC on an RP column, where ascorbic acid is derived to dehydroascorbic acid by iodine oxidation [detection limit (S/N = 3), 46 fmol on-column] [225].

*DDB Application: Determination of Total Ascorbic Acid in Human Serum [225]*

Freshly drawn blood is allowed to stand for 30 min at room temperature (15–25 °C) and centrifuged at 1000 g at 4 °C for 10 min. The serum (2 μl) is diluted with 5 μl of 10.0 mM 6,7-dimethoxy-3-propyl-2(1*H*)-quinoxaline (internal standard) in methanol and 0.1 ml of 0.6 M trichloroacetic acid. The mixture is allowed to stand at 0 °C for approximately 5 min and then centrifuged at 1000 g for 5 min. To 50 μl of the supernatant, 50 μl of 0.3 M sodium carbonate in 0.1 M acetate buffer (pH 4.5) and 10 μl of 1.0 mM iodine in 1.5 mM potassium iodide are successively added, and the mixture is allowed to stand at room temperature for 10–20 s. The excess of iodine is decomposed by the addition of 10 μl of 50 mM sodium thiosulfate solution. To the mixture are added 50 μl of the DDB solution and the resulting solution is warmed at 37 °C for 30 min in the dark. The reaction mixture (10 μl) is applied to RP-HPLC with fluorescence detection (Ex. 371 nm, Em. 458 nm).

The DDB solution is prepared by dissolving DDB monohydrochloride (16.4 mg) in 5.0 ml of 0.1 M sodium thiosulfate (stabilizer of DDB) and diluted to 40 ml with 0.1 M acetate buffer (pH 4.5). The solution is stored in the dark and used within a day.

## 3.4 REAGENTS FOR ALCOHOLS

Bioactive compounds with hydroxyl moiety exist in nature, with steroids and carbohydrates being representative examples of the important compounds. In order to determine hydroxy compounds in various complex mixtures of biological importance at high sensitivity and selectivity, many fluorescence reagents for alcohols have been reported. Most of these reagents are of the fluorescence-labeling type. They can be classified as follows by the function groups of the reaction sites; carbonyl or sulfonyl chlorides, carbonyl azides, carbonylnitriles and chloroformates.

*(1) Carbonyl and Sulfonyl Chloride*

Figure 3.43 shows fluorescence labeling reagents for alcohols with carbonyl or sulfonyl chloride: DBD-COCl [226], OMB-COCl [227], 7MC-COCl [228,229], 3MC-COCl [230], DNS-Cl [231,232] (Em. 520 nm, Ex. 355 nm) and NBI-SOCl (Em. 475 nm, Ex. 365 nm) [233,234]. They react with primary and secondary alcohols in aprotic solvents such as benzene and chloroform to give

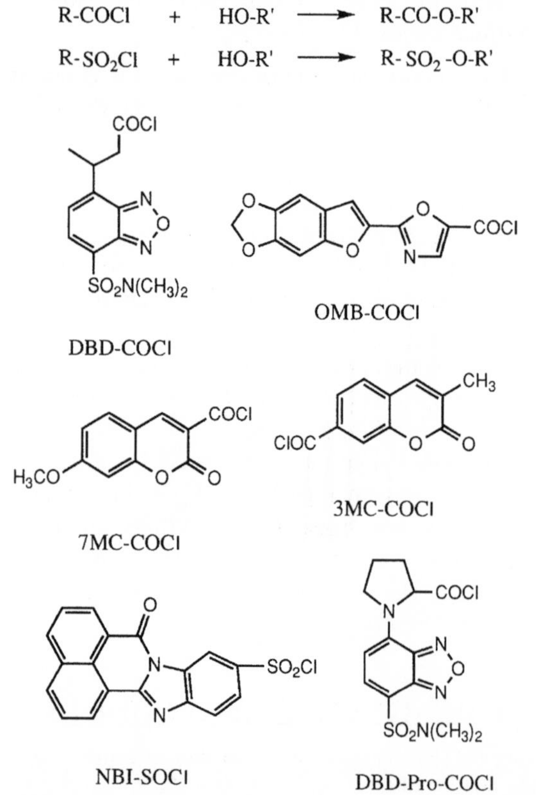

$$\text{R-COCl} + \text{HO-R'} \longrightarrow \text{R-CO-O-R'}$$
$$\text{R-SO}_2\text{Cl} + \text{HO-R'} \longrightarrow \text{R-SO}_2\text{-O-R'}$$

DBD-COCl

OMB-COCl

7MC-COCl

3MC-COCl

NBI-SOCl

DBD-Pro-COCl

**Fig. 3.43.** Fluorescence labeling reagents with carbonyl and sulfonyl chloride and their reaction with alcohols.

corresponding fluorescent esters. Most of them can also react with amines under different reaction conditions.

Of the reagents, DBD-COCl reacts in anhydrous benzene solution at room temperature or 60 °C with androsterone (a representative of hydroxyls), (−)-mandelic and D,L-lactic acid (hydroxy acid), estrone (phenol), benzylamine (primary amine), phenetidine (aromatic amine) and $\alpha$-mercapto-$N$,2-naphthylacetamide (thiol) to give fluorescent products bearing fluorescence wavelengths at between 543 nm and 555 nm (excitation at 437 nm and 445 nm). The base catalyst, quinuclidine, is required to complete the reaction with the nucleophiles except aromatic amines. The reaction solution is subjected to an RP-HPLC and the detection limits of the derivatives are at the fmol range on-column [226].

OMB-COCl reacts with primary and secondary alcohols in anhydrous benzene at 100 °C in the presence of pyridine to produce the corresponding fluorescent esters, which can be separated on an RP column with aqueous acetonitrile. The reagent can also react with tertiary alcohols when 140 °C is selected as the reaction temperature. The detection limits (S/N = 3) are 3–12 fmol for primary and secondary alcohols and 900 fmol for tertiary alcohols [227].

DBD-Pro-COCl [235] is applied to the determination of alcohols with LIF detection. DBD-Pro-COCl derivatives have their excitation and emission maxima at around 450 nm and 560 nm, respectively. The detection limits (S/N = 2) with conventional fluorescence detection using a xenon arc lamp are 10–500 fmol, while those with LIF detection using argon ion at 488 nm are in the range 2–10 fmol.

The hydrolysis products of these types of reagent (carboxylic acids or sulfonic acids), produced during the reaction, often show strong fluorescence. Therefore, the extraction of those compounds from the reaction media prior to the column separation is sometimes very important.

*(2) Carbonyl Azides*

Fluorescence labeling reagents with carbonyl azide change to isocyanate by heating, which can react

R-$CON_3$ + HO-R' ⟶ R-NHCOO-R'

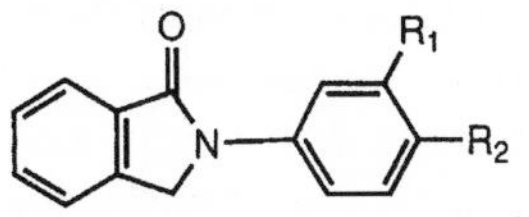

| | $R_1$ | $R_2$ |
|---|---|---|
| *m*-Phibyl-$N_3$ : | $CON_3$ | H |
| *p*-Phibyl-$N_3$ : | H | $CON_3$ |

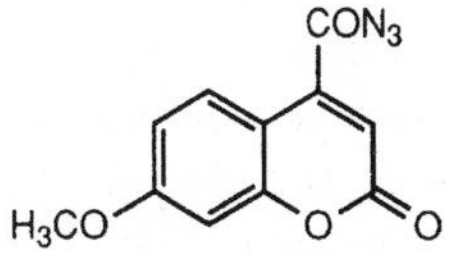

7MC-4-$CON_3$

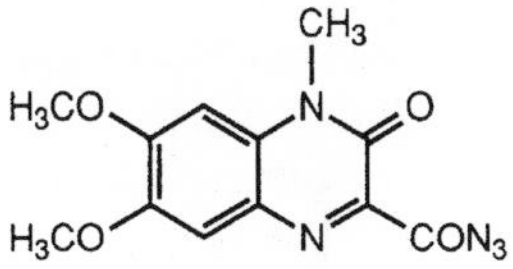

DMEQ-$CON_3$

**Fig. 3.44.** Fluorescence labeling reagents with azide and their reaction with alcohols.

with alcohols to give fluorescent urethanes. *m*-Phibyl-$N_3$ and *p*-Phibyl-$N_3$ [236], 7MC-4-$CON_3$ [237] and DMEQ-$CON_3$ [238] react with primary and secondary alcohols in benzene (Fig. 3.44).

*m*- and *p*-Phibyl-$N_3$, prepared from OPA and corresponding aminobenzoic acids in several steps, react with alcohols at 125 °C for 30 min to produce the fluorescent derivatives, which can be separated on an RP column with aqueous acetonitrile as eluent. The fluorescence intensity is monitored at 440 nm emission and 302 nm excitation for *m*-Phibyl-$N_3$, and at 460 nm emission and 305 nm excitation for *p*-Phibyl-$N_3$, respectively. The detection limit for the alcohols using *m*-Phibyl-$N_3$ is 0.1–0.4 pmol on-column. *m*-Phibyl-$N_3$ is applied to the determination of cholesterol in human serum by RP-HPLC. The extent of conversion of cholesterol into fluorescent derivative is 100%.

*DMEQ- $CON_3$ Application: Determination of 7-dehydrochlesterol (7-DHC) in Human Skin Surface Lipids (Fig. 3.45) [239]*

7-DHC is present in human skin and is converted photochemically into vitamin $D_3$ via previtamin

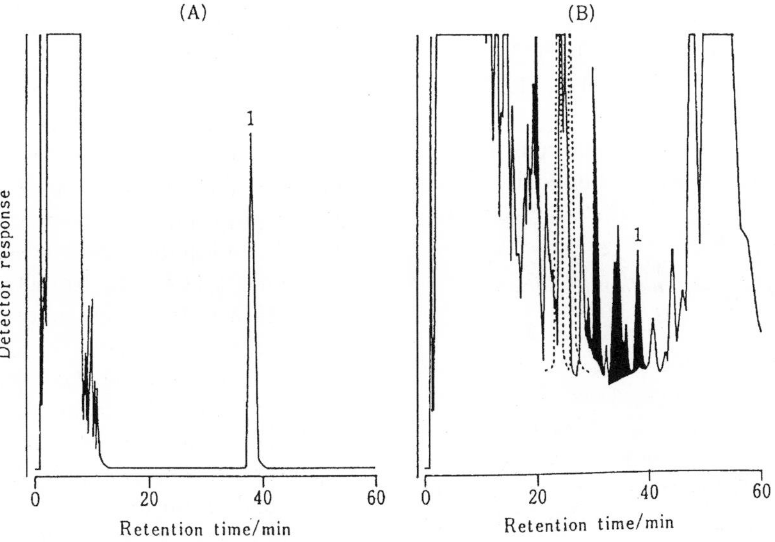


**Fig. 3.45.** Chromatograms obtained with: (A) a standard solution; and (B) an extract of human skin (abdomen) surface using DMEQ-$CON_3$. Peaks and concentrations: 1 = 7-DHC [A, 80.0 pmol (30.7 ng); 13.8 pmol (5.29 ng) $cm^{-2}$ skin surface]; others = A, the reagent blank; B, the reagent blank and endogenous substances in human skin surface. The shaded areas disappeared upon oxidation by 3-chloroperbenzoic acid and two peaks (in broken line) appeared. [Reproduced from ref. 239, p. 672, Fig. 2.].

$D_3$ *in vivo*. 7-DHC is extracted with 30 ml of hexane-ethanol (1:1, v/v; the flow rate, about 1 ml/min) from human skin surface (2.25 $cm^2$) using an extraction device. The extract is evaporated to dryness under a nitrogen stream, and the residue, after dissolving in acetone, is placed in a PTFE screw-capped reaction vial; the solvent is evaporated to dryness in the same manner. The residue is dissolved in 0.2 ml of 2 mM DMEQ- $CON_3$ in benzene. The vial is tightly closed and heated at 120 °C for 60 min in the dark. After cooling, the reaction mixture is diluted with 1.8 ml of methanol, and an aliquot (10 µl) of the resulting solution is injected in the RP-HPLC. The fluorescence is monitored at 440 nm emission and at 360 nm excitation. The detection limits are 18 fmol on-column (S/N = 3).

*(3) Carbonyl Nitriles*

Carbonyl nitriles react with alcohols in aprotic solvents in the presence of the bases as catalysts to give fluorescent derivatives. 9-AN (Em. 460 nm, Ex. 360 nm) reacts selectively with equatorial hydroxyl groups of bile acids and 6$\beta$-hydroxycortisol, follows by separation on a silica gel column and detection at the sub-pmol level [240]. PCN [241], 1-AN (Em. 470 nm, Ex. 370 nm) [242], DNN (Em. 530 nm, Ex. 350 nm) [243] have also been reported (Fig. 3.46). For the simultaneous determination of $\beta$-adrenergic blockers in biological fluids, MBC has been developed [244].

R-COCN + HO-R' ⟶ R-CO-O-R'

|  | $R_1$ | $R_2$ |
|---|---|---|
| 1-AN: | COCN | H |
| 9-AN: | H | COCN |

PCN

DNN

MBC

**Fig. 3.46.** Fluorescence labeling reagents with carbonyl nitrile and their reaction with alcohols.

*1-AN Application: Determination of Serum Bile Acids (Fig. 3.47) [242]*

To a serum sample (100 µl) are added free and glycine- and taurine-conjugated deoxycholate 12-propionates (250 ng each) as internal standard, and the mixture is diluted with 0.17 M phosphate buffer (pH 7.0) (1 ml) and applied to a BondElut cartridge. After successive washing with water (2 ml) and 1.5% ethanol (1 ml), bile acids are eluted with 90% ethanol (2 ml), A 400 µl aliquot of the effluent is evaporated, 1-AN (200 µg) in acetonitrile (100 µl) and 0.16% quinuclidine in acetonitrile (100 µl) are added and the mixture

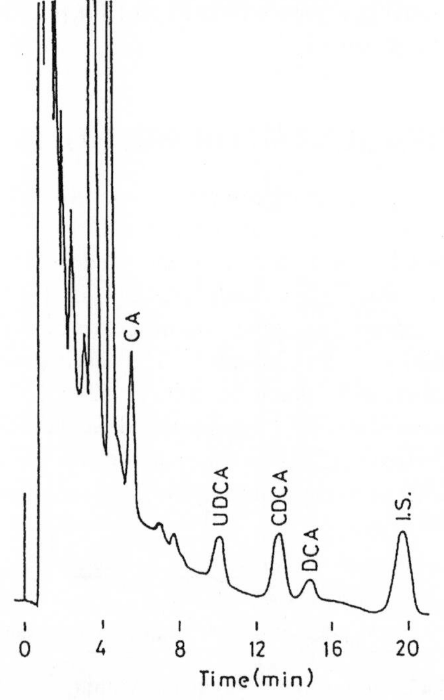


**Fig. 3.47.** Chromatogram of free bile acids in serum of a healthy subject. [Reproduced from ref. 242, p. 298, Fig. 7.].

is heated at 60 °C for 20 min. After addition of methanol (50 µl) to decompose excess of 1-AN, the mixture is evaporated under nitrogen. The residue is dissolved in 90% ethanol (1 ml) and applied to a PHP-LH-20 column (100 mg). Elution is carried out at a flow-rate of 0.2 ml/min. After washing with 90% ethanol (1 ml), free and glycine- and taurine-conjugated bile acids are fractionally separated by stepwise elution with 0.1 M acetic acid in 90% ethanol (5 ml), 0.2 M formic acid in 90% ethanol (5 ml) and 0.3 M acetic acid-potassium acetate (pH 6.3) in 90% ethanol (5 ml). Each fraction is evaporated and the residue is dissolved in methanol (100–200 µl). A 5–10 µl aliquot of the solution is injected into the HPLC system with an RP column.

*(4) Chloroformates*

Chloroformates such as DNS-ECF [245] (Em. 534 nm, Ex. 342 nm) react in alkaline media with alcohols such as cholesterol, estradiol and hydrocortisone to give fluorescent derivatives. These derivatives are separated on RP columns and detected sensitively.

## 3.5 REAGENTS FOR PHENOLS

A few reagents have been reported for the selective fluorescent derivatization of phenols.

DNS-Cl, used for primary and secondary amines, also reacts with phenols in an alkaline medium to give fluorescent derivatives (Em. 470 nm, Ex. 360 nm). The derivatives of the drugs with phenolic moiety [246,247,248], estrogens [249,250], vanillylmandelic acid [251] and thyroxin [252] are separated on reversed [249,252] or normal-phase [250,251] columns and determined with fluorescence and chemiluminescence detection. The detection limits for estrogens are about 1 pmol by fluorescence [249] and 0.1 pmol by oxalate chemiluminescence detection, respectively.

FBIBT [253] has been reported as a fluorescence labeling reagent for estrogens. FBIBT reacts with estrogens very quickly at room temperature in an alkaline medium containing DMSO (Fig. 3.48). The detection limit ($S/N = 2$) of the authentic derivative obtained from the reaction of estron with FBIBT is 10 fmol per injection.

*General Procedure for Fluorescence Labeling Reaction of Estrogens with FBIBT [253]*

To 10 µl of a standard mixture of estrogens in DMSO is added 50 µl of 0.4 mM FBIBT in DMSO, followed by the addition of 10 µl of 20 mM sodium hydroxide. After standing at room temperature for 5 s, 10 µl of 52 mM acetic acid is added; the components are mixed throughly. Then, 10 µl of the final solution is injected in the HPLC with an RP column. The fluorescence is monitored at 560 nm emission and at 540 nm excitation.

PSCL is used for the labeling reagent of estrogens in human serum by HPLC with LIF detection [254]. The fluorescence labeling is carried out using a two-phase system with tetrapentylammonium bromide as a phase transfer catalyst. Labeling reaction is completed in 30 s at room temperature. The derivatives are stable for over a 24 h period. The method is applied to the determination of $\beta$-estradiol at low pg/ml concentrations in 1 ml of human serum. The PSCL derivatives of estrogens demonstrate excellent potential for excitation by LIF using the 325 nm line of an He-Cd laser. Furthermore, a

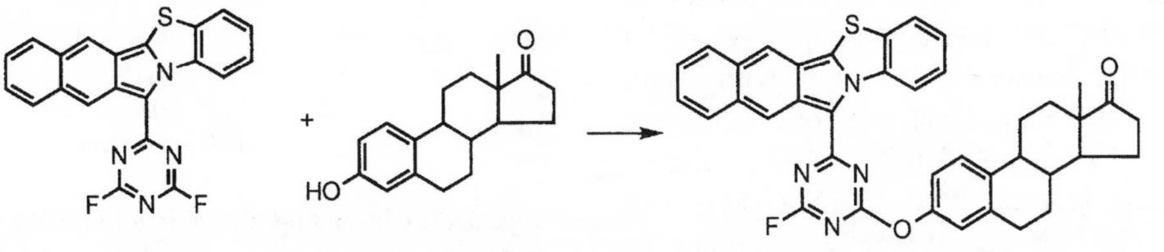

**Fig. 3.48.** Fluorescence labeling reaction of estron with FBIBT.

column-switching technique is used to separate excess reagent and to get a simple chromatogram. The detection limit is 0.2 fmol on-column.

Tyrosine-containing peptides can be converted to highly fluorescent derivatives with DDB [255]. a fluorescence derivatization reagent for aromatic aldehydes. The derivatization method is based on the formylation of the phenolic moiety in the tyrosyl residue by means of the Reimer–Tiemann reaction and the resulting aldehydes converted into the fluorescent derivatives by reaction with DDB (Fig. 3.49). The method is applicable to the HPLC determination of bioactive peptides containing tyrosine residue in their molecules such as angiotensin I, II and III and methionine- and leucine-enkephalins [256,257].

### *DDB Application: HPLC Determination of Leucine-enkephalin-like Peptide in Rat Brain [256]*

A portion (about 0.2 g) of the brain tissue is homogenized at 0–4 °C with 3 ml of 0.1 M hydrochloric acid, then a 50-μl portion of 1.0 nmol/ml [Ala2,Ala3]methionine-enkephalin as an internal standard, is added to the homogenate. The homogenate is transferred into a centrifuge tube, rinsing with 1 ml of 0.1 M hydrochloric acid. The homogenate is deproteinized with 0.5 ml of 2 M perchloric acid. After centrifugation at 800 g for 10 min, the precipitate is suspended with 2 ml of 0.2 M perchloric acid and centrifuged again. The combined supernatant is then neutralized at pH 7–8 with about 2 ml of 1 M sodium hydrogen carbonate solution. The deproteinized solution is applied to a BondElut C18 cartridge. In advance, the cartridge is washed with 3 ml each of water and methanol. After loading the sample solution, 1 ml of water, 2 ml of dichloromethane to remove strongly hydrophobic substances, 1 ml of water, 3 ml of 0.1 M hydrochloric acid, 1 ml of water, 3 ml of 0.1 M borate buffer (pH 8.5) and 1 ml of water are successively passed through the cartridge. Finally, the leucine-enkephalin-rich fraction is obtained by elution with 1 ml of aqueous 90% methanol. After evaporation *in vacuo* at about 30 °C, the residue is dissolved in 200 μl of water. To a 100 μl portion of the sample solution are added 50 μl of chloroform and 25 μl of 3.0 M potassium hydroxide solution. The mixture is warmed at 60 °C for 10 min, then cooled in ice-water for about 1 min. To the mixture, 25 μl of 14 M acetic acid and 25 μl of 4.6 mM DDB are added. The mixture is warmed at 60 °C for 18 min to derivatize the formylated residue, then cooled. A 100 μl portion of the final reaction mixture is injected into the RP-HPLC system. The fluorescence detection is performed at 425 nm for emission with irradiation at 350 nm for excitation. This method permits the determination of the endogenous leucione-enkephalin at concentrations as low as 5.6 pmol/g present in rat brain tissues.

Opioid peptides and their precursors can be derivatized into the corresponding fluorescent compounds by reaction with hydroxylamine, cobalt(II) ion and borate at 100 °C for 1–5 min in a weakly alkaline solution (pH 8–9) [258]. This derivatization method is highly selective for the opioid peptides which have tyrosyl residue at the *N*-terminal, and applied to both pre and post-column HPLC measurement. Methionine and leucine-enkephalins, methionine-enkephalin-Arg-Phe, methionine-enkephalin-Arg-Gly-Leu and kyotorphin in rat brain are determined by HPLC with pre-column fluorescent derivatization [259,260,261,262]. In the method,

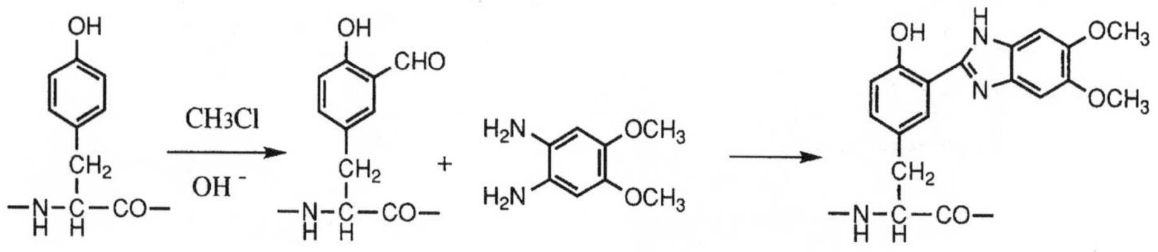

**Fig. 3.49.** Fluorescence derivatization reaction of tyrosine-containing peptides with DDB.

the detection limits (S/N = 3) for the peptides are 0.33–1.2 pmol on-column.

Seven endogenous opioid peptides in rat brain can be simultaneously determined by the post-column HPLC method [263]. Three opioid peptide precursors, proopiomeranocortine and proenkephalins A and B in rat brain are determined after enzymatic reaction by tryptic enzyme [264]. The post-column method can be also applicable to the assay of enkephalin-related enzymes such as enkephalinase A and B [265], and enkephalin-generating enzymes [266]. In these methods, the detection limits (S/N = 3) for the peptides are 0.5–1.5 pmol on-column.

## 3.6 REAGENTS FOR THIOLS

Because thiolic compounds are often found in industrial and urban wastes, and drugs, the development of sensitive and selective methods including fluorescence derivatization reagents for the determination of thiols plays an important role in environmental and pharmaceutical analytical chemistry. Thiols can undergo many types of chemical reactions including disulfide bond formation, oxidation and metal chelation. In the pretreatment of actual samples, therefore, disodium EDTA is added to the sample solution to mask trace metals which catalyze the oxidation of thiols, especially when very small amounts of thiols are to be analyzed. 1,4-dithiothreitol (DTT) is often used to protect thiols against oxidation or to reduce disulfides to thiols. Furthermore, thiols are stablized by lowering the pH to less then 2. *N*-substituted maleimide, alkyl halides, bimanes and other types of reagent have been reported.

*(1) N-substituted Maleimides*

Maleimides react with thiols at the 3,4-double bond. By introduction of a fluorescent substituent on the imide nitrogen, many fluorescent maleimides have been reported: DPM [267], DBPM [268,269], NAM (Em. 465 nm, Ex. 355 nm) [270,271] and BPM (Em. 368 nm, Ex. 318 m) [272] (Fig. 3.50). The reaction proceeds in a few minutes at pH 5–8 and room temperature (or

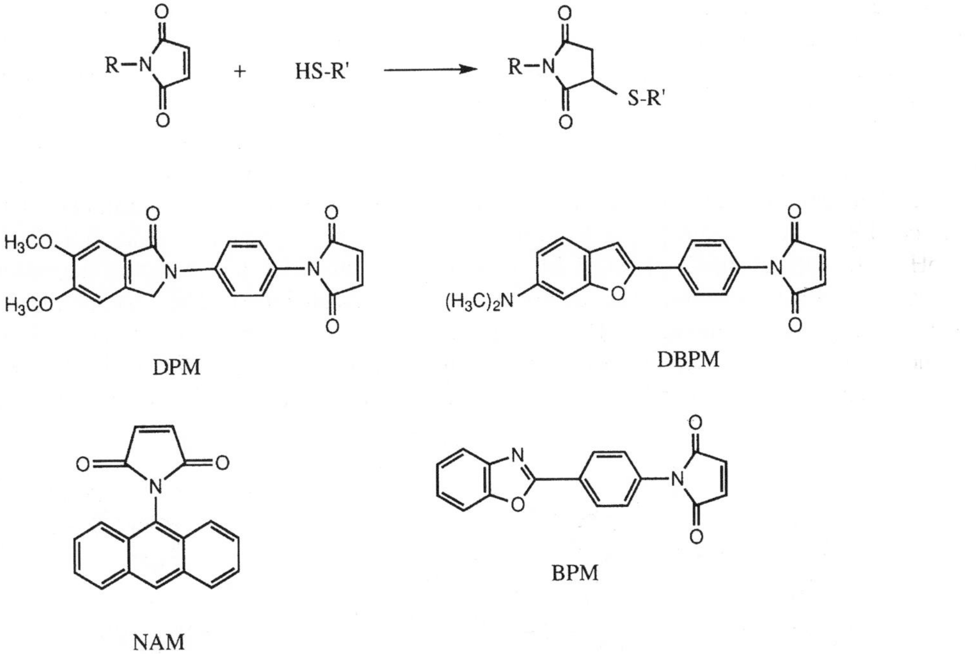

R–N + HS-R' → R–N S-R'

DPM

DBPM

NAM

BPM

**Fig. 3.50.** Fluorescence derivatization reagents with *N*-substituted maleimide and their reaction with thiols.

37 °C). The fluorescent adducts are unstable and are converted into two ring-cleaved fluorophores [273]. Thus, more than one fluorescent derivatives sometimes appear in the chromatogram. The site of scission depends on the steric hindrance of the $N$-substituted moiety of the reagents.

*General Procedure for Fluorescence Labeling Reaction of Thiols with DPM [267]*

To a mixture of 0.01 M phosphate buffer (pH 8.0) (0.7 ml) and acetonitrile (1.2 ml) are added 0.5 mM DPM in acetonitrile (50 μl) and an aqueous solution of thiols (50 μl). The resulting mixture is allowed to stand at room temperature for 15 min. A 20 μl portion of the final reaction mixture is injected into the RP-HPLC system. The fluorescence detection is performed at 422 nm for emission with irradiation at 312 nm for excitation. The detection limits of the method are 10–100 fmol (S/N = 2).

*DBPM Application: Determination of Glutathione (GSH) and Cysteine in Human Serum (Fig. 3.51) [268]*

Human serum is diluted in 20 mM EDTA to adjust 10% (v/v). To 1 ml of this solution is added

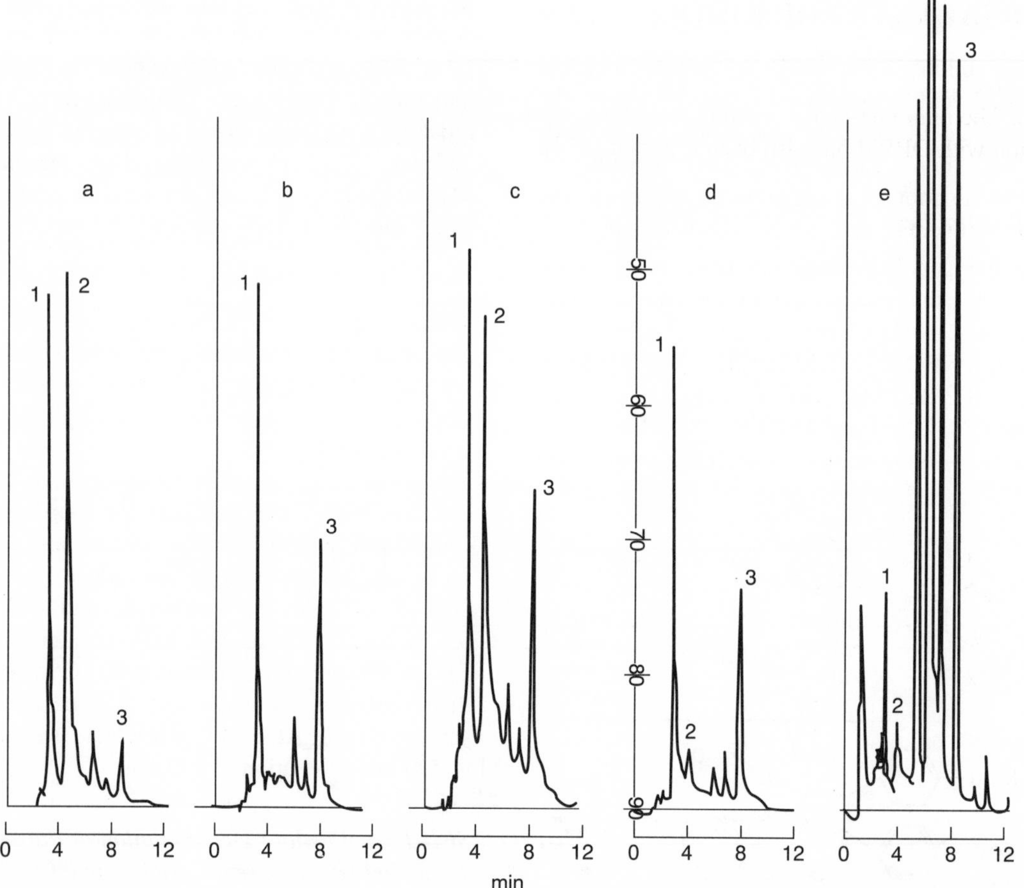


**Fig. 3.51.** Chromatograms of thiols following derivatization with DBPM in rat homogenates, rat plasma and normal human serum: (a) rat liver; (b) rat kidney; (c) rat spleen; (d) rat plasma; (e) normal human serum. Peaks: 1 = ANS; 2 = GSH; 3 = cysteine. [Reproduced from ref. 268, p. 14, Fig. 5.].

0.2 ml of 30% (w/v) metaphosphoric acid. After centrifugation at 2000 g at 4 °C for 20 min, 0.5 ml of the supernatant is mixed with 0.24 ml of 2 M potassium hydroxide. The resulting solution is used as a sample. To 0.1 ml of a sample solution are added 0.4 ml of 0.1 M borate buffer (pH 8.5), 0.3 ml of DBPM-acetonitrile solution (0.24 mM) and 0.2 ml of disodium 6-amino-1,3-naphthalene disulfonate (ANS) solution as an internal standard. The mixture is incubated at 60 °C for 30 min, then cooled for 5 min in tap water. The mixture is centrifuged at 2000 g at 4 °C for 15 min. The supernatant is filtered off through a membrane filter (Millipore, 2 μm) and 20 μl of the filtrate are injected into the HPLC with an RP column. The fluorescence is monitored at 457 nm for emission against at 355 nm for excitation. The detection limits for GSH and cysteine are 17 and 20 fmol, respectively.

The chemiluminescence method based on the reaction with DBPM has also been reported [274].

### *(2) Alkyl Halides*

Many halogenobenzofurazans have been developed for the fluorescence derivatization of the thiol compounds: NBD-Cl (Em. 510 nm, Ex. 425 nm) [275,276], NBD-F [277,278], SBD-Cl [279], SBD-F [280,281,282,283], ABD-F (Em. 510 nm, Ex. 380 nm) [284,285] and DBD-F [286] (Fig. 3.52). NBD-thiols fluoresce less than NBD-amines and

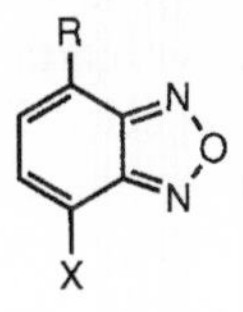

| | X | R |
|---|---|---|
| NBD-Cl : | Cl | $NO_2$ |
| NBD-F : | F | $NO_2$ |
| SBD-Cl : | Cl | $SO_3^- NH_4^+$ |
| SBD-F : | F | $SO_3^- NH_4^+$ |
| ABD-F : | F | $SO_2NH_2$ |
| DBD-F : | F | $SO_2N(CH_3)_2$ |

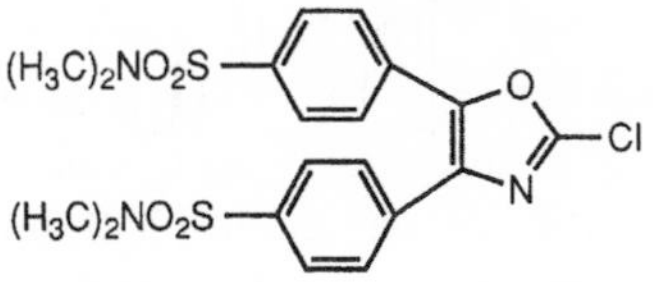

$(H_3C)_2NO_2S$   $(H_3C)_2NO_2S$   Cl

SAOX-Cl

**Fig. 3.52.** Fluorescence derivatization reagents with active halide for thiols.

tend to undergo an SN rearrangement if an excess of amines is present in the medium. Thus, NBD-F and NBD-Cl are not suitable for the detection of thiols.

SBD-F reacts 30–100 times faster than SBD-Cl with thiols [280]. An antihypertensive drug, captopril, is determined by being pre-labeled with SBD-F [287]. SBD-F can also be applicable to the determination of thiols in CE [288]. A more electron-withdrawing group such as $SO_2NH_2$ at the *p*-position of the fluorine moiety affords a more reactive reagent ABD-F than SBD-F. ABD-F is used for differentiating one thiol from the other thiols in a complex matrix of egg white albumin [285].

The reaction rates DBD-F with thiols are several times higher than those with ABD-F; it is suggested that the electron withdrawing effect of the dimethylsulfonamide group is larger than that of the sulfonamide group. DBD-F does not react with amino acids under the derivatization conditions. The detection limits of DBD-F derivatives are in the range 0.13–0.92 pmol on-column.

### *DBD-F Application: HPLC Determination of Thiols in Rat Tissues (Fig. 3.53) [286]*

Tissues (liver, kidney, spleen, testis, lung and pancreas) are exenterated from an etherized rat, washed with physiological saline solution and stored in a freezer at −80 °C prior to analysis. Aliquots (4 ml/g of tissue) of 5% TCA containing 5 mM EDTA are added to the tissues with cooling in ice-water. The samples are homogenized, centrifuged at 3500 rpm for 10 min and to each 10 μl of the supernatant are added 150 μl of DBD-F (27 mM) in acetonitrile and 440 μl of 0.1 M borax solution (pH 8.0) containing 1 mM EDTA. The solution is heated at 50 °C for 10 min, then cooled in ice-water and a 20 μl aliquot is subjected to the RP-HPLC. The eluate is monitored at 520 nm (excitation at 390 nm).

### *SBD-F Application: Determination of Plasma Total Homocysteine and Cysteine by HPLC with Fluorescence Detection [283]*

Whole blood is collected from apparently healthy individuals into a lithiumheparin tube. Samples are

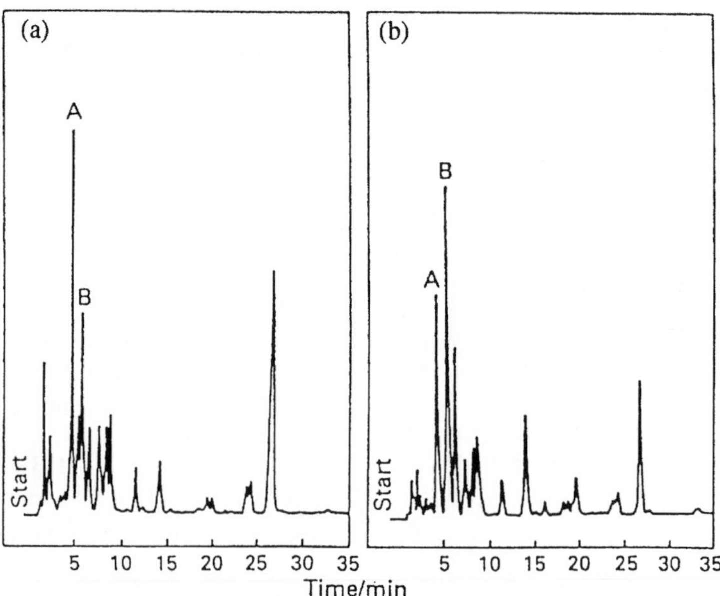


**Fig. 3.53.** Chromatograms obtained from rat tissues: (a) kidney and (b) pancreas using DBD-F. Peaks: A = cysteine; B = GSH. [Reproduced from ref. 286, p. 417, Fig. 6.].

immediately centrifuged at 2000 g for 10 min at room temperature and plasma stored at −20 °C.

To 100 µl of plasma are added 50 µl of 12.5 µM 2-mercaptoethylamine as an internal standard and freshly prepared 10% tributylphosphine. Samples are then incubated at 4 °C for 30 min to reduce all thiols. TCA (150 µl) is then added and the samples are left at room temperature for exactly 10 min. After centrifuging the samples at 4000 g for 10 min, 50 µl of the clear supernatant is taken and added to a mixture containing 125 µl of 0.125 M borate buffer (pH 9.5), 15 µl of 1.25 mM sodium hydroxide and 50 µl of SBD-F (1.0 mg/ml). The mixture is incubated at 60 °C for 60 min. The final solution (20 µl) is injected into the RP-HPLC. The fluorescence is monitored at 515 nm against at 385 nm.

*SBD-F Application: CE Determination of Glutathione in Human Whole Blood with Fluorescence Detection [288]*

To a 1 ml portion of human whole blood is added an equal volume of a chilled solution of 10% (w/v) TCA containing 1.0 mM disodium EDTA. The sample is vortex mixed and centrifuged at 3500 rpm for 5 min at 0 °C. A 40 µl volume of the supernatant is mixed successively with 40 µl of an internal standard solution (50 µg/ml mercaptoethanol in 2.5 M sodium borate buffer, pH 9.5), 100 µl of SBD-F solution (2 mg/ml in the same 2.5 M borate buffer) and 220 µl of the 2.5 M borate buffer. The reaction mixture is then mixed vigorously, heated an a water-bath at 60 °C for 60 min and cooled in an ice-bath. Disposable 4 mm syringe filters (0.45 µm pore size) are used to ensure that clear particle-free derivatized sample solutions are injected into the CE system. Samples are injected on the cathodic side by electromigration, using different loading times and voltages. Relative standard deviations ($n = 10$) are 1.9% for the migration times and 4.4% for the peak areas. The fluorescence is measured by a conventional LC fluorescence detector equipped with a laboratory-made square quartz cell at 510 nm emission and at 380 nm excitation. The detection limits are 0.2–1.0 µg/ml for various kinds of thiols.

An oxazole-based labeling reagent for thiols, SAOX-Cl, has been developed [289]. SAOX-Cl

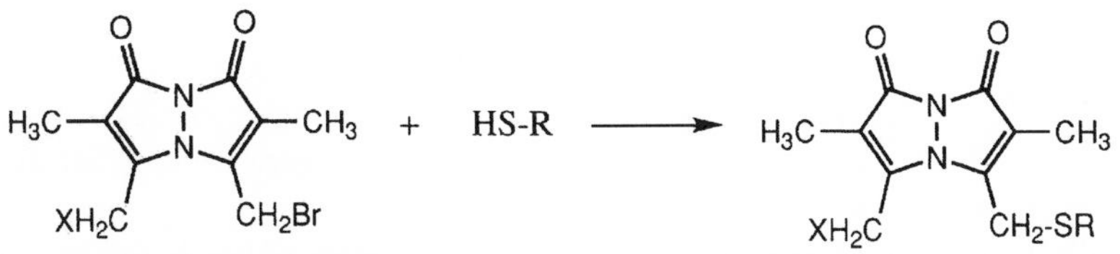

Fig. 3.54. Fluorescence labeling reaction of thiols with bimanes.

reacts with thiols in slightly alkaline media by heating at 60 °C for 90 min. The fluorescence wavelengths of the derivatives are around 425 nm against 330 nm. The fluorescence intensities of the derivatives are constant over a wide pH range (pH 4–12). The derivatives are separated by RP-HPLC and detected fluorimetrically. The detection limits (S/N = 2) are 1.2–1.9 fmol on-column.

*(3) Bimanes*

Monobromotrimethylammoniobimane bromide, monobromobimane and dibromobimane have been reported for pre-column labeling reagents of thiols (Fig. 3.54).

Since the derivatives are not stable at alkaline pH and are sensitive to light, fluorescence labeling reaction is carried out as close as possible to neutral pH in the dark. Excess amounts of the reagents, which are fluorescent, sometimes interfere the sensitive determination. Therefore, they have to be removed by the addition of thiol agarose after the reaction [290]. The detection limits are in the pmol range.

*General Procedure for Fluorescence Labeling Reaction of Thiols with Monobromobimane [291]*

One molar equivalent of DTT is added per equivalent of thiol and the mixture is allowed to stand for 5 min at room temperature. Six equivalents of monobromobimane is then added and the reaction is allowed to proceed for 15 min. The excess of the reagent is scavenged with thiol agarose. Acetic acid (final concentration 3%) is added to the supernantant. The fluorescence wavelengths of the derivatives are at around 480 nm emission and at 398 nm excitation.

*(4) Other Reagents*

DNS-aziridine reacts with thiols as strong nucleophiles with the opening of the aziridine ring and produces stable fluorescent thioethers [292]. Other weak nucleophiles such as phenols, alcohols and amines do not react with the reagent. DNS-aziridine is prepared by reaction of DNS-Cl with equimolar amounts of ethyleneimine in benzene in the presence of triethylamine [292]. The labeling reaction with thiols is carried out by mixing the sample solution in slightly alkaline solution with the same volume of DNS-aziridine in methanol and heating at 60 °C for 30 min [293]. Thiols can be detected at the pmol level with excitation at 338 nm and emission at 540 nm.

OPA and taurine (Em. 420 nm, Ex. 350 nm) [294,295,296] reagents are used for the fluorescent determination of thiols. Nucleophiles (thiols) such as $\beta$-mercaptoethanol are required for the fluorescence derivatization of primary amines using OPA.

CY5.4a-IA, which has a cyanine as a basic chemical structure, is used for the visible diode LIF detection of thiols (Em. 689 nm, Ex. 665 nm) [297] (Fig. 3.55). The derivative of 2-mercaptobenzothiazole is detected sensitively with

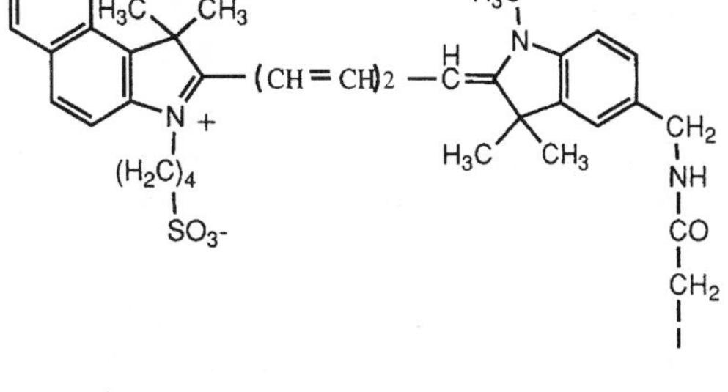

Fig. 3.55. CY5.4a-IA.

a visible diode laser (670 nm) after separation by HPLC. The detection limit is 200 amol on-column (S/N = 3).

## 3.7 REAGENTS FOR ALDEHYDES AND KETONES

Aldehydes and ketones are major constituents of food aromas, and widely distributed all over the environment. Many pharmaceuticals and biological compounds also have aldehydes and ketones in their molecules. Fig. 3.56 shows various kinds of fluorescence reagents for aldehydes and ketones.

The fluorescence labeling reagents with hydrazine as a reaction site have been developed for the determination of aldehydes and ketones. DNS-hydrazine (Em. 505 nm, Ex. 365 nm) [298,299] and NBD-H (Em. 530–570 nm, Ex. 470 nm) [300] are applied

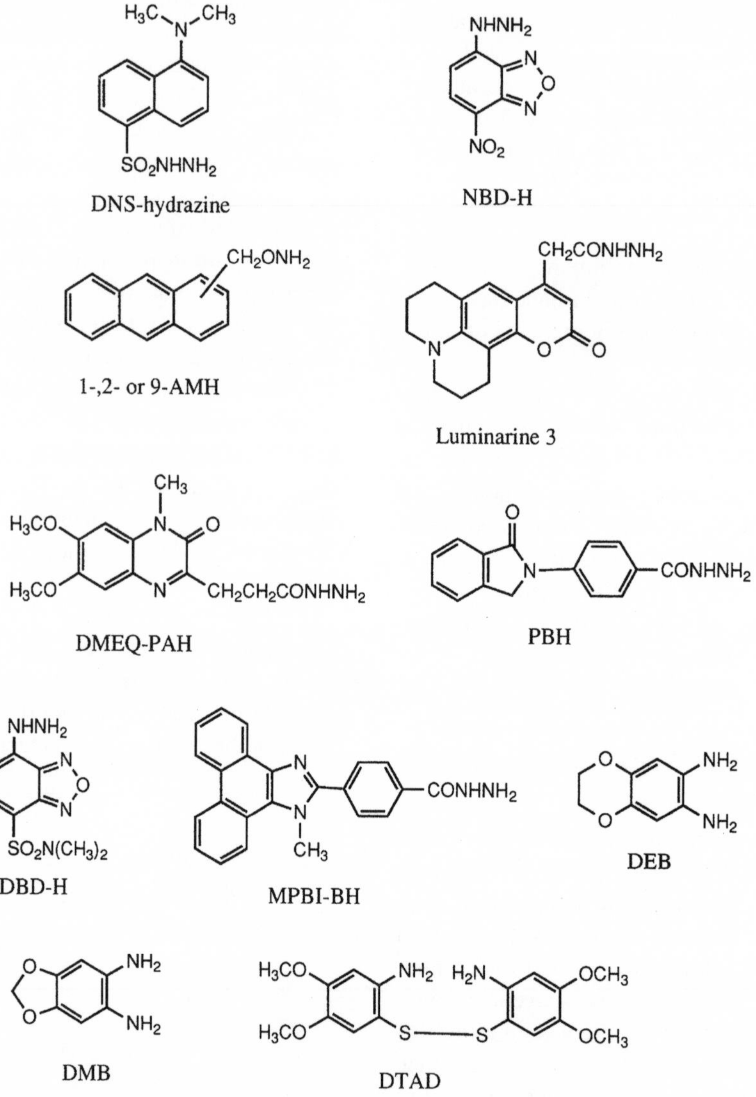

**Fig. 3.56.** Fluorescence reagents for aldehydes and ketones.

for the measurement of cortisol [298] and oligosaccharides [299]. NBD-H reacts with aldehydes and ketones in the presence of 0.1% trifluoroacetic acid for 1 h (aldehydes) or for 5 h (ketones) at room temperature. NBD-H is not fluorescent, however, the reaction products with aldehydes and ketones fluoresce at wavelengths from 548 to 580 nm with excitation from 450 to 470 nm. The detection limits are in the sub-pmol to pmol range.

Bile acids are also derivatized by this reaction following enzymatic conversion of the acids into 3-oxo-bile acids [301]. 3-oxo-bile acids also react with 1-, 2- or 9-AMH in a similar manner to form the corresponding fluorescent oximes [302]. The detection limit is 20 fmol on-column, respectively.

Luminarin 3 [303] reacts with aliphatic, unsaturated and aromatic carbonyls, dicarbonyls and other difunctional carbonyls in about 50 mM sulfuric acid at room temperature for 60 min (aldehydes) or 240 min (ketones) in the dark. After the reaction, the pH of the solution is adjusted to 7.0 by adding sodium hydrogencarbonate and the fluorophore are extracted with dichloromethane. The dichloromethane layer is evaporated under a stream of nitrogen and the resulting residue is dissolved in acetonitrile or ethyl acetate. The derivatives are separated by reversed and normal-phase columns and detected at around 450–470 nm with excitation at around 380–400 nm (detection limits; 60–1950 fmol).

DMEQ-PAH [304] reacts with aliphatic and aromatic aldehydes in the presence of TCA at 60 °C to produce the corresponding fluorescent derivatives, which are separated on RP-HPLC by isocratic elution with aqueous methanol as eluent. The fluorescence detection is performed at 442 nm with excitation at 362 nm. The detection limits are 13–18 fmol for an injection volume.

PBH has been developed for the fluorescent labeling reagent of aliphatic aldehydes. The labeling reaction is carried out at 25 °C for 5 min in the presence of acetic acid [305,306]. The method is applied for the determination of malondialdehyde in rat hepatocyates.

DBD-H [307,308] is developed for the sensitive fluorescence labeling reagent of aldehydes and ketones.

*PBH Application: Measurement of Malondialdehyde Generated from Isolated Rat Hepatocytes Stimulated with Tert-butyl Hydroperoxide [306]*

Rat liver cells, isolated according to the method of Berry and Friend, are suspended in 10 mM HEPES-buffered Hanks' medium (pH 7.0) to give a concentration of $0.5 \times 10^5$ cells/ml. Aliquots (2 ml) of the cell suspension are transferred into 50 ml round-bottomed centrifuge tubes and then *tert*-butyl hydroperoxide is added at various concentrations (0.025–0.5 mM in the suspension medium). After incubation for 2 h at 37 °C, the suspensions are centrifuged for 5 min at 2000 r.p.m. To the supernatant (40 µl) placed in a screw-capped vial, PBH reagent solution (160 µl) is added and the contents are mixed well. After the mixture had been allowed to stand for 60 min at 50 °C, an aliquot (20 µl) is injected into the RP-HPLC system. The fluorescence is detected at 385 nm with excitation wavelength of 320 nm. The detection limit (S/N = 3) is 8 fmol on-column.

*DBD-H Application: Determination of Acetaldehyde in Human Plasma [308]*

To 50 µl of human plasma in a 1.5-ml plastic tube are added 10 µl of aqueous propionaldehyde solution as internal standard, 40 µl of 10 mM DBD-H in acetonitrile, and 20 µl of 1% TFA in acetonitrile. After vortex-mixing for 30 s, the mixture is left for 30 min at room temperature. After centrifugation at 400 g for 10 min (4 °C), the upper layer is passed through a membrane filter (0.45 µm), and a 20 µl aliquot of the filtrate is injected onto the RP-HPLC system. Analytes are detected at an emission wavelength of 560 nm and an excitation wavelength of 445 nm. Submicromolar levels of formaldehyde, acetaldehyde, propionaldehyde and butylaldehyde can be determined. The method is applied to the measurement of acetaldehyde levels in normal human plasma before and 30 min after ingestion of ethanol (Fig. 3.57).

MPBI-BH [309] has been reported as a labeling reagent for aldehydes in HPLC with LIF

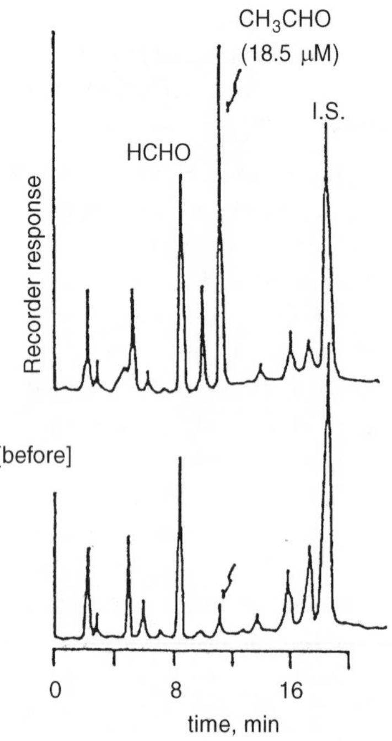


**Fig. 3.57.** Chromatogram of plasma before and 30 min after injection of ethanol using DBD-H. [Reproduced from ref. 308, p. 210, Fig. 4.].

detection. The derivatives have their fluorescence excitation wavelength at 325 nm and emission wavelength at 460 nm, respectively. The detection limits using He-Cd in laser are several hundred amol levels.

Aromatic 1,2-diamino compounds have been found to react selectively with aromatic aldehydes in an acidic medium to give 2-substituted imidazole derivatives. 1,2-diaminonaphthalene, *o*-phenylenediamine, DDB [310] and DEB [311] have been developed as fluorogenic reagents for aromatic aldehydes, which give the corresponding fluorescent imidazole derivatives.

DDB also reacts with arylaliphatic aldehydes such as phenylacetaldehydes and cinnamaldehyde and has been applied to a pre-column fluorescence derivatization of bestatin [(2*S*,3*R*)-3-amino-2-hydroxy-4-phenylbutanoyl-*S*-leucine], an aminopeptidase inhibitor, in human and mouse sera and mouse muscle, which can be derived to phenylacetaldehyde by periodate oxidation [312].

DMB [313] has been developed for the fluorescence derivatization reagent of 1,2-dicarbonyl compounds. Corticosteroids and their related drugs in human serum and urine can be derivatized to fluorescent compounds by reaction with DMB [314,315], after oxidative conversion of the steroids into the corresponding glyoxal compounds by Porter-silver reaction. The derivatives are separated on an RP column, and their detection limits are 0.5–1 pmol per injection volume (S/N = 3).

#### *DMB Application: Determination of 6β-hydroxycortisol and Cortisol in Urine (Fig. 3.58) [316]*

A urine specimen is centrifuged at 1000 g for 30 min to remove insoluble matter. To the resulting urine sample (1 ml), 200 μl of the fludrocortisone (internal standard) and 6 ml of ethyl acetate are added. The mixture is vortex-mixed for about 3 min and the ethyl acetate extract is washed with 1 ml of sodium hydroxide, and subsequently with 2 ml of ultra pure water. After a brief centrifugation at 1000 g for about 3 min, 4.5 ml of the extract are evaporated to dryness under a flow of nitrogen. The residue, dissolved in 100 μl of methanol, is used as a sample solution. A 100 μl volume of the sample solution, placed in a screw-capped vial, is mixed with 20 μl of the copper (II) acetate solution. The mixture is allowed to stand at room temperature for 1 h, and then 100 μl of the DMB solution [1 mg/ml in water containing sodium hydrosulfite (50 mg/ml) and β-mercaptoethanol (50 μl/ml)] is added. After heating at 60 °C for 40 min, the mixture is cooled in ice-water and filtered through a membrane filter (0.45 μm pore size). A 100 μl aliquot of the filtrate is injected to the HPLC system. An excitation wavelength of 350 nm and an emission wavelength of 390 nm are used for fluorescence detection. The lower limits of detection (S/N = 3) for 6β-hydroxycortisol and cortisol are 1.8 pmol (680 pg) and 2.4 pmol (950 pg)/ml urine (0.6 pmol and 0.8 pmol on-column), respectively.

DTAD has been developed for highly sensitive and selective fluorescence derivatization reagents of aromatic aldehydes [317]. DTAD

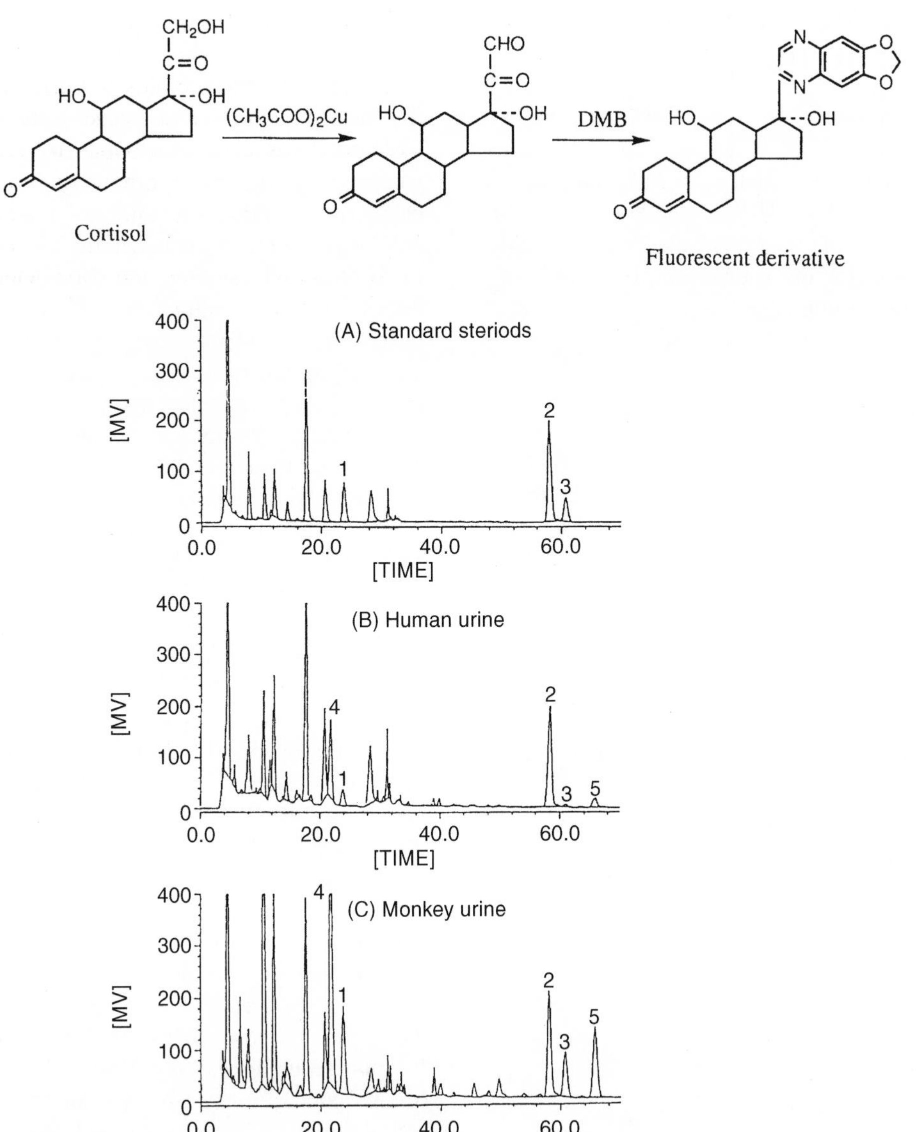


**Fig. 3.58.** Fluorescence derivatization reaction of cortisol with DMB and chromatograms obtained with: (A) standard steroids; (B) human urine and (C) monkey urine. Peaks: 1 = $6\beta$-hydroxycortisol; 2 = IS; 3 = cortisol; 4 = unknown urinary component; 5 = cortisone; others = reagent blank components. [Reproduced from ref. 316, p. 18, Fig. 1.].

reacts selectively with aromatic aldehydes in acidic media in the presence of *tri-n*-butylphosphine, sodium sulfite and disodium hydrogenphosphate to give corresponding 2-substituted thiazole compounds. Arylaliphatic aldehydes (e.g., phenylacetaldehyde and cinnamaldehyde) and aliphatic aldehydes (e.g., formaldehyde, acetaldehyde and *n*-butylaldehyde) do not fluoresce under the procedure. The derivatization reaction is complete within 60 min at 37 °C. The derivatives are separated by RP-HPLC and their detection limits (S/N = 3) are 8–20 fmol on-column.

## 3.8 REAGENTS FOR CARBOHYDRATES

There is interest in the determination of various carbohydrates in both clinical samples and food products which are the components of oligosaccharides. HPLC methods for the determinations of carbohydrates are greatly improved in sensitivity and selectivity by utilizing post or pre-column fluorescence derivatization reagents.

*(1) Postcolumn Derivatization Reagents*

Many fluorogenic reagents such as ethylenediamine [318], 2-ethanolamine [319,320], 2-cyanoacetamide [321,322], arginine [323], arylamidines [324] and *meso*-1,2-diarylethylenediamines [325], have been developed for reducing carbohydrates (Fig. 3.59). These reagents require an alkaline or neutral medium during heating for the derivatization reaction. Of these reagents, 2-cyanoacetamide, BA, *p*-MBA and *p*-MOED have good sensitivity and reactivity towards various reducing carbohydrates, including amino sugars, uronic acids and sialic acids.

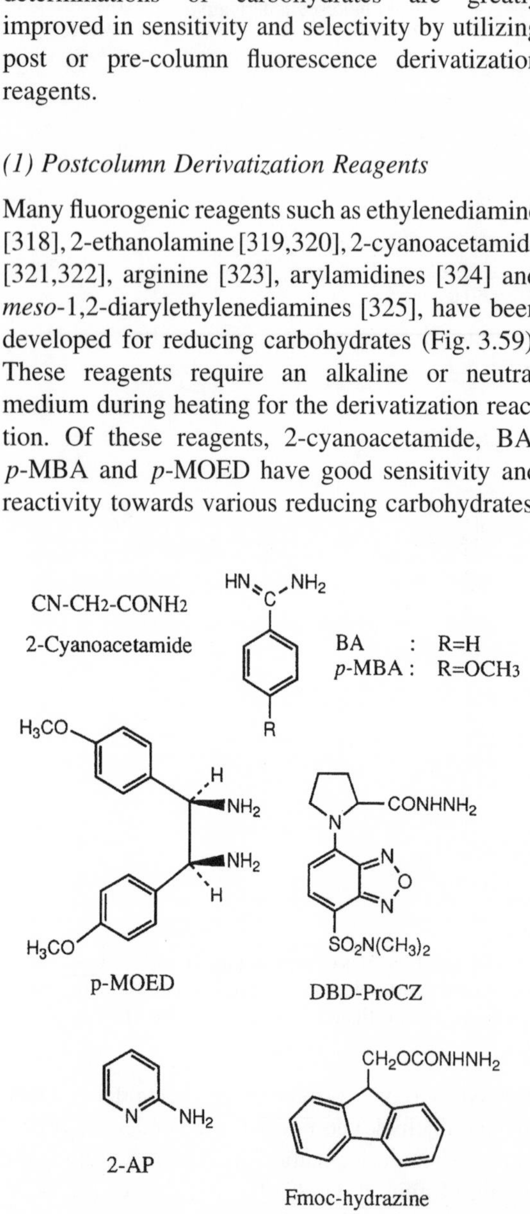

2-cyanoacetamide has been widely used for the determination of various carbohydrates in biological samples. The reagent requires the participation of the hydroxyl group at the 2-position of the carbohydrate in the reaction [326] and cannot give derivatives for 2-deoxy sugars. In the method, the detection limits (S/N = 2) are 0.1–4.2 nmol on-column [321,322].

BA and *p*-MBA react by reducing sugars rapidly: the reaction is completed within 3 min at 100 °C [324]. The reagents also give fluorescent derivatives for 2-deoxy sugars. Reducing carbohydrates in biological matrices such as human serum, mustard plants and wines are determined by the post-column HPLC methods using BA [327,328]. The detection limits (S/N = 3) for the compounds are 2–63 pmol on-column.

On the other hand, *p*-MOED reacts with all kinds of reducing carbohydrates, especially 2-deoxy sugars such as 2-deoxyribose and 2-deoxyglucose [325]. The post-column derivatization HPLC method is applied to the determination of reducing sugars in human urine and serum [329] and of 2-deoxyglucose in rat serum with periodate oxidation [330]. *p*-MOED is also used for the determination of glycated albumin in human serum [331].

Guanidine has been able to react with not only reducing but also non-reducing carbohydrates to give fluorescent derivatives [332]. In the case of post-column HPLC determination for non-reducing carbohydrates using guanidine, 0.1 M potassium tetraborate solution (pH 11.5) containing 50 mM of guanidine hydrochloride and 0.5 mM sodium metaperiodate is used for the reagent solution [333]. The fluorescence is monitored at 433 nm emission and at 314 nm excitation. The detection limits are 5–10 pmol on-column.

*(2) Pre-column Derivatization Reagents*

DBD-ProCZ [334], ANS [335], 2-AP [336,337], DNS-hydrazine [338] and Fmoc-hydrazine [339] are reported for pre-column fluorescence labeling reagents of reducing carbohydrates in HPLC (Fig. 3.59). The derivatization reactions are all



**Fig. 3.59.** Fluorescence reagents for carbohydrates.

based on the condensation of the hydrazine or amine moiety of the reagents with the aldehyde or keto group of the carbohydrates.

### *General Procedure for Fluorescence Labeling Reaction of Reducing Carbohydrate with DBD-ProCZ (Fig. 3.60) [334]*

A 50 μl volume of 0.5 mM reducing carbohydrates (water-acetonitrile = 7:3, v/v), 50 μl of 10 mM DBD-ProCZ in acetonitrile and 100 μl of 0.5% TCA in acetonitrile solution are mixed in a 1.5 ml mini-vial. The vials are capped tightly and heated at 65 °C with a dry heat block for 180 min under protection from light. After derivatization, 200 μl each of water and ethyl acetate are added to the reaction solution (50 μl), and then the excess reagent in the solution is extracted out after centrifugation at 3000 r.p.m. for 2 min. The same extraction procedure is repeated three times. The combined aqueous phase is dried under reduced pressure and dissolved in 200 μl of acetonitrile. An aliquot of the solution is injected into the HPLC. The excitation and emission wavelengths are fixed at 450 nm and 540 nm, respectively. The detection limits are at the low pmol levels.

CBQAC has been utilized as a pre-column derivatization reagent for various amino sugars

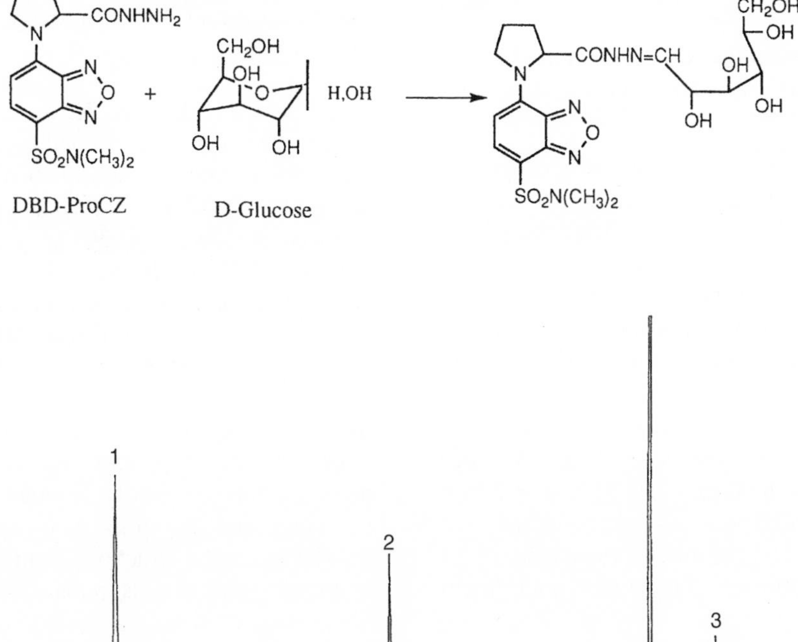


**Fig. 3.60.** Fluorescence labeling reaction of carbohydrate with DBD-ProCZ and chromatograms obtained from hexose and hexosamine by Amide-80 column after elimination of excess reagent. Peaks: 1 = Glu; 2 = Xyl; 3 = GluNAC. [Reproduced from ref. 334, p. 134, Fig. 7.].

[340]. Constituents of various biological mixtures can be converted to highly fluorescent isoindole derivatives (Ex. 456 nm, Em. 552 nm), separated by CE and determined at amol levels by a LIF (He-Cd laser, 442-nm line) detector. The method has been applied to the analysis of monosaccharides and acid-hydrolyzed polysaccharides. Carbohydrate moieties derived from a glycoprotein are also determined.

## 3.9 REAGENTS FOR DIENES

Cookson-type reagents such as P-TAD [341], DMEQ-TAD [342,343] and A-TAD [344] are introduced for the sensitive determination of vitamin D3 and their related conjugates (Fig. 3.61).

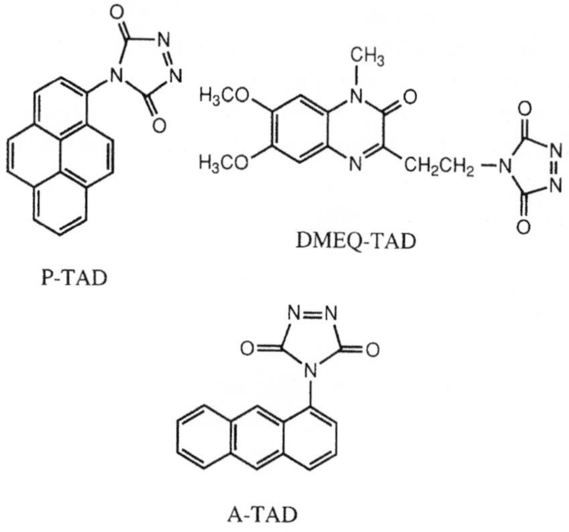

P-TAD

DMEQ-TAD

A-TAD

**Fig. 3.61.** Cookson-type fluorescence labeling reagents for dienes.

*DMEQ-TAD Application: Determination of 25-hydroxyvitamin D3 in Human Plasma (Fig. 3.62) [343]*

To 0.2 ml of human plasma is added a radioactive standard (25-hydroxy[*H3*]vitamin-D3) to monitor the recovery of the metabolite. After the plasma had stood for 30 min in the dark, dichloromethane-methanol (1:2, 1.6 ml) is added, and the whole is vigorously vortexed for 1 min and then centrifuged (3000 rpm) at 20 °C for 10 min. The precipitate is discarded and 0.2 M KOH (0.5 ml) and then dichloromethane (0.5 ml) are added to the supernatant. The mixture is vortexed for 10 s and centrifuged (2000 rpm) at 20 °C for 5 min. The lower organic phase is taken, mixed with methanol-water (1:1, 2 ml), vortexed for 10 s, centrifuged (2000 rpm) for 10 min, and finally evaporated to dryness *in vacuo*. A Sep Pak classic silica cartridge is conditioned successively with ethylacetate-hexane (60:40, 10 ml) and ethyl acetate-hexane (7:93, 10 ml), and then the plasma extracts, dissolved in 300 µl of ethyl acetate-hexane (7:93), are loaded. The column is eluted with 30 ml of ethyl acetate-hexane (7:93) (non-polar lipid fraction to be discarded), 40 ml of ethyl acetate-hexane (15:85) (monohydroxyvitamin D fraction), and 30 ml of ethyl acetate-hexane (60:40) (dihydroxyvitamin D fraction).

For the assay of 25-hydroxyvitamin D3, the monohydroxyvitamin D fraction is evaporated *in vacuo* and the residue is applied to silica gel HPLC. The fraction (1 ml) is placed in a micro-V-vial and is pumped up to complete

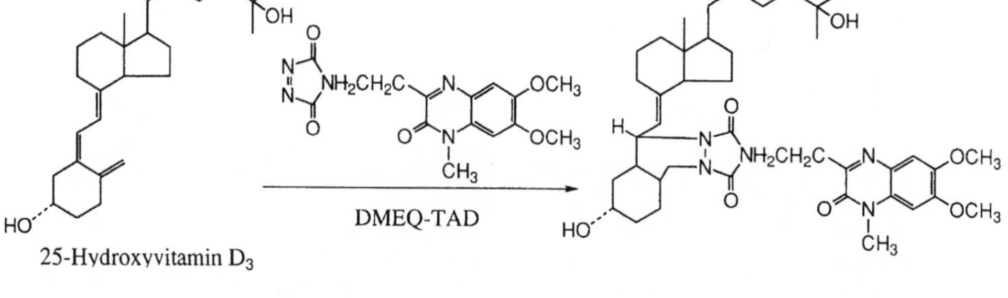

25-Hydroxyvitamin $D_3$

DMEQ-TAD

**Fig. 3.62.** Fluorescence labeling reaction of 25-hydroxyvitamin $D_3$ with DMEQ-TAD.

dryness. A solution of DMEQ-TAD (2.6 μg) in dichloromethane (100 μl) is added to the residue in the vial and the mixture is stirred at room temperature. After 15 min, the same amount of DMEQ-TAD solution is added again and the whole is stirred further for 15 min. Water (20 μl) is added, and after 30 min the mixture is evaporated *in vacuo*. The residue is dissolved in 200 μl of methanol and a portion (10 μl) is injected to the RP-HPLC with gradient elution. The fluorescence detection is performed at 440 nm with excitation wavelength at 370 nm. 24*R*,25-dihydroxyvitamin D3 is also determined with some modification of sample pretreatment.

*A-TAD Application: Determination of 7-DHC in Human Skin [344]*

Human skin (11.34 $cm^2$) is extracted with hexane-ethanol (1:1, v/v) and after concentration of the solvent to 0.5 ml, the sample is filtered through a membrane filter (0.5 μm). The filtrate is evaporated under a stream of nitrogen and the residue obtained is dissolved in ethyl acetate (0.2 ml). The ethyl acetate solution of the reagent (about 65 μg) is added to the above sample and kept at 4 °C for 10 min. An aliquot of the final solution is subjected to HPLC.

## 3.10 REAGENTS FOR NUCLEIC ACID RELATED COMPOUNDS

Nucleic acids have complex chemical structures being formed of sugars (pentoses), phosphoric acid, and nucleobases (purines and pyrimidines). The derivatization reagents and methods for the determination of nucleobases, nucleosides and nucleotides in physiological fluids by HPLC have been developed for biochemical and biomedical studies.

*(1) Fluorescence Labeling of Phosphoric Acid Moiety of Nucleotides*

Fluorescein isothiocyanate is used for the fluorescence labeling of 5′-phosphate group of four major deoxynucleotide-5′-monophosphates [345]. 5′-Phosphate is first labeled with ethylenediamine via a phosphorimidazolide intermidiate in methanol with *N*,*N*-dicyclohexylcarbodiimide (DCC). The resulting ethylenediaminephosphoramidate products are reacted in turn with fluorescein isothiocyanate at room temperature for 18 h. The fluorescence detection is performed at 520 nm with excitation wavelength at 480 nm.

DNS-EDA has been reported as a pre-column fluorescence labeling reagent for mono, di and trinucleotides, and 2′-deoxynucleoside-5′-monophosphate [346,347,348]. The method is based on the reaction of DNS-EDA with the phosphoric acid moiety of nucleotide. The fluorescent DNS-EDA derivatives of nucleotides are separated by RP-HPLC. The detection limits for the nucleotides are between 5 and 20 pmol on-column. This sensitivity permits the determination of RNA of more than 40 ng by one analysis.

*DNS-EDA Application: Determination of Ribonucleoside 5′-monophosphate Produced by Nuclease P1 Digestion of Yeast RNA (Fig. 3.63) [346]*

Yeast RNA in 10 μl of distilled water is heated in a boiling bath for 5 min, then rapidly cooled on ice. To this solution is added 10 μl of nuclease *P1* (Yamasa, EC 3.1.30.1.) solution prepared by dissolving 2 units of nuclease *P1* in 1 ml of sodium acetate buffer (40 mM, pH 5.3) containing 0.2 mM $ZnCl_2$. After incubation at 50 °C for 1 h, ribonucleoside 5′-monophosphate, which is generated by nuclease *P1* digestion, is used for the labeling reaction. To this solution is added 180 μl of 1-methylimidazole buffer (0.1 M, pH 7.5), 10 μl of 0.1 M 1-ethyl-3-(3-dimethylaminopropyl)carbodiimide hydrochloride (EDC) in 1-methylimidazole buffer (0.1 M, pH 7.5) and 40 μl of 40 mM DNS-EDA in DMSO, successively. The reaction is carried out in dark for 18 h at 27 °C. An aliquot of the final solution is subjected to RP-HPLC. The fluorescence from DNS-EDA derivatives is monitored at an excitation wavelength of 270 nm and at an emission wavelength of 546 nm.

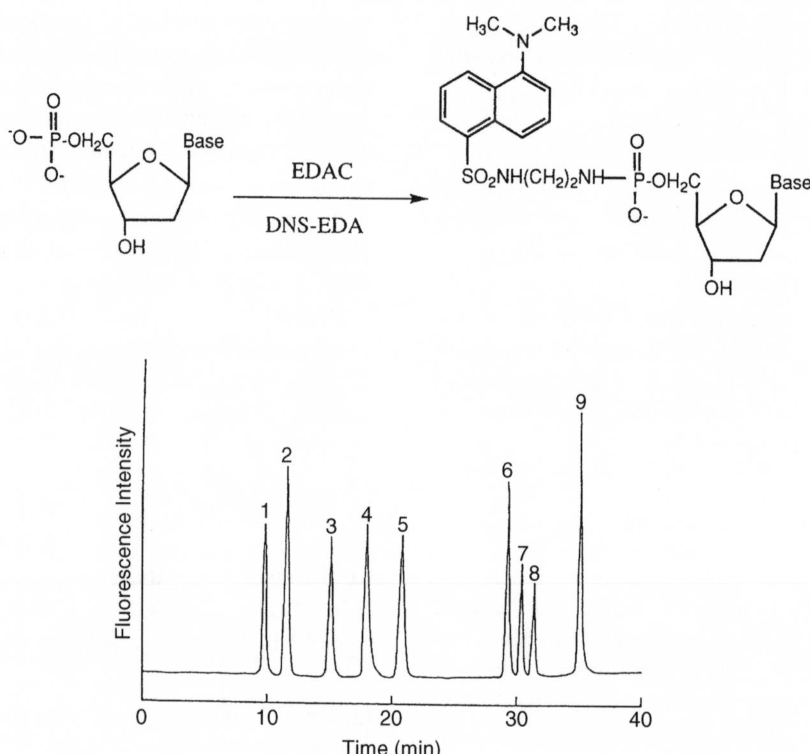


**Fig. 3.63.** Fluorescence labeling reaction of 2′-deoxynucleoside 5′-monophosphate with DNS-EDA and chromatogram of DNS-EDA derivatives of mono-, di- and trinucleotides. Peaks: 1 = GTP; 2 = ATP; 3 = GDP; 4 = by-product; 5 = ADP; 6 = GMP; 7 = UMP; 8 = CMP; 9 = AMP. [Reproduced from ref. 346, p. 1061, Fig. 2.].

### *(2) Fluorescence Labeling of Sugars of Nucleosides and Nucleotides*

Fluorescence labeling reagents for the sugar moiety of nucleosides or nucleotides have been reported for the determination of nucleic acid related compounds in the HPLC.

OMB-COCl, a fluorescence-labeling reagent for alcohols [227], can also react with the 5′-hydroxyl group of the sugar moiety in nucleosides and nucleotides to produce the corresponding fluorescent esters. The fluorescent derivatives of ten ribo- and deoxyribonucleosides and deoxyribonucleotides are separated by RP HPLC with isocratic elution [349,350]. This labeling method is applied to the HPLC determination of 2′,3′-dideoxyinosine and 2′,3′-dideoxyadenosine, which have potent activity against human immunodeficiency virus, in rat plasma for pharmacokinetic studies [351]. Mono to decanucleotides can also be derivatized with OMB-COCl in the presence of sodium azide to give corresponding OMB-carbamates of the 5′-terminal hydroxyl groups, and the resulting derivatives are determined sensitively by RP and size exclusion HPLC [352].

#### *OMB-COCl Application: Determination of 2′,3′-dideoxyinosine (ddI) and 2′,3′-Dideoxyadenosine (ddA) in Rat Plasma by Precolumn HPLC (Fig. 3.64) [351]*

To a plasma sample (0.1 ml) are added 85 nmol/ml 3′-deoxythymidine (dT; IS) solution (10 μl) and 0.1 M phosphate buffer (pH 4.5; 0.89 ml). The mixture is passed through a Toyopak ODS M cartridge at a flow rate of 0.12–0.15 ml/min. The cartridge is washed with 0.1 M phosphate buffer

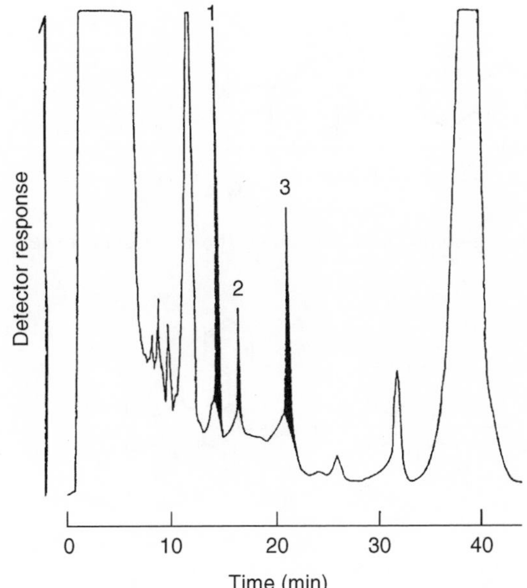


**Fig. 3.64.** Chromatogram obtained with rat plasma added with ddI, ddA and dT. Peaks: 1 = ddI; 2 = ddA; 3 = dT (IS). Amount (pmol on column): 17 pmol each. [Reproduced from ref. 351, p. 2203, Fig. 1.].

(pH 4.5) (2 ml) and water (2 ml). The adsorbed analytes are eluted with methanol (1 ml). The eluate, placed in a screw-capped vial (3.5 ml), is evaporated to dryness under a stream of nitrogen at 37 °C. To the residue dissolved in dried pyridine (0.1 ml) is added 3 mM OMB-COCl in dried benzene (0.9 ml). The vial is tightly closed and heated at 100 °C for 50 min in the dark. After cooling, the reaction mixture is concentrated to dryness under a stream of nitrogen at 60 °C. The residue is dissolved in the mobile phase for HPLC (1 ml). The resulting solution (20 μl) is subjected to HPLC with isocratic elution. The fluorescence is monitored at 475 nm with excitation wavelength at 360 nm. The detection limits (S/N = 3) for 2′,3′-dideoxyinosine and 2′,3′-dideoxyadenosine are 1.3 and 5.4 pmol on-column, respectively.

*p*-MOED has been applied to the selective derivatization of the ribose moiety in ribonucleosides and nucleotides [353]. The fluorescence derivatization of the ribose moiety is performed by heating with *p*-MOED at 140 °C for 15 min in 10 mM hydrochloric acid in the presence of 0.85 mM sodium periodate. This derivatization reaction is used for the post-column HPLC determination of ribonucleosides and nucleotides [354,355]. Pseudouridine (a modified nucleoside as a tumor marker) in human urine and serum [354], and several ribonucleotides in human erythrocytes [355] are determined by this HPLC method (detection limits for the nucleosides and nucleotides are 4–67 pmol on-column).

### *(3) Purine Nucleosides and Nucleotides*

Chloroacetaldehyde [356] and bromoacetaldehyde [357] have been reported for the derivatization reagents of adenine moiety in the nucleosides and nucleotides to give the corresponding stable fluorescent ethenoderivatives (Fig. 3.65). Chloroacetaldehyde is synthesized from chloroacetaldehyde dimethyl acetal [358]. Adenine and its nucleosides and nucleotides in biological samples have been determined by pre [359,360] and post-column [361] HPLC with fluorescence detection. The detection limits (S/N = 3) are 0.5–10 pmol on-column.

#### *Chloroacetaldehyde Application: Simultaneous Determination of Nerve-induced Adenine Nucleotides and Nucleosides Released from Rabbit Pulmonary Artery [360]*

The rabbit pulmonary artery suspended in an oxygenated Krebs solution at 37 °C is stimulated electrically. The Krebs solution in the organ bath is replaced prior to stimulation and again following stimulation of the tissues. To 1 ml of sample solution (Krebs solution) are added 40 μl of chloroacetaldehyde reagent, 100 μl of 0.6 mM 9-$\beta$-D-arabinofuranosyladenine (IS) and 360 μl of 0.1 M citrate-phosphate buffer (pH 4.0). The mixture is then heated to 80 °C for 40 min. The derivatization reaction is terminated by placing the sample on ice. The resulting ethenopurine derivatives are separated by RP-HPLC. The fluorescence is monitored at 420 nm with excitation at 305 nm.

PGO reacts selectively with guanine-containing compounds to provide corresponding single

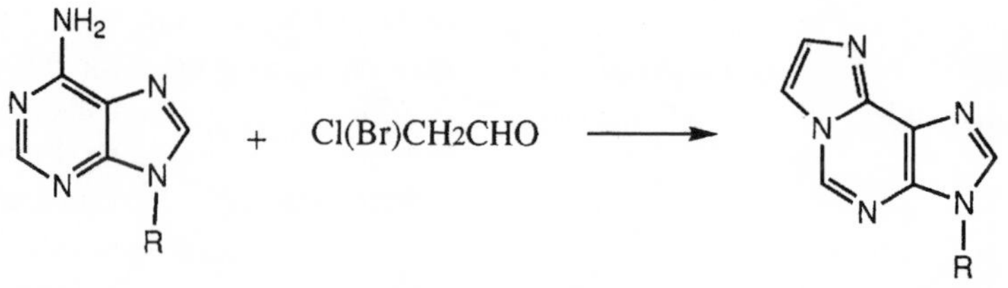

+ $Cl(Br)CH_2CHO$ →

**Fig. 3.65.** Fluorescence derivatization reaction of adenine-containing compounds with chloro(bromo)acetaldehyde.

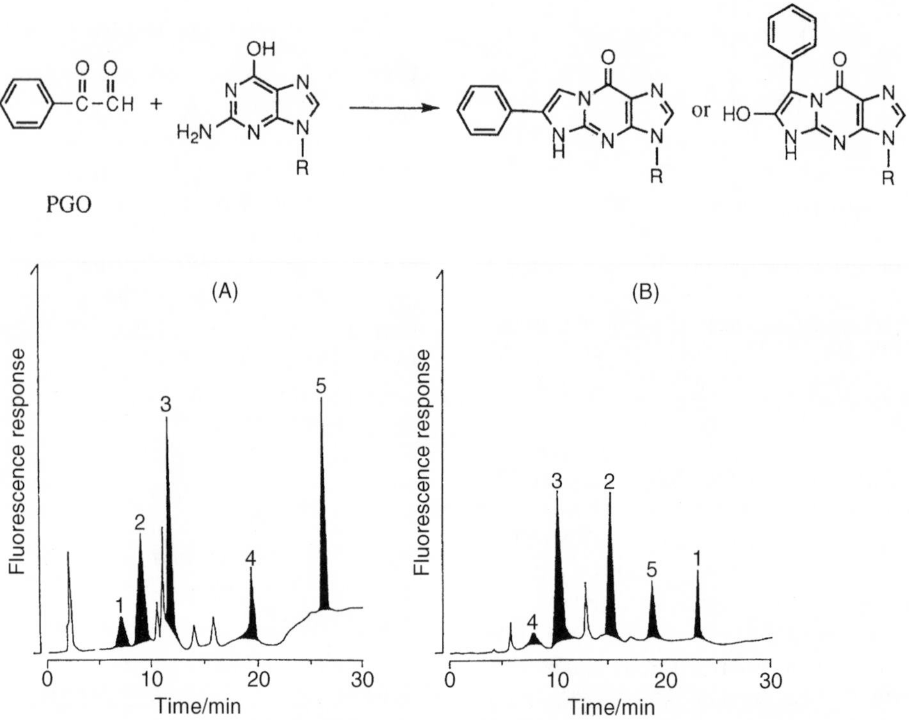


**Fig. 3.66.** Fluorescence derivatization reaction of guanine-containing compounds with PGO and chromatograms of PGO derivatives obtained with a tissue sample of rat brain cortex by: (A) reversed-phase HPLC; and (B) ion-pair reversed-phase HPLC. Peaks(endogenous concentrations, nmol/g tissue): 1 = GTP (72); 2 = GDP (69); 3 = GMP (158); 4 = guanosine (60); 5 = IS; other = unknown. [Reproduced from ref. 364, p. 250, Fig. 2.].

fluorescent derivatives which are separated by RP-HPLC. The pre-column fluorescence derivatization reaction is performed by warming at 37–60 °C for 15–30 min in a slightly acidic solution (pH 4.0–6.0) [362,363,364].

The method is also applicable to the post-column HPLC measurement of guanine and its nucleosides and nucleotides in human erythrocytes [365]. The post-column HPLC method gives detection limits (S/N = 3) of 3.2–10 pmol on-column.

Alkoxyphenylglyoxals such as DMPG are reported for more highly sensitive fluorescent reagents of guanine and its nucleosides and nucleotides [366]. The derivatization reaction is carried out by warming at 37 °C for 5–7 min in a phosphate buffer (pH 7.0), which is much milder than those for phenylglyoxal. In pre-column derivatization HPLC using this reagent, the detection limits (S/N = 3) for guanine and its nucleosides and nucleotides are 0.04–0.4 pmol on-column.

*PGO Application: HPLC Determination of Guanine and its Nucleosides and Nucleotides in Rat Brain (Fig. 3.66) [364]*

Rat brain tissue (100 ± 20 mg) is homogenized at 0–4 °C with 200 μl of 0.1 M hydrochloric acid and then mixed with 200 μl of methylcellosolve, and 50 μl of 0.2 μm ol/ml 9-ethylguanine as internal standard. The homogenate is neutralized with 40 μl of 0.5 M disodium hydrogenphosphate to a pH of approximately 6.0. The homogenate is centrifuged at 1400 g for 10 min. The supernatant (100 μl) is mixed with 100 μl each of 25 mM phosphate buffer (pH 6.0), 0.1 M phenylglyoxal in DMSO and water. The mixture is warmed at 37 °C for 15 min. A 100 μl portion of the final reaction mixture is used for RP-HPLC analysis. The fluorescence is monitored at 515 nm emission and 365 nm excitation. GTP, GDP, GMP and guanosine in the biospecimens are quantified. The detection limits (S/N = 3) for the compounds are 0.11–2.54 pmol on-column.

*(4) Pyrimidine Nucleosides and Nucleotides*

Phenacyl bromide has been reported for the selective derivatization reagent for the HPLC determination of cytosine-containing compounds (Fig. 3.67) [367]. Highly fluorescent 2-phenyl-3, *N*4-ethenocytosine derivatives are produced by a reaction of cytosine-containing compounds with phenacyl bromide in weakly acidic acetonitrile solution at 80 °C for 45 min. The fluorescence is monitored at 370 nm emission and at 305 nm excitation. The method is applied to the determination of cidofovir, an acyclic cytidine monophosphate analog which has potent antiviral activity, in cynomolgus monkey plasma. The limit of detection for cidofovir in cynomolgus monkey plasma is 5 ng/ml (ca. 100 fmol on-column).

Br-DMEQ and Br-MMC give fluorescent products from nucleosides with active imino hydrogens such as uridine, deoxyuridine and thymidine (Fig. 3.68). The derivatives of the pyrimidine nucleosides with Br-MMC [368,369] can be determined by RP-HPLC. The fluorescence is monitored at 395 nm emission and at 345 nm excitation. The detection limits (S/N = 2) are 0.5–1 pmol on-column. Br-DMEQ is applied to the determination of an antitumour agent, 5-FU, and its pro-drug, 5-fluoro-2′-deoxyuridine, in human serum [370,371].

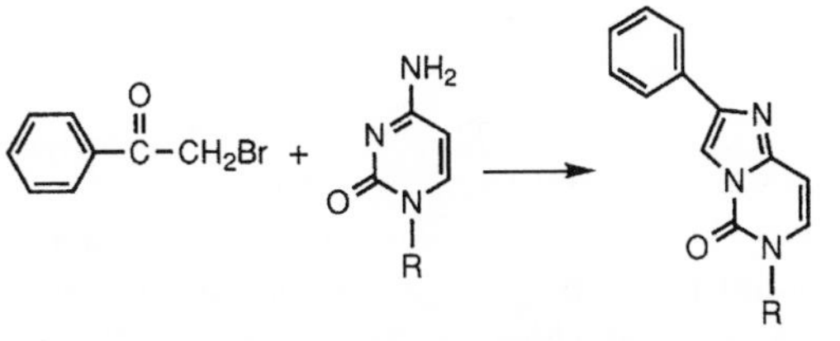

**Fig. 3.67.** Fluorescence derivatization reaction of cytosine-containing compounds with phenacyl bromide.

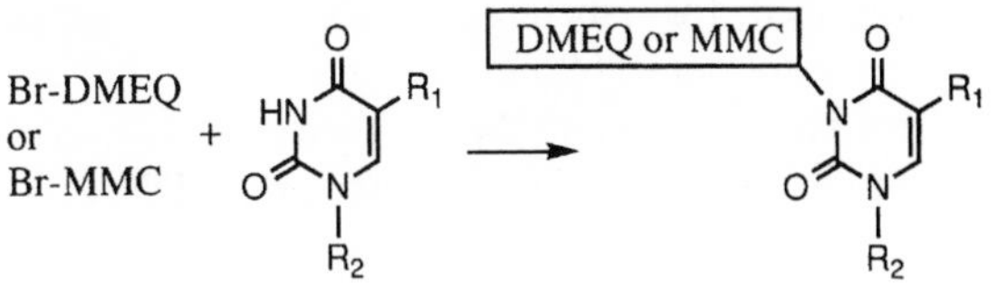

**Fig. 3.68.** Fluorescence labeling reaction of pyrimidine related compounds with Br-DMEQ or Br-MMC.

*Br-DMEQ Application: HPLC Determination of 5-FU and 5-fluoro-2′-deoxyuridine in Human Serum [371]*

A 100 μl aliquot of human serum is added to 5 μl of the 5-chlorouracil (IS) solution, and the mixture is allowed to stand at room temperature for about 3 min. To the resulting solution, 0.5 ml of 0.1 M potassium phosphate buffer (pH 3.5) and 2.0 ml of ethyl acetate are added. the mixture is vortexed for about 2 min and centrifuged at 1000 g for 5 min. The organic layer (about 1.4 ml) is placed in a screw-capped test tube (10 ml) and evaporated to dryness *in vacuo*. To the tube are added 20 mg of a mixture of potassium hydrogen carbonate and anhydrous sodium sulfate (1:7, w/w) and 50 μl each of 1.3 mM of Br-DMEQ in acetone and 1.5 mM of 18-crown-6 in acetone. The tube is tightly closed and heated at 50 °C for 20 min. The final reaction mixture (10 μl) is applied to the RP-HPLC with fluorescence detection (excitation, 370 nm and emission, 455 nm). The detection limits (S/N = 5) are 0.35–0.55 pmol on-column.

Semiconductor LIF detection of DNA has been accomplished by free-solution CE using an oligonucleotide labeled with a cyanine dye as a DNA probe [372]. Cyanine dye which has succinimidyl moiety as a reaction site is synthesized for labeling the DNA. The molar absorptivity at 788 nm of the cyanine dye derivatives are about 245 000 $cm^{-1}$ $mol^{-1}$. The detection limit is $8 \times 10^{-9}$ M for probe DNA.

## REFERENCES


[1] Roth, M. (1971) *Anal. Chem.*, **43**, 880.
[2] Benson, J.R. and Hare, P.E. (1975) *Proc. Nat. Acad. Sci. U.S.A.*, **72**, 619.
[3] Schuster, R. (1988) *J. Chromatogr.*, **431**, 271.
[4] Kamisaki, Y., Takao, Y., Itoh, T., Shimomura, T., Takahashi, K., Uehara, N. and Yoshino, Y. (1990) *J. Chromatogr.*, **529**, 417.
[5] Graser, T., Godel, H., Albers, S., Foldi, P. and Furst, P. (1985) *Anal. Biochem.*, **151**, 142.
[6] Liu, H. and Worthen, H.G. (1992) *J. Chromatogr.*, **579**, 215.
[7] Fleury, M.O. and Ashley, D.V. (1983) *Anal. Biochem.*, **133**, 330.
[8] Godel, H., Graser, T., Foldi, P., Pfaender, P. and Furst, P. (1984) *J. Chromatogr.*, **297**, 49.
[9] Buck, R.H. and Krummen, K. (1984) *J. Chromatogr.*, **315**, 279.
[10] Nimura, N. and Kinoshita, T. (1986) *J. Chromatogr.*, **352**, 169.
[11] Uhe, A., Collier, G., McLenna, E., Tucker, D. and O'Dea, K. (1991) *J. Chromatogr.*, **564**, 81.
[12] Miller, R., Grone, J.L., Skolnick, P. and Boje, K.M. (1992) *J. Chromatogr.*, **578**, 103.
[13] De Montigny, P., Stobaugh, J.F., Givens, R.S., Carlson, R.G., Srinivachar, K, Sternson, L.A. and Higuchi, T. (1987) *Anal. Chem.*, **59**, 1096.
[14] Koning, H., Wolf, H., Venoma, K. and Korf, J. (1990) *J. Chromatogr.*, **533**, 171.
[15] Lunte, S. and Wong, O. (1989) *LC-GC Intern.*, **2**, 20.
[16] Beale, S.C., Hsieh, Y.-Z., Wiesler, D. and Novotny., M. (1990) *J. Chromatogr.*, **409**, 579.
[17] Beale, S.C., Savege, J., Hsieh, Wiesler, D., Wiestock, S. and Novotny, M. (1988) *Anal. Chem.*, **60**, 1765.
[18] Liu, J., Hsieh, Y.-Z., Wiesler, D. and Novotny, M. (1991) *Anal. Chem.*, **63**, 408.
[19] Roach, M.C. and Harmony, M.D. (1987) *Anal. Chem.*, **59**, 411.
[20] Udenfriend, S., Stein, S., Bohlen, P., Dairman, W., Leimgruber, W. and Wieble, M. (1972) *Sciences*, **178**, 871.
[21] Samejima, K. (1974) *J. Chromatogr.*, **96**, 250.
[22] Samejima, K., Kawase, M., Sakamoto, S., Okada, M and Endo, Y. (1976) *Anal. Biochem.*, **60**, 78.
[23] Felix, A.M. and Terkelsen, G. (1974) *Anal. Biochem.*, **60**, 78.
[24] Frei, R.W., Michel, L. and Santi, W. (1976) *Anal. Biochem.*, **126**, 665.
[25] Weigle, M., de Bernard, S., Leimgruber, W., Cleeland, R. and Grunberg, E. (1973) *Biochem. Biophys. Res. Commun.*, **54**, 899.
[26] Sakai, T. and Nagasawa, T. (1992) *J. Chromatogr.*, **576**, 155.
[27] Gray, W.R. and Hartley, B.S. (1963) *Biochem. J.*, **89**, 371.
[28] Tapuhi, Y., Schmidt, D.E., Linder, W. and Karger, B.L. (1981) *Anal. Biochem.*, **115**, 123.
[29] Ishida, J., Abe, K., Nakamura, M. and Yamaguchi, M. (1995) *Anal. Sci.*, **11**, 743.
[30] Nachtmann, F., Spitzy, H. and Frei, R.W. (1975) *Anal. Chim. Acta*, **76**, 57.
[31] Kamimura, H., Sasaki, H and Kawamura, S. (1981) *J. Chromatogr.*, **225**, 115.
[32] Osborne, N.N., Stahl, W.L. and Neufoff, V. (1976) *J. Chromatogr.*, **123**, 212.
[33] Tsuruta, Y., Date, Y. and Kohashi, K. (1990) *J. Chromatogr.*, **502**, 178.
[34] Hara, S., Aoki, J., Yoshikuni, K., Tatsuguchi, Y. and Yamaguchi, M. (1997) *Analyst*, **122**, 475.
[35] Hendlich, C.E., Berdecia-Rodriguez, J., Wiedmeier, V.T. and Porterfield, S.P. (1992) *J. Chromatogr.*, **577**, 19.
[36] Einarsson, S., Josefsson, B. and Lagerkvist, S. (1983) *J. Chromatogr.*, **282**, 606.
[37] Gustavsson, B. and Betner, I. (1990) *J. Chromatogr.*, **507**, 67.
[38] Haynes, P.A., Sheumack, D., Kibby, J. and Redmond, J.W. (1991) *J. Chromatogr.*, **540**, 177.
[39] Haynes, P.A., Sheumack, D., Greig, J., Kibby, J. and Redmond, J.W. (1991) *J. Chromatogr.*, **588**, 107.
[40] Blankenship, D., Krivanek, M., Ackermann, B. and Cardin, A. (1989) *Anal. Biochem.*, **178**, 227.
[41] Worthen, H. and Liu, H. (1992) *J. Liq. Chromatogr.*, **15**, 3323.
[42] Gubitz, G., Wintersteiger, R. and Hartinger, H. (1981) *J. Chromatogr.*, **218**, 51.
[43] Faulkner, A.J., Veening, H., Becker, H.-D. (1991) *Anal. Chem.*, **63**, 292.

[44] Ishida, J., Yamaguchi, M., Iwata, T. and Nakamura, M. (1989) *Anal. Chim. Acta*, **223**, 319.
[45] Yoshida, T., Moriyama, Y., Nakamura, K. and Taniguchi, H. (1993) *Analyst*, **118**, 29.
[46] Fujino, H. and Goya, S. (1990) *Anal. Sci.*, **6**, 465.
[47] Ishida, J., Yamaguchi, M. and Nakamura, M. (1990) *Anal. Biochem.*, **184**, 86.
[48] Ishida, J., Yamaguchi, M. and Nakamura, M. (1991) *Anal. Biochem.*, **195**, 168.
[49] Iwata, T., Fujino, H., Sonoda, J. and Yamaguchi, M. (1997) *Anal. Sci.*, **13**, 467.
[50] Kawauchi, H., Tujimura, T., Maeda, H. and Ishida, N. (1969) *J. Biochem.*, **66**, 783.
[51] Ichikawa, H., Tanimura, T., Nakajima, T. and Tamura, Z. (1979) *Chem. Pharm. Bull.*, **18**, 1493.
[52] Jin, S.-W., Chen, G.-X., Palacz, Z. and Wittmann-Liebold, B. (1986) *FEBS Lett.*, **198**, 150.
[53] Imakyure, O., Kai, M., Mitsui, T., Nohta. H. and Ohkura, Y. (1993) *Anal. Sci.*, **9**, 647.
[54] Yamaguchi, M., Hara, S., Aoki, J., Yoshikuni, K. and Iwata, T. (1998) *Anal. Sci.*, **14**, 425.
[55] Imakyure, O., Kai, M. and Ohkura, Y. (1994) *Anal. Chim. Acta*, **291**, 197.
[56] Steinert, J., Khalaf, H., Keese, W. and Rimpler, K. (1996) *Anal. Chim. Acta*, **327**, 153.
[57] Nimura, N., Iwaki, K., Takeda, K. and Ogura, H. (1986) *Anal. Chem.*, **58**, 2372.
[58] Iwaki, K., Nimura, N., Hiraga, K., Kinoshita, T., Takeda, K. and Ogura, H. (1987) *J. Chromatogr.*, **407**, 273.
[59] Hirai, T., Umezawa, S., Kitamura, M. and Inoue, Y. (1991) *Bunseki Kagaku*, **40**, 233.
[60] Cohen, S.A. and Michaud, D.P. (1993) *Anal. Biochem.*, **211**, 179.
[61] Ghosh, P.B. and Whitehouse, M.W. (1968) *Biochem. J.*, **108**, 155.
[62] Watanabe, Y. and Imai, K. (1981) *Anal. Chim. Acta*, **130**, 377.
[63] Watanabe, Y. and Imai, K. (1981) *Anal. Biochem.*, **116**, 471.
[64] Watanabe, Y. and Imai, K. (1984) *J. Chromatogr.*, **309**, 279.
[65] Toyo'oka, T., Suzuki, T., Saito, Y., Uzu, S. and Imai, K. (1989) *Analyst*, **114**, 413.
[66] Kato, M., Fukushima, T., Santa, T. and Imai, K. (1995) *Biomed. Chromatogr.*, **9**, 193.
[67] Toyo'oka, T., Suzuki, T., Saito, Y., Uzu, S. and Imai, K. (1989) *Analyst*, **114**, 1233.
[68] Hamachi, Y., Tsujiyama, T., Nakashima, K. and Akiyama, S. (1995) *Biomed. Chromatogr.*, **9**, 216.
[69] Toyo'oka, T., Chokushi, H., Carlson, G., Givens, R. and Lunte, S. (1993) *Analyst*, **118**, 257.
[70] Mank, A.J.C., Molenaar, E.J., Lingeman, H., Gooijer, C., Brinkman, U.A.Th. and Velthorst, N.H. (1993) *Anal. Chem.*, **65**, 2197.
[71] Gao, C.-X., Chou, T.-Y., Colgan, S.T., Krull, I.S., Dorschel, C. and Bidilingmeyer, B.A. (1988) *J. Chromatogr. Sci.*, **26**, 449.
[72] Yu, J.H., Li, G.D., Krull, I.S. and Cohen, S. (1994) *J. Chromatogr. B.*, **658**, 249.
[73] Higashijima, H., Fuchigami, T., Imasaka, T. and Ishibashi, N. (1992) *Anal. Chem.*, **64**, 711.
[74] Fuchigami, T., Imasaka, T. and Shiga, M. (1993) *Anal. Chim. Acta*, **282**, 209.
[75] Mank, A.J.G. and Yeng, E.S. (1995) *J. Chromatogr.*, **708**, 309.
[76] Williams, R.J., Lipowska, K., Patonay, G. and Strekowski, L. (1993) *Anal. Chem.*, **65**, 601.
[77] Yui, Y., Fujita, T., Yamamoto, T., Itokawa, Y. and Kawai, C. (1980) *Clin. Chem.*, **26**, 194.
[78] Yamatodani, A. and Wada, H. (1981) *Clin. Chem.*, **27**, 1983.
[79] Seki, T. (1978) *J. Chromatogr.*, **155**, 415.
[80] Mori, K. and Imai, K. (1985) *Anal. Biochem.*, **146**, 283.
[81] Mitsui, A., Nohta, H. and Ohkura, Y. (1985) *J. Chromatogr.*, **344**, 61.
[82] van der Hoorn, F.A.J., Boomsma, F., Man in't Veld, A.J. and Schalekamp, M.A.D.H. (1991) *J. Chromatogr.*, **563**, 348.
[83] Husek, P., Malikova, J. and Herzogova, G. (1990) *J. Chromatogr.*, **553**, 166.
[84] Kamahori, M., Taki, M., Watanabe, Y. and Miura, J. (1991) *J. Chromatogr.*, **553**, 166.
[85] Denckla, W.D. and Dewey, H.K. (1967) *J. Lab. Clin. Ed.*, **69**, 160.
[86] Iizuka, H. and Yajima, T. (1981) *Chem. Pharm. Bull.*, **33**, 2591.
[87] Iizuka, H. and Yajima, T. (1993) *Biol. Pharm. Bull.*, **16**, 103.
[88] Kojima, E., Kai, M. and Ohkura, Y. (1991) *Anal. Chim. Acta*, **248**, 213.
[89] Kojima, E., Kai, M. and Ohkura, Y. (1993) *J. Chromatogr.*, **612**, 187.
[90] Ishida, J., Yamaguchi, M. and Nakamura, M. (1991) *Analyst*, **116**, 301.
[91] Ishida, J., Iizuka, R. and Yamaguchi, M. (1993) *Analyst*, **118**, 165.
[92] Ishida, J., Iizuka, R. and Yamaguchi, M. (1993) *Clin. Chem.*, **39**, 2355.
[93] Iizuka, R., Ishida, J., Yoshitake, T., Nakamura, M. and Yamaguchi, M. (1996) *Biol. Pharm. Bull.*, **19**, 762.
[94] Ishida, J., Yoshitake, T., Fujino, K., Kawano, K., Kehr, J. and Yamaguchi, M. (1998) *Anal. Chim. Acta*, **365**, 227.
[95] Nohta, H., Yukizawa, T., Ohkura, Y., Yoshimura, M., Ishida, J. and Yamaguchi, M. (1997) *Anal. Chim. Acta*, **344**, 233.
[96] Yamaguchi, M., Ishida, J. and Yoshimura, M. (1998) *Analyst*, **123**, 307.

[97] Yamada, S. and Itano, H.A. (1966) *Biochim. Biophys. Acta*, **130**, 538.
[98] Ohkura, Y. and Kai, M. (1979) *Anal. Chim. Acta*, **106**, 89.
[99] Kai, M., Miura, T., Kohashi, K. and Ohkura, Y. (1981) *Chem. Pharm. Bull.*, **29**, 1115.
[100] Nakazawa, H. (1987) *J. Chromatogr.*, **417**, 409.
[101] Chow, J., Orenberg, J.B. and Nugent, K.D. (1987) *J. Chromatogr.*, **386**, 243.
[102] Mifune, M., Krehbiel, D.K., Stohbaugh, J.F. and Riley, C.M. (1989) *J. Chromatogr.*, **496**, 55.
[103] Boppana, V.K., Miller-Stein, S., Politowsky, J.F. and Rhodes, G.R. (1991) *J. Chromatogr.*, **548**, 319.
[104] Ohno, M., Kai, M. and Ohkura, Y. (1987) *J. Chromatogr.*, **392**, 309.
[105] Ohno, M., Kai, M. and Ohkura, Y. (1989) *J. Chromatogr.*, **490**, 301.
[106] Kai, M., Miyazaki, T., Sakamoto, S. and Ohkura, Y. (1985) *J. Chromatogr.*, **322**, 473.
[107] Kemp, R. (1991) *LC-GC Intern.* **4**, 40.
[108] Dunges, W., (1977) *Uvspect. Group Bull.*, **5**, 38.
[109] Wolf, J.H. and Korf, J. (1992) *J. Pharm. Biol. Analysis*, **10**, 99.
[110] van der Horst, F.A.L. (1989) *Trends Anal. Chem.*, **8**, 268.
[111] Wolf, J.H. and Korf, J. (1990) *J. Chromatogr.*, **502**, 423.
[112] Jansen, E.H.J.M. and de Fluiter, P. (1992) *J. Liq. Chromatogr.*, **15**, 2247.
[113] Gueddour, R.B., Matt, M. and Nicolas, A. (1993) *Anal. Lett.*, **26**, 429.
[114] Lloyd, J.B.F. (1979) *J. Chromatogr.*, **178**, 249.
[115] Wolf, J.H. and Korf, J. (1992) *J. Pharm. Biomed. Anal.*, **10**, 9 9.
[116] Yoo, J.S. and McGuffin, V.L. (1992) *J. Chromatogr.*, **627**, 87.
[117] Farinotti, R., Siard, P., Bourson, J., Kirkiacharian, S., Valeur, B. and Mahuzier, G. (1983) *J. Chromatogr.*, **269**, 81.
[118] Yoshida, S., Toshihary, A. and Hirose, S. (1988) *J. Chromatogr.*, **430**, 156.
[119] Naganuma, H. and Kawahara, Y. (1989) *J. Chromatogr.*, **478**, 149.
[120] Tuchiya, H., Hayashi, T., Naruse, H. and Takagi, N. (1982) *J. Chromatogr.*, **234**, 121.
[121] Yamaguchi, M., Matsunaga, R., Hara, S., Nakamura, M. and Ohkura, Y. (1986) *J. Chromatogr.*, **375**, 27.
[122] Nakashima, K., Okamoto, K., Yoshida, K., Kuroda, N., Akiyama, S. and Yamaguchi, M. (1992) *J. Chromatogr.*, **584**, 275.
[123] Chao, E.-F., Kai, M. and Ohkura, Y. (1990) *J. Chromatogr.*, **526**, 77.
[124] Yamaguchi, M., Matsunaga, Fukuda, K. and Nakamura, M. (1987) *J. Chromatogr.*, **414**, 275.
[125] Yamaguchi, M. and Nakamura, M. (1987) *Chem. Pharm. Bull.*, **35**, 3740.
[126] Yamaguchi, M. Nakamura, M., Kuroda, N. and Ohkura, Y. (1987) *Anal. Sci.*, **3**, 75.
[127] Yamaguchi, M., Takehiro, O., Hara, S., Nakamura, M. and Ohkura, Y. (1988) *Chem. Pharm. Bull.*, **36**, 2263.
[128] Yamaguchi, M., Hara, S., Takemori, Y. and Nakamura, M. (1986) *Anal. Sci.*, **5**, 35.
[129] Alekseev, S.M., Konkin, E.E., Sarycheva, I.K., Pomoinitskii, V.D., Zolotukhin, S.V. and Evstigneeva, R.P. (1982) *Khim. Farm. Zh.*, **17**, 619.
[130] Akasaka, K., Suzuki, T., Ohrui, H., Meguro, H., Shindo, Y. and Takahashi, H. (1987) *Anal. Lett.*, **20**, 1581.
[131] Takadate, A., Masuda, T., Tajima, C., Murata, M., Irikura, M. and Goya, S. (1992) *Anal. Sci.*, **8**, 663.
[132] Takadate, A., Masuda, C., Murata, M., Haratake, C., Isobe, A., Irikura, M. and Goya, S. (1992) *Anal. Sci.*, **8**, 695.
[133] Goto, J., Shamsa, F. and Nambara, T. (1983) *J. Liq. Chromatogr.*, **16**, 1977.
[134] Kamada, S., Maeda, M. and Tsuji, A. (1983) *J. Chromatogr.*, **272**, 29.
[135] Kwakman, P.J.M., van Schaik, H.-P., Brinkman, U.A.Th. and de Jong, C.J. (1991) *Analyst*, **116**, 1385.
[136] Wadkins, W.D. and Peterson, M.B. (1982) *Anal. Biochem.*, **125**, 30.
[137] Allenmark, S., Chelminska-Bertilsson, M. and Thompson, R.A. (1990) *Anal. Biochem.*, **185**, 279.
[138] Gatti, R., Cavrini, V. and Roveri, P. (1992) *Chromatographia*, **33**, 13.
[139] Roach, M.C., Ungar, L.W., Zare, R.N., Reimer, L.M., Pompliano, D.L. and Frost, J.W. (1987) *Anal. Chem.*, **59**, 1056.
[140] van der Horst, F.A.L., Post, M.H., Holthuis, J.J.M. and Brinkman, U.A.Th. (1990) *J. Chromatogr.*, **500**, 443.
[141] Nimura, N. and Kinoshita, T. (1980) *Anal. Lett.*, **13**, 191.
[142] Nimura, N., Kinoshita, T., Yoshida, T., Uetake, A. and Nakai, C. (1988) *Anal. Chem.*, **60**, 2067.
[143] Yoshida, T., Uetake, A., Nakai, C., Nimura, N. and Kinoshita, T. (1988) *J. Chromatogr.*, **456**, 421.
[144] Takadate, A., Tahara, T., Fujino, H. and Goya, S. (1982) *Chem. Pharm. Bull.*, **30**, 4120.
[145] Remens, Th.H.M. and van der Vusse, G.J. (1991) *J. Chromatogr.*, **570**, 243.
[146] Tagawa, K., Hayashi, K., Mizobe, M. and Noda, K. (1993) *Anal. Biochem.*, **60**, 2067.
[147] Yoshida, T., Uetake, A., Yamaguchi, H., Nimura, N. and Kinoshita, T. (1988) *Anal. Biochem.*, **173**, 70.

[148] Ikeda, M., Shimada, K., Sakaguchi, T. and Matsumoto, U. (1984) *J. Chromatogr.*, **305**, 261.
[149] van Den Beld, C.M.B. and Lingeman, H. (1991) in *Luminescence Techniques in Chemical and Biochemical Analysis* Bayens, W.R.G., Keukeleire, D.D. and Korkidis, K., (eds), p. 237. Marcel Dekker, Inc., New York.
[150] Lingeman, H., Tjaden, U.R., Van Den Beld, C.M.B. and Van Der Greef, J. (1988) *J. Pharm. Biomed. Anal.*, **6**, 687.
[151] Narita, S. and Kitagawa, T. (1989) *Anal. Sci.*, **5**, 31.
[152] Lee, Y.-M. Nakamura, H. and Nakajima, T. (1989) *Anal. Sci.*, **5**, 209.
[153] Lee, Y.-M. Nakamura, H. and Nakajima, T. (1989) *Anal. Sci.*, **5**, 681.
[154] Yanagisawa, I., Yamane, M. and Urayama, T. (1985) *J. Chromatogr.*, **345**, 229.
[155] Yamaguchi, M., Iwata, T., Inoue, K., Hara, S. and Nakamura, M. (1990) *Analyst*, **115**, 1363.
[156] Iwata, T., Ishimaru, T., Nakamura, M. and Yamaguchi, M. (1994) *Biomed. Chromatogr.*, **8**, 283.
[157] Iwata, T., Nakamura, M. and Yamaguchi, M. (1992) *Anal. Sci.*, **8**, 889.
[158] Yamaguchi, M., Hara, S. and Obata, K. (1995) *J. Liq. Chromatogr.*, **18**, 2991.
[159] Nakashima, K., Taguchi, Y., Kuroda, N., Akiyama, S. and Duan, G. (1993) *J. Chromatogr.*, **619**, 1.
[160] Saito, Y., Ushijima, T., Sasamoto, K., Takata, K., Ohkura, Y. and Ueno, K. (1995) *Anal. Chim. Acta*, **300**, 243.
[161] Saito, M., Chiyoda, Y., Ushijima, T., Sasamoto, K. and Ohkura, Y. (1994) *Anal. Sci.*, **10**, 669.
[162] Iwata, T., Hirose, T., Nakamura, M. and Yamaguchi, M. (1994) *Analyst*, **119**, 1747.
[163] Nakajima, R., Shimada, K., Fujii, Y., Yamamoto, A. and Hara, T. (1991) *Bull. Chem. Soc. Jpn.*, **64**, 3173.
[164] Baty, J.D., Pazouki, S. and Dolphin, J. (1987) *J. Chromatogr.*, **395**, 403.
[165] Goya, S., Takadate, A. and Fujino, H. (1981) *Yakugaku Zasshi*, **101**, 1164.
[166] Goya, S., Takadate, A., Fujino, H. and Irikura, M. (1981) *Yakugaku zasshi*, **101**, 1064.
[167] Jones, O.T. (1985) *Biochemistry*, **24**, 2195.
[168] Lingeman, H., Hulshoff, A., Underberg, W.J.M. and Offermann, F.B.J.M. (1984) *J. Chromatogr.*, **290**, 215.
[169] Mukherjee, P.S., DeSilva, K.H. and Karnas, H.T. (1996) *Anal. Chim. Acta*, **124**, 99.
[170] Yasaka, Y., Tanaka, M., Shono, T., Tetsumi, T. and Katakawa, J. (1990) *J. Chromatogr.*, **508**, 133.
[171] Akasaka, K., Ohrui, H. and Meguro, H. (1993) *Analyst*, **118**, 765.
[172] Toyo'oka, T., Ishibashi, M., Takeda, Y., Nakashima, K., Akiyama, S., Uzu, S. and Imai, K. (1991) *J. Chromatogr.*, **588**, 61
[173] Toyo'oka, T., Ishibashi, M., Takeda, Y., Nakashima, K., Akiyama, S., Uzu, S. and Imai, K. (1991) *Analyst*, **116**, 609.
[174] Toyo'oka, T., Ishibashi, M. and Terao, T. (1992) *Analyst*, **117**, 727.
[175] Toyo'oka, T., Ishibashi, M. and Terao, T. (1992) *J. Chromatogr.*, **625**, 357.
[176] Toyo'oka, T., Ishibashi, M. and Terao, T. (1993) *Anal. Chim. Acta*, **278**, 71.
[177] Toyo'oka, T., Ishibashi, M. and Terao, T. (1992) *J. Chromatogr.*, **627**, 75.
[178] Toyo'oka, T. and Liu, Y.-M. (1994) *Anal. Proc.*, **31**, 265.
[179] Toyo'oka, T. and Liu, Y.-M. (1995) *J. Chromatogr. A*, **695**, 11.
[180] Toyo'oka, T., Takahashi, M., Suzuki, A. and Ishii, Y. (1995) *Biomed. Chromatogr.*, **9**, 162.
[181] Toyo'oka, T., Ishibashi, M., Terao, T. and Imai, K. (1992) *Biomed. Chromatogr.*, **6**, 143.
[182] Toyo'oka, T., Ishibashi, M., Terao, T. and Imai, K. (1991) *Analyst*, **116**, 609.
[183] Santa, T., Kimoto, K., Hukushima, T., Homma, H. Imai, K. (1996) *Biomed. Chromatogr.*, **10**, 83.
[184] Prados, P., Fukushima, T., Santa, T., Honma, H., Tsunoda, M., Al-Kindy, S., Mori, S., Yokosu, H. and Imai, K. (1997) *Anal. Chim. Acta*, **344**, 227.
[185] Mank, A.J.B., Beekman, M.C., Velthorst, N.H., Brinkman, U.A.T., Lingeman, H. and Gooijer, C. (1995) *Anal. Chim. Acta*, **315**, 209.
[186] Rahavendran, V. and Karnes, H.T. (1996) *J. Pharm. Biomed. Anal.*, **15**, 83.
[187] Voelter, W. and Huber, R. (1981) *J. Chromatogr.*, **217**, 491.
[188] Wolf, J.H., Veenma vd Duin, L. and Korf, J. (1989) *J. Chromatogr.*, **487**, 496.
[189] Wolf, H. and Korf, J. (1988) *J. Chromatogr.*, **436**, 437.
[190] Thuchiya, H., Hayashi, T., Naruse, H. and Takagi, N. (1982) *J. Chromatogr.*, **234**, 121.
[191] Yamaguchi, M., Matsunaga, R., Hara, S., Nakamura, M. and Ohkura, Y. (1986) *J. Chromatogr.*, **375**, 27.
[192] Yamaguchi, M., Matsunaga, R., Fukuda, K., Nakamura, M. and Oh kura, Y. (1986) *Anal. Biochem.*, **155**, 256.
[193] Yamaguchi, M., Ishida, J. and Nakamura, M. (1989) *Chem. Pharm. Bull.*, **37**, 2846.
[194] Shimomura, Y., Taniguchi, K., Suie, T., Murakami, M., Sugiyama, S. and Ozawa, T. (1984) *Clin. Chim. Acta*, **143**, 361.
[195] Kargas, G., Rudy, T., Spennetta, T., Takayama, K., Ouerishi, N. and Shrago, E. (1990) *J. Chromatogr.*, **256**, 331.

[196] Yanagisawa, I., Yamane, M. and Urayama, T. (1985) *J. Chromatogr.*, **345**, 229.
[197] Sein, J., Milovic, V., Zeu Zem, S. and Caspary, W.F. (1993) *J. Liq. Chromatogr.*, **16**, 2915.
[198] Nakashima, K., Taguchi, Y., Kuroda, N., Akiyama, S. and Duan, G. (1993) *J. Chromatogr.*, **619**, 1.
[199] Iwata, T., Inoue, K., Nakamura, M. and Yamaguchi, M. (1992) *Biomed. Chromatogr.*, **6**, 120.
[200] Tsuchiya, H., Hayashi, T., Naruse, H. and Takagi, N. (1982) *J. Chromatogr.*, **231**, 247.
[201] Engels, W., Kamps, M.A.F., Lemmens, P.J.M.R., Van Der Vusse, G.J. and Reneman, R.S. (1988) *J. Chromatogr.*, **427**, 209.
[202] Yamaguchi, M., Fukuda, K., Hara, S., Makamura, M. and Ohkura, Y. (1986) *J. Chromatogr.*, **380**, 257.
[203] Hatsumi, M., Kimata, S. and Hirosawa, K. (1982) *J. Chromatogr.*, **253**, 271.
[204] Yanagisawa, I., Yamane, M. and Urayama, T. (1985) *J. Chromatogr.*, **345**, 229.
[205] Toyo'oka, T., Ishibashi, M., Terao, T. and Imai, K. (1992) *Biomed. Chromatogr.*, **6**, 143.
[206] Stein, T.A., Angus, S., Borrero, E., Anguste, L.J. and Wise, L. (1987) *J. Chromatogr.*, **395**, 591.
[207] Salari, H., Yeung, M., Douglas, S. and Morozowich, W. (1987) *Anal. Biochem.*, **165**, 220.
[208] Andreolini, F., Beale, S.C. and Novotony, M. (1988) *J. High Resolut. Chromatogr. Chromatogr. Commun.*, **11**, 24.
[209] Shimada, K., Komine, Y. and Mitamura, K. (1991) *J. Chromatogr.*, **565**, 111.
[210] Allenmark, S., Chelminska-Bertilsson, M. and Thompson, R.A. (1990) *Anal. Biochem.*, **185**, 279.
[211] Cavrini, V., Gatti, R., Roda, A., Cerre, C. and Roveri, P. (1993) *J. Pharm. Biomed. Anal.*, **11**, 761.
[212] Gatti, R., Roda, A., Cerre, C., Bonazzi, D. and Cavrini, V. (1997) *Biomed. Chromatogr.*, **11**, 11.
[213] Leroy, P., Chakir, S. and Nicolas, A. (1986) *J. Chromatogr.*, **351**, 267.
[214] Iwata, T., Hirose, T., Nakamura, M. and Yamaguchi, M. (1994) *J. Chromatogr. B.*, **654**, 171.
[215] Iwata, T., Hirose, T. and Yamaguchi, M. (1997) *J. Chromatogr. B.*, **695**, 201.
[216] Hara, S., Takemori, Y., Iwata, T., Yamaguchi, M., Nakamura, M. and Ohkura, Y. (1985) *J. Chromatogr.*, **344**, 33.
[217] Keen, R.E., Nissensoon, C.H. and Barrio, J.R. (1993) *Anal. Biochem.*, **213**, 23.
[218] Hara, S., Takemori, Y., Yamaguchi, M. and Nakamura, M. (1985) *J. Chromatogr.*, **344**, 33.
[219] Nakamura, M., Hara, S., Yamaguchi, M., Takemori, Y. and Ohkura, Y. (1987) *Chem. Pharm. Bull.*, **35**, 687.
[220] Hara, S., Yamaguchi, M., Takemori, Y., Nakamura, M. and Ohkura, Y. (1986) *J. Chromatogr.*, **377**, 111.
[221] Hara, S., Takemori, Y., Yamaguchi, M., Nakamura, M. and Ohkura, Y. (1987) *Anal. Biochem.*, **164**, 138.
[222] Hara, S., Yamaguchi, M., Takemori, Y., Furuhata, K., Ogura, H. and Nakamura, M. (1987) *Anal. Biochem.*, **179**, 162.
[223] Lagana, A., Mario, A., Fago, G. and Martinez, B.P. (1995) *Anal. Chim. Acta*, **306**, 65.
[224] Iwata, T., Hara, S., Yamaguchi, M. and Nakamura, M. (1985) *Chem. Pharm. Bull.*, **33**, 3499.
[225] Iwata, T., Yamaguchi, M. and Nakamura, M. (1985) *J. Chromatogr.*, **344**, 351.
[226] Imai, K., Fukushima, T. and Yokosu, H. (1994) *Biomed. Chromatogr.*, **8**, 107.
[227] Nagaoka, H., Nohta, H., Kaetsu, Y., Saito, M. and Ohkura, Y. (1989) *Anal. Sci.*, **5**, 525.
[228] Hamada, C., Iwasaki, M., Kuroda, N. and Ohkura, Y. (1985) *J. Chromatogr.*, **341**, 426.
[229] Matsuoka, C., Nohta, H., Kuroda, N. and Ohkura, Y. (1985) *J. Chromatogr.*, **341**, 342.
[230] Karlsson, K.E., Wiesler, D., Alasandro, M. and Novotory, M. (1985) *Anal. Chem.*, **57**, 229.
[231] Werhoven-Goewie, C.E., Brinkman, U.A.Th. and Frei, R.W. (1980) *Anal. Chim. Acta*, **114**, 147.
[232] Sommandossi, J.P., Lemar, M., Necciari, J., Sumirtapura, Y., Cano, J.P. and Gaillot, J. (1982) *J. Chromatogr.*, **228**, 205.
[233] Tocksteinova, D., Slosar, J., Urbanek, J. and Churacek, J. (1979) *Microchim. Acta II*, **193**.
[234] Jandera, P., Pechova, H., Tocksteinova, D., Churacek, J. and Kralovsky, J. (1982) *Chromatographia*, **16**, 275.
[235] Toyo'oka, T., Liu, Y., Hanioka, N., Jinno, H. and Ando, M. (1994) *Anal. Chim. Acta*, **285**, 343.
[236] Tsuruta, Y., Date, Y. and Kohashi, K. (1991) *Anal. Sci.*, **7**, 411.
[237] Takadate, A., Irikura, M., Suehiro, T., Fujino, H. and Goya, S. (1985) *Chem. Pharm. Bull.*, **33**, 1164.
[238] Yamaguchi, M., Iwata, T., Nakamura, M. and Ohkura, Y. (1987) *Anal. Chim. Acta*, **193**, 209.
[239] Iwata, T., Hanazono, H., Yamaguchi, M., Nakamura, M. and Ohkura, Y. (1989) *Anal. Sci.*, **5**, 671.
[240] Goto, J., Shamsa, F. and Nambara, T. (1983) *J. Liq. Chromatogr.*, **6**, 1977.
[241] Goto, J., Komatsu, S., Inada, M. and Nambara, T. (1986) *Anal. Sci.*, **2**, 585.
[242] Goto, J., Saito, M., Chikai, T., Goto, N. and Nambara, T. (1983) *J. Chromatogr.*, **276**, 289.
[243] Goto, J., Komatsu, S., Goto, N. and Nambara, T. (1981) *Chem. Pharm. Bull.*, **29**, 899.
[244] Goto, J., Shao, G., Fukusawa, M., Nambara, T. and Miyano, S. (1991) *Anal. Sci.*, **7**, 645.

[245] Takadate, A., Iwai, M., Fujino, H., Tahara, K. and Goya, S. (1983) *Yakugaku Zasshi*, **103**, 962.
[246] Schultz, B. and Hansen, S.H. (1982) *J. Chromatogr.*, **228**, 279.
[247] Williams, A.T.R., Winfield, S.A. and Balloli, R.C. (1982) *J. Chromatogr.*, **240**, 224.
[248] Lawrence, J.F., Renault, C. and Frei, R.W. (1976) *J. Chromatogr.*, **121**, 343.
[249] Schmidt, G.L.P.O., Vandemark, F.L. and Slavin, W. (1978) *Anal. Biochem.*, **91**, 636.
[250] Roos, R.W. and Medwick, T. (1980) *J. Chromatogr. Sci.*, **18**, 626.
[251] Yamada, K., Kayama, E., Aizawa, Y., Oka, K. and Hara, S. (1981) *J. Chromatogr.*, **223**, 176.
[252] Bongiovanni, R., Burman, K.D., Garis, R.K. and Boehm, T. (1981) *J. Liq. Chromatogr.*, **4**, 813.
[253] Fujino, H. and Goya, S. (1992) *Anal. Sci.*, **8**, 715.
[254] deSilva, K.H., Vest, F.B. and Karnes, H.T. (1996) *Biomed. Chromatogr.*, **10**, 318.
[255] Ishida, J., Kai, M. and Ohkura, Y. (1986) *J. Chromatogr.*, **356**, 171.
[256] Kai, M., Ishida, J. and Ohkura, Y. (1988) *J. Chromatogr.*, **430**, 271.
[257] Ishida, J., Kai, M. and Ohkura, Y. (1985) *J. Chromatogr.*, **344**, 267.
[258] Kai, M. and Ohkura, Y. (1986) *Anal. Chim. Acta*, **182**, 177.
[259] Nakano, M., Kai, M., Ohno, M. and Ohkura, Y. (1987) *J. Chromatogr.*, **411**, 305.
[260] Kai, M., Nakano, M., Zhang, G.Q. and Ohkura, Y. (1987) *Anal. Sci.*, **5**, 289.
[261] Zhang. G.Q., Kai, M. and Ohkura, Y. (1991) *Chem. Pharm. Bull.*, **39**, 126.
[262] Kai, M., Nakashima, A. and Ohkura, Y. (1997) *J. Chromatogr. B*, **688**, 205.
[263] Zhang, G.Q., Kai, M. and Ohkura, Y. (1990) *Anal. Sci.*, **6**, 671.
[264] Zhang, G.Q., Kai, M. and Ohkura, Y. (1991) *Chem. Pharm. Bull.*, **39**, 2369.
[265] Ohno, M., Kai, M. and Ohkura, Y. (1988) *J. Chromatogr.*, **430**, 291.
[266] Zhang, G.Q., Kai, M. and Ohkura, Y. (1991) *Anal. Sci.*, **7**, 561.
[267] Tsuruta, Y., Moritani, K., Date, Y. and Kohashi, K. (1992) *Anal. Sci.*, **8**, 393.
[268] Nakashima, K., Umekawa, C., Yoshida, H., Nakatsuji, S. and Akiyama, S. (1987) *J. Chromatogr.*, **414**, 11.
[269] Nakashima, K., Nishida, K., Nakatsuji, S. and Akiyama, S. (1986) *Chem. Pharm. Bull.*, **34**, 1678.
[270] Hatakeyama, E., Matsumoto, N., Ochi, T., Suzuki, T., Ohrui, H. and Meguro, H. (1989) *Anal. Sci.*, **5**, 657.
[271] Schafer, J., Turnbull, D.M. and Reichmann, H. (1993) *Anal. Biochem.*, **209**, 53.
[272] Miners, J.O., Fearnly, I., Smith, K.J., Brikett, D.J., Brooks, P.M. and Whitehouse, M.W. (1983) *J. Chromatogr.*, **275**, 89.
[273] Yamamoto, K., Sekine, T. and Kanaoka, Y. (1977) *Anal. Biochem.*, **79**, 83.
[274] Nakashima, K., Umekawa, C., Nakatsuji, S., Akiyama, S. and Givens, R.S. (1989) *Biomed. Chromatogr.*, **3**, 39.
[275] Frank, H., Thiel, D. and Langer, K. (1984) *J. Chromatogr.*, **309**, 261.
[276] Nishikawa, Y. and Kuwata, K. (1985) *Anal. Chem.*, **57**, 1864.
[277] Watanabe, Y. and Imai, K. (1983) *Anal. Chem.*, **55**, 1786.
[278] Toyo'oka, T., Miyano, H. and Imai, K. (1986) *Biomed. Chromatogr.*, **1**, 15.
[279] Andrew, J.L., Ghosh, P., Ternai, B. and Whitehouse, M.W. (1982) *Arch. Biochem. Biophys.*, **214**, 386.
[280] Sueyoshi, T., Miyata, T., Iwanaga, S., Toyo'oka, T. and Imai, K. (1985) *J. Biochem.*, **97**, 1811.
[281] Toyo'oka, T. and Imai, K. (1983) *J. Chromatogr.*, **282**, 495.
[282] Toyo'oka, T. and Imai, K. (1984) *Analyst*, **109**, 1003.
[283] Daskakakis, I., Lucock, M.D., Anderson, A., Schorah, C.J. and Levene, M.I. (1996) *Biomed. Chromatogr.*, **10**, 205.
[284] Toyo'oka, T. and Imai, K. (1984) *Anal. Chem.*, **56**, 2461.
[285] Toyo'oka, T. and Imai, K. (1985) *Anal. Chem.*, **57**, 1931.
[286] Toyo'oka, T., Suzuki, T., Saito, Y. and Imai, K. (1989) *Analyst*, **114**, 413.
[287] Toyo'oka, T., Imai, K. and Kawahara, Y. (1984) *J. Pharm. Biomed. Anal.*, **2**, 473.
[288] Ling, B.L. and Baeyens, W.R.G. (1991) *Anal. Chim. Acta*, **255**, 283.
[289] Toyo'oka, T., Chokshi, H.P., Carlson, R.G., Givens, R.S. Lunte, S.M. (1993) *Analyst*, **118**, 257.
[290] Fahey, R.C., Newton, G.L., Dorian, R. and Kosower, E.M. (1981) *Anal. Biochem.*, **111**, 357.
[291] Burton, N.K., Aherne, G.W. and Marks, V. (1984) *J. Chromatogr.*, **309**, 409.
[292] Machida, M., Ushijima, N., Machida, M.I. and Kanaoka, Y. (1975) *Chem. Pharm. Bull.*, **23**, 1385.
[293] Machida, M., Machida, M.I., Sekine, T. and Kanaoka, Y. (1977) *Chem. Pharm. Bull.*, **25**, 1678.
[294] Nakamura, H. and Tamura, Z. (1981) *Anal. Chem.*, **53**, 2190.
[295] Sano, A., Takitani, S. and Nakamura, H. (1995) *Anal. Sci.*, **11**, 299.
[296] Mopper, K. and Delmas, D. (1984) *Anal. Chem.*, **56**, 2557.

[297] Mank, A.J.G., Molenaar, E.J., Lingeman, H., Gooijer, C., Brinkman, U.A.Th. and Velthort, N.H. (1993) *Anal. Chem.*, **65**, 2197.

[298] Kawasaki, T., Maeda, M. and Tsuji, A. (1979) *J. Chromatogr.*, **163**, 143.

[299] Hull, S.R. and Turco, S.J. (1985) *Anal. Biochem.*, **146**, 143.

[300] Gubitz, G., Wintersteiger, R. and Frei, R.W. (1984) *J. Liq. Chromatogr.*, **7**, 839.

[301] Kawasaki, T., Maeda, M. and Tsuji, A. (1983) *J. Chromatogr.*, **272**, 261.

[302] Goto, J., Saito, Y. and Nambara, T. (1989) *Anal. Sci.*, **5**, 399.

[303] Traore, F., Tod, M., Chalom, J., Farinotti, R. and Mahuzier, G. (1992) *Anal. Chim. Acta*, **269**, 211.

[304] Iwata, T., Hirose, T., Nakamura, M. and Yamaguchi, M. (1993) *Analyst*, **118**, 517.

[305] Tsuruta, Y., Tonogaito, H., Date, Y., Sugino, E. and Kohashi, K. (1993) *Anal. Sci.*, **9**, 311.

[306] Tsuruta, Y., Date, Y., Tonogaito, H., Sugihara, N., Furuno, K. and Kohashi, K. (1994) *Analyst*, **119**, 1047.

[307] Uzu, S., Kanda, S., Imai, K., Nakashima, K. and Akiyama, S. (1990) *Analyst*, **115**, 1477.

[308] Nakashima, K., Hidaka, Y., Yoshida, T., Kuroda, N. and Akiyama, S. (1994) *J. Chromatogr. B*, **661**, 205.

[309] Iwata, T., Ishimaru, T. and Yamaguchi, M. (1997) *Anal. Sci.*, **13**, 501.

[310] Nakamura, M., Toda, M., Saito, M. and Ohkura, Y. (1982) *Anal. Chim. Acta*, **134**, 39.

[311] Chao, W.F., Kai, M., Ishida, J., Ohkura, Y., Hara, S. and Yamaguchi, M. (1988) *Anal. Chim. Acta*, **215**, 259.

[312] Ishida, J., Yamaguchi, M., Kai, M., Ohkura, Y. and Nakamura, M. (1984) *J. Chromatogr.*, **305**, 381.

[313] Hara, S., Yamaguchi, M., Takemori, Y., Yoshitake, T. and Nakamura, M. (1988) *Anal. Chim. Acta*, **215**, 267.

[314] Yamaguchi, M., Yoshitake, T., Ishida, J. and Nakamura, M. (1989) *Chem. Pharm. Bull.*, **37**, 3022.

[315] Yoshitake, T., Ishida, J., Sonezaki, S. and Yamaguchi, M. (1992) *Biomed. Chromatogr.*, **6**, 217.

[316] Inoue, S., Inokuma, M., Harada, T., Shibutani, Y., Yoshitake, T., Charles, B., Ishida, J. and Yamaguchi, M. (1994) *J. Chromatogr. B*, **661**, 15.

[317] Hara, S., Nakamura, M., Sakai, F., Nohta, H., Ohkura, Y. and Yamaguchi, M. (1994) *Anal. Chim. Acta*, **291**, 189.

[318] Mopper, K., Dawson, R., Liebezeit, G. and Hansen, H.P. (1980) *Anal. Chem.*, **52**, 2018.

[319] Kato, T. and Kinoshita, T. (1980) *Anal. Biochem.*, **106**, 238.

[320] DelNozal, M.J., Bernal, J.L., Gomez, F.J., Antolin, A. and Toribio, L. (1992) *J. Chromatogr.*, **607**, 191.

[321] Honda, S., Matsuda, Y., Takahashi, M., Kakehi, K. and Ganno, S. (1980) *Anal. Chem.*, **52**, 1079.

[322] Honda, S., Suzuki, S., Takahashi, M., Kakehi, K. and Ganno, S. (1983) *Anal. Biochem.*, **134**, 34.

[323] Mikami, H. and Ishida, Y. (1983) *Bunscki Kagaku*, **32**, E207.

[324] Kai, M., Tamura, K., Yamaguchi, M. and Ohkura, Y. (1985) *Anal. Sci.*, **1**, 59.

[325] Umegae, Y., Nohta, H. and Ohkura, Y. (1989) *Anal. Chim. Acta*, **217**, 263.

[326] Honda, S., Kakehi, K., Fujikawa, K., Oka, Y. and Takahashi, M. (1988) *Carbohydr. Res.*, **103**, 59.

[327] Coquet, A., Veuthey, J.L. and Haerdi, W. (1991) *J. Chromatogr.*, **553**, 255.

[328] Coquet, A., Veuthey, J.L. and Haerdi, W. (1991) *Anal. Chim. Acta*, **252**, 173.

[329] Umegae, Y., Nohta, H. and Ohkura, Y. (1989) *Anal. Sci.*, **5**, 675.

[330] Umegae, Y., Nohta, H. and Ohkura, Y. (1990) *Chem. Pharm. Bull.*, **38**, 963.

[331] Zhang, G.Q., Kai, M., Nohta, H., Umegae, Y. and Ohkura, Y. (1993) *Anal. Sci.*, **9**, 9.

[332] Yamauchi, S., Nakai, C., Nimura, N., Kinoshita, T. and Hanai, T. (1993) *Analyst*, **118**, 773.

[333] Yamauchi, S., Nakai, C., Nimura, N., Kinoshita, T. and Hanai, T. (1993) *Analyst*, **118**, 769.

[334] Toyo'oka, T. and Kuze, A. (1997) *Biomed. Chromatogr.*, **11**, 132.

[335] Hong, S., Nakamura, H. and Nakajima, T. (1994) *Anal. Sci.*, **10**, 647.

[336] Coles, E., Reinhold, V.N. and Carr, S.A. (1985) *Carbohydr. Res.*, **139**, 1.

[337] Maness, N.O., Miranda, E.T. and Mort, A.J. (1991) *J. Chromatogr.*, **587**, 177.

[338] Takeda, M., Maeda, M. and Tsuji, A. (1982) *J. Chromatogr.*, **244**, 347.

[339] Zhang, R.E., Cao, Y.L. and Hearn, M.W. (1991) *Anal. Biochem.*, **195**, 160.

[340] Liu, J., Shirota, O. and Novotony, M. (1991) *Anal. Chem.*, **63**, 413.

[341] Shimada, K. and Oe, T. (1990) *Anal. Sci.*, **6**, 461.

[342] Shimizu, M., Kamachi, S., Nishii, Y. and Yamada, S. (1991) *Anal. Biochem.*, **194**, 77.

[343] Shimizu, M., Gao, Y., Aso, T., Nakatsu, K. and Yamada, S. (1992) *Anal. Biochem.*, **204**, 258.

[344] Shimada, K., Oe, T. and Mizuguchi, T. (1991) *Analyst*, **116**, 1393.

[345] Al-Deen, A.N., Cecchini, D.C., Abdel-Baky, A., Moneam, N.M.A. and Giese, R.W. (1990) *J. Chromatogr.*, **512**, 409.

[346] Sonoki, S., Sanda, A. and Hisamatsu, S. (1994) *J. Liq. Chromatogr.*, **17**, 1057.

[347] Sonoki, S., Kadoike, Y., Kiyokawa, M. and Hisamatsu, S. (1993) *J. Liq. Chromatogr.*, **16**, 2731.

[348] Sonoki, S., Hisamatsu, S. and Kiuchi, A. (1993) *Nucleic Acids Res.*, **21**, 2776.

[349] Nagaoka, H., Nohta, H., Saito, M. and Ohkura, Y. (1992) *Anal. Sci.*, **8**, 345.

[350] Nagaoka, H., Nohta, H., Saito, M. and Ohkura, Y. (1992) *Anal. Sci.*, **8**, 565.

[351] Nagaoka, H., Nohta, H., Saito, M. and Ohkura, Y. (1992) *Chem. Pharm. Bull.*, **40**, 2202.

[352] Nagaoka, H., Nohta, H., Saito, M. and Ohkura, Y. (1992) *Chem. Pharm. Bull.*, **40**, 2559.

[353] Umegae, Y., Nohta, H. and Ohkura, Y. (1990) *Chem. Pharm. Bull.*, **38**, 452.

[354] Umegae, Y., Nohta, H. and Ohkura, Y. (1990) *J. Chromatogr.*, **515**, 495.

[355] Umegae, Y., Nohta, H. and Ohkura, Y. (1990) *Anal. Sci.*, **6**, 519.

[356] Kochetkov, N.K., Shibaev, V.N. and Kost, A.A. (1971) *Tetrahedron Lett.*, **1993**.

[357] Yoshioka, Y., Nishidate, K., Iizuka, H., Nakamura, A., Ei-Merzabani, M.M., Tamura, Z. and Miyazaki, T. (1984) *J. Chromatogr.*, **309**, 63.

[358] Levitt, B., Head, R.J. and Westfall, D.P. (1984) *Anal. Biochem.*, **137**, 93.

[359] Yoshioka, M., Yamada, K., Abu-Zeid, M.M., Fujimori, H., Fuke, A., Hirai, K., Goto, A., Ishii, M., Sugimoto, T. and Parvez, H. (1987) *J. Chromatogr.*, **400**, 133.

[360] Mohri, K., Takeuchi, K., Shinozuka, K., Bjur, R.A. and Westfall, D.P. (1993) *Anal. Biochem.*, **210**, 262.

[361] Fujimori, H., Sasaki, T., Hibi, K., Senda, M. and Yoshioka, M. (1990) *J. Chromatogr.*, **515**, 363.

[362] Kai, M., Ohkura, Y., Yonekura, S. and Iwasaki, M. (1988) *Anal. Chim. Acta*, **207**, 243.

[363] Yonekura, S., Iwasaki, M., Kai, M. and Ohkura, Y. (1993) *J. Chromatogr.*, **641**, 235.

[364] Yonekura, S., Iwasaki, M., Kai, M. and Ohkura, Y. (1994) *Anal. Sci.*, **10**, 247.

[365] Yonekura, S., Iwasaki, M., Kai, M. and Ohkura, Y. (1994) *J. Chromatogr.*, **654**, 19.

[366] Ohba, Y., Kai, M., Nohta, H. and Ohkura, Y. (1994) *Anal. Chim. Acta*, **287**, 215.

[367] Eisenberg, E.J. and Cundy, K.C. (1996) *J. Chromatogr. B*, **679**, 119.

[368] Iwamoto, M., Yoshida, S. and Hirose, S. (1984) *J. Chromatogr.*, **310**, 51.

[369] Yoshida, S. and Hirose, S. (1986) *J. Chromatogr.*, **383**, 61.

[370] Yamaguchi, M., Hara, S., Matsunaga, R., Nakamura, M. and Ohkura, Y. (1985) *J. Chromatogr.*, **346**, 227.

[371] Yamaguchi, M., Nakamura, M., Kuroda, N. and Ohkura, Y. (1987) *Anal. Sci.*, **3**, 75.

[372] Kaneta, T., Okamoto, T. and Imasaka, T. (1996) *Anal. Sci.*, **12**, 875.

# 4

# Reagent for CL Detection

**Naotaka Kuroda and Kenichiro Nakashima**
Nagasaki University, Nagasaki, Japan

| | |
|---|---|
| 4.1. Introduction | 167 |
| 4.2. Label of amines ($-NH_2$, $-NH$) | 171 |
| 4.3. Label of carboxyl ($-COOH$) | 176 |
| 4.4. Label of hydroxyl ($-OH$) and thiol ($-SH$) | 178 |
| 4.5. Label of other functional groups | 179 |
| 4.6. Application | 183 |
| References | 187 |

## 4.1 INTRODUCTION

Chemiluminescence (CL) is the emission of light as the result of electronic excitation of the luminescing species by a chemical reaction of a precursor of that species. The process of emission of light in CL, i.e., decay of excited molecules to the electronic ground state, is the same as in photoluminescence except for its excitation process. Therefore, the resultant spectra of CL are identical to the fluorescence spectra of the emitter. Bioluminescence is the kind of CL produced by certain living organisms (e.g., firefly, jellyfish, and squid), which involves an oxidation of luciferin substrates catalyzed by luciferase enzymes, or a reaction of photoprotein aequorin with $Ca^{2+}$ [1–4].

A CL reaction doesn't need an exciting light source, and thus is not accompanied by any light scattering. This permits an increase in the detector's sensitivity, and the attainment of a large signal-to-noise ratio (S/N). Therefore, the CL detection system has been applied to the determination of a wide range of analytes, e.g., ultra trace metal ions, environmental pollutants, pharmaceuticals, and biological components.

The phenomenon of CL has been known for a long time. In 1877, Radziszewski [5] found that lophine (2,4,5-triphenylimidazole) emitted a

Edited by Toshimasa Toyo'oka: *Modern Derivatization Methods for Separation Sciences* 

green light when it reacted with oxygen in the presence of a base. This is the first example of CL using a synthetic organic compound. Since then, a number of chemiluminescent compounds have been synthesized and studied for their chemiluminescent properties; luminol (5-amino-2,3-dihydro-1,4-phthalazinedione), lucigenin (*N*,*N*′-dimethyl-9,9′-bisacridinium dinitrate), tetrakis(dimethylamino)ethylene [6], indole derivatives [7], Schiff bases [8], and others have been reported. Various CL reactions were utilized in analytical chemistry, and the luminol CL reaction discovered by Albrecht [9] in 1928 is one of the best known examples. The luminol CL reaction is catalyzed with hemin and thus has been utilized to identify blood in many criminal investigations. CL of lucigenin was first observed by Glue and Petsch [10] in 1935, and it has been utilized for the assay of phagocytic leukocyte oxygenation activities [11]. An acridinium ester synthesized by Weeks *et al.* [12] has also been applied to an immunoassay of human $\alpha$-fetoprotein. In 1968, McCapra showed that 1,2-dioxetanes produce emission of light when cleaved concertedly to form two carbonyl compounds [13]. Many thermally stable dioxetanes were synthesized and studied on their chemically or enzymatically triggered chemiluminescent decompositions [14]. Recently, adamantyl-1,2-dioxetane derivatives have been applied to enzyme immunoassays to detect alkaline phosphatase [15] and $\beta$-D-galactosidase [16]. These compounds so far described were directly able to produce emission of light by themselves due to oxidation reactions with oxygen or hydrogen peroxide.

However, another chemiluminescent reaction involving an energy transfer reaction has also become well known. The investigation on this area originated from the observation of visible light upon reaction of oxalyl chloride with hydrogen peroxide reported by Chandross in 1963 [17]. Rauhut *et al.* reviewed the oxalyl chloride chemiluminescent system in 1965 [18], and showed that aryloxalates could also be used for this system instead of oxalyl chloride. They have synthesized numerous oxalates and oxamides, and established a new strongly luminescing system, namely, the peroxyoxalate CL (PO-CL) system. In this system, oxalates and oxamides react with hydrogen peroxide to yield intermediate peroxides which produce light by energy transfer to a co-existing fluorophore. Therefore, a suitable combination of oxalates or oxamides with strongly fluorescent compounds yields an intense emission. For example, when 9,10-bis(phenylethynyl)anthracence was used as a fluorophore, oxamide showed a quantum yield of 34% [19].

The emission of light from aromatic hydrocarbons such as 9,10-diphenylanthracene and rubrene was observed in the presence of diphenoylperoxide by Koo and Schuster [20,21] in 1977/78. Diphenoylperoxide was decomposed to yield the non-fluorescent benzocoumarin. The mechanism of this phenomenon was considered to be that the aromatic hydrocarbons, having low oxidative electric potentials, transfer an electron to diphenoylperoxide to form a charge-transfer complex from which benzocoumarin and the excited hydrocarbon are produced: this mechanism is known as the CIEEL mechanism.

Most CL reactions are oxidation reactions except for that of lucigenin. Organic chemiluminescent compounds are excited by oxidation with oxygen or hydrogen peroxide, and then emit light in the visible region of the emission spectrum (400–800 nm). The efficiency of the CL reaction can be determined according to the following equation,

$$\Phi_{CL} = \Phi_C \times \Phi_E \times \Phi_F$$

where $\Phi_{CL}$ is the total CL quantum efficiency; $\Phi_C$, the efficiency of the chemical reaction; $\Phi_E$, the electronic excitation energy; and $\Phi_F$, the fluorescence efficiency. The total light corresponds to the amounts of substrates in the CL reaction. Therefore, these substrates can be determined quantitatively by measuring the light produced in the reaction.

In spite of the fact that many kinds of CL reactions have been so far reported, few of them are used as practical analytical tools. Some reviews on the reagents for analyses with CL detection have already been published [22–24]. Among many kinds of separation techniques, CL detection is most frequently employed in combination with high-performance liquid chromatography (HPLC).

As the reagents that show direct CL reactions (chemiluminogenic compounds), luminol and its derivatives such as *N*-(4-aminobutyl)-*N*-ethylisoluminol (ABEI) are exclusively used. Luminol is a powerful detection tool for determining hydrogen peroxide in biological samples by HPLC. ABEI was synthesized by Schroeder *et al.* [25] and applied to the CL immunoassay of biotin and thyroxine [26]. ABEI possessing an alkyl amine moiety reacts with amines and carboxylic acids in the presence of appropriate condensing reagents. Luminol derivatives produce CL by oxidation with hydrogen peroxide in an alkaline medium. Metal ions (i.e., iron(II), cobalt(II), etc), halogen, hemin, hemoglobin, etc. catalyze the luminol CL (Fig. 4.1).

A PO-CL system is one of the most efficient and versatile CL systems, and thus has been most widely used for post-column detection in HPLC [27]. As described above, PO-CL is the emission of light produced by a chemical reaction of aryloxalate, peroxide and a fluorophore (Fig. 4.2). The reaction is catalyzed with some bases such as imidazole and triethylamine to enhance the CL intensity. Imidazole has been proven to be superior to all other catalysts tested [28] and is most widely used. In this CL reaction, the total emission of light is proportional to the concentration of the substrates, i.e., oxalates, hydrogen peroxide, fluorophores or bases. Consequently, these substrates and bases could be determined by using the PO-CL reaction as a detection system.

The kind of aryloxalate used in the PO-CL system is a very important factor in CL intensity. Several aryloxalates [e.g., bis(2,4,6-trichlorophenyl)oxalate (TCPO), bis(2,4-dinitrophenyl)oxalate (DNPO), bis(2,6-difluorophenyl)oxalate (DFPO), bis(pentafluorophenyl)oxalate (PFPO), and bis[2-(3,6,9-trioxadecyloxycarbonyl)-4-phenyl]oxalate (TDPO)] were evaluated on their properties in terms of intensity, rate of CL decay, solubility in different solvents, and stability in the presence of hydrogen peroxide [29–31]. Among them, TCPO is the most frequently used. TDPO and DNPO also have excellent features of solubility in organic solvents such as acetonitrile and methanol. Recently, 1,1′-oxalyldiimidazole (ODI) has been applied to determine hydrogen peroxide, and the results showed that ODI was about 10 times more sensitive than TCPO [32]. The detection of fluorophores by PO-CL was first reported by Curtis and Seitz who detected dansylated amino

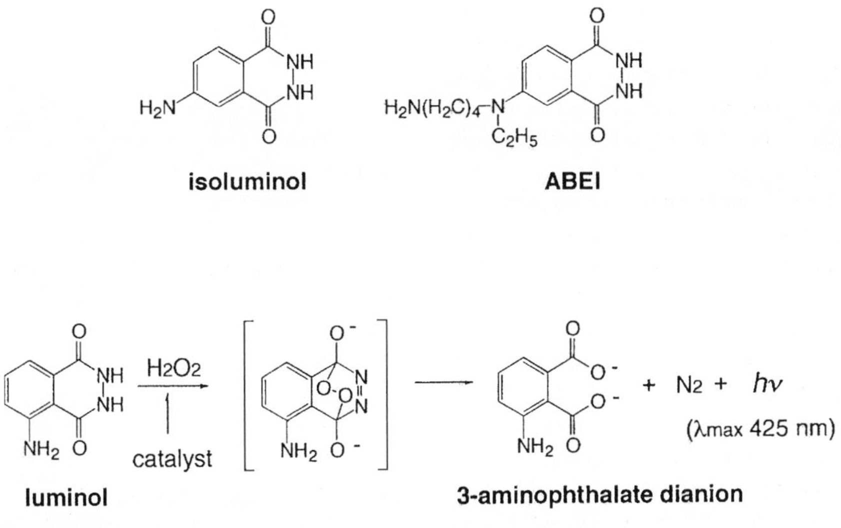

**Fig. 4.1.** Luminol and its derivatives, and CL reaction of luminol.

$Ar-O-C(=O)-C(=O)-O-Ar \xrightarrow{H_2O_2} Ar-O-C(=O)-C(=O)-O-OH \longrightarrow$ [ 1,2-dioxetane intermediates ] $\xrightarrow{F}$ [ $F^{\dot{+}}$ radical anion intermediates ] $\longrightarrow F^* \longrightarrow F + h\nu$

aryloxalate

F: fluorophore

Ar :

**TCPO** **DNPO** **DFPO** **TDPO**

**Fig. 4.2.** PO-CL reaction and aryloxalates.

acids on a thin-layer chromatographic (TLC) plate [33]. When chemically excited, not all the fluorophores are more efficiently detected than when using photoexcitation. Several characteristics of the fluorophore, including high fluorescence efficiency, low oxidation and low singlet energy, contribute to efficient chemical excitation [34].

A CL detection system for HPLC should be constructed by considering the conditions for HPLC separation, efficiency of CL reaction, and stabilities of reagents. Typical systems are illustrated in Fig. 4.3. System B is generally used for the CL detection of luminol derivatives, in which two pumps are required for delivering the solutions of hydrogen peroxide and a catalyst such as potassium hexacyanoferrate(III), respectively. In PO-CL detection, system A is the most simple one and is used in certain cases in which separated hydrogen peroxide is determined by using a single solution containing an aryloxalate and a fluorophore for post-column CL reaction. Systems B and C are widely used for PO-CL detection after HPLC separation. System B is employed when the solutions for CL reaction are mixed successively with an eluent. In the case when the CL reaction needs to be optimized before mixing with the CL reagent, system C is

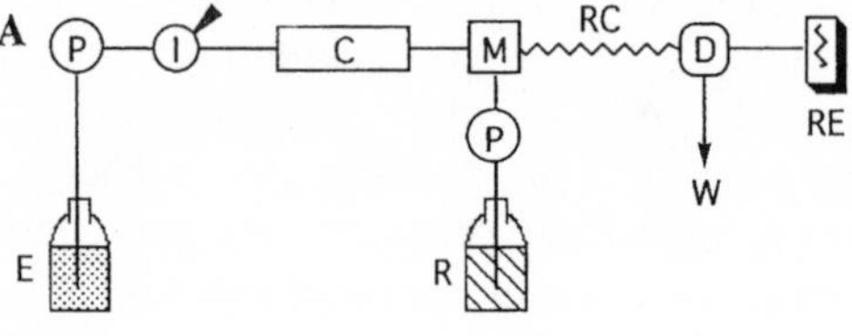


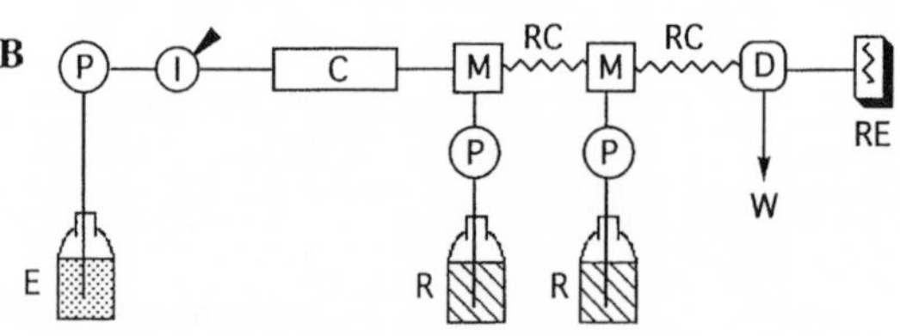


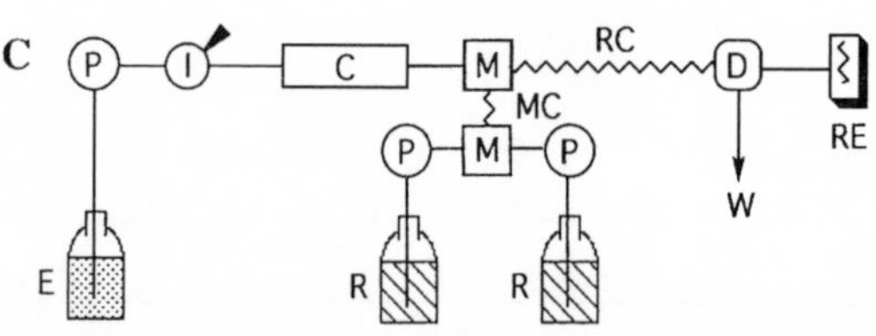


**Fig. 4.3.** CL detection systems in combination with HPLC: P, pump; I, injector; C, column; M, mixing tee; D, detector; RC, reaction coil; MC, mixing coil; RE, recorder; E, eluent; R, reagent; W, waste.

convenient. A more complicated system will be required when a gradient elution is selected for the separation. A simple system is preferred from the point of view of operation and economy.

## 4.2 LABEL OF AMINES ($-NH_2$, $-NH$)

Generally, amino groups are so reactive that labeling reactions proceed under fairly mild conditions. Many CL labeling reagents for amines have been developed and applied to the highly sensitive determination of a wide variety of analytes.

Among luminol derivatives, ABEI has been used as a labeling reagent for primary and secondary amines using *N*,*N*′-disuccinimidyl carbonate (DSC) as a condensing reagent [35]. Labeling of amines with ABEI is performed in two steps; at first, the ABEI-DSC intermediate is prepared and then amines react with ABEI-DSC in the presence of triethylamine as a catalyst (Fig. 4.4). ABEI derivatives of amines were separated and detected by their CL which was produced by reaction with hydrogen peroxide and potassium hexacyanoferrate(III). This ABEI-CL detection technique in HPLC has been applied to determine abuse drugs such as methamphetamine and amphetamine in serum or urine of methamphetamine addicts [36,37]. Twenty fmol of methamphetamine in serum and urine, and 100 fmol amphetamine in urine could be detected.

Luminol derivatives having isothiocyanate as a reactive functional group for amines, 4-isothiocyanatophthalhydrazide [38] and 6-isothiocyanobenzo[*g*]phthalazine-1,4-(2*H*,3*H*)-dione (IPO) have been synthesized [39] as highly sensitive chemiluminescence labeling reagents. Twelve amino acids labeled with 4-isothiocyanatophthalhydrazide have been separated and detected with an average detection limit of 10 fmol per injection [38]. Various primary and secondary amines reacted with IPO within ca. 10 min at 80 °C in the presence of triethylamine, and separated by a reversed-phase HPLC (Fig. 4.5). The detection limits ($S/N = 3$) for primary and secondary amines obtained with IPO are in the range 30–120 and 0.8–3 fmol per injection, respectively; the sensitivities for secondary amines, which are 10–100 times more sensitive than those for primary amines, are 7–20 times higher than those obtained with ABEI

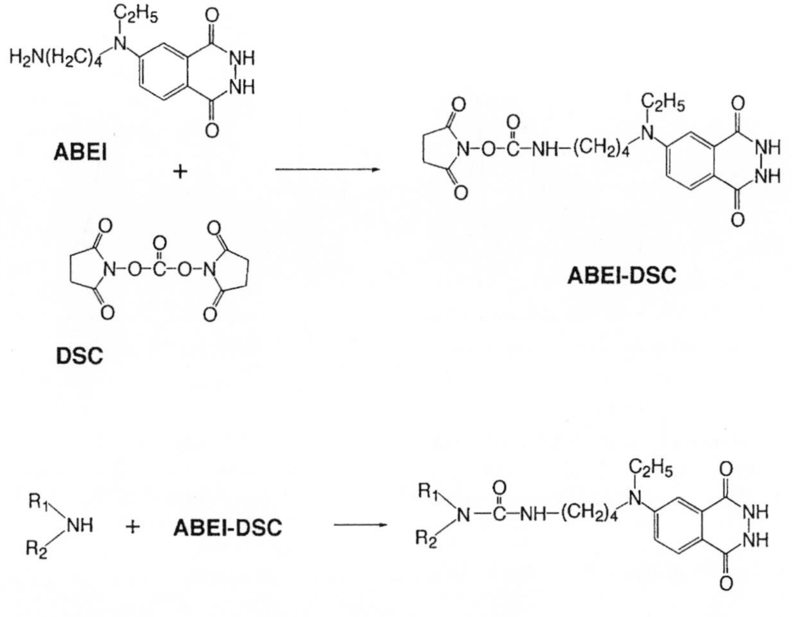

**Fig. 4.4.** Labeling reaction of primary and secondary amines with ABEI.

**IPO**

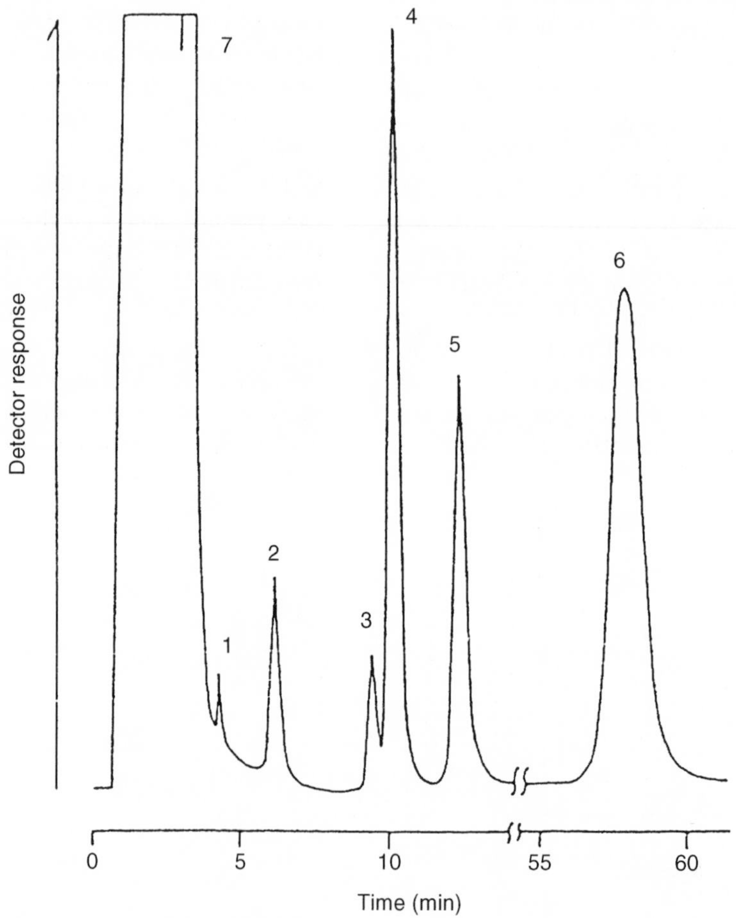


**Fig. 4.5.** IPO and chromatogram of IPO derivatives of amines. Peaks (pmol on column): 1 = *n*-hexylamine (5.0); 2 = *n*-heptylamine (5.0); 3 = *n*-octylamine (5.0); 4 = *N*-methyloctylamine (0.5); 5 = di-*n*-amylamine (0.5); 6 = di-*n*-hexylamine (0.5); 7 = reagent blank. Reprinted from *Analytical Chimica Acta*, 302, J. Ishida, N. Horike and M. Yamaguichi, 6-Isothiocyanatobenzo[g]phthalazine-1.4 (2H,3H)-dione as a highly sensitive chemiluminescence derivatization reagent for amines in liquid chromatography, 61–67, 1995, with kind permission of Elsevier Science—NL, Sara Burgerhartstraat 25, 1055 KV Amsterdam, The Netherlands.

and 4-isothiocyanatophthalhydrazide. IPO has also been successfully applied to the analysis of maprotiline, a widely used antidepressant, in human plasma [40].

A PO-CL detection system in HPLC was utilized for a highly sensitive determination of amines labeled with various fluorescent compounds. Fluorescamine was used for the determination of catecholamines and the sensitivity of the method, with 25 fmol of detection limit, was about 20 times higher than that of the conventional fluorescence detection system [41]. Catecholamines in human urine could be determined with less than 10 μl of urine sample. Fluorescamine was also used for the derivatization of histamine by Walters, *et al* [42]. They compared PO-CL detection, using TCPO, with fluorescence detection, and obtained the results that fluorescence detection was superior (the detection limits of fluorescamine-histamine were 13 pg for fluorescence and 1.0 ng for CL detection on column) due to the limitations of the mobile phase needed for the chromatographic separation and the spectral and chemical characteristics of the derivative itself. For regulatory purposes in the daily screening of residual sulfamethazine in edible animal tissues, that in samples of chicken serum and egg was determined using fluorescamine by PO-CL with TDPO and hydrogen peroxide as post-column reagents; the detection limit in the standard solution was 1 ng/ml, and the calibration curve was linear between 1 and 100 ng/ml [43].

Dansyl chloride (Dns-Cl) is often utilized for PO-CL detection. Dns-amino acids (i.e., Dns-alanine, -glutamic acid, -methionine and -norleucine) were determined by using TCPO. These Dns-amino acids were separated within 30 min on a reversed-phase column with a detection limit of 10 fmol [44]. By using a linear gradient elution, sixteen kinds of Dns-amino acids were separated and detected with detection limits of 2–5 fmol (S/N = 2) [45] (Fig. 4.6). To increase

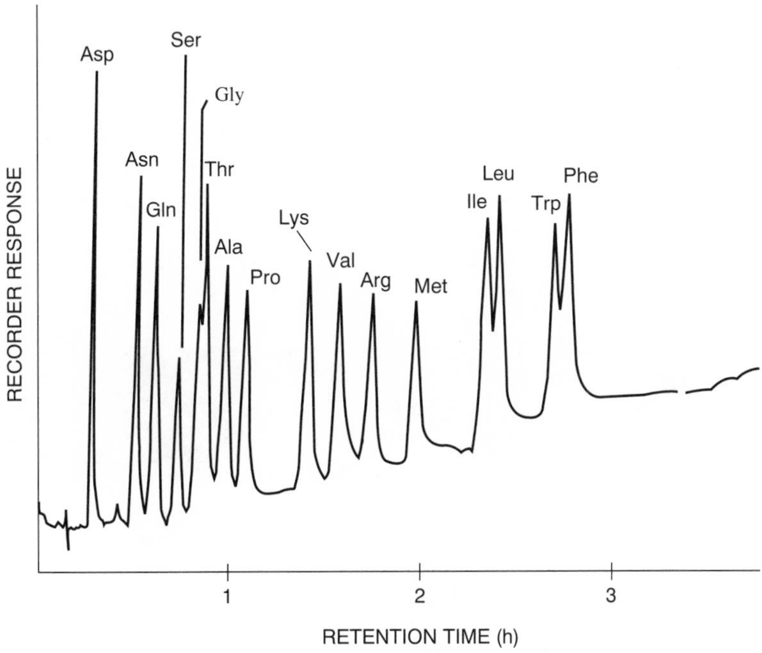


**Fig. 4.6.** Chromatogram of Dns-amino acids standards. Conditions: gradient linear program, from 30% B (70% A) to 99% B (1% A) in 256 min. Each peak corresponds to 100 fmol Dns-amino acid injected. Reprinted from *Journal of Chromatography*, 303, K. Miyaguchi, K. Honda and K. Imai, Sub-picomol chemiluminescence detection of Dns-amino acids separated by high performance liquid chromatography with gradient elution, 173–176, 1984, with kind permission of Elsevier Science—NL, Sara Burgerhartstraat 25, 1055 KV Amsterdam, The Netherlands.

the sensitivity of this method, a microbore column (250 × 1 mm, i.d.) chromatography was introduced. As a result, the injected samples are not subjected to dilution as much as in conventional columns; consequently, a higher peak can be obtained after the CL reaction. Four Dns-amino acids (Ala, Val, Ileu and Phe) were separated and detected with detection limits of the order of 0.2 fmol (S/N = 3) [46], almost ten times lower than those with the ordinary-bore HPLC-CL detection system. A microbore HPLC-PO-CL detection technique for Dns-amino acids was also investigated to determine the *N*-terminal amino acid of bradykinin (Arg-Pro-Pro-Gly-Phe-Ser-Pro-Phe-Arg) as a model peptide. By the improved CL detection system, 20 fmol of each of sixteen Dns-amino acids was separated and detected; the detection limits were 0.8–1.7 fmol. The amount of Dns-Arg liberated after hydrolysis of Dns-bradykinin was about 10% of the starting amount of bradykinin; 40–80 fmol of the starting materials would be sufficient for the analysis of *N*-terminal amino acids [47].

Dns-Cl was successfully applied to the determination of amphetamine, methamphetamine and piperidine in human urine. Methamphetamine and amphetamine are widely abused drugs and these three compounds are usually found in human urine samples of abusers. The corresponding peaks obtained from urine were defined as Dns-derivatives by EI-MS. Methamphetamine levels as low as $2 \times 10^{-10}$ M in urine were determined [48,49]. The non-fluorescent anti-arrhythmic drug, mexiletine, was sensitively determined in rat plasma by an HPLC-PO-CL system with TDPO after being labeled with Dns-Cl. The calibration curve for mexiletine in rat plasma was linear over the range 20–100 ng/ml plasma (20.6–103 fmol/injection) with the detection limit (S/N = 2) of 1.0 fmol [50].

Hayakawa *et al.* investigated HPLC-PO-CL detection of trace levels of amphetamine-related compounds (APs) after derivatization with Dns-Cl, 4-fluoro-7-nitrobenzoxadiazole (NBD-F) and naphthalene-2,3-dicarboxaldehyde (NDA) [51] (Fig. 4.7). TCPO and hydrogen peroxide in acetonitrile was used as a post-column CL reagent. Dns-Cl was the most suitable of the three for the simultaneous determination of both primary and secondary amino APs, while NDA gave the most sensitive derivatives (cyanobenz[*f*]isoindole, CBI derivatives) but only with primary amino APs. The on-column detection limits were 3–4 fmol for Dns-APs and as low as 0.2 fmol for CBI-APs; the method was more sensitive than GC-MS by a factor of 70 for CBI-APs and 3.5 for Dns-APs. Methamphetamine and its metabolites in the urine samples from methamphetamine abusers were successfully determined by the method using Dns-Cl and NDA [52]. As CBI derivatives are usually so stable, NDA has been used for

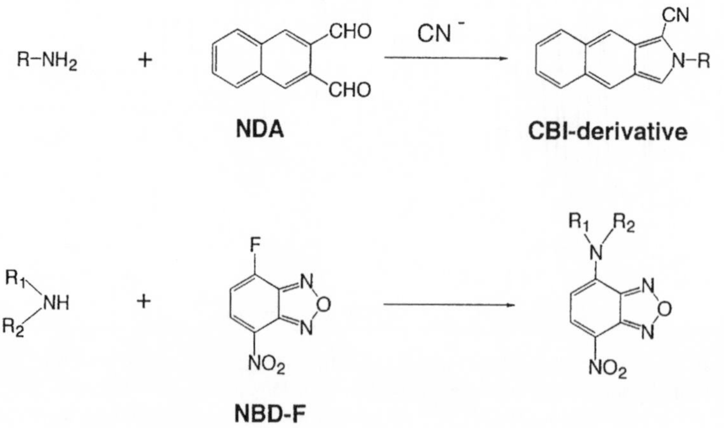

**Fig. 4.7.** Labeling reactions of amines with NDA and NBD-F.

fluorescence labeling of primary amines [53]. NDA was also used for the labeling of dopamine and norepinephrine [54], and the anti-depressant, fluvoxamine [55]; fmol amounts of these CBI derivatives separated by HPLC were detected by PO-CL systems.

The fluorogenic reagent for amines and amino acids, 4-(*N*,*N*-diemethylaminosulfonyl)-7-fluoro-2,1,3-benzoxadiazole (DBD-F) has been used, as well as NBD-F. DBD-F labeled amino acids and epinephrine were separated on a reversed-phase column and detected at the fmols level by PO-CL detection with TDPO [56]. The method was also applied to the determination of metoprolol (a $\beta$-blocker having an isopropylamino group) in serum [57]. The lower detection limit of the drug spiked in serum was 0.8 ng/ml (3 nmol/l) using 20 µl of serum (S/N = 5).

Luminarin 1 is a labeling reagent with quinolizino coumarin structure and an *N*-hydroxysuccinimide ester reactive function, which reacts with primary amines in 20 min at 50 °C, and with secondary amines in 180 min at 70 °C (Fig. 4.8). Several amines (i.e., pentylamine, pyrrolidine, tyramine and proline) were labeled and separated by HPLC [58]. The limit of detection (S/N = 3) with TCPO were between 15 and 100 fmol/injection, which were 3–10 times lower than those with fluorescence detection.

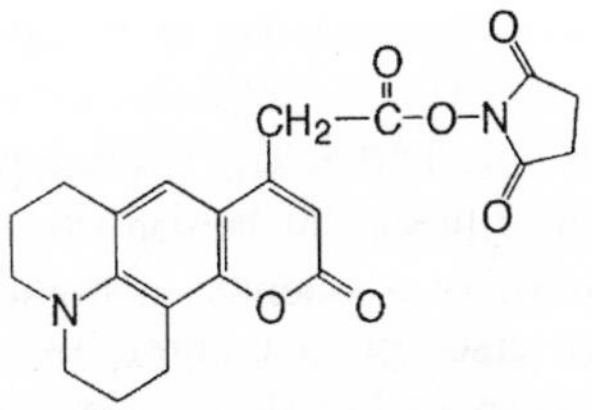

**luminarin 1**

**Fig. 4.8.** Luminarin 1.

A unique CL reaction, tris(2,2′-bipyridyl)ruthenium(II) [$Ru(bpy)_3^{2+}$] for post-column CL reaction, was used for the detection of amino acids [59], Dns-amino acids and oxalate [60,61] after HPLC separation. The oxidative-reduction reaction scheme of CL from $Ru(bpy)_3^{2+}$ has been postulated by Rubinstein *et al.* [62] (Fig. 4.9). The initial oxidation of $Ru(bpy)_3^{2+}$ to $Ru(bpy)_3^{3+}$ is performed at the electrode surface, and then $Ru(bpy)_3^{3+}$ reacts with analytes to emit light at 620 nm. The electrogenerated CL (ECL) intensity is directly proportional to the amount of the reductant, which is analyte. The method was applied to

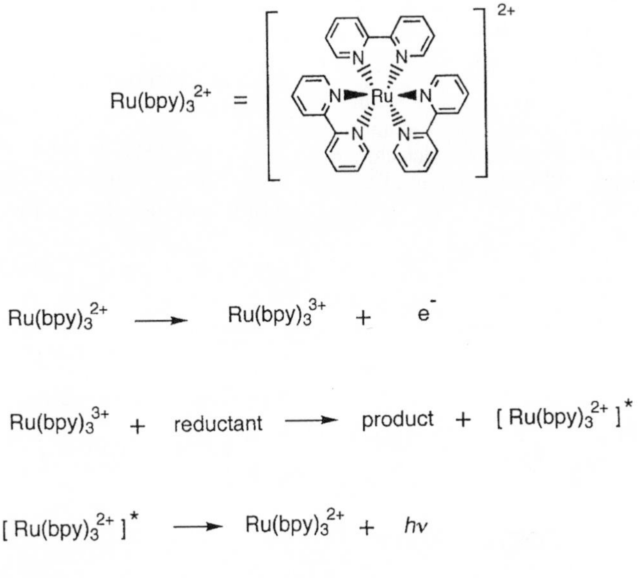

$Ru(bpy)_3^{2+}$ =

$$Ru(bpy)_3^{2+} \longrightarrow Ru(bpy)_3^{3+} + e^-$$

$$Ru(bpy)_3^{3+} + \text{reductant} \longrightarrow \text{product} + [Ru(bpy)_3^{2+}]^*$$

$$[Ru(bpy)_3^{2+}]^* \longrightarrow Ru(bpy)_3^{2+} + h\nu$$

**Fig. 4.9.** $Ru(bpy)_3^{2+}$ CL reaction.

the detection of D- and L-Trp after separation by a ligand-exchange HPLC [59]. The detection limits (S/N = 2) for D- and L-Trp were both 0.2 pmol per injection. Dns-amino acids were determined using $Ru(bpy)_3^{2+}$ CL reaction after HPLC separation with a reversed-phase column [60]. The HPLC system is shown in Fig. 4.10. Since the order of increasing CL intensities for amines reacted with $Ru(bpy)_3^{3+}$ was tertiary > secondary > primary amines, Dns derivatives make the CL intense due to the secondary and tertiary amine groups. The calibration curve for Dns-Glu covered three orders of magnitude with a detection limit of 0.1 μM (2 pmol/injection, S/N = 2). Although underivatized amino acids could be detected with $Ru(bpy)_3^{3+}$ CL, the Dns-derivatives had detection limits improved by three orders of magnitude. $Ru(bpy)_3^{2+}$ CL has the advantages of reagent stability and greater compatibility with common reversed-phase HPLC solvent systems. Oxalate in urine and blood plasma samples has also been determined by a reversed-phase ion-pair HPLC [61]. Direct addition of $Ru(bpy)_3^{2+}$ to the mobile phase was investigated and compared with the conventional post-column $Ru(bpy)_3^{2+}$ addition. The detection limit using oxalate standards with $Ru(bpy)_3^{2+}$ in the mobile phase was below 0.1 μM, which was significantly better than the post-column technique. The mobile-phase addition method allowed the instrumentation to be simplified and reduced band broadening from post-column mixing.

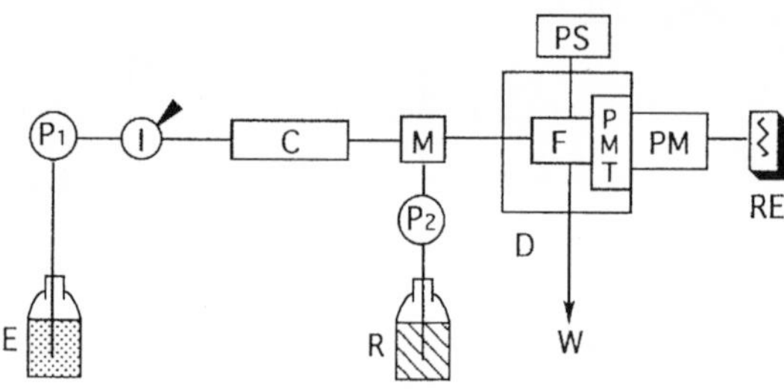


**Fig. 4.10.** $Ru(bpy)_3^{2+}$ CL detection system with HPLC: $P_1$ and $P_2$, pump; I, injector; C, column; M, mixing tee; E, eluent; R, reagent; D, detector; F, flow cell; PS, potentiostat; PMT, photomutiplier; PM, photometer; RE, recorder; W, waste.

## 4.3 LABEL OF CARBOXYL (−COOH)

ABEI described in Section 4.2 can also be used as a labeling reagent for carboxylic acids utilizing 2-chloro-1-methylpyridinium iodide (CMPI) and 3,4-dihydro-2*H*-pyrido[1,2-*a*]pyrimidin-2-one (DPP) as condensing reagents [35] (Fig. 4.11). A variety of ABEI derivatives of carboxylic acids (i.e., free fatty acids or unconjugated bile acids) were separated by a reversed-phase column and sensitively detected using the hydrogen peroxide-potassium hexacyanoferrate(III) system as a post-column reaction (Fig. 4.12). The detection limit of cholic acid was 20 fmol (S/N = 3.5) [35]. This ABEI CL detection technique in HPLC has been applied to the determination of eicosapentaenoic acid in serum. The calibration curve was linear in the range 2 pmol to 2 nmol with a detection limit of 200 fmol (S/N = 2) [63]. Eleven fatty acids were detected in human serum samples by this method. Recently, ABEI

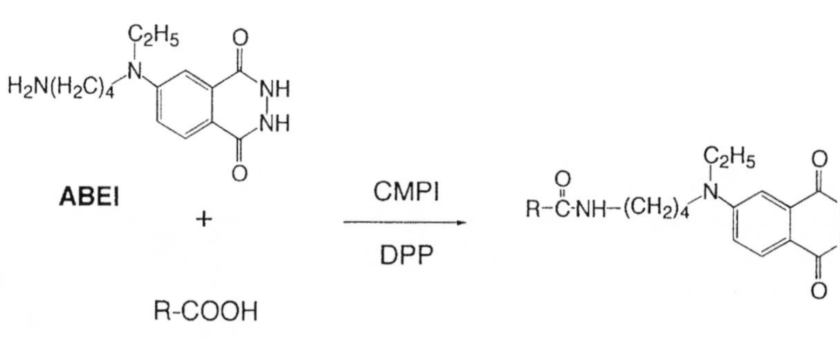

**Fig. 4.11.** Labeling reaction of carboxylic acids with ABEI.

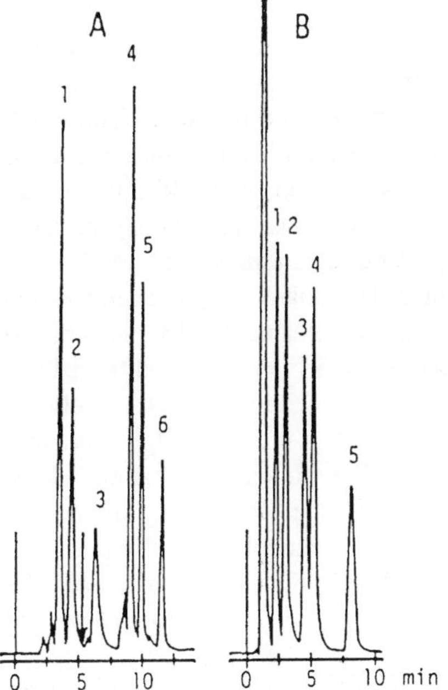


**Fig. 4.12.** Chromatograms of ABEI derivatives of a standard mixture of carboxylic acids. (A) Free fatty acids. Peaks: 1 = acetic acid; 2 = propionic acid; 3 = *n*-butyric acid; 4 = caproic acid; 5 = caprylic acid; 6 = capric acid. (B) Bile acids. Peaks: 1 = ursodeoxycholic acid; 2 = cholic acid; 3 = chenodeoxycholic acid; 4 = deoxycholic acid; 5 = lithocholic acid. Reprinted from, *Journal of Chromatography*, 328, T. Kawasaki, M. Maeda and A. Tsuji, Chemiluminescence liquid chromatography using *N*-(4-aminobutyl)-*N*-ethylisoluminol as a precolumn labelling reagent, 121–126, 1985, with kind permission of Elsevier Science—NL, Sara Burgerhartstraat 25, 1055 KV Amsterdam, The Netherlands and The Pharmaceutical Society of Japan.

has been applied to the determination of ibuprofen in saliva [64]. The labeling was performed with 1-hydroxybenzotriazole (HOBT) as a pre-activator of the carboxylic acid function and *N*-ethyl-*N*′-(3-dimethylaminopropyl)carbodiimide (EDC) as a condensing reagent. CL detection was carried out using a post-column on-line electrochemical hydrogen peroxide generation system and the addition of microperoxidase as a catalyst. The detection limit of labeled ibuprofen in human saliva was 0.7 ng/0.5 ml saliva, with a recovery of 96.1 ± 1.3%.

Several fluorogenic labeling reagents were used for the determination of carboxylic acids with HPLC followed by PO-CL detection (Fig. 4.13). 4-(Bromomethyl)-7-methoxycoumarin (Br-Mmc), 7-(diethylamino)coumarin-3-carbohydrazide (DCCH), and 7-(diethylamino)-3-[4-((iodoacetyl)amino)phenyl]-4-methylcoumarin (DCIA) were evaluated as carboxylic acid labeling reagents to be detected using PO-CL with HPLC [65]. TCPO was used for the evaluation. The labeling procedure for Br-Mmc and DCIA requires only one step as opposed to two for DCCH derivatization. No CL was observed from the Br-Mmc derivatives; this result suggests that the singlet excitation energy for Br-Mmc exceeds the maximum energy of 105 kcal/mol produced by the peroxyoxalate reaction [66]. The detection limits of straight chain

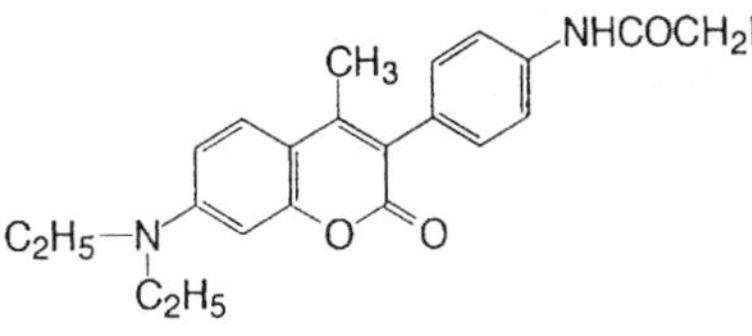

**DCIA**

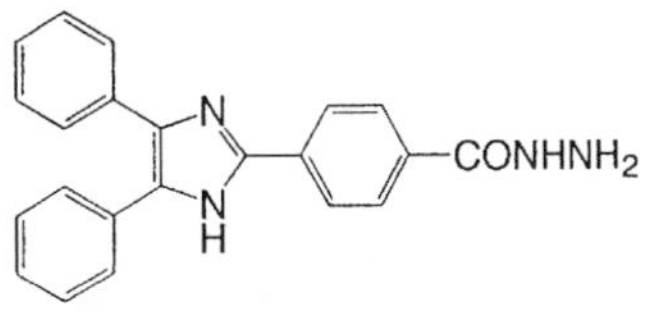

**HCPI**

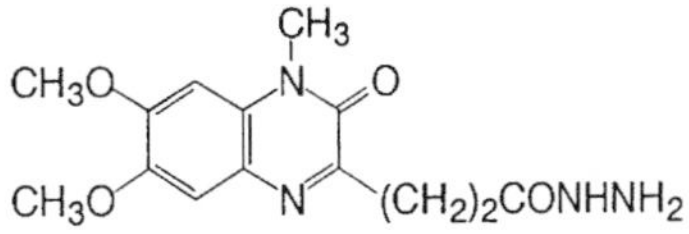

**6,7-dimethoxy-1-methyl-2(1*H*)-quinoxalinone-3-proprionylcarboxylic acid hydrazide**

**Fig. 4.13.** Fluorescence labeling reagents for carboxylic acids used for PO-CL detection.

carboxylic acids labeled with DCIA by PO-CL detection were in the low fmol range. A quinoxalinone fluorescent label, 6,7-dimethoxy-1-methyl-2(1*H*)-quinoxalinone-3-proprionyl-carboxylic acid hydrazide [67], was synthesized and evaluated as a carboxylic acid labeling reagent for PO-CL detection. EDC and *N*,*N*′-dicyclohexylcarbodiimide (DCC) were used as condensing reagents. The post-column CL reaction conditions with TCPO were optimized using several arachidonic acid metabolites to give detection limits of 500 amol/injection (S/N = 3). 2-(4-Hydrazinocarbonylphenyl)-4,5-diphenylimid-azole (HCPI) developed as a fluorescent labeling reagent of carboxylic acids has been applied to the HPLC determination of the saturated fatty acids (margaric and arachidic acid) via the PO-CL detection using TCPO [68]. The detection limits (S/N = 2) were 12 and 18 fmol per injection for margaric and arachidic acid, respectively. The method was applied to the determination of these acids in normal human serum.

## 4.4 LABEL OF HYDROXYL (−OH) AND THIOL (−SH)

Not so many reagents for CL detection were so far reported for hydroxyl and thiol groups. For phenolic hydroxyl group, a derivatization scheme employing 10-methyl-9-acridinium carboxylate as a CL label for sensitive detection of environmentally important chlorophenols was investigated using HPLC [69]. A two-step derivatization scheme was used to generate the CL derivatives (Fig. 4.14). Following the separation under reversed-phase conditions, the CL reaction was performed by base-catalyzed post-column oxidation. The quantum efficiency was dependent on the species of analyte. The detection limits of chlorophenols (S/N = 3) ranged from 1.25 fmol to 300 amol per injection, approximately one order of magnitude less than those for approaches using capillary GC with electron capture detection or HPLC with CL detection using the peroxyoxalate reaction. Chlorophenols have also been determined by using PO-CL detection. Lissamine Rhodamine B sulfonyl chloride (laryl chloride) was employed as a pre-column labeling reagent for phenolic compounds [70]. The advantages of rhodamine labels are the high quantum yield, the long wavelength emission (>550 nm) which permits a considerable reduction of the CL background, and the fact that electronegative heavy atom substituents do not quench the CL. The labeling procedure for chlorophenols was made quantitative in 1 min at room temperature. The detection limits for several chlorophenols in both the reversed- and normal-phase HPLC system were

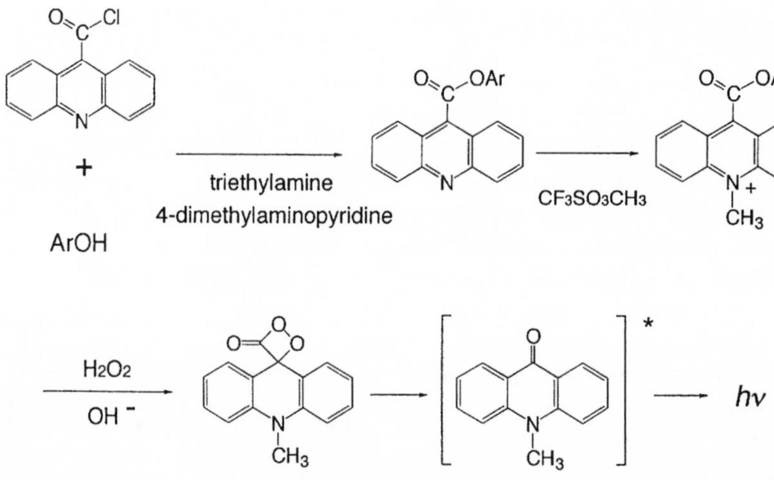

**Fig. 4.14.** Labeling reaction of chlorophenols and CL reaction of the acridinium salts.

in the low pg range. The derivatization of pentachlorophenol spiked in river water (0.8 ppb) was carried out directly on the sample without any preconcentration or clean-up step(s).

For the labeling of SH groups, several kinds of maleimide-type fluorescent reagents have been known. Among them, *N*-[4-(6-dimethylamino-2-benzofuranyl)phenyl]maleimide (DBPM) was applied to the PO-CL determination of biological thiols including glutathione, cysteine, *N*-acetylcysteine, cysteamine and D-penicillamine [71]. The labeling reaction was carried out at 60 °C for 30 min (pH 8.5). Five kinds of labeled thiols were separated within 12 min on an ODS column and detected in the ranges from 500 fmol to 2 pmol (cysteamine and *N*-acetylcysteine), to 3 pmol (cysteine) and to 5 pmol (glutathione and D-penicillamine) on column (Fig. 4.15). The lower detection limits (S/N = 2) were from 7 fmol (cysteamine) to 113 fmol (glutathione). The method was applied to biological samples, and the amounts of glutathione and cysteine in rat liver were successfully determined.

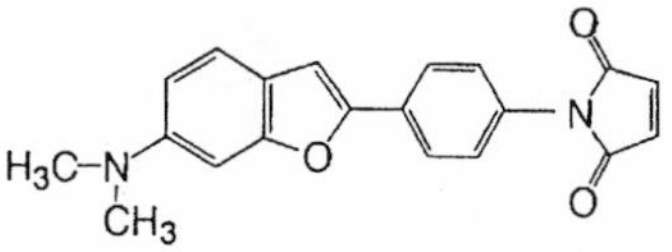

**DBPM**

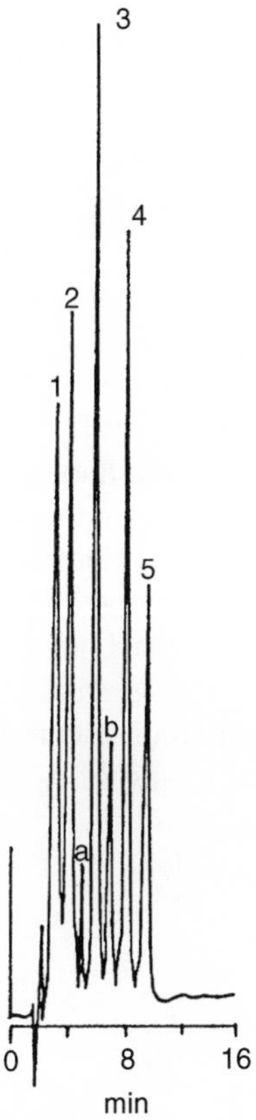

**Fig. 4.15.** DBPM and chromatogram of DBPM derivatives of thiols. Peaks (pmol on column): 1 = glutathione (5); 2 = *N*-acetylcysteine (2); 3 = cysteine (2); 4 = D-penicillamine (10); 5 = cysteamine (0.8); a = reagent blank; b = unknown. [Reproduced from ref. 71 with permission from Heyden & Son Limited.].

## 4.5 LABEL OF OTHER FUNCTIONAL GROUPS

For the labeling of the carbonyl group, 5-*N*,*N'*-dimethylaminonaphthalene-1-sulfonohydrazide (Dns-H) has been widely used in combination with PO-CL detection. Oxo-steroids and oxo-bile acid ethyl esters were derivatized with Dns-H to Dns-hydrazone, purified by a high-performance gel-permeation chromatography, separated on an ODS column and detected via the PO-CL reaction using TDPO [72]. The detection limits (S/N = 2) for corticosterone, testosterone and progesterone were 3, 2 and 4 fmol, respectively. The unusual oxo-bile acid, 7$\alpha$-hydroxy-3-oxo-5$\beta$-cholanic acid, has also been detected at the nmol/l level in urine from a patient with cholestatic liver disease. 3$\alpha$- or 3$\beta$-Hydroxysteroids, such as bile acids (free and glycine and taurine conjugates), 3$\beta$-hydroxy-5-cholenic acid, pregnanediol, 5-pregnene-3$\beta$, 20$\beta$-diol and 5-pregnene-3$\beta$, 20$\alpha$-diol were enzymatically converted to 3-oxosteroids using immobilized hydroxysteroid dehydrogenase and labeled with Dns-H [73]. The Dns-hydrazones of these were separated on an ODS column and detected at the levels of a few fmol (S/N = 2).

Dns-H has been used for the labeling of sugars. HPLC-PO-CL methods for the determination of hyaluronic acid (HA) in blood plasma [74], and HA, chondroitin sulfhate (CS) and dermatan sulfhate (DS) [75] using Dns-H were

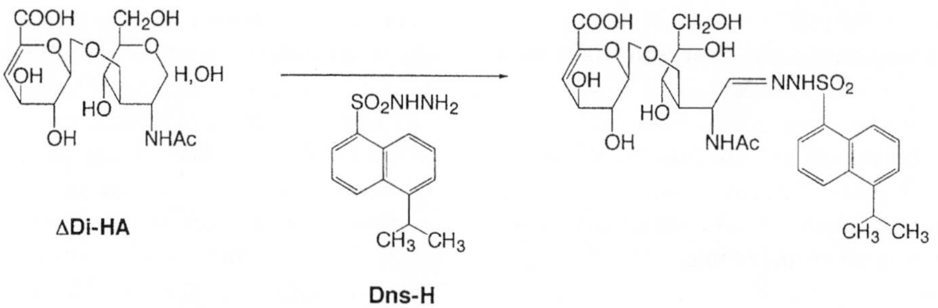

ΔDi-HA

Dns-H

Fig. 4.16. Labeling reaction of unsaturated disaccharide with Dns-hydrazine.

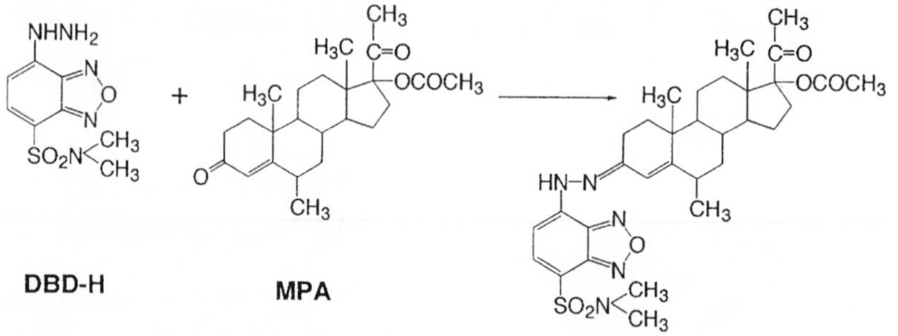

DBD-H

MPA

Fig. 4.17. Labeling reaction of medroxyprogesterone acetate (MPA) with DBD-H.

reported. For determining HA, unsaturated disaccharide, {2-acetamide-2-deoxy-3-*O*-(*β*-D-gluco-4-enepyranosyluronic acid)-D-glucose (ΔDi-HA)}, derived from HA by enzymatic digestion with hyaluronidase SD was labeled with Dns-H (Fig. 4.16) [74]. The detection limit of ΔDi-HA derived from HA was 100 fmol (S/N = 3). The Dns-hydrazone derivatives of the unsaturated disaccharides derived from HA, CS and DS by chondroitinase ABC and/or chondroitinase ACII were separated by a reversed-phase chromatography. The calibration curves for Dns-unsaturated disaccharides were linear in the range from 500 fmol to 5 nmol with a detection limit of 100 fmol (S/N = 3) [75]. The method was applicable to the determination of the levels of HA, CS and DS in rat peritoneal mast cells.

Medroxyprogesterone acetate (MPA), a synthetic progesterone possessing high progestational activity, was determined using 4-(*N*,*N*-dimethylaminosulfonyl)-7-hydrazino-2,1,3-benzoxadiazole (DBD-H) as a label (Fig. 4.17). The linear range of the standard curve, in serum, was 15.6–96.6 ng/ml with a detection limit of 9 ng/ml using only 100 μl of serum [76].

3-Aminofluoranthene, a labeling reagent for aldehydes and ketones, was developed for PO-CL detection [77]. Reductive amination reaction by the use of a borane-pyridine complex was incorporated with the labeling reaction. Analytes were separated by HPLC and detected using PO-CL with fmol-amount detection limits, the detectability of which was thirty times better than fluorescence.

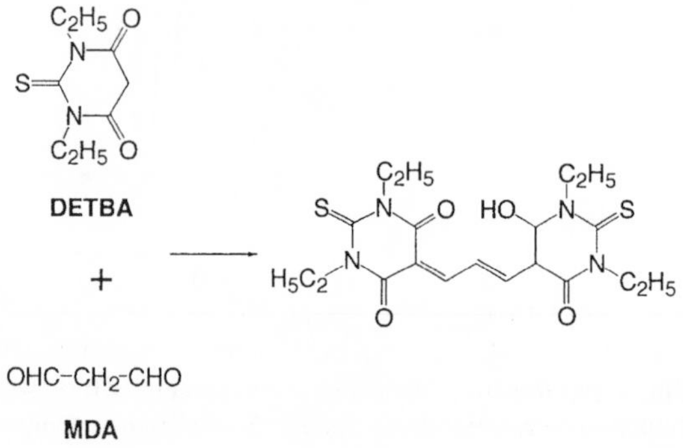

DETBA

MDA

Fig. 4.18. Labeling reaction of MDA with DETBA.

Malondialdehyde (MDA) derived from peroxidized polyunsaturated lipid was derivatized with 1,3-diethyl-2-thiobarbituric acid (DETBA), and the derivative was determined after separation with a reversed-phase column by PO-CL detection. Detection limit was achieved at 20 fmol (S/N = 2). The method was applied to the determination of MDA in rat brain (Fig. 4.18) [78].

Highly sensitive methods with PO-CL detection using 7-(diethylamino)-3-[4-((iodoacetyl)amino)phenyl]-4-methylcoumarin (DCIA) as a label were employed for the determination of fluoropyrimidines [79,80] which are widely used in the chemotherapy for a variety of human carcinomas. As described in Section 4.3, DCIA is a labeling reagent for carboxylic acids. However, DCIA can also be used for the alkylation at the nitrogen atom of acidic imide group in the pyrimidine compounds by using the crown ether-potassium complex as a catalyst. The typical detection limits were 40 fmol for 5-fluorouracil and 20 fmol for 5-fluorouridine, 5-fluoro-5′-deoxyuridine and 1-(tetrahydro-2-furanyl)-5-fluorouracil per 4-μl injection (S/N = 2) [80].

A unique luminol derivative, 4,5-diaminophthalhydrazide (DPH, Fig. 4.19), has been synthesized

DPH + phenylpyruvic acid →

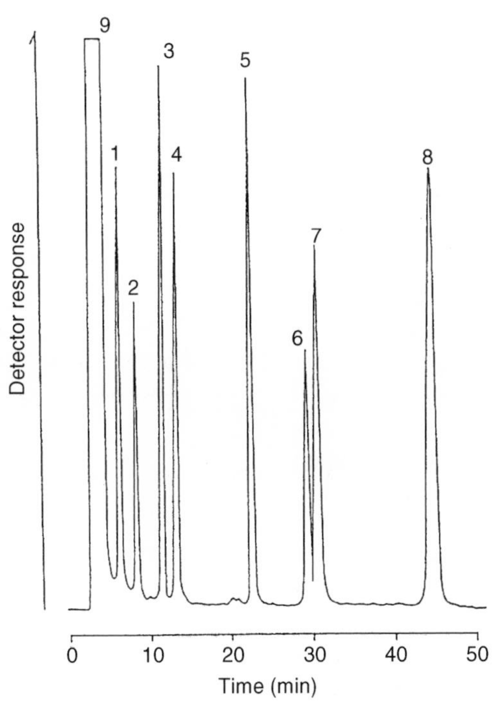


**Fig. 4.19.** Reaction of phenylpyruvic acid with DPH and a chromatogram of DPH derivatives of $\alpha$-keto acids. Peaks: 1 = $\alpha$-ketobutylic acid; 2 = $p$-hydroxyphenylpyruvic acid; 3 = $\alpha$-ketovaleric acid; 4 = $\alpha$-ketoisovaleric acid; 5 = $\alpha$-ketoisocaproic acid; 6 = $\alpha$-keto-$\beta$-methylvaleric acid; 7 = $\alpha$-ketocaproic acid; 8 = phenylpyruvic acid; 9 = reagent blank. [Reproduced from ref. 82 with permission from Academic Press Inc.].

and used as a highly sensitive CL derivatization reagent for $\alpha$-keto acids which are important intermediates in the biosynthesis of amino acids, carboxylic acids and sugars [81]. DPH reacts selectively with $\alpha$-keto acids to give quinoxaline derivatives which produce CL by the reaction with hydrogen peroxide and potassium hexacyanoferrate(III). DPH derivatives of eight biologically important $\alpha$-keto acids were separated within 50 min by a reversed-phase column (Fig. 4.19). The detection limits for these acids were in the range 4–50 fmol per injection. The determination of $\alpha$-keto acids including phenylpyruvic acid in human plasma has also been achieved [82]. The detection limits were in the range 9–92 pmol/ml in plasma (S/N = 3); this sensitivity allowed the first determination of phenylpyruvic acid in normal human plasma. DPH was adapted to the determination of *N*-acetylneuraminic acid (NANA) [83]. For the measurement of NANA, serum and urine were hydrolyzed to release NANA from glycoproteines and glycolipids. NANA and *N*-glycolylneuraminic acid (internal standard) were converted into chemiluminescent derivatives with DPH, and separated by a reversed-phase column. This method is very sensitive with a detection limit of 9 fmol (S/N = 3), and thus requires only small volumes of serum and urine (10 and 50 µl, respectively).

DPH was found to react with $\alpha$-dicarbonyl compounds under different conditions from those for $\alpha$-keto acid to give quinoxaline derivatives. The detection limits of $\alpha$-dicarbonyl compounds, phenylglyoxal, diacetyl, 2,3-pentanedione and 2,3-hexanedione, were in the range 1.1–8.7 fmol on column, except for 3,4-hexanedione (300 fmol on column) [84]. The method has been applied to the measurement of $3\alpha,5\beta$-tetrahydroaldosterone [85] and dexamethasone [86]. These compounds have a $\alpha$-ketol moiety in their structure, which is converted to $\alpha$-dicarbonyl moiety by the reaction with copper(II) acetate. The resultant glyoxal compounds are derivatized with DPH into chemiluminescent derivatives.

A new luminol-type reagent, 6-aminomethylphthalhydrazide (6-AMP), was synthesized as a highly sensitive and selective CL derivatization reagent for 5-hydroxyindoles [87]. The reagent reacts selectively with the indoles in the presence of potassium hexacyanoferrate(III) to give highly CL derivatives (Fig. 4.20). The derivatives of three 5-hydroxyindoles (i.e., 5-hydroxytryptophan, serotonin, and 5-hydroxyindole-3-acetic acid) were separated within 35 min by a reversed-phase HPLC, followed by CL detection. The detection limits for 5-hydroxyindoles were in the range 0.7–4 fmol per injection (S/N = 3).

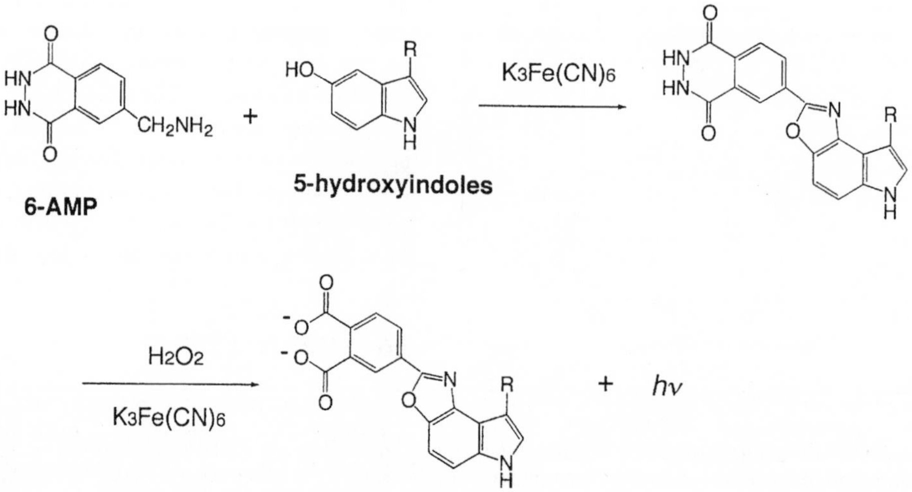

**Fig. 4.20.** Labeling reaction of 5-hydroxyindoles with 6-AMP, and CL reaction of the labeled 5-hydroxyindoles.

A highly sensitive method has been developed for the determination of plasma catecholamines, norepinephrine, epinephrine and dopamine. Catecholamines were derivatized in-line with ethylenediamine (ED) to the corresponding intensely fluorescent compounds, and then detected by PO-CL reaction utilizing TDPO (Fig. 4.21). The detection limit for all the catecholamines obtained was 1 fmol (S/N = 2) [88]. Concentrations of catecholamines in 100 μl of rat plasma were successfully determined by this method. In order to exclude the tedious manual extraction step of catecholamines, a fully-automated analyser was constructed [89]. An automatic in-line extraction of catecholamines using an ion-exchange pre-column, in-line derivatization with ED, and HPLC-PO-CL detection were combined in the analyser (Fig. 4.22). For the PO-CL determination of catecholamines, fourteen 1,2-diarylethylenediamines wereevaluated as a fluorescence derivatization reagent [90]. Among them, 1,2-bis(3-chlorophenyl)ethylenediamine, 1,2-bis(3,4-dichlorophenyl)ethylenediamine and 1,2-bis(4-chlorophenyl)ethylenediamine were found to be the most sensitive for all catecholamines on the basis of the PO-CL reaction. The derivatization and PO-CL reaction conditions using TDPO were optimized for 1,2-bis(3-chlorophenyl)ethylenediamine, and the detection limits for catecholamines obtained were approximately 40–120 amol per injection (S/N = 3).

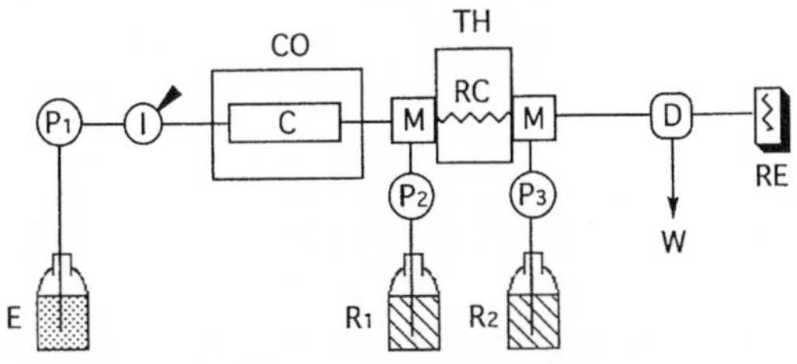


**Fig. 4.21.** HPLC-PO-CL detection system for catecholamines: $P_1$, $P_2$ and $P_3$, pump; I, injector; C, column; CO, column oven; M, mixing device; RC, reaction coil; TH, thermostat; E, eluent; $R_1$, fluorogenic reagent; $R_2$, CL reagent; D, CL detector; RE, recorder; W, waste.

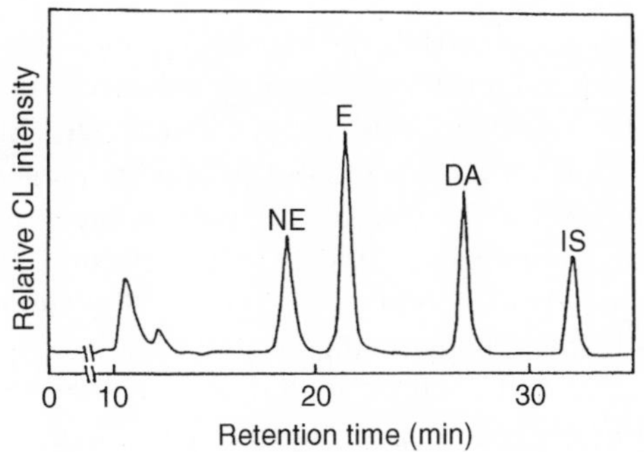


**Fig. 4.22.** Chromatogram of a standard catecholamines with in-line extraction and HPLC-PO-CL detection. Peaks: NE = norepinephrine; E = epinephrine; DA = dopamine; IS = internal standard (*N*-methyldopamine). [Reproduced from ref. 89 with permission from John Wiley & Sons, Ltd.].

CL methods for the determination of adenine and its nucleos(t)ides [91,92], and guanine and its nucleos(t)ides have been reported [93]. These methods are based on derivatization techniques that can selectively convert non-luminescent analytes to chemiluminescent derivatives by using glyoxal derivatives. The individual derivatization conditions are developed for adenine- and guanine-containing compounds. Structures of the chemiluminescent species of both nucleic acid bases are still unknown, but the possible pathways of the derivatization reaction with phenylglyoxal to produce CL are considered in the manner shown in Fig. 4.23. The derivatization products exhibit intense CL in an alkaline medium in the presence of the aprotic polar solvent, *N*,*N*′-dimethylformamide. The detection limits for adenine and guanine were 37 and 53 pmol/ml in the reaction mixture, respectively. These methods might be applicable to a post-column CL detection system in HPLC for the sensitive determination of these nucleic acid bases and nucleos(t)ides.

## 4.6 APPLICATION

In order to increase the sensitivity of CL detection in HPLC, several options have been attempted. In the view to obtain a stable base line, reproducible peak height and less band broadening, some mixing devices (e.g., a rotating flow mixing device

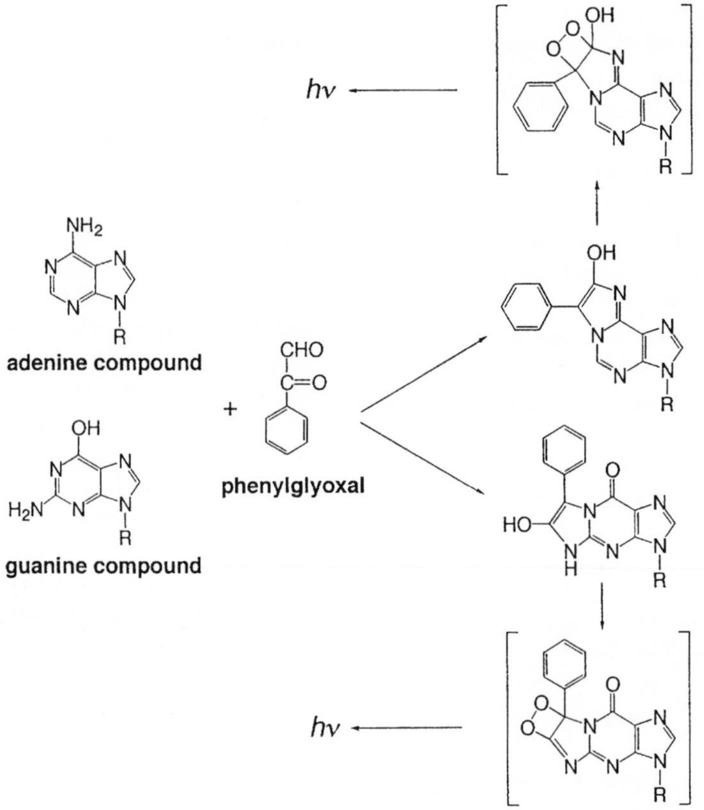

Fig. 4.23. Possible pathways of the CL reactions between adenine and guanine compounds, and phenylglyoxal.

having two or three directional inlets and an outlet at the center of the top of the vessel) have been proposed for PO-CL detection [94–96].

The recent development of liquid chromatography (LC) with small-bore columns has led to an increase in the sensitivity, since a higher solute concentration is obtained in the peak maximum. This type of column has other advantages: lower solvent consumption and small requirement of sample amounts. A PO-CL detection system for packed capillary columns (0.32 mm i.d.) has been demonstrated, in which a zero dead volume system for mixing of the LC eluate and the reagents was developed [97]. For a few dansylated compounds, detection limits of about 100 fg were obtained. Weber and Grayeski evaluated PO-CL as a detection mode for packed capillary LC [98]. The use of relatively large volume flow cells (>1 μl) based on a sheathing flow of CL reagents around the column effluent resulted in enhanced sensitivity; detection limits in the fmol range were possible for certain fluorophores.

Another approach, to find fluorophores having efficient chemical excitation as enhancers in the PO-CL reaction, has been carried out. Eighteen kinds of pyrimido[5,4-*d*]pyrimidines together with several fluorescent compounds were evaluated as enhancers for PO-CL detection [99]. Among them, 2,6-bis[di-(2-hydroxyethyl)amino]-4,8-dipiperidinopyrimido[5,4-*d*]pyrimidine(Dipyridamole) and 2,4,6,8-tetrathiomorpholinopyrimido [5,4-*d*]pyrimidine gave intense CL which were larger than those of any other commercially available fluorescent compounds tested (e.g., 10 times larger than that of perylene). The highly sensitive detection of near-infrared (near-IR) fluorescent dyes using HPLC with PO-CL detection was examined [100]. Due to the fact that compounds having low singlet excitation energy are excited effectively by the intermediate, near-IR fluorescent dyes are assumed to be suitable for PO-CL detection. The detection limits for methylene blue, pyridine 1, oxazine 1 and 3,3′-diethylthiadicarbocyanine iodide (DTDCI) were 120, 27, 31, 0.19 fmol on column, respectively (S/N = 2). DTDCI was found to be the preferable structure for PO-CL detection and its sensitivity was 250 times that obtained by HPLC with the conventional fluorescence detection. The development of fluorescence labeling reagents having pyrimido[5,4-*d*]pyrimidine and DTDCI structures are to be expected.

Lophine derivatives, in which CL was enhanced with Co(II) and hydroxylamine hydrochloride, are known as CL compounds. The CL intensities of eighteen kinds of the derivatives were evaluated by a flow injection method [101]; the results suggest that the derivatives having COCl or $CONHNH_2$ as a reactive group at the 4-position of the phenyl ring of the 2-position of the imidazole skeleton might be usable as CL labeling reagents for amines or carboxylic acids in HPLC. Reagents for CL detection to increase the sensitivity and selectivity are still under active development. The development of CL reagents showing intense light emission under mild conditions in solvents similar to those used for separation is being pursued.

Recently, capillary electrophoresis (CE) has been explored as a powerful tool of a wide range of analytes because of its advantage for the simultaneous determination of a number of analytes with high resolution and speed. UV/visible absorption is most commonly used as a detection method. However, the improvement in limits of detection arise from the small sample volume which is expected. To meet this requirement, the detection methods utilizing laser-induced fluorescence (LIF), electrochemistry and mass spectrometry, etc. have been demonstrated. Some applications of CL detection to CE have also been attempted, although difficulties in coupling the separation technique with the CL detector still exist [102].

PO-CL detection in CE using a two-step approach, involving switching off the CE power supply at an appropriate time and connecting the capillary to a syringe pump to effect dynamic flow, was demonstrated [103]. The scheme for the post-column reactor is shown in Fig. 4.24. The average lower limit of three Dns-amino acids obtained by this method was about 1.2 fmol (about 85 nM) which is approximately 35-fold lower than UV absorption methods. On-line PO-CL detection of proteins utilizing Eosine Y by CE has been developed [104,105]. The methods are based on the finding that Eosine Y comigrates with protein as its complex in a capillary tube containing phosphate buffer (pH 3.5). The detection limit of bovine serum albumin was 1.7 fmol ($6.0 \times 10^{-8}$ mol/l) [105]. A detection interface has been designed to allow the addition of post-column reagents to assess CL as a method of detection for CE [106]. The interface utilized a reactor that introduces the reagent, which consisted of a base and hydrogen peroxide, into the electrophoretic system in a sheathing flow profile. The reaction conditions including pH, concentration, and flow rates of the reagents for acridinium CL have been studied. The detection limit for the interface was found to be in the low-fmol to upper-amol range for acridiniums. Dadoo *et al.* reported an end-column CL detector for CE, in which the signal is generated at the column outlet [107]. As the analytes emerge from the column, they react with CL reagents to produce visible light that is transported by a fiber optic to a photomultiplier tube (Fig. 4.25). The luminol CL and firefly luciferase bioluminescence reaction were adapted for use with this system. The limits of detection (S/N = 3) measured for luminol and ATP are 500 amol ($2 \times 10^{-8}$ M) and 100 amol ($5 \times 10^{-9}$ M) per injection, respectively. These values are approximately three orders of magnitude lower than those obtained with absorbance. Although the end-column CL detector provided relatively high detection sensitivity, the separation efficiencies obtained were lower than those typically found in CE. Band broadening as a result of the mixing at the column outlet, slow CL reaction kinetics, and the relatively large detection zone (1 mm) may explain the relatively low number of theoretical plates. CL detection offers several advantages

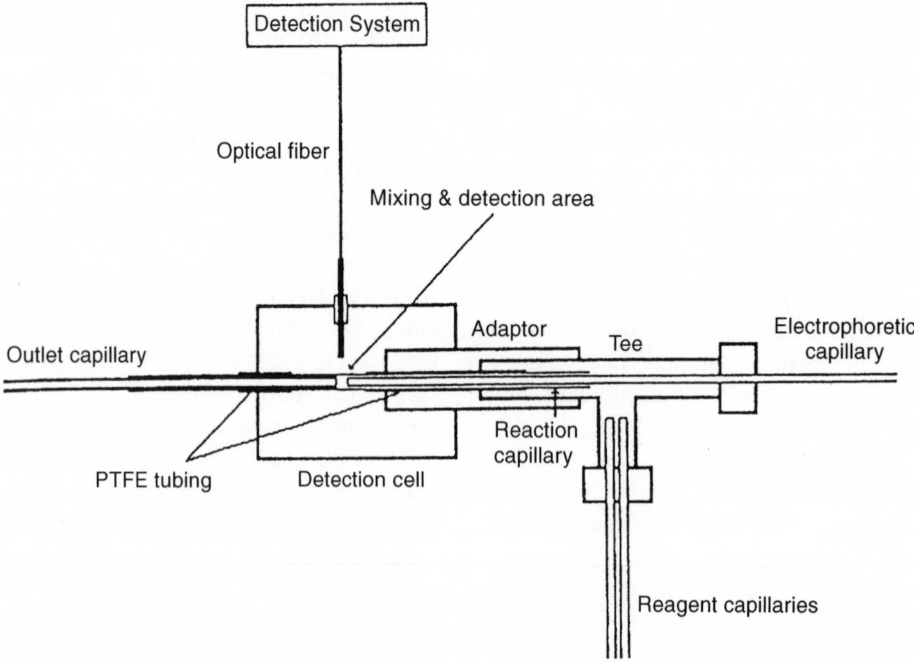


**Fig. 4.24.** Scheme of the post-column CL reactor for CE. Reprinted from *Journal of Chromatography*, 634, N. Wu and C.W. Huie, Peroxyoxalte chemiluminescence detection in capillary electrophoresis, 309–315, 1993, with kind permission of Elsevier Science—NL, Sara Burgerhartstraat 25, 1055 KV Amsterdam, The Netherlands.

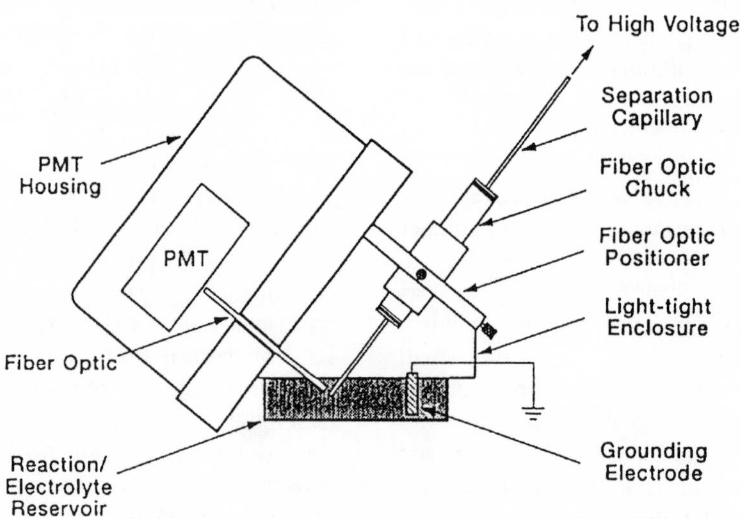


**Fig. 4.25.** Scheme of the end-column CL detector for CE. [Reproduced from ref. 107 with permission from the American Chemical Society.].

as a method of detection for CE. Increased sensitivity can be achieved by improving interfaces to obtain larger detection volumes without the band broadening.

Research on the application of CL detection to diverse separation techniques involving HPLC is progressing very fast owing to its great sensitivity and simplicity of instrumentation. These advantages are well suited for miniaturized separation techniques such as capillary liquid chromatography, CE, and capillary electrochromatography, which will be further extended. In the near future, CL detection techniques will have a wider range of applications in the field of life science, environmental science, and other areas.

## REFERENCES


[1] Harvey, E.N. (1952) *Bioluminescence*, Academic Press, New York.

[2] Deluca, M.A. (Ed.), (1978) *Methods in Enzymology*, Vol. 57, Academic Press, New York.

[3] Deluca, M.A. and McElroy, W.D. (Eds), (1981). *Bioluminescence and Chemiluminescence*, Academic Press, New York.

[4] Deluca, M.A. and McElroy, W.D. (Eds), (1986). *Methods in Enzymology*, Vol. 133, Academic Press, New York.

[5] Radziszewski, B. (1877) *Chem. Ber.*, **10**, 70.

[6] Pruett, R.L. (1950) *J. Am. Chem. Soc.*, **72**, 3646.

[7] Philbrook, G.F., Qyers, J.B. and Totter, J.R. (1965) *Photochem. Photobiol.*, **4**, 869.

[8] McCapra, F. and Burford, A. (1976) *J. Chem. Soc., Chem. Commun.*, 607.

[9] Albrecht, H.D. (1928) *Z. Phys. Chem.*, **136**, 321.

[10] Glue, K. and Petsch, W. (1935) *Angew. Chem.*, **48**, 57.

[11] Allen, R.C. (1986) *Methods Enzymol.*, **133**, 449.

[12] Weeks, I., Beheshti, I., McCapra, F., Campbell, A.K. and Woodhead, J.S. (1983) *Clin. Chem.*, **29**, 1474.

[13] McCapra, F. (1968) *J. Chem. Soc., Chem. Commun.*, 155.

[14] Schaap, A.P., Handley, R.S. and Giri, B.P. (1987) *Tetrahedron Lett.*, **28**, 935.

[15] Bronstein, I., Edward, B. and Voyta, J.C. (1989) *J. Biolumin. Chemilumin.*, **4**, 99.

[16] Schaap, A.P., Akhavan, H. and Romano, L.J. (1989) *Clin. Chem.*, **35**, 1863.

[17] Chandross, E.A. (1963) *Tetrahedron Lett.*, 761.

[18] Rauhut, M.M., Sheehan, D., Clarke, R.A. and Semsel, A.M. (1965) *Photochem. Photobiol.*, **4**, 1097.

[19] Tseng, S.S., Mohan, A.G., Haines, L.G., Vizcarra, L.S. and Rauhut, M.M. (1979) *J. Org. Chem.*, **44**, 4113.

[20] Koo, J.Y. and Schuster, G.B. (1977) *J. Am. Chem. Soc.*, **99**, 6107.

[21] Koo, J.Y. and Schuster, G.B. (1978) *J. Am. Chem. Soc.*, **100**, 4496.

[22] Nakashima, K. and Imai, K., in Schulman, S.G. (Ed.), (1993) *Molecular Luminescence Spectroscopy*, Part 3, Chemical Analysis Series, Vol. 77, John Wiley & Sons, Inc., Chichester, West Sussex.

[23] Calokerions, A.C., Deftereos, N.T. and Baeyens, W.R.G. (1995) *J. Pharm. Biomed. Anal.*, **13**, 1063.

[24] Nakashima, K. and Imai, K., in Hanai, T. and Hatano, H. (Eds), (1996) *Advances in Liquid Chromatography*, Methods in Chromatography, Vol. 1, World Scientific Publishing Co. Ltd., Singapore.

[25] Schroeder, H.R., Boguslaski, R.C., Carrico, R.J. and Buckler, R.T. (1978) *Method Enzymol.*, **57**, 424.

[26] Schroeder, H.R. and Yeager, F.M. (1978) *Anal. Chem.*, **50**, 1114.

[27] Robards, K. and Worsfold, P.J. (1992) *Anal. Chim. Acta*, **266**, 147.

[28] Hanaoka, N., Givens, R.S., Schowen, R.L. and Kuwana, T. (1988) *Anal. Chem.*, **60**, 2193.

[29] Honda, K., Miyaguchi, K. and Imai, K. (1985) *Anal. Chim. Acta*, **177**, 103.

[30] Imai, K., Nawa, H., Tanaka, M. and Ogata, H. (1986) *Analyst*, **111**, 209.

[31] Nakashima, K., Maki, K., Akiyama, S., Wang, W.H., Tsukamoto, Y. and Imai, K. (1989) *Analyst*, **114**, 1413.

[32] Stigbrand, M., Ponten, E. and Irgum, K. (1994) *Anal. Chem.*, **66**, 1766.

[33] Curtis, T.G., Seitz, W. (1977) *J. Chromatogr.*, **134**, 343.

[34] Sigvardson, K.W., Kennish, J.M. and Birks, J.W. (1984) *Anal. Chem*, **56**, 1096.

[35] Kawasaki, T., Maeda, M. and Tsuji, A. (1985) *J. Chromatogr.*, **328**, 121.

[36] Nakashima, K., Suetsugu, K., Akiyama, S. and Yoshida, K. (1990) *J. Chromatogr.*, **530**, 154.

[37] Nakashima, K., Suetsugu, K., Yoshida, K., Imai, K. and Akiyama, S. (1991) *Anal. Sci.*, **7**, 815.

[38] Spurlin, S.R. and Cooper, M.M. (1986) *Anal. Lett.*, **19**, 2277.

[39] Ishida, J., Horike, N. and Yamaguchi, M. (1995) *Anal. Chim. Acta*, **302**, 61.

[40] Ishida, J., Horike, N. and Yamaguchi, M. (1995) *J. Chromatogr. B*, **669**, 390.

[41] Kobayashi, S., Sekino, J., Honda, K. and Imai, K. (1981) *Anal. Biochem.*, **112**, 99.

[42] Walters, D.L., James, J.E., Vest, F.B. and Karnes, H.T. (1994) *Biomed. Chromatogr.*, **8**, 207.
[43] Tsai, C-E., Kondo, F., Ueyama, Y. and Azama, J. (1995) *J. Chromatogr. Sci.*, **33**, 365.
[44] Kobayashi, S. and Imai, K. (1980) *Anal. Chem.*, **52**, 424.
[45] Miyaguchi, K., Honda, K. and Imai, K. (1984) *J. Chromatogr.*, **303**, 173.
[46] Miyaguchi, K., Honda, K. and Imai, K. (1984) *J. Chromatogr.*, **316**, 501.
[47] Miyaguchi, K., Honda, K., Toyo'oka, T. and Imai, K. (1986) *J. Chromatogr.*, **352**, 255.
[48] Hayakawa, K., Imaizumi, N., Ishikura, H., Minogawa, E., Takayama, N., Kobayashi, H. and Miyazaki, M. (1990) *J. Chromatogr.*, **515**, 459.
[49] Takayama, N., Hayakawa, K., Kobayashi, H. and Miyazaki, M. (1991) *Eisei Kagaku*, **37**, 14.
[50] Nishitani, A., Kanda, S. and Imai, K. (1992) *Biomed. Chromatogr.*, **6**, 124.
[51] Hayakawa, K., Hasegawa, K., Imaizumi, N., Wong, O.S. and Miyazaki, M. (1989) *J. Chromatogr.*, **464**, 343.
[52] Hayakawa, K., Miyoshi, Y., Kurimoto, H., Matsushima, Y., Takayama, N., Tanaka, S. and Miyazaki, M. (1993) *Biol. Pharm. Bull.*, **16**, 817.
[53] de Montigny, P., Stobaugh, J.F., Givens, R.S., Carlson, R.G., Srinivasachar, K., Sternson, L.A. and Higuchi, T. (1987) *Anal. Chem.*, **59**, 1096.
[54] Kawasaki, T., Imai, K., Higuchi, T. and Wong, O.S. (1990) *Biomed. Chromatogr.*, **4**, 113.
[55] Kwakman, P.J.M., Koelewijn, H., Kool, I., Brinkman, U.A.Th. and de Jong, G.J. (1990) *J. Chromatogr.*, **511**, 155.
[56] Uzu, S., Imai, K., Nakashima, K. and Akiyama, S. (1991) *Biomed Chromatogr.*, **5**, 184.
[57] Uzu, S., Imai, K., Nakashima, K. and Akiyama, S. (1991) *Analyst*, **116**, 1353.
[58] Kouwatli, H., Chalom, J., Tod, M., Farinotti, R. and Mahuzier, G. (1992) *Anal. Chim. Acta*, **266**, 243.
[59] Uchikura, K. and Kirisawa, M. (1991) *Anal. Sci.*, **7**, 971.
[60] Lee, W-Y. and Nieman, T.A. (1994) *J. Chromatogr., A*, **659**, 111.
[61] Skotty, D.R., Lee, W-Y., and Nieman, T.A. (1996) *Anal. Chem.*, **68**, 1530.
[62] Rubinstein, I., Martin, C.R. and Bard, A.J. (1983) *Anal. Chem.*, **55**, 1580.
[63] Yuki, H., Azuma, Y., Maeda, N. and Kawasaki, H. (1988) *Chem. Pharm. Bull.*, **36**, 1905.
[64] Steijger, O.M., Lingeman, H., Brinkman, U.A. Th., Holthuis, J.J.M., Smilde, A.K. and Doornbos, D.A. (1993) *J. Chromatogr.*, **615**, 97.
[65] Grayeski, M.L. and DeVasto, J.K. (1987) *Anal. Chem.*, **59**, 1203.
[66] Lechtken, P. and Turro, N.J. (1974) *Mol. Photochem.*, **6**, 95.
[67] Sandmann, B.W. and Grayeski, M.L. (1994) *J. Chromatogr. B*, **653**, 123.
[68] Duan, G-L., Nakashima, K., Kuroda, N. and Akiyama, S. (1995) *J. Chin. Pharm. Sci.*, **4**, 22.
[69] Novak, T.J. and Grayeski, M.L. (1994) *Microchemical Journal*, **50**, 151.
[70] Kwakman, P.J.M., Mol, J.G.J., Kamminga, D.A., Frei, R.W., Brinkman, U.A. Th. and de Jong, G.J. (1988) *J. Chromatogr.*, **459**, 139.
[71] Nakashima, K., Umekawa, C., Nakatsuji, S., Akiyama, S. and Givens, R.S. (1989) *Biomed. Chromatogr.*, **3**, 39.
[72] Imai, K., Higashidate, S., Nishitani, A., Tsukamoto, Y., Ishibashi, M., Shoda, J. and Osuga, T., (1989) *Anal. Chim. Acta*, **227**, 21.
[73] Higashidate, S., Hibi, K., Senda, M., Kanda, S. and Imai, K. (1990): *J. Chromatogr.*, **515**, 577.
[74] Akiyama, H., Toida, T. and Imanari, T. (1991) *Anal. Sci.*, **7**, 807.
[75] Akiyama, H., Shidawara, S., Mada, A., Toyoda, H., Toida, T. and Imanari, T. (1992) *J. Chromatogr.*, **579**, 203.
[76] Uzu, S., Imai, K., Nakashima, K. and Akiyama, S., (1992) *J. Pharm. Biomed. Anal.*, **10**, 979.
[77] Mann, B. and Grayeski, M.L. (1987) *J. Chromatogr.*, **386**, 149.
[78] Nakashima, K., Nagata, M., Takahashi, M. and Akiyama, S. (1992) *Biomed. Chromatogr.*, **6**, 55.
[79] Yoshida, S., Urakami, K., Kito, M., Takeshima, S. and Hirose, S. (1990) *J. Chromatogr.*, **530**, 57.
[80] Yoshida, S., Urakami, K., Kito, M., Takeshima, S. and Hirose, S. (1990) *Anal. Chim. Acta*, **239**, 181.
[81] Ishida, J., Yamaguchi, M., Nakahara, T. and Nakamura, M. (1990) *Anal. Chim. Acta*, **231**, 1.
[82] Nakahara, T., Ishida, J., Yamaguchi, M. and Nakamura, M. (1990) *Anal. Biochem.*, **190**, 309.
[83] Ishida, J., Nakahara, T. and Yamaguchi, M. (1992) *Biomed. Chromatogr.*, **6**, 135.
[84] Ishida, J., Sonezaki, S. and Yamaguchi, M. (1992) *J. Chromatogr.*, **598**, 203.
[85] Ishida, J., Sonezaki, S., Yamaguchi, M. and Yoshitake, T. (1992) *Analyst*, **117**, 1719.
[86] Ishida, J., Sonezaki, S., Yamaguchi, M. and Yoshitake, T. (1993) *Anal. Sci.*, **9**, 319.
[87] Ishida, J., Yakabe, T., Nohta, H. and Yamaguchi, M. (1997) *Anal. Chim. Acta*, **346**, 175.
[88] Higashidate, S. and Imai, K. (1992) *Analyst*, **117**, 1863.
[89] Prados, P., Higashidate, S. and Imai, K. (1994) *Biomed. Chromatogr.*, **8**, 1.
[90] Ragab, G.H., Nohta, H., Kai, M., Ohkura, Y. and Zaitsu, K. (1995) *J. Pharm. Biomed. Anal.*, **13**, 645.
[91] Kuroda, N., Nakashima, K. and Akiyama, S., (1993) *Anal. Chim. Acta*, **278**, 275.

[92] Sato, N., Shirakawa, K., Sugihara, K. and Kanamori, T., (1997) *Anal. Sci.*, **13**, 59.
[93] Kai, M., Ohkura, Y., Yonekura, S. and Iwasaki, M., (1994) *Anal. Chim. Acta*, **287**, 75.
[94] Kobayashi, S. and Imai, K. (1980) *Anal. Chem.*, **52**, 1548.
[95] Sugiura, M., Kanda, S. and Imai, K. (1993) *Biomed. Chromatogr.*, **7**, 149.
[96] Baeyens, W., Bruggeman, J., Dewaele, C., Lin, B. and Imai, K. (1990) *J. Biolumin. Chemilumin.*, **5**, 13.
[97] de Jong, G.J., Lammers, N., Spruit, F.J., Dewaele, C. and Verzele, M. (1987) *Anal. Chem.*, **59**, 1458.
[98] Weber, A.J. and Grayeski, M.L. (1987) *Anal. Chem.*, **59**, 1452.
[99] Nakashima, K., Maki, K., Akiyama, S. and Imai, K. (1990) *Biomed. Chromatogr.*, **4**, 105.
[100] Kimoto, K., Gohda, R., Murayama, K., Santa, T., Fukushima, T., Homma, H. and Imai, K. (1996) *Biomed. Chromatogr.*, **10**, 189.
[101] Nakashima, K., Yamasaki, H., Kuroda, N. and Akiyama, S. (1995) *Anal. Chim. Acta*, **303**, 103.
[102] Baeyens, W.R.G., Lin Ling, B., Imai, K., Calokerions, A.C. and Schlman, S.G. (1994) *J. Microcol. Sep.*, **6**, 195.
[103] Wu, N. and Huie, C.W. (1993) *J. Chromatogr.*, **634**, 309.
[104] Hara, T., Okamura, S., Katou, S., Yokogi, J. and Nakajima, R. (1991) *Anal. Sci.*, **7** (supplement), 261.
[105] Hara, T., Kayama, S., Nishida, H. and Nakajima, R. (1994) *Anal. Sci.*, **10**, 223.
[106] Ruberto, M.A. and Grayeski, M.L. (1992) *Anal. Chem.*, **64**, 2758.
[107] Dadoo, R., Seto, A.G., Colon, L.A. and Zare, R.N. (1994) *Anal. Chem.*, **66**, 303.

# 5

# Reagents for Electrochemical Detection

**Kenji Shimada,* Tomokazu Matsue[†] and Kazutake Shimada[‡]**
*Niigata College of Pharmacy, [†]Graduate School of Engineering, Tohoku University,
[‡]Faculty of Pharmaceutical Sciences, Kanazawa University

5.1 Introduction 192
5.1.1 Amperometric Analysis in Flowing Streams 192
5.1.2 Electrochemical Detectors 192
5.1.3 Electrode Material 193
5.1.4 Electrode Configuration 194
5.1.5 Mobile Phase 195
5.1.6 Chemical Derivatization 196
5.2 Labeling of Primary and Secondary Amines 198
5.2.1 *o*-phthalaldehyde 198
5.2.2 Naphthalene-2,3-dicarbaldehyde 199
5.2.3 Ferrocene 200
5.2.4 Isocyanate and Isothiocyanate 202
5.2.5 Salicylic Acid Chloride 203
5.2.6 2,4-Dinitrofluorobenezene 204
5.2.7 1,2-Diphenylethylenediamine 204
5.2.8 Bolton and Hunter Type Reagent 204
5.2.9 Others 205
5.3 Labeling of Carboxy Groups 205
5.4 Labeling of Hydroxy Groups 207
5.5 Labeling of Thiol Groups 208

5.6 Labeling of Other Functional Groups 208
5.7 Applications 211
References 213

## 5.1 INTRODUCTION

### 5.1.1 Amperometric Analysis in Flowing Streams

Electroanalytical methods are categorized by their mode of measurement into amperometric (measurement of current), potentiometric (measurement of voltage), and conductimetric (measurement of conductivity) methods. This chapter deals with amperometric electroanalysis in flowing streams [1–3]. The major advantages of amperometric detection over classical spectroscopic measurements are its high selectivity, low detection limits, and the simplicity of the instrument.

Amperometric reaction (or electrode reaction) is composed of a series of complicated steps; but the amperometric response is usually limited by either or both of the mass transport from the bulk solution to the electrode surface (or electrode surface to bulk solution) and/or the surface reaction including electron transfer at the electrode. When the surface reaction is very rapid, the amperometric response is governed by mass transport. The driving forces of mass transport are: migration for charged species under the influence of an electric field; diffusion caused by a concentration gradient; and convection caused by stirring or hydrodynamic flow [4]. The contribution of migration to mass transport can be minimized by adding an excess of a non-electroactive salt (supporting electrolyte) to the solution containing the redox species. Because flowing paths greatly affect the mass transport of the analyte, the amperometric response in flowing streams is very sensitive to the structure of the detectors. The detectors proposed to date can be classified into thin-layer, wall-jet, tubular and coulometric detectors (Fig. 5.1). The next section describes the properties of these electrochemical detectors.

### 5.1.2 Electrochemical Detectors

A thin-layer cell is the most common configuration of amperometric detector in flowing streams. In many cases, the cell utilizes a planar electrode in a rectangular flow-through cell. For the theoretical and practical analyses of the flow-through cell it is important to clarify the type of fluid flow which is characterized by a dimensionless parameter, the Reynolds number (Re):

$$\mathrm{Re} = v_{\mathrm{av}} d / \nu \tag{1}$$

where $v_{\mathrm{av}}$ is the average velocity, $d$ is the flow length, and $\nu$ is the kinematic viscosity. The

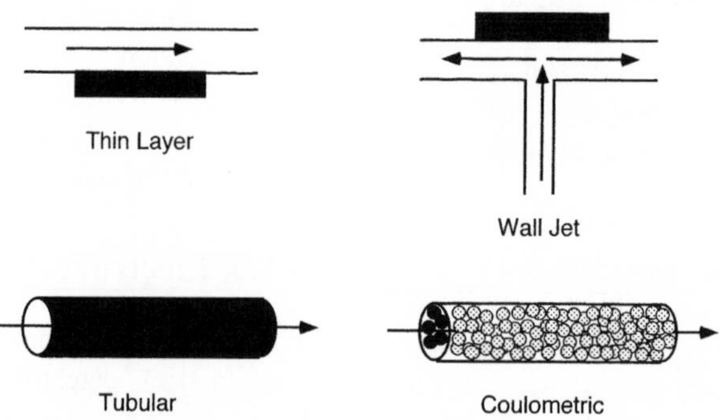


**Fig. 5.1.** Amperometric cells for flow analysis.

quantitative analysis of the amperometric response requires the flowing stream to be laminar, in which the flow is smooth and steady. The flowing stream becomes laminar in a channel flow cell when $Re < 10^4$. Thus, for a dilute aqueous solution, the average velocity should be less than 100 cm/s to develop laminar flow in a thin-layer cell of 1 cm flow length because the kinetic viscosity of water at 20 °C is about 0.01 $cm^2/s$.

In laminar flow, the fluid moves parallel to the surface of the electrode and, therefore, the flow shows no direct motion of the redox species toward the electrode surface. However, the flow refreshes the solution with depleted redox species near the electrode and increases the flux of redox species to the surface. If we neglect longitudinal mass transport, the steady state current ($i_{SS}$) in a rectangular flow-through thin-layer cell under steady-state, laminar flow conditions is given by [5]:

$$i_{SS} = 1.467nFC^*(AD/b)^{2/3}U^{1/3} \quad (2)$$

where $C^*$ is the bulk concentration of the analyte, $A$ is the electrode area, $b$ is the channel height, and $U$ is the volume flow rate. It should be noted here that the amperometric response in thin-layer cells is proportional to the concentration of redox analyte and to the cube root of the flow rate.

In a wall-jet cell, the mobile solution was introduced from a small inlet as a jet of flow perpendicular to the wall with an electrode. The mobile solution then flows radially over the wall. The amperometric response in this case is proportional to the concentration of redox analyte and to the three-fourths power of the flow rate [6]:

$$i_{SS} = 0.898nFC^*D^{2/3}\nu^{-5/12}a^{-1/2}A^{3/8}U^{3/4} \quad (3)$$

where $a$ is the diameter of inlet conduit. Wall-jet cells, in general, provide efficient mass transfer to the electrode surface and thereby provide higher sensitivity.

Tubular cells have also been used as a detector for electrochemical flow analysis. When a solution containing a redox analyte is continuously passed through a tubular electrode at a constant laminar flow, we observe a steady-state current which is proportional again to the concentration of the analyte and to the cube root of the flow rate [5,7]:

$$i_{SS} = 1.61nFC(DA/r)^{2/3}U^{1/3} \quad (4)$$

where $r$ is the radius of the tubular electrode. The demerits of this cell are that some electrode materials cannot be used and that mechanical polishing of the electrode surface is difficult.

The amperometric detectors described above electrolyze only a part of the analyte in flowing to the electrode. Some detectors are designed to electrolyze almost all of the analyte. The detectors of this kind are commonly called ‘coulometric detectors’ to discriminate them from ordinary amperometric detectors [8]. The integral of the amperometric response of a coulometric detector with respect to time is directly proportional to the number of moles ($N$) of the analyte:

$$Q = \int i\,dt = nFN \quad (5)$$

Therefore, unlike many amperometric detectors, calibration is not necessary. The peak area corresponds directly to the amount of analyte injected if the value of $n$ is known. More importantly, because the peak area is not influenced by fluctuation of the flow rate, reproducibility is, in general, better than that with the commonly-used amperometric detectors. The complete electrolysis of the analyte requires a large area/volume ratio. Thus, a typical coulometric detector consists of an electrode with a large surface area such as screens of fine mesh metal, sintered metal, packed beds of conducting particles, or carbon fibers. The drawbacks of coulometric detectors are that they need relatively large cell volumes which cause a peak broadening. In addition, the large surface area increases area-dependent noise and lowers signal-to-noise ratio (S/N). The detection limit of coulometric cells is generally inferior to that of thin-layer amperometric cells.

### 5.1.3 Electrode Material

The above arguments on the amperometric (and coulometric) detectors in flowing streams are based on the mass transport of the analyte to the electrode surface. Another important aspect in amperometric detection is the surface reaction including electron transfer. Because the surface reaction is greatly dependent on the physical and chemical properties of the electrode material, the selection of the electrode is crucial in order to obtain reliable results. The potential window, the range of accessible potentials with low background currents, is also an important factor in choosing the electrode material. In protic solutions, the negative window is usually limited by hydrogen evolution. The positive window is mainly limited by dissolution of the electrode and/or oxidation of the solvent. The electrode materials used frequently for flow analysis are carbon, platinum, gold, and mercury [9].

Carbon is the most commonly used material for amperometric detection in flow systems. Several different forms of carbon are available from commercial sources to make electrodes including glassy carbon, carbon paste, and graphite. Glassy carbon has excellent mechanical stability and electrical conductivity. It is highly resistant to chemical attack, gas impermeable, and free of chemical impurities. The potential window of glassy carbon is relatively large because hydrogen evolution and oxidation of many protic solvents proceeds very slowly at glassy carbon surfaces. Despite the wide potential window, the electron transfer rates of many redox species have been found to be large. Carbon paste electrodes prepared by dispersing graphite powder in an organic liquid such as Nujol have also been used for flow-through amperometric detectors. Carbon paste electrodes show good reproducibility and low residual current in the positive potential range. However, carbon paste contains oxygen which is reduced at the electrode, resulting in a large residual current in the negative potential region. Carbon paste tends to dissolve in organic solvents, therefore, it is not suitable for organic mobile phases.

Noble metals such as platinum and gold have also been used in the amperometric detectors. These metals are chemically stable, readily obtained in high purity, and can be fabricated readily into a variety of geometric configurations. However, the potential windows of platinum and gold are narrow compared with carbon electrodes. Platinum has extremely small over potentials for hydrogen evolution and oxygen reduction because the electrode itself is a catalyst for these reactions. At positive potentials, platinum and gold form oxide films at the surfaces which sometimes induce undesired chemical reactions.

Mercury has a wide potential window in the negative potential region because of its high overpotential for hydrogen evolution. A clean surface is easy to form due to its liquid-metallic nature at room temperature. Thus, mercury has been thought to be the best electrode material for investigation in the negative potential region. At the positive potential region, mercury is easily oxidized, particularly in the presence of some species which can form mercury complexes. This property can be used for selective detection of thiols such as cysteine and homocysteine [10]. However, it is technically difficult to fabricate a well-defined detector with mercury electrode for flow analysis. Because mercury contamination in the environment has been a matter of social interest, the applicability of mercury in flow analysis is now limited.

Amperometric detectors are, in general, based on a three-electrode system which consists of a working (or indicator) electrode as described above, a reference electrode to which the potential of working electrode is referenced, and a counter (or auxiliary) electrode. The commonly-used reference electrodes are an Ag/AgCl electrode and a saturated calomel electrode (SCE) for aqueous solutions and an $Ag/Ag^{+}$ electrode for non-aqueous solutions. Platinum, gold, or carbon has been used as the counter electrode because these materials are chemically inert.

## 5.1.4 Electrode Configuration

Recent progress in microfabrication technology enables us to use a microarray electrode, an assembly of well-defined microelectrodes, for electrochemical measurements [11]. To date, various kinds of microarray electrodes have been fabricated for flow analysis; including a split disk electrode [12] for a wall-jet detector, arrays of microdisks [13–16], microholes [17], and

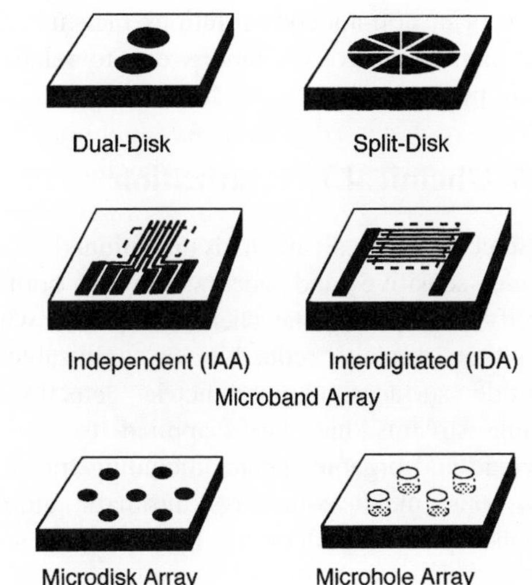


**Fig. 5.2.** Electrode arrays for flow amperometric detectors.

microbands [11] for thin-layer amperometric detectors (Fig. 5.2). When the electrode elements are located close together, chemical and physical processes occurring at an electrode element affect the reaction proceeding at adjacent elements. This 'cross-talking' effect brings about unique electrochemical behavior which is never expected for ordinary electrodes and single microelectrodes.

The origin of the microarray electrodes for flow analysis is a thin layer detector with a dual-disk electrode [18]. There are two types of configuration of placements of the dual-disk electrode with respect to the direction of solution flow; i.e., parallel and series configurations. In a parallel configuration, one can monitor the amperometric responses at two different potentials. In a series configuration, the upstream electrode is generally used as a generator and the electrogenerated products are monitored at the downstream electrode. This configuration has substantial advantages in selectivity because the downstream electrode can detect only reversible redox species.

The above characteristics of a dual-disk electrode are further emphasized using interdigitated microarray (IDA) electrodes which consist of two metal-microband arrays opposing each other on insulating substrates. When an IDA electrode is placed in series configuration, the amperometric response is amplified by redox cycling [11] occurring at adjacent microelectrode elements. The redox cycling is operative only on reversible redox species and amplifies the redox current. The amplification factor for amperometric response by redox cycling increases with decreasing electrode and gap widths. As is true for an ordinary flow-through cell, the peak current at a thin-layer cell with an IDA electrode increases linearly with the cube root of the flow rate [19,20].

When a microband array electrode is placed in parallel configuration with respect to the flow direction, one can obtain three-dimensional (3D) information (time, potential, and current) [21,22]. Acquisition of 3D information usually requires a rapid potential scan to detect species flowing in and out of the detector, resulting in an increases in the background due to charging current and also in distortion of the voltammetric shape. Such undesired features can be eliminated using multichannel electrochemical detection (ECD) systems with independently addressable microarray (IAA) electrodes. An IAA electrode consisting of an assembly of 16 independent microband electrodes held at different potentials was used to obtain 16 and 80 channel 3D information. Hydrodynamic voltammograms of analytes can easily be obtained from a cross-section of the 3D result at various time domains on the time axis.

### 5.1.5 Mobile Phase

The mobile phase for ECD should be adequately conductive, be chemically and electrochemically inert in the potential range investigated, and dissolve the analyte of interest. To support electrical conductivity, non-electroactive salts (supporting electrolytes) are added to the mobile solution. Purification of the solvent and the supporting electrolyte is of crucial importance to reduce residual current and therefore to attain high sensitivity.

Non-polar organic solvents cannot be used because of their inability to dissolve supporting electrolytes. Polar aprotic solvents such as acetonitrile and DMF with tetraalkylammoium salts as a supporting electrolyte have been frequently used for electrochemical flow analysis. The use of a non-aqueous mobile phase has a substantial advantage over an aqueous mobile phase in terms of accessible potentials: for example, the potential window in dry acetonitrile expands as wide as from −3 to +3 V *vs* an Ag/AgCl reference electrode whereas the typical potential window in aqueous solution is from −1 to +1 V *vs* as SCE. However, amperometric detection at very negative or very positive potentials in non-aqueous solutions is severely influenced by the presence of trace water. Detectors in non-aqueous solutions pick up more noise than in aqueous solutions due to relatively high impedance.

### 5.1.6 Chemical Derivatization

The electrochemical methods mentioned above provide sensitive and accurate ways both to quantify and qualify the chemical species which are either directly reducible or oxidizable at electrode surfaces. Amperometric detection in flowing streams has been applied to various inorganic and organic species including transition metal ions, metal complexes, aromatic amines, phenols, quinones, alcohols, thiols, halogenated

**Table 5.1.** Electrochemical derivatization reagents

| Labeling reagent | Substrate | References |
|---|---|---|
| *o*-phthalaldehyde (OPA) | amine (amino acid) | 2,28,34–45 |
| | amine | 2,29–33 |
| Naphthalene-2,3-dicarbaldehyde (NDA) | amine | 46–48 |
| *N*-succinimidyl 3-ferrocenylpropionate | amine | 50 |
| Ferrocenylisothiocyanate | amine | 51 |
| Ferrocenylethylisothiocyanate | amine | 51 |
| Ferrocene carboxaldehyde | amine | 52 |
| Ferrocene carboxylic acid chloride (FAC) | amine | 52,53 |
| *p*-*N*,*N*-dimethylaminophenylisothiocyanate (DMAPI) | amine (amino acid) | 54 |
| Phenylisothiocyanate (PITC) | amine (amino acid) | 55 |
| *R*-(+)-α-methylbenzylisocyanate | amine | 56 |
| *O*-acetylsalicyloyl chloride | amine | 57 |
| Salicylic acid chloride | amine | 58 |
| 2,4-dinitrofluorobenzene | amine (amino acid, amino alcohol) | 59,60 |
| 1,2-diphenylethylenediamine | amine | 61–63 |
| Homovanillic acid-*N*-hydroxysuccinimide | amine | 64 |
| Sulfosuccinimidyl-3-(4-hydroxyphenyl) propionate | amine | 65 |
| 3-(4-hydroxyphenyl)propionic acid-*N*-hydroxysuccinimide | amine | 66 |
| *N*-(4-anilinophenyl)isomaleimide (APIM) | amine | 67 |
| Polymeric anhydride (containing *O*-acetylsalicyl) | amine | 68 |
| 5-(dimethylamino)-1-naphthalene sulfonyl chloride (Dansyl chloride) | amine | 69 |
| *p*-aminophenol | carboxylic acid | 71 |

**Table 5.1.** *(continued)*

| Labeling reagent | Substrate | References |
|---|---|---|
| 2,4-dimethoxyaniline | carboxylic acid (PG) | 72 |
| *p*-nitrobenzyloxyamine | carboxylic acid (PG) | 72 |
| 2-bromo-2′-nitroacetophenone | carboxylic acid (PG) | 73 |
| 2,4-dinitrophenylhydrazine | carboxylic acid (PG) | 73 |
| 4-bromomethyl-7-methoxy-6-nitrocoumarin | carboxylic acid | 74 |
| 3-bromoacetyl-1,1′-dimethylferrocene | carboxylic acid | 75 |
| *R*-(−)-1-ferrocenylethylamine | carboxylic acid | 76 |
| *S*-(+)-1-ferrocenylpropylamine | carboxylic acid | 76 |
| 1-(4-hydroxyphenyl)-2-bromoethanone (4-HBE) | carboxylic acid | 77 |
| 1-(2,4-dihydroxyphenyl)-2-bromoethanone (2,4-DBE) | carboxylic acid | 77 |
| 1-(2,5-dihydroxyphenyl)-2-bromoethanone (2,5-DBE) | carboxylic acid | 77 |
| Trimethylaminomethyl ferrocene iodide | carboxylic acid | 52 |
| Ferrocenoyl azide | alcoholic-OH | 78 |
| 3-ferrocenylpropionyl azide | alcoholic-OH | 78 |
| 4-aminoantipyrine | phenolic-OH | 79 |
| Ferroceneboronic acid (FBA) | alcoholic-OH | 80,81 |
| 2-[2-(azidocarbonyl)ethyl]-3-methyl-1,4-naphthoquinone (AMQ) | alcoholic-OH | 82 |
| 2-[2-(isocyanate)ethyl]-3-methyl-1,4-napthoquinone (IMQ) | alcoholic-OH | 83,84 |
| *N*-(4-anilinophenyl)maleimide (APM) | thiol | 86 |
| *N*-(ferrocenyl)maleimide | thiol | 87 |
| Phenylhydrazine | carbonyl | 88 |
| *p*-nitrophenylhydrazine | carbonyl | 88,89,94 |
| 2,5-dihydroxybenzohydrazide (DBH) | carbonyl | 91 |
| | aldehyde | 95 |
| 2,4-dinitrophenylhydrazine | carbonyl | 92,93 |
| 2-cyanoacetamide | carbohydrate | 96 |
| Silver picrate | alkyl halide | 97 |
| Hydroxylamine | hydroxyphenyl pyruvate | 99 |
| 2-ferrocenylethylamine | carboxylic acid (steroid glucuronide) | 100 |
| Tryptamine | isocyanate | 101 |
| 2-mercaptobenzothiazole | chloramine | 102 |
| Ferrocencarboxyhydrazide | aldehyde | 52 |
| *R*-(−)-2-amino-1-propanol | aldehyde | 103 |

compounds, nitro derivatives, aldehydes, and ketones [1]. However, there are many compounds which are analytically important but do not show redox behavior in the available potential windows. Thus, the attachment of a redox moiety to the species of interest has become commonplace in ECD. Several recent reviews on the electrochemical derivatization have been published [2,23–25] and the reagents of the derivatization for several functional groups are summarized in Table 5.1. The next section describes the derivatization in detail.

## 5.2 LABELING OF PRIMARY AND SECONDARY AMINES

### 5.2.1 *o*-phthalaldehyde

*o*-phthalaldehyde (OPA), which was originally developed by Roth [26] as a fluorescent derivatization reagent for primary amines in the presence of a reducing agent, 2-mercaptoethanol, as the post-column reaction reagent, is the most widely used ECD labeling reagent. The derivatized isoindoles from amines also show favorable electroactive properties and a great number of applications have been reported for the derivatization of amines and amino acids [2]. In 1983, Joseph and Davies demonstrated that the isoindoles produced from the amino acids in the OPA reaction are electroactive [27]. The reaction formula is shown in Fig. 5.3, the details of which are described in the last paragraph of this section. Hoskins *et al.* showed that analysis of amino acids using OPA derivatization and ECD is possible and satisfactory using samples containing nanogram quantities of amino acids. Detection was electrochemical at +0.6 V *vs* an Ag/AgCl electrode [28].

Direct oxidation of histamine requires a much higher potential (peak potential +1.05 V with Pd electrode). HPLC with ECD was adopted to determine histamine following pre-column derivatization with OPA and 2-mercaptoethanol. The isoindole derivative obtained as the reaction product of histamine and OPA was electroactive at moderate potentials (peak potential +0.4 V with Pd electrode) [29].

In order to assay sympathomimetic drugs (heptaminol, amphetamine, phenylpropanolamine, $\beta$-phenethylamine (PEA) and 2-heptylamine), a derivatization procedure with the use of OPA and various thiols (2-mercaptoethanol, ethanethiol or *t*-butanethiol) was developed. PEA is also an endogenous amine present in several mammalian tissues. PEA lacks hydroxy groups in the aromatic ring. Therefore, PEA itself gives no signal in the ECD system even at high oxidation potentials. The electroactive properties of these substituted isoindolic products were investigated by amperometry (+0.9 V *vs* a SCE). Picomole detection limits were achieved [30].

Baclofen, a chemical analogue of $\gamma$-aminobutyric acid (GABA), used for the symptomatic relief of muscular spasm, was derivatized with OPA and *t*-butanethiol at room temperature and subjected to HPLC analysis with ECD. The detection limit was 2.5 ng/ml. By combining solid phase extraction with highly sensitive ECD, the optimum performance for the pharmacokinetic studies of baclofen was achieved. Refer to the detailed descriptions in Section 5.7 on applications [31].

Amines in wines were analyzed by HPLC with ECD (coulometric detection) after pre-column derivatization with OPA. The coulometric detector consisted of four cell packs in series. For histamine, tryptamine, tyramine, PEA, putrescine, cadaverine, 1,6-diaminohexane and tryptophan, the calibration graphs were rectilinear from 10 ng/ml to 100 μg/ml and the detection limits were 12–25 ng/ml. This method yields a high precision without preliminary separation of the different families of compounds which is necessary in other methods [32]. Polyamines (spermidine, putrescine, cadaverine and spermine) in mouse brain tissues were also derivatized by treatment

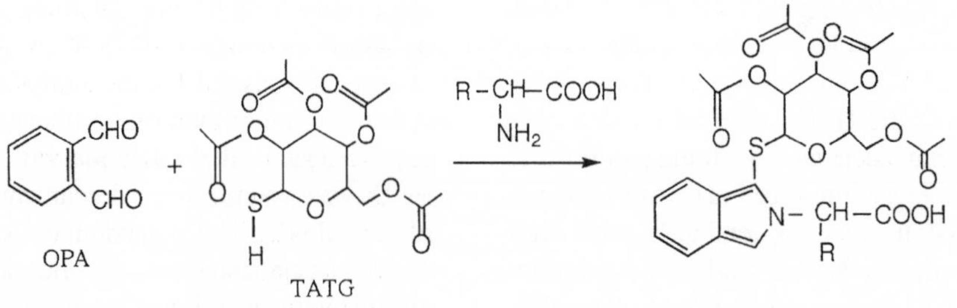

**Fig. 5.3.** Reaction of amino acid and OPA/TATG.

with methanolic OPA-ethanethiol reagent by pre-column derivatization [33].

For the measurement of GABA in rat brain microdialysis, OPA-sulfite derivatization and HPLC with ECD were introduced by Smith and Sharp [34]. They overcome the poor stability of the electroactive GABA derivative by substitution of the OPA-alkylthiol reaction with an OPA-sulfite reaction. The derivatives were analyzed with ECD at +0.85 V *vs* an Ag/AgCl electrode, and the limit of detection was 25–50 fmol. Turiak and Volicer examined the stability of OPA-sulfite derivatives of five amino acids (alanine, arginine, glutamic acid, serine, tyrosine) and their methyl esters with respect to electrochemical and chromatographic properties [35]. Simultaneous determination of GABA and polyamines (putrescine, spermine, spermidine, cadaverine) by OPA-2-mercaptoethanol pre-column derivatization and gradient elution HPLC with ECD was proposed by Zambonin *et al.* [36]. HPLC separation of reduced and oxidized glutathione ratio in tumor cells was performed with pre-column derivatization with OPA and ECD with a glassy carbon working electrode at +0.75 V *vs* an Ag/AgCl [37]. Twenty five standard amino acids could be separated with OPA-2-mercaptoethanol derivatization by automated pre-column derivatization with ECD [38]. Very rapid and simple assay of taurine in the brain tissue samples was achieved within 2 min by HPLC with ECD after derivatizing with OPA [39]. Various cysteine derivatives were labeled with OPA and *t*-butanethiol and separated by HPLC on a column of Spherisorb ODS. ECD was set at +0.75 V *vs* an Ag/AgCl electrode with a detection limit of 130–160 fmol [40]. Determination of amino acids in rat brain was carried out utilizing pre-column derivatization with OPA-*t*-butylthiol and ECD with a glassy carbon working electrode in the oxidation mode (+0.7 V *vs* an Ag/AgCl). The validity and usefulness of the HPLC method for detecting neurotoxicity-related changes in brain amino acid metabolism were established [41]. Amino acids were derivatized in aqueous methanol with OPA and 2-methylpropane-2-thiol, and the derivatives were separated on a column of LiChrospher ODS. The derivatives were detected at +0.5 V *vs* a SCE [42]. Boomsma *et al.* applied an OPA derivatization method to the determination of D,L-threo-3,4-dihydroxyphenylserine (D,L-threo-DOPS) and of the D- and L-enantiomer in human plasma and urine in the presence of *N*-acetyl-L-cysteine. Good separation was achieved with this procedure (resolution factor 2.33, Fig. 5.4) [43]. Pre-column derivatization with OPA-2-mercaptoethanol and the ECD method was developed for the determination of the mucolytic drug *S*-carboxymethyl-L-cysteine (SCMC) and its major metabolites *S*-methyl-L-cysteine (SMC), SCMC-sulfoxide and SMC-sulfoxide. Good resolution was obtained for urine [44].

Separation of amino acid enantiomers using pre-column derivatization with OPA and 2,3,4,6-tetra-*O*-acetyl-1-thio-$\beta$-glucopyranoside (TATG) was developed by Einarsson *et al.*; the reaction was completed in a few minutes at room temperature and the derivatives were quite stable (Fig. 5.3). The formed diastereomers were separated by reversed-phase chromatography and selectivity was generally good, except for lysine and ornithine. The electrochemical half-wave potential ($E_{1/2}$) was +0.65 ~ +0.75 V *vs* an Ag/AgCl. The detection limits (for L-leucine) were 23 pmol and 1 pmol for fluorescence and ECD, respectively [45].

## 5.2.2 Naphthalene-2,3-dicarbaldehyde

Naphthalene-2,3-dicarbaldehyde (NDA), in the presence of cyanide ion ($CN^-$) as a nucleophile, reacts with primary amines to produce fluorescent cyano[*f*]benzoisoindole (CBI) derivatives which are similar to those produced with OPA (Fig. 5.5). The CBI derivatives have been found to be electroactive and are oxidized at a modest oxidation potential (+0.75 V *vs* Ag/AgCl). ECD is especially useful for the analysis of compounds containing more than one primary amine site. The combination of derivatization with NDA/$CN^-$ and ECD was found to be linear over three orders of magnitude. The detection limit for CBI-lysine and CBI-desmosine was 100 fmol. Amino acids in elastin were quantitated using this method. Refer to the detailed descriptions in Section 5.7 [46].

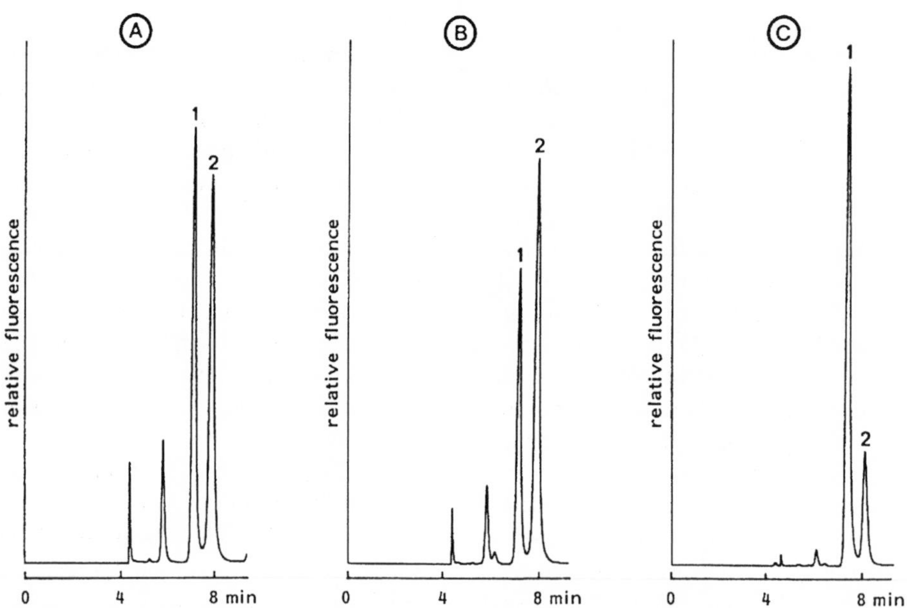


**Fig. 5.4.** Separation of D- and L-threo-DOPS after derivatization with OPA and *N*-acetyl-L-cysteine [43]. (A) Standard mixture; (B) plasma sample; (C) urine sample. Peaks: 1 = D-threo-DOPS; 2 = L-threo-DOPS. Reprinted from *Journal of Chromatography*, 427, F. Boomsma *et al.*, Separation of D- and L-threo-DOPS, 222, 1988, with kind permission of Elsevier Science-NL, Sara Burgerhartstraat 25, 1055 KV Amsterdam, The Netherlands.

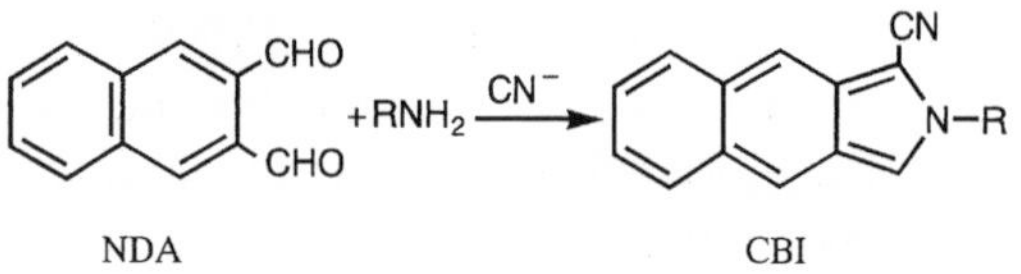

CHO
CHO
$+ RNH_2 \xrightarrow{CN^-}$
CN
N–R
NDA
CBI

**Fig. 5.5.** Structure of NDA and reaction with amine.

Alendronic acid (4-amino-1-hydroxybutane-1, 1-bisphosphonic acid), previously developed as a potent inhibitor of bone resorption by the researcher, was determined in urine and plasma by HPLC with ECD as its cyanide derivative. The assay was based on pre-column derivatization of alendronate with NDA in the presence of $CN^-$ to produce the electroactive *N*-substituted CBI derivative. The HPLC with ECD assay in human urine was validated in the concentration range of 2.5–50.0 ng/ml [47,48].

### 5.2.3 Ferrocene

Highly selective and sensitive derivatization reagents for amine or alcoholic hydroxy compounds for HPLC with ECD, which possess the ferrocenyl group as an electrophore, have been developed by Tanaka *et al.* [49]. The ferrocenyl moiety can be easily oxidized and can be detected selectively in the presence of other electroactive compounds such as phenols and aromatic amines. Fig. 5.6 a shows the structure of *N*-succinimidyl 3-ferrocenylpropionate, a labeling reagent for amines. Quantitative condensation of arylalkylamines such as PEA and tryptamine with *N*-succinimidyl 3-ferrocenylpropionate was effected at room temperature for 20 min in acetonitrile-0.05 M borate buffer (pH 8.0) (1:1). The maximum sensitivity could be obtained with a glassy carbon working electrode at +0.40 V *vs* an Ag/AgCl reference electrode and the detection limit was 0.2 pmol [49]. Of six reagents having ferrocene as an electrophore evaluated with respect to stability and reactivity with the use of PEA, *N*-succinimidyl 3-ferrocenylpropionate was the best with respect to reactivity, stability and electrochemical properties [50].

Glycine (as a model compound) was derivatized with ferrocenylisothiocyanate (Fig. 5.6b) or

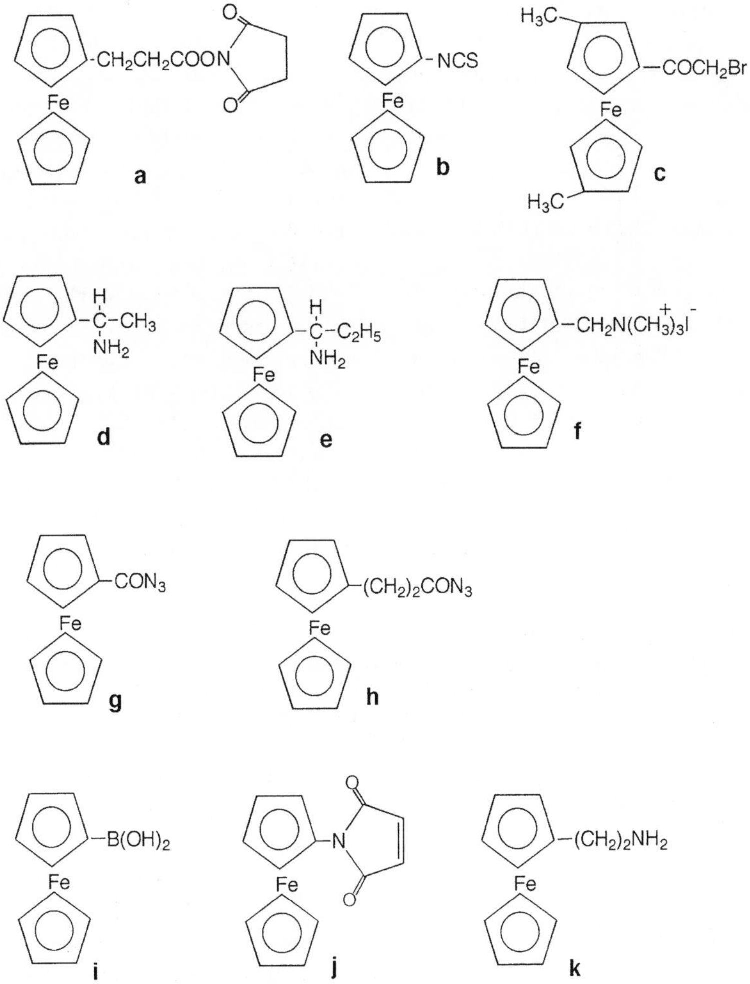

**Fig. 5.6.** Structures of ferrocene derivatization reagents.

ferrocenylethylisothiocyanate in aqueous acetonitrile containing triethylamine at 70 °C for 30 min and 90 min. The derivatives were determined by HPLC on a column of TSKgel ODS-80TM with ECD with a glassy carbon working electrode *vs* an Ag/AgCl reference electrode. The ferrocenylisothiocyanate was applied in the determination of GABA in the brain, and the applied potential was +0.50 V. The detection limit was 0.05 pmol [51]. Kubab *et al.* developed a derivatization method using ferrocene reagents for the determination of carboxylic acid, amino acid and aldehyde by HPLC with ECD. Among the reagents examined, ferrocene carboxaldehyde and ferrocene carboxylic acid chloride (FAC) proved to be satisfactory reagents for amines with respect to reactivity, stability and electrochemical properties. Maximum sensitivities were observed at +0.50 V and +0.75 V *vs* an Ag/AgCl, respectively, with a detection limit of 0.5 pmol [52]. FAC reacts readily with both primary and secondary amines and amino acids to yield stable amides, which are reversibly oxidized in mixed aqueous acetonitrile solution at moderate potentials (+0.22 V *vs* Ag/AgCl). Ferrocenesulfonylchloride was also investigated as a ferrocene tagging reagent, but gives low sulfonamide yields.

A mixture of 14 amino acids was derivatized and separated by gradient elution in HPLC with ECD within 45 min. The HPLC with ECD determination of the FAC derivative yielded a detection limit of 500 fmol [53].

### 5.2.4 Isocyanate and Isothiocyanate

Amino acids react with *p*-*N*,*N*-dimethylaminophenylisothiocyanate (DMAPI) to form the corresponding substituted phenylhydantoin, which can be reversibly oxidized with a glassy carbon working electrode at pH 2 at $E_{1/2} = +0.68 \pm 0.01$ V *vs* an Ag/AgCl reference electrode. The derivatized amino acids were separated on a C8 column in 0.01 M phosphate buffer-25% acetonitrile (pH 2 or 6) and detected at a glassy carbon working electrode set at +0.85 V *vs* an Ag/AgCl reference electrode. A mixture of 21 amino acids has been separated with 80–90% recovery and with a linear response from 1 to greater than 150 ng and detection limits of 0.5–1.0 ng [54].

An HPLC method was proposed for the separation of amino acids in blood or urine, using precolumn derivatization with phenylisothiocyanate (PITC), gradient elution and ECD. The use of PITC derivatives for ECD virtually eliminates interference and enables secondary amino acids to be measured. Detection was performed with a glassy carbon working electrode at +1.1 V *vs* an Ag/AgCl reference electrode. The derivatives were stable for long periods when stored dry at −20 °C [55].

Chiral derivatization with *R*-(+)-α-methylbenzylisocyanate of the four stereoisomers of a new

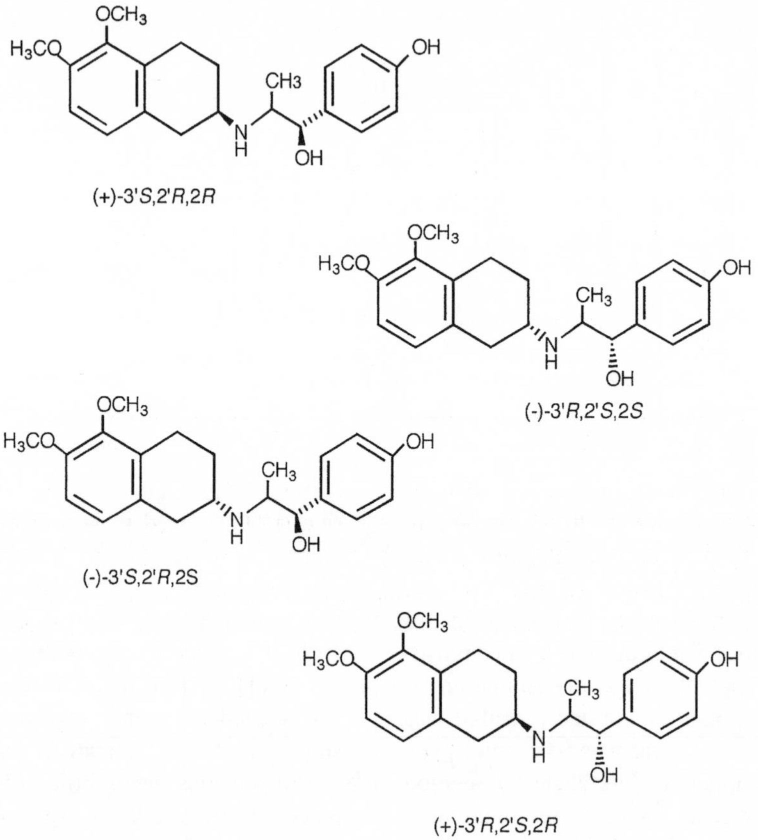

**Fig. 5.7.** Structures of stereoisomers of aminotetralin.

aminotetralin, (+)-3′*S*, 2′*R*, 2*R*-, (−)3′*R*, 2′*S*, 2*S*-, (−)-3′*S*, 2′*R*, 2*S*- or (+)-3′*R*, 2′*S*, 2*R*-5,6-dimethoxy-2-[3′-(*p*-hydroxyphenyl)-3′-hydroxy-2′-propyl]aminotetralin was introduced. The use of an electrochemical detector, operating in the oxidative mode, allows the quantitation in plasma of all four urea derivatives at a nanogram level. The method is precise, reproducible and applicable to pharmacokinetic studies after administration of the two epimeric racemates. The structures of the stereoisomers of a new aminotetralin are shown in Fig. 5.7. A chromatogram obtained from rat plasma after oral administration of the drug (50 mg/kg) as the racemate is also shown in Fig. 5.8 [56].

Ferrocenylisothiocyanate or ferrocenylethylisothiocyanate, a labeling reagent for amines, is described as a ferrocene reagent (5.2.3).

### 5.2.5 Salicylic Acid Chloride

*O*-acetylsalicyloyl chloride was developed as a pre-column derivatization reagent for aliphatic amines. Initially this compound reacts with an amino group to form an *o*-acetylsalicylamide, which will be hydrolyzed spontaneously under the alkaline conditions of the synthesis to yield a 2-hydroxybenzamide (Fig. 5.9). This free phenolic group can then be detected electrochemically at low concentrations in aqueous solution following separation by HPLC [57].

Aliphatic amines, such as propylamine, butylamine, 2-pentylamine, dimethylamine, diethylamine and dipropylamine react readily with salicylic acid chloride to produce electroactive amide derivatives with good chemical stability. The

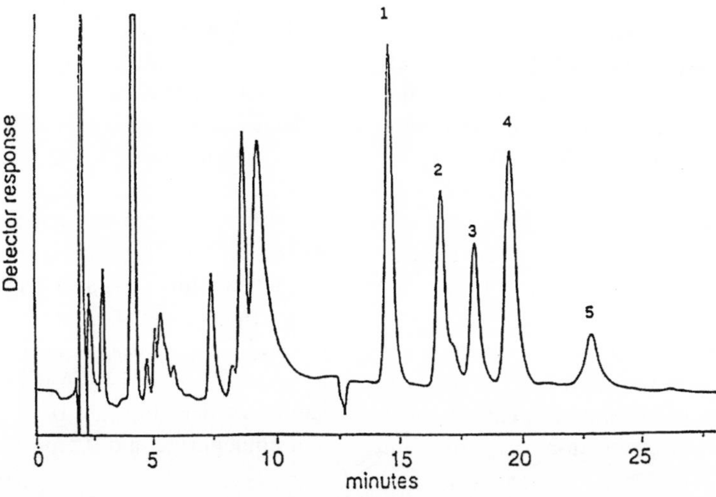


**Fig. 5.8.** Chromatogram obtained from rat plasma after oral administration of the drug as racemates [56]. Samples were submitted to the extraction procedure for the determination of total stereoisomers of aminotetralin. Peaks: 1 = (+)-3′*S*, 2′*R*, 2*R*; 2 = (−)-3′*R*, 2′*S*, 2*S*; 3 = (−)-3′*S*, 2′*R*, 2*S*; 4 = (+)-3′*R*, 2′*S*, 2*R*; 5 = IS. Reprinted from *Journal of Chromatography*, 612, I. Rondelli *et al.*, Chromatograms obtained from rat plasma, 103, 1993, with kind permission of Elsevier Science-NL, Sara Burgerhartstraat 25, 1055 KV Amsterdam, The Netherlands.

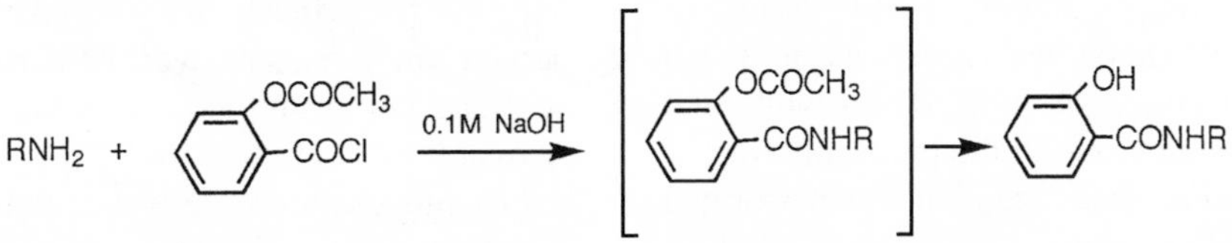

$RNH_2$ + ($OCOCH_3$, $COCl$) —0.1M NaOH→ [($OCOCH_3$, $CONHR$)] → ($OH$, $CONHR$)

**Fig. 5.9.** Reaction of *O*-acetylsalicyloyl chloride with an amine in alkaline solution.

method was sensitive enough to detect the derivatives in the picomole range with ECD (+1.2 V *vs* Ag/AgCl). The method was also applied for the determination of amines in plasma and urine using 2-propylamine and dipropylamine as model substances. Following a solid phase extraction and derivatization, the compounds could be detected in the lower nanogram range/ml of biological material [58].

## 5.2.6 2,4-Dinitrofluorobenzene

The nitrophenyl group is ideally suited to ECD operated in the reduction mode. The electrochemistry of the nitrophenyl group is not well understood in all of its details.

In 1987, Chang *et al.* developed the pre-column derivatization for improved detection in HPLC-photolysis-electrochemistry [59]. Pre-column, homogeneous chemical derivatization with 2,4-dinitrofluorobenzene (Sanger's reagent) was utilized to improve the chromatographic and detection properties of amino alcohols and amino acids. The 2,4-dinitrophenyl derivatives were separated using reversed-phase HPLC, and the column eluate was passed through a photochemical reactor to produce electroactive species and thence to an electrochemical detector containing a single or dual electrode system. The derivatization-detection approach provides limits of detection in the low parts-per-billion range with a linearity of roughly three orders of magnitude. The method was applied to the determination of serine in beer [59].

In 1995, Leube and Fischer reported pre-column derivatization and post-column irradiation for the ECD method [60]. Renin inhibitor I has been found to be very potent, thereby necessitating a sensitive method for the evaluation of its pharmacokinetics in man. Because the renin inhibitor contains no structural group allowing highly sensitive detection, derivatization was necessary to reach the required quantification limit. Renin inhibitor I possesses an amine group in the imidazole ring of its histidine moiety. Derivatization was performed as an automated sequential process with a Gilson sample processor, which added 0.2 M sodium borate buffer (pH 8) and 2,4-dinitrofluorobenzene in acetonitrile solution. After 25 min at 80 °C, an aliquot (50 μl) was analyzed by HPLC. Irradiation was performed at 254 nm with a Beam Boost using a Teflon coil, and ECD was set at +0.55 V [60].

## 5.2.7 1,2-Diphenylethylenediamine

Catecholamines are converted to trihydroxyindole with $[Fe(CN)_6]^{3-}$ as oxidant, and the fluorescence intensities are sensitively detected. The post-column reaction based on trihydroxyindole formation has frequently been used in the analysis of catecholamines. However, dopamine could not be detected by this method. 1,2-diphenylethylenediamine has been found to react in neutral medium with catecholamines and other catechol compounds in the presence of $[Fe(CN)_6]^{3-}$ to give fluorescent diphenylquinoxaline derivatives [61]. By an alternative ECD with pre-column derivatization with 1,2-diphenylethylenediamine, dopamine and the other catecholamines could be easily determined. The detection limit was 2 fmol [62].

The fluorometric method for catecholamine determination with the selective pre-column derivatization with a fluorescent agent, 1,2-diphenylethylenediamine, has also been evaluated by Hoorn *et al.* by comparison with an HPLC with ECD method. The electrochemical detector was operated at +0.6 V *vs* an Ag/AgCl reference electrode. Limits of detection are 0.3 pg for norepinephrine and epinephrine and 0.5 pg for dopamine [63].

## 5.2.8 Bolton and Hunter Type Reagent

In 1979, Shimada *et al.* reported the *N*-hydroxysuccinimide esters (Bolton and Hunter type reagent) of vanillic and homovanillic acid as highly sensitive electroactive derivatization reagents for primary and secondary amines. The derivatization reagents were prepared from vanillic acid and homovanillic acid by condensation with *N*-hydroxysuccinimide [64].

For the chemical derivatization of PEA with a readily oxidizable group, sulfosuccinimidyl-3-(4-hydroxyphenyl)propionate (sulfo Bolton and

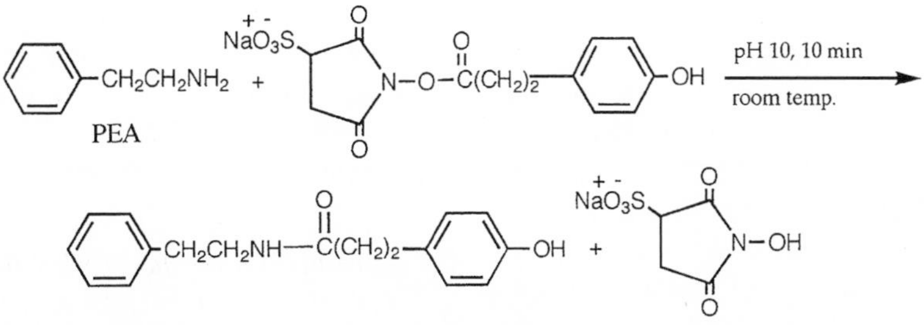

**Fig. 5.10.** Derivatization of PEA with the sulfo Bolton and Hunter type reagent.

Hunter type reagent) was selected. As shown in Fig. 5.10, the reaction of PEA with an excess of the reagent at pH 10 was essentially complete at room temperature in 10 min. The Bolton and Hunter type derivative of PEA gave a single symmetrical peak in the HPLC system. The detector response for a single concentration of PEA increased with the electrode voltage, and +0.55 V produced the optimal S/N [65].

Cyclohexylamine and benzylamine were derivatized with 3-(4-hydroxyphenyl)propionic acid-*N*-hydroxysuccinimide ester to produce the corresponding electroactive amines. For cyclohexylamine, the calibration graph was rectilinear for 1 to 10 μM and the detection limit was 0.10 μM [66].

### 5.2.9 Others

PEA and piperidine as the primary and secondary amines reacted with *N*-(4-anilinophenyl)isomaleimide (APIM) prepared from pre-column derivatization reagent in acetonitrile-0.05 M borate buffer solution (pH 9.0) (1:1). Derivatization of amines with the reagent was complete within 20 min at room temperature. The derivative formed with APIM was responsive to an electrochemical detector. The detection limit for the PEA-APIM adduct was about 0.1 pmol. The electrochemical response was found to be linear in the range 0.1–1.0 ng of PEA [67].

A polymeric anhydride containing *O*-acetylsalicyl group was utilized as the labeling moiety for primary and secondary amines. The derivatives were detected by an electrochemical detector in the oxidative mode and oxidative ECD after post-column photolysis (HPLC-$h\nu$-ECD). Derivatization improves the detection limits by 3–4 orders of magnitude [68].

The use of ECD in the HPLC of 5-(dimethylamino)-1-naphthalene sulfonyl (dansyl) derivatives of biogenic amines was reported by Chiavari *et al.* [69]. Putrescine, cadaverine, 1,6-diaminohexane, tryptamine, histamine, tyramine, spermine and spermidine were derivatized with dansyl chloride. The detection limits of the ECD were compared with those of ultraviolet and spectrofluorimetric detectors. Detection limits with ECD (+0.55 V) ranged from 0.18 to 1.0, by fluorescence from 0.52 to 1.25, and 4 to 8 pmol by UV absorption, respectively.

## 5.3 LABELING OF CARBOXY GROUPS

The derivatizations of carboxy groups on the electroactive compounds have been described by several reviewers [23–25] [70].

The fatty acids, bile acids, and prostaglandins (PGs) were converted to the corresponding *p*-hydroxyanilides by reaction with *p*-aminophenol in the presence of 2-bromo-1-methylpyridinium iodide and triethylamine. *p*-hydroxyanilide derivatives of these compounds were stable for more than one week at 4 °C. The anilides were oxidized in a two-electron process and analyzed by HPLC with ECD as shown in Fig. 5.11. The detection limits were 0.5, 2 and 2 ng for stearic acid, chenodeoxycholic acid and $PGF_2\alpha$, respectively. Fig. 5.12 shows the chromatogram of *p*-hydroxyanilides of fatty acids [71].

$$\mathrm{RCONH{-}C_6H_4{-}OH} \xrightarrow[+\,\mathrm{H_2O}]{-\,2\,e^-} \mathrm{RCONH_2} + \mathrm{O{=}C_6H_4{=}O} + 2\,\mathrm{H^+}$$

**Fig. 5.11.** Electrochemical reaction of *p*-hydroxyanilide.

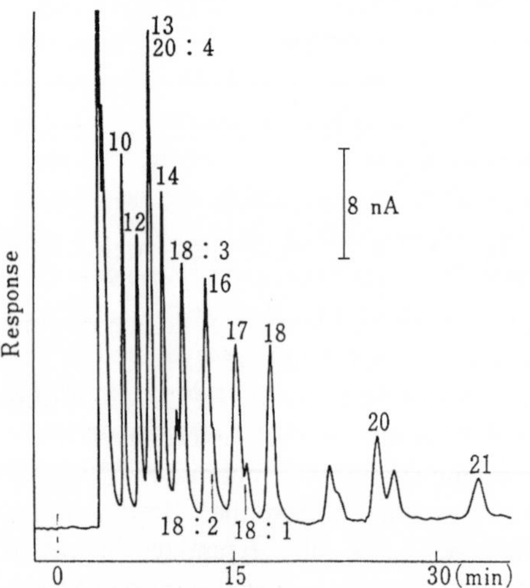


**Fig. 5.12.** Chromatogram of *p*-hydroxyanilides of fatty acids [71]. The numbers in the figure present the ratio (number of carbon atoms): (number of double bonds) for fatty acids. Reprinted from *Chemical & Pharmaceutical Bulletin*, 28, S. Ikenoya *et al.*, Chromatogram of *p*-Hydroxyanilides, 2945, 1980, with kind permission of The Pharmaceutical Society of Japan.

Similarly, Knospe *et al.* developed the ECD method for PGs as their 2,4-dimethoxyanilides. The derivatization reactions were performed for $PGE_2$, $PGF_{2}\alpha$, $PGD_2$ and thromboxane $B_2$ as their 2,4-dimethoxyanilides. The ECD potential was set at +1.10 V *vs* Ag/AgCl. The detection limit on the column was 50 pg [72].

Conventional reversed-phase HPLC conditions have been optimized for resolution of a mixture containing $PGE_1$, $PGE_2$, $PGF_{1}\alpha$ and $PGF_{2}\alpha$. Electroactive derivative-forming reagents, such as *p*-nitrobenzyloxyamine, 2-bromo-2′-nitroacetophenone, and 2,4-dinitrophenylhydrazine have been evaluated for use as pre-column reagents for forming PG derivatives. The results indicate that detection limits of 120 pg are achievable with ECD. The use of different derivatizing reagents allows for choice of the electroactive moiety to be introduced and this can be used to differentiate the classes of PGs by their functional groups [73].

On the other hand, carboxylic acids can be alkylated with 4-bromomethyl-7-methoxy-6-nitrocoumarin. ECD was achieved at −0.75 V with the detection limit of 100 pmol [74].

3-bromoacetyl-1,1′-dimethylferrocene (Fig. 5.6c), in a series of ferrocene reagents, was developed for the determination of fatty acids. Condensation of fatty acids with the ferrocene reagent was effected in the presence of 18-crown-6 and potassium fluoride. The resulting esters showed a satisfactorily electroactive response at +0.60 V *vs* an Ag/AgCl reference electrode with a detection limit of 0.5 pmol [75].

Electrochemical derivatization reagents for optical resolution of carboxylic acids were also developed by Shimada *et al.* *R*-(−)-1-Ferrocenylethylamine (Fig. 5.6d) and *S*-(+)-1-ferrocenylpropylamine (Fig. 5.6e) were synthesized and used as derivatization reagents to resolve 2-arylpropionic acids (ibuprofen or naproxen racemates) or an *N*-acetylamino acid. The reagent was mixed with the sample solution in the presence of 1-ethyl-3-(3-dimethylaminopropyl)carbodiimide and 1-hydroxybenzotriazole at 4 °C, and the resulting solution was analyzed. HPLC for the corresponding diastereoisomeric amides was carried out on a column of Develosil ODS-5 with a mobile phase of acetonitrile-1.5% sodium acetate buffer and ECD at +0.45 V *vs* an Ag/AgCl reference electrode. The detection limit was 0.5 pmol [76].

Munns *et al.* synthesized the electrochemical derivatizing reagents for carboxylic acids and investigated their suitability for drugs and metabolites. Quinoxaline-2-carboxylic, benzoic and salicylic acids each was derivatized with 1-(4-hydroxyphenyl)-2-bromoethanone (4-HBE), 1-(2,4-dihydroxyphenyl)-2-bromoethanone (2,4-DBE) and 1-(2,5-dihydroxyphenyl)-2-bromoethanone (2,5-DBE). Conditions of derivatization are relatively mild at 60 °C for 60 min or less, and the reaction is 76% complete. The detection limits are very low for ECD determinations of the esters of 4-HBE and 2,5-DBE, respectively. The oxidation of the derivatives of 2,5-DBE at the lower potentials eliminates those substances that are oxidized above +0.4 ~ +0.6 V *vs* an Ag/AgCl [77].

Carboxylic acids were determined by derivatization with trimethylaminomethyl ferrocene iodide (Fig. 5.6f), followed by HPLC with ECD. The best sensitivity with the reagent was observed at +0.5 V *vs* an Ag/AgCl. The detection limit was 0.5 pmol [52].

## 5.4 LABELING OF HYDROXY GROUPS

Ferrocenoyl azide and 3-ferrocenylpropionyl azide, as shown in Fig. 5.6g,h, condense readily with alcoholic hydroxy compounds under mild conditions to provide urethanes, which can be detected electrochemically with a glassy carbon working electrode at +0.4 V *vs* an Ag/AgCl reference electrode with a detection limit of 0.5 pmol. The method was applied to the determination of hydroxysteroids such as dehydroepiandrosterone (DHEA), methyl lithocholate, 5$\beta$-cholane-3$\alpha$, 24-diol, estrone, digitoxigenin, 3-epidigitoxigenin, digoxigenin and digitoxigenin monodigitoxoside [78].

Phenols were derivatized by reaction with 4-aminoantipyrine in the presence of $K_3Fe(CN)_6$ to form the corresponding quinoneimines prior to reversed-phase HPLC (Fig. 5.13). In contrast to the conventional oxidative detection of phenols with a high applied potential of over +0.95 V, a low potential of −0.20 V *vs* an Ag/AgCl reference electrode was used successfully to detect the derivatized phenols with a glassy carbon working electrode. The method has the advantage that the applied potential is so low that oxygen removal from the mobile phase is not required [79].

Ferroceneboronic acid (FBA, Fig. 5.6i) was used in the derivatization of brassinosteroids with respect to reactivity and sensitivity. The steroids, which have two sets of vicinal diol functional groups in the A ring and in the side chain, were readily condensed with FBA under mild conditions to provide the corresponding boronates, exhibiting maximum sensitivity at +0.6 V *vs* an Ag/AgCl reference electrode with a detection limit of 25 pg for brassinolide, and calibration graphs were rectilinear from 50 pg to 5 ng. Ferroceneboronate was found to be stable for two months in solution in pyridine-acetonitrile at 0 °C [80,81].

Nakajima *et al.* developed derivatization reagents for alcohols with fluorescence and ECD. The reagents are constituted from 3-methyl-1,4-naphthoquinone structure, in which 2-[2-(azidocarbonyl)ethyl] is attached to the basic structure. The reactivity and sensitivity of 2-[2-(azidocarbonyl)ethyl]-3-methyl-1,4-naphthoquinone (AMQ) were examined with 1-eicosanol as a model compound. Alcohols were reacted with the reagent in acetone at 100 °C for 15 min to produce the corresponding carbamic acid esters. The detection limit for 1-eicosanol with ECD was 16 fmol/10 μl [82]. By modifying the AMQ reagent with 2-[2-(isocyanate)ethyl]-3-methyl-1,4-napthoquinone (IMQ) as shown in Fig. 5.14,

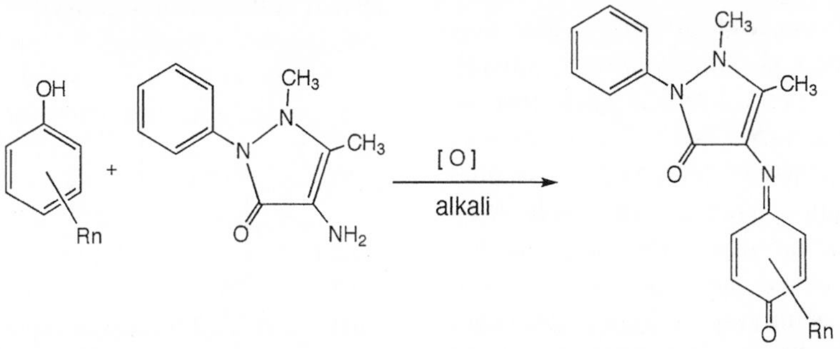

**Fig. 5.13.** Derivatization of phenols to quinoneimines employing 4-aminoantipyrine. Rn: substituents.

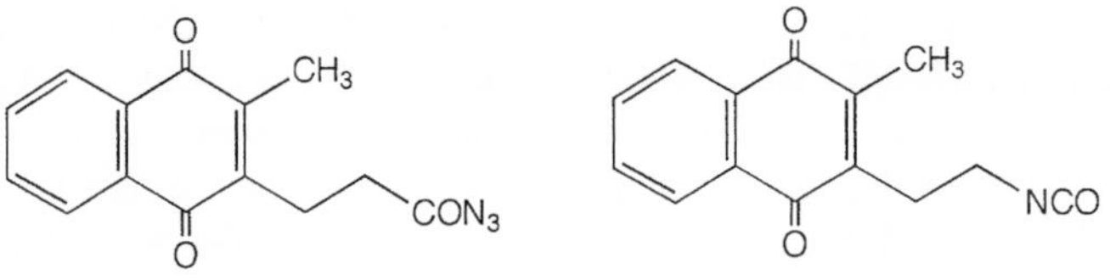

AMQ IMQ

**Fig. 5.14.** Structures of AMQ and IMQ.

a highly sensitive labeling reagent for hydroxysteroids in HPLC with ECD was achieved. The steroids labeled with these reagents are once reduced in a post-column reduction with a platinum catalyst, then detected oxidatively with an Ag/AgCl reference electrode at +0.7 V at a detection limit of 17 fmol for cholesterol [83,84].

## 5.5 LABELING OF THIOL GROUPS

Review of the derivatization of thiol-containing compounds used in ultraviolet, visible, fluorometric, chemiluminescence and ECD has been discussed by Shimada and Mitamura [85].

The derivatization of thiols with *N*-(4-anilinophenyl)maleimide (APM) is normally performed in the pre-chromatographic mode. A chilled solution of the maleimide is mixed with thiol. The reaction mixture is kept for 10–90 min, and subsequently the excess of reagent is removed by diethylether extraction. The aqueous phase is heated for 20 min, and an aliquot is subjected to isocratic reversed-phase chromatography. Detection is performed with a glassy carbon working electrode and an Ag/AgCl reference electrode. The detection limits of glutathione and penicillamine derivatives are about 20 pg [86]. The details are described in the application section (5.2.7).

*N*-(ferrocenyl)maleimide (Fig. 5.6j) as a pre-column derivatization reagent for thiols was prepared and determined by Shimada and co-workers. *N*-(ferrocenyl)maleimide is the most favorable reagent with respect to its reactivity, stability and electrochemical properties. The dual-electrode coulometric detection of the adduct showed high selectivity and sensitivity with a detection limit of 0.06 pmol [87].

## 5.6 LABELING OF OTHER FUNCTIONAL GROUPS

Derivatization of 17-oxosteroids with hydrazine having an aromatic group for HPLC with ECD was carried out. 17-oxosteroids, DHEA and androsterone (AND), were derivatized with the hydrazine to form their hydrazones, as shown in Fig. 5.15. The *p*-nitrophenylhydrazones exhibited the highest chemically irreversible oxidation response in methanol-buffer which have peak potentials (glassy carbon working electrode) of +0.8 V *vs* an Ag/AgCl reference electrode [88]. 17-oxosteroid, particularly DHEA was derivatized with *p*-nitrophenylhydrazine in trichloroacetic acid-benzene solution to produce its hydrazone. The *p*-nitrophenylhydrazones were separated by HPLC on a C18 column. The method provides a quantitation limit of 80 ng/ml. A good correlation was observed between the values obtained by the proposed method and radioimmunoassay for DHEA sulfate in serum [89].

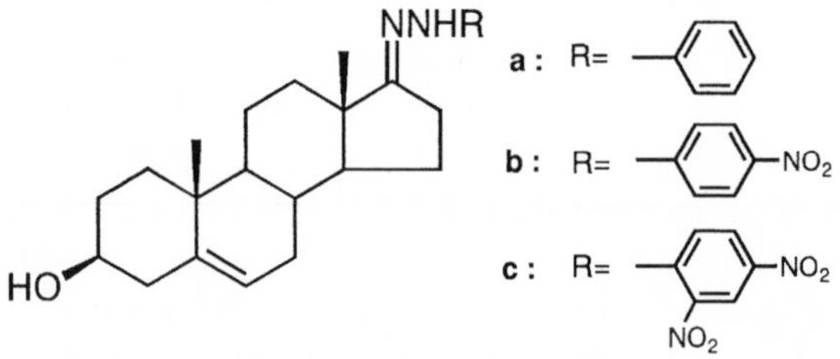

**Fig. 5.15.** Structures of DHEA derivatives.

For the determination of carbonyl compounds, a highly sensitive and selective derivatization reagent was developed by Ueno and Umeda [90]. Of several hydrazines and hydrazides, 2,5-dihydroxybenzohydrazide (DBH) was selected as the most suitable reagent. The hydrazones were sensitively detected at a low oxidative potential (+0.20 V *vs* Ag/AgCl). The detection limits were in the range 60–500 fmol. A good linear relationship was observed between the peak area and the concentration in the range 0.4–200 pmol of corticosterone and DHEA.

The usefulness of DBH was proved by applying the reagent to the derivatization of new antiandrogenic modified steroidal anthrasteroids, 3,8-dione and 8-one. The derivatives were detected electrochemically at +0.4 V *vs* an Ag/AgCl reference electrode. The detection limits for 3,8-dione and 8-one in 0.5 ml of human serum were 0.5 and 1 ng/ml, respectively [91].

In 1972, Fitzpatrick *et al.* reported that the 2,4-dinitrophenylhydrazine derivatives of 4-epimeric forms of AND and DHEA can be separated by HPLC [92]. Samples of urine and blood hydrolysates have been derivatized with the hydrazine and analyzed by the proposed technique.

Jacobs and Kissinger developed 2,4-dinitrophenylhydrazine as an electrochemical derivatization reagent for aldehydes and ketones [93]. Formaldehyde, acetaldehyde, acetone and acrolein were derivatized with the reagent. Optimum detection was achieved at an operating potential of −0.75 V *vs* an Ag/AgCl reference electrode. Detection limits for those derivatives were 54–99 pg. These detection limits were approximately 20 times lower than obtainable with UV absorbance at 254 nm. The increasing acidity of atmospheric cloud, fog and precipitation is of environmental concern. The major acids responsible for 'acid rain' are formed in the troposphere by various chemical reactions. Some organic compounds, such as aldehydes and ketones, play an important role in such chemical reactions, and the development of sensitive methods for the determination of aldehydes and other carbonyls is important for the elucidation of processes that control the acidity and composition of atmospheric precipitation. The possibility of using HPLC for the separation and quantitation of nanogram amounts of some carbonyl compounds considered ubiquitous in the air, were examined. HPLC with ECD has been applied for the separation of the carbonyl compounds as 2,4-dinitrophenylhydrazine derivatives. Nine carbonyl compounds, such as methanal, ethanal, propenal, propanal, acetone, *n*-butanal, 2-butanal, 3-methylbutanal, and benzaldehyde were separated as its derivatives by a C18 column, and detected by ECD at the optimal potential of +1.10 V (Fig. 5.16). The detection limit ranged from 30 to 212 pg [94].

DBH was used for an electrochemical labeling reagent for aliphatic aldehydes by Bousquet *et al.* [95]. The hydrazone derivatives of octanal, decanal and cyclohexanecarboxaldehyde (internal standard: IS) were detected at a graphite electrode set at an oxidation potential of +0.3 V. The derivatization was shown to be quantitative and the response linear between 1 and 15 ng/ml. The method is rapid, reproducible and the detection limit is 130 fmol for an injection volume of 5 ml.

Reducing carbohydrates were derivatized post-column with 2-cyanoacetamide to readily oxidizable compounds. ECD allowed sensitive and reproducible monitoring of glucose with a detection limit and a linear range of 20 pmol and 50 pmol–2 nmol, respectively. Similar results were obtained with other reducing carbohydrates. The diene-monool isolated intermediate compounds were readily oxidized under the same conditions as those for ECD [96].

Colgan *et al.* developed the use of picric acid salts as a labeling reagent for alkyl halides and epoxides. Alkyl, pentyl and octyl halides and cyclohexene oxide were derivatized with silver picrate. The labeled products are allyl picryl ether, pentyl picryl ether, octyl picryl ether, and 1-hydroxy-2-picrylcyclohexane, respectively. The labeled derivatives can be monitored with an electrochemical detector in the reductive mode or with an electrochemical detector in the oxidative mode after post-column photolysis [97]. Colgan *et al.* also extended silver picrate to determine the cited compound, ethylene dibromide. Two

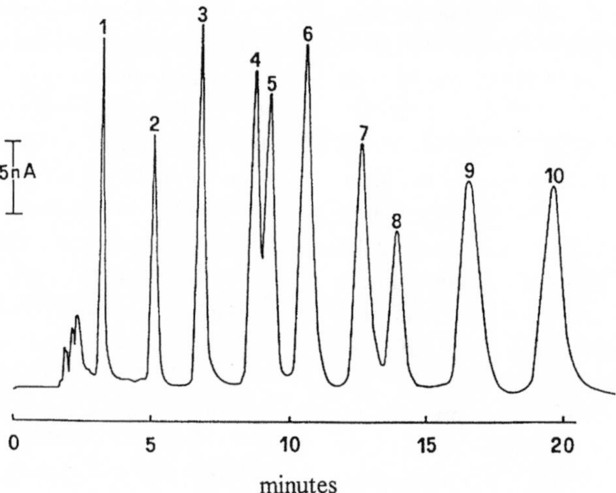


**Fig. 5.16.** HPLC separation of carbonyl compounds as 2,4-dinitrophenylhydrazin derivatives [94]. 1 = Reagent; 2 = methanal; 3 = ethanal; 4 = propenal; 5 = propanal; 6 = acetone; 7 = *n*-butanal; 8 = 2-butenal; 9 = 3-methylbutanal; 10 = benzaldehyde. Reprinted from *Journal of Chromatography*, 318, G. Chiavari *et al.*, HPLC separation of carbonyl compounds, 428, 1985, with kind permission of Elsevier Science-NL, Sara Burgerhartstraat 25, 1055 KV Amsterdam, The Netherlands.

products, 2-bromoethyl picrate and/or ethane-1,2-diyl dipicrate, were formed. Sub-parts-per-million detection limits were obtained [98].

An electrochemical derivatization technique has been developed for the determination of *p*-tyrosine aminotransaminase activity. *p*-hydroxyphenylpyruvate obtained from the *p*-tyrosine transamination reaction was derivatized with hydroxylamine to form a stable oxime. This oxime derivative is readily separated from *p*-hydroxyphenylpyruvate, followed by ECD at oxidation potential (+0.9 V *vs* Ag/AgCl). *p*-hydroxyphenylpyruvate and its oxime derivative in the low picomole range can be detected. This is a sensitive method for *p*-tyrosine aminotransaminase assay [99].

A derivatization reagent for steroid glucuronides was developed by Shimada *et al.* Estrogen glucuronide was derivatized with 2-ferrocenylethylamine (Fig. 5.6k) and then subjected to HPLC with ECD (+0.45 V *vs* an Ag/AgCl), with a detection limit of 0.5 pmol [100].

Tryptamine was developed as a derivatizing agent for the determination of isocyanates. Methyl isocyanate is easily vaporized, resembling most airborne isocyanates in the workplace, when simulated air sampling was conducted. Tryptamine contains an indoyl moiety and undergoes fluorescence and ECD (amperometric oxidation). Reversed-phase HPLC separation was employed for the isocyanate derivative with ECD. The detection method was highly specific and sensitive. This method provides an extremely low level of quantification. The detection limit was 1 ng [101].

Inorganic and organic chloramines were derivatized with 2-mercaptobenzothiazole. The detection limit was 1 pmol. The method was applied to determine chloramines in water. The derivatization pathway is shown in Fig. 5.17. The resulting sulfenamides are stable and can be conveniently analyzed by HPLC using ECD [102].

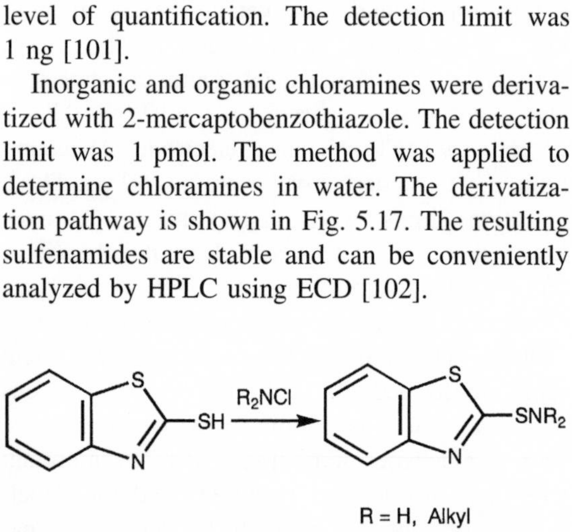

$R_2NCl$

R = H, Alkyl

**Fig. 5.17.** The reaction of 2-mercaptobenzothiazole with chloramine.

Ferrocenecarboxyhydrazide has been applied to the determination of anisic aldehyde in the oxidation mode and proved to be satisfactory with respect to reactivity, stability and electrochemical properties. Maximum sensitivity was observed at +0.75 V *vs* an Ag/AgCl [52].

## 5.7 APPLICATIONS

The HPLC method combines solid phase extraction and ECD and reaches the optimum performance for pharmacokinetic studies of medicinal drugs.

Baclofen was extracted from plasma using SCX Bond Elut columns. Plasma (1 ml) was added to 100 μl of water and 1 ml of citrate buffer (pH 2.6) and applied to a column. Baclofen was eluted with 1.5 ml borate hydroxide buffer of pH 10.4. An aliquot of the extract (200 μl) was derivatized with 50 μl of derivatization reagent (prepared daily by mixing 75 mg OPA/5 ml methanol/80 ml *t*-butanethiol/5 ml borate hydroxide buffer of pH 9.3). A portion (20 μl) was applied to a C18 Novapak column (4 μm; 15 cm × 3.9 mm, i.d.). The flow rate of the mobile phase, phosphate buffer-methanol (18:37), was 0.8 ml/min., and ECD (amperometric detection) was at +0.7 V (screening electrode at +0.2 V) with a sensitivity of 50 nA. Calibration graphs were linear from 10 to 500 ng/ml baclofen; the detection limit was 2.5 ng/ml. Both the within- and between-day relative standard deviation and inaccuracy are less than 10% and 7%, respectively. The method was used to monitor the plasma concentration of baclofen in volunteers after a single 20 mg oral dose. Fig. 5.18 shows a chromatogram of baclofen in a plasma sample [31].

Desmosine and isodesmosine were determined in hydrolyzed elastin after derivatization with NDA in the presence of $CN^-$ (Fig. 5.5). The derivatives were determined on a column of Supelco LC-18 DB (3 μm; 15 cm × 4.6 mm, i.d.). For ECD (amperometric detection) with a glassy carbon working electrode at +0.75 V *vs* an Ag/AgCl reference electrode, the mobile phase was methanol-citrate buffer (7:3) containing 50 mM $NaClO_4$. With ECD, calibration graphs were rectilinear for 8–160 pmol of amino acid injected and the detection limit for desmosine and isodesmosine was 100 fmol. In derivatization for

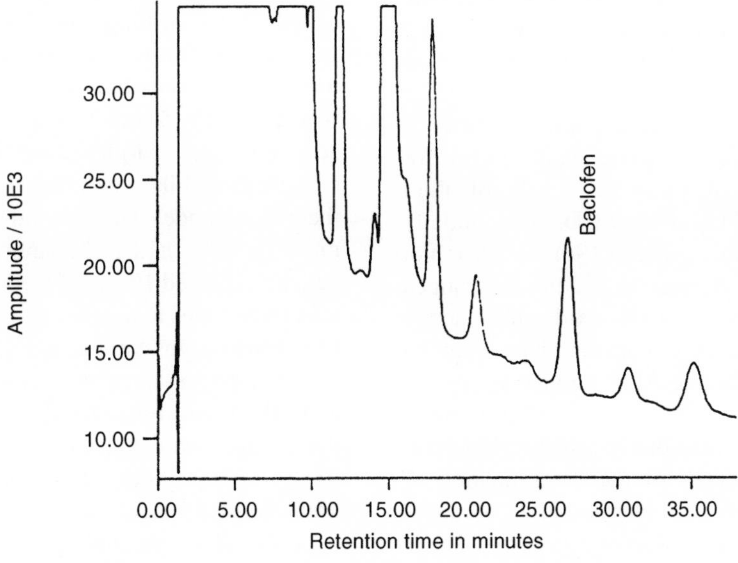


**Fig. 5.18.** Chromatogram of baclofen in a plasma sample from a volunteer 8 hr after a single 20 mg oral dose of baclofen. Reprinted from *Journal of Chromatography A*, 729, L. Millerious *et al.*, 312, 1996, with kind permission of Elsevier Science-NL, Sara Burgerhartstraat 25, 1055 KV Amsterdam, The Netherlands.

amino acid analysis, an appropriate aliquot of an amino acid mixture (containing 0.16 mM of each amino acid) was dissolved in approximately 5 ml of borate buffer. NDA and $CN^-$ (the amount of both materials added depends on the total concentration of primary amines present in the derivatization solution) were added. The final solution was diluted with borate buffer to a 10 ml volume. The reaction was allowed to proceed for 30 min before the injection was made [46].

Shimada *et al.* examined three *N*-substituted maleimides as derivatizing reagents for thiols. APM was the most favorable reagent in terms of sensitivity and reactivity. *N*-acetyl-L-cysteine, glutathione, L-cysteine and D-penicillamine were readily converted into the adducts with APM. Picogram levels of these thiol compounds were separated and quantified. A mixture of glutathione, L-cysteine, *N*-acetyl-L-cysteine and D-penicillamine was derivatized with APM at 0 °C for 90 min, and excess reagent was removed by extraction with ether. The aqueous layer was heated at 50 °C for 20 min to complete the rearrangement of the APM adducts with L-cysteine and D-penicillamine. When the applied potential of a glassy carbon working electrode was set at +0.7 V *vs* an Ag/AgCl reference electrode, L-cysteine and D-penicillamine were detected selectively with almost the same sensitivity as that obtained at +1.0 V *vs* an Ag/AgCl reference electrode (amperometric detection). The use of APM for derivatization seems very useful for the determination of picogram levels of thiol compounds by HPLC with ECD. The initially formed cysteine adduct (Fig. 5.19a), which is not detectable at +0.7 V, can be readily rearranged to provide the more easily oxidizable lactam (Fig. 5.19b). A similar phenomenon was observed with the D-penicillamine-APM adduct, as shown in Fig. 5.19 [86].

An HPLC with pre-column chemical derivatization was developed for the determination of gossypol enantiomers in plasma, after administration of the racemate. Racemic gossypol acetic acid in plasma was extracted into acetonitrile and analyzed using a reversed-phase column and a coulometric detector in the redox mode. To separate the enantiomers, 30 μl of the chiral derivatizing reagent, (*R*)-(−)-2-amino-1-propanol (50 mg/ml) and 15 μl of 20% acetic acid were added. After heating the mixture at 60 °C for 100 min, the derivatized enantiomers (Fig. 5.20) were determined by HPLC on a column of Novapak C18 (5 μm; 15 cm × 4.6 mm, i.d.) with acetonitrile-0.2 M-phosphate buffer of pH 3.5 (31:19) as mobile phase (1.5 ml/min) and ECD in the redox mode. The method successfully separated the (+)- and (−)-enantiomers. Two cancer patients received 10 mg of racemic gossypol acetic acid three times a day. In one patient, the racemic, (+)- and (−)-gossypol acetic acid plasma concentration after 65 days of therapy were 317, 213 and 104 ng/ml, respectively. In the other patient, these values were 362, 210 and 152 ng/ml, respectively, after a week of therapy. The calibration graph was rectilinear from 25 μg/l (detection limit) to 1.6 mg/l of racemic gossypol acetic acid in plasma, and the intra-assay coefficient of variation was 6.6% ($n = 6$) at 200 μg/l. This represents the first determination of the individual enantiomer

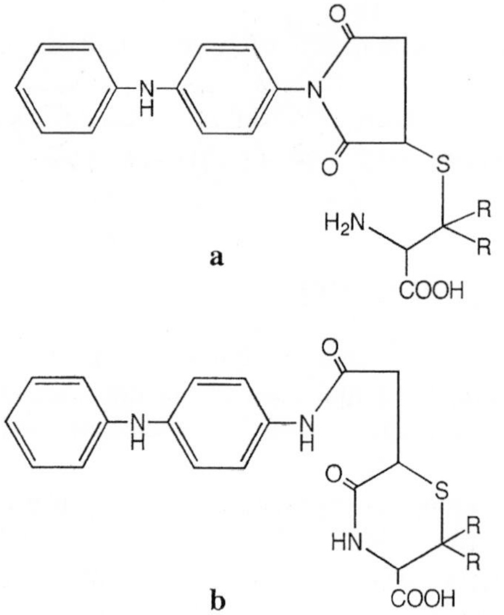

**Fig. 5.19.** Structure of APM adduct.

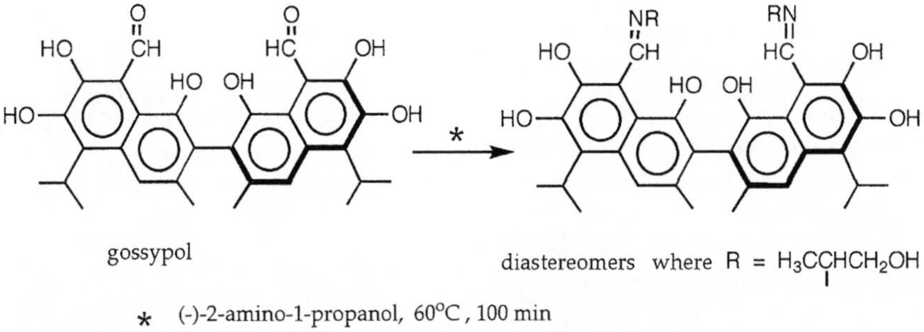

**Fig. 5.20.** Chemical structures of racemic gossypol and diastereomeric derivatives.

levels of gossypol after administration of the racemate [103].

ECD is selective, sensitive and also produces satisfactory recovery, precision and accuracy. ECD is comparable to fluorescence detection in sensitivity, accuracy and selectivity. In order to improve the sensitivity, ECD detection methods are widely utilized. Several labeling reagents for ECD detection have been developed for the microdetermination of biological materials, medicinal drugs or environmental pollutants, which often lack sensitivity or selectivity. ECD derivatization techniques have recently become more and more important.

## REFERENCES


[1] Porthault, M. (1988) *High-performance Liquid Chromatography in Electrochemical Detection Techniques in Applied Biosensors*, Junter, G. (Ed.) John Wiley & Sons, New York, pp. 53–84.

[2] Johnson, D.C., Weber, S.G., Bond, A.M., Wightman, R.W., Shoup, R.E. and Krull, I.S. (1986) *Anal. Chim. Acta* **180**, 187.

[3] Kissinger, P.T. (1984) *Electrochemical Detection in Liquid Chromatgraphy and Flow Injection Analysis in Laboratory Techniques in Electroanalytical Chemistry*, Marcel Dekker, New York, pp. 611–635.

[4] Bard, A.J. and Faulkner, L.R. (1980) *Electrochemical Methods, Fundamentals and Applications*, John Wiley & Sons, New York, pp. 1–43.

[5] Matsuda, H. (1967) *J. Electroanal. Chem.* **15**, 325.

[6] Yamada, J. and Matsuda, H. (1973) *J. Electroanal. Chem.* **44**, 189.

[7] Blaedel, W.J. and Klatt, L.N. (1966) *Anal. Chem.* **38**, 879.

[8] Stulik, K. and Pacakova, V. (1981) *J. Chromatogr.* **208**, 269.

[9] Sawyer, D.T. and Boberts, J.L. Jr (1974) *Experimental Electrochemistry for Chemists*, John Wiley & Sons, New York, pp. 11–115.

[10] Bond, A.M., Thomson, S.B., Tucker, D.T. and Briggs, M.H. (1984) *Anal. Chim. Acta.* **156**, 33.

[11] Matsue, T. (1993) *Tr. Anal. Chem.* **12**, 100.

[12] Iwasaki, Y., Niwa, O., Morita, M., Tabei, H. and Kissinger, P.T. (1996) *Anal. Chem.* **68**, 3797.

[13] Thormann, W., Bosch, P. and Bond, A.M. (1985) *Anal. Chem.* **57**, 2764.

[14] Strohben, W.E., Smith, D.K. and Evans, D.H. (1990) *Anal. Chem.* **62**, 1709.

[15] Caudill, W.L., Howell, J.O. and Wightman, R.M. (1982) *Anal. Chem.* **54**, 2532.

[16] Kasai, N., Matsue, T., Uchida, I., Horiuchi, T., Morita, M. and Niwa, O. (1996) *J. Electrochem. Soc. Jp.* **64**, 1269.

[17] Tokuda, K., Morita, K. and Shimizu, Y. (1989) *Anal. Chem.* **61**, 1763.

[18] Roston, D.A., Shoup, R.E. and Kissinger, P.T. (1982) *Anal. Chem.* **54**, 1417A.

[19] Matsue, T., Aoki, A., Abe, T. and Uchida, I. (1989) *Chem. Lett.* 133.

[20] Aoki, A., Matsue, T. and Uchida, I. (1990) *Anal. Chem.* **62**, 2206.

[21] Matsue, T., Aoki, A., Ando, E. and Uchida, I. (1990) *Anal. Chem.* **62**, 407.

[22] Aoki, A., Matsue, T. and Uchida, I. (1992) *Anal. Chem.* **64**, 44.

[23] Lisman, J.A., Underberg, W.J.M. and Lingeman, H. (1990) *Electrochemical Derivatization in Chromatographic Science Series* Vol. 48 *Detection-Oriented Derivatization Techniques in Liquid Chromatography*, Lingeman, H. and Underberg, W.L.M. (Eds) Marcel Dekker, New York, pp. 283–322.

[24] Imai, K. and Toyo'oka, T. (1988) *Selective sample handling and detection in high-performance liquid chromatography*, Frei, R.W. and Zech, K. (Eds), Elsevier, New York, pp. 265–288.
[25] Toyo'oka, T. (1995) *J. Chromatogr. B* **671**, 91.
[26] Roth, M. (1971) *Anal. Chem.* **43**, 880.
[27] Joseph, M.H. and Davies, P. (1983) *J. Chromatogr.* **277**, 125.
[28] Hoskins, J.A., Holliday, S.B. and Davies, F.F. (1986) *J. Chromatogr.* **375**, 129.
[29] Harsing, L.G. Jr., Nagashima, H., Vizi, E.S. and Duncalf, D. (1986) *J. Chromatogr.* **383**, 19.
[30] Leroy, P., Nicolas, A. and Moreau, A. (1983) *J. Chromatogr.* **282**, 561.
[31] Millerioux, L., Brault, M., Gualano, V. and Mignot, A. (1996) *J. Chromatogr. A* **729**, 309.
[32] Achilli, G., Cellerino, G.P. and d'Eril, G.M. (1994) *J. Chromatogr. A* **661**, 201.
[33] M-Teissier, E., Drieu, K. and Rips, R. (1988) *J. Liq. Chromatogr.* **11**, 1627.
[34] Smith, S. and Sharp, T. (1994) *J. Chromatogr. B* **652**, 228.
[35] Turiak, G. and Volicer, L. (1994) *J. Chromatogr. A* **668**, 323.
[36] Zambonin, P.G., Guerrieri, A., Rotunno, T. and Palmisano, F. (1991) *Anal. Chim. Acta* **251**, 101.
[37] M-Teissier, E., Mestdagh, N., Bernier, J-L and Henichart, J-P (1993) *J Liq. Chromatogr.* **16**, 573.
[38] Schmidt, J. and MaClain, C.J. (1991) *J. Chromatogr.* **568**, 207.
[39] Murai, S., Saito, H., Masuda, Y. and Itoh, T. (1990) *J. Pharmacol. Meth.* **23**, 195.
[40] Ziegler, S.J. and Sticher, O. (1988) *HRC-CC J. High Resolut. Chromatogr. Chromatogr. Commun.* **11**, 639.
[41] Hikal, A.H., Lipe, G.W., Slikker, W. Jr., Scallet, A.C., Ali, S.F. and Newport, G.D. (1988) *Life Sci.* **42**, 2029.
[42] Dutrieu, J., Miller, A.O.A. and Delmotte, Y.A. (1988) *Fresenius' Z. Anal. Chem.* **330**, 398.
[43] Boomsma, F., van der Hoorn, F.A.J., Man in't Veld, A.J. and Schalekamp, M.A.D.H. (1988) *J. Chromatogr.* **427**, 219.
[44] Woolfson, A.D., Millership, J.S. and Karim, E.F.I.A. (1987) *Analyst* **112**, 1421.
[45] Einarsson, S., Folestad, S. and Josefsson, B. (1987) *J. Liq. Chromatogr.* **10**, 1589.
[46] Lunte, S.M., Mohabbat, T., Wong, O.S. and Kuwana, T. (1989) *Anal. Biochem.* **178**, 202.
[47] Kline, W.F., Matuszewski, B.K. and Bayne, W.F. (1990) *J. Chromatogr.* **534**, 139.
[48] Kline, W.F. and Matuszewski, B.K. (1992) *J. Chromatogr.* **583**, 183.
[49] Tanaka, M., Shimada, K. and Nambara, T. (1984) *J. Chromatogr.* **292**, 410.
[50] Shimada, K., Oe, T., Tanaka, M. and Nambara, T. (1989) *J. Chromatogr.* **487**, 247.
[51] Shimada, K., Kawai, Y., Oe, T. and Nambara, T. (1989) *J. Liq. Chromatogr.* **12**, 359.
[52] Kubab, N., Farinotti, R., Montes, C., Chalom, J. and Mahuzier, G. (1989) *Analusis* **17**, 559.
[53] Cox, R.L., Schneider, T.W. and Koppang, M.D. (1992) *Anal. Chim. Acta* **262**, 145.
[54] Mahachi, T.J., Carlson, R.M. and Poe, D.P. (1984) *J. Chromatogr.* **298**, 279.
[55] Sherwood, R.A., Titheradge, A.C. and Richards, D.A. (1990) *J. Chromatogr.* **528**, 293.
[56] Rondelli, I. Mariotti, F., Acerbi, D., Redenti, E., Amari, G. and Ventura, P. (1993) *J. Chromatogr.* **612**, 95.
[57] Smith, R.M., Ghani, A.A., Haverty, D.G., Bament, G.S., Chamsi, A.Y. and Fogg, A.G. (1988) *J. Chromatogr.* **455**, 349.
[58] Wintersteiger, R., Barary, M.H., El-Yazbi, F.A., Sabry, S.M. and Wahbi, A-AM. (1995) *Anal. Chim. Acta* **306**, 273.
[59] Chang, M.Y., Chen, L.R., Ding, X.D., Selavka, C.M., Krull, I.S. and Bratin, K. (1987) *J. Chromatogr. Sci.* **25**, 460.
[60] Leube, J. and Fischer, G. (1995) *J. Chromatogr. B* **665**, 373.
[61] Nohta, H., Mitsui, A. and Ohkura, Y. (1984) *Anal. Chim. Acta* **165**, 171.
[62] Mori, K. (1987) *Life Sci.* **41**, 901.
[63] van der Hoorn, F.A.J., Boomsma, F., Man in 't Veld, A.J. and Schalekamp, M.A.D.H. (1989) *J. Chromatogr.* **487**, 17.
[64] Shimada, K., Tanaka, M. and Nambara, T. (1979) *Chem. Pharm. Bull.* **27**, 2259.
[65] Gusovsky, F., Jacobson, K.A., Kirk, K.L., Marshall, T. and Linnoila, M. (1987) *J. Chromatogr.* **415**, 124.
[66] Smith, R.M. and Ghani, A.A. (1990) *Electroanalysis* **2**, 167.
[67] Shimada, K., Tanaka, M. and Nambara, T. (1983) *J. Chromatogr.* **280**, 271.
[68] Chou, T.Y., Colgan, S.T., Kao, D.M., Krull, I.S., Dorschel, C. and Bidlingmeyer, B. (1986) *J. Chromatogr.* **367**, 335.
[69] Chiavari, G., Galletti, G.C. and Vitali, P. (1989) *Chromatographia* **27**, 216.
[70] Yasaka, Y. and Tanaka, M. (1994) *J. Chromatogr. B* **659**, 139.
[71] Ikenoya, S., Hiroshima, O., Ohmae, M. and Kawabe, K. (1980) *Chem. Pharm. Bull.* **28**, 2941.
[72] Knospe, J., Steinhilber, D., Herrmann, T. and Roth, H.J. (1988) *J. Chromatogr.* **442**, 444.
[73] Beck, G.M., Roston, D.A. and Jaselskis, B. (1989) *Talanta* **36**, 373.
[74] Kubab, N., Farinotti, R. and Mahuzier, G. (1986) *Analusis* **14**, 125.
[75] Shimada, K., Sakayori, C. and Nambara, T. (1987) *J. Liq. Chromatogr.* **10**, 2177.

[76] Shimada, K., Haniuda, E., Oe, T. and Nambara, T. (1987) *J. Liq. Chromatogr.* **10**, 3161.
[77] Munns, R.K., Roybal, J.E., Shimoda, W. and Hurlbut, J.A. (1988) *J. Chromatogr.* **442**, 209.
[78] Shimada, K., Orii, S., Tanaka, M. and Nambara, T. (1986) *J. Chromatogr.* **352**, 329.
[79] Li, C.-Y. and Kemp, M. W. (1988) *J. Chromatogr.* **455**, 241.
[80] Gamoh, K., Sawamoto, H., Takatsuto, S., Watabe, Y. and Arimoto, H. (1990) *J. Chromatogr.* **515**, 227.
[81] Gamoh, K. and Takatsuto, S. (1994) *J. Chromatogr. A* **658**, 17.
[82] Nakajima M., Wakabayashi H., Yamato S. and Shimada K. (1991) *Anal. Sci. 7 suppl.* 173.
[83] Nakajima, M., Wakabayashi, H., Yamato, S. and Shimada, K. (1993) *J. Chromatogr.* **641**, 176.
[84] Nakajima, M., Yamato, S., Wakabayashi, H. and Shimada, K. (1995) *Biol. Pharm. Bull.* **18**, 1762.
[85] Shimada, K. and Mitamura, K. (1994) *J. Chromatogr. B* **659**, 227.
[86] Shimada, K., Tanaka, M. and Nambara, T. (1983) *Anal. Chim. Acta* **147**, 375.
[87] Shimada, K., Oe, T. and Nambara, T. (1987) *J. Chromatogr.* **419**, 17.
[88] Shimada, K., Tanaka, M. and Nambara, T. (1980) *Anal. Lett.* **13**, 1129.
[89] Shimada, K., Tanaka, M. and Nambara, T. (1984) *J. Chromatogr.* **307**, 23.
[90] Ueno, K. and Umeda, T. (1991) *J. Chromatogr.* **585**, 225.
[91] Ueno, K., Morimoto, A. and Umeda, T. (1992) *Anal. Sci.* **8**, 13.
[92] Fitzpatrick, F.A., Siggia, S. and Dingman, J. Sr. (1972) *Anal. Chem.* **44**, 2211.
[93] Jacobs, W.A. and Kissinger, P.T. (1982) *J. Liq. Chromatogr.* **5**, 669.
[94] Chiavari, G. and Bergamini, C. (1985) *J. Chromatogr.* **318**, 427.
[95] Bousquet, E., Tirendi, S., Prezzavento, O. and Tateo, F. (1995) *J. Liq. Chromatogr.* **18**, 1933.
[96] Honda, S., Konishi, T. and Suzuki, S. (1984) *J. Chromatogr.* **299**, 245.
[97] Colgan, S.T., Krull, I.S., Neue, U., Newhart, A., Dorschel, C., Stacey, C. and Bidlingmeyer, B. (1985) *J. Chromatogr.* **333**, 349.
[98] Colgan, S.T., Krull, I.S., Dorschel, C. and Bidlingmeyer, B. (1986) *Anal. Chem.* **58**, 2366.
[99] Yu, P.H. and Bailey, B.A. (1986) *J. Chromatogr.* **362**, 55.
[100] Shimada, K., Nagashima, E., Orii, S. and Nambara, T. (1987) *J. Pharm. Biomed. Anal.* **5**, 361.
[101] Wu, W.S., Nazar, M.A., Gaind, V.S. and Calovini, L (1987) *Analyst* **112**, 863.
[102] Lukasewycz, M.T., Bieringer, C.M., Liukkonen, R.J., Fitzsimmons, M.E., Corcoran, H.F., Lin, S. and Carlson, R.M. (1989) *Environ. Sci. Technol.* **23**, 196.
[103] Wu, D.-F., Reidenberg, M.M. and Drayer, D.E. (1988) *J. Chromatogr.* **433**, 141.

# 6

# Derivatization for Resolution of Chiral Compounds

**Toshimasa Toyo'oka**
University of Shizuoka, Shizuoka, Japan


| | |
|---|---|
| 6.1 Introduction | 218 |
| 6.1.1 Fundamentals of Stereochemistry | 218 |
| 6.1.2 Chirality and Biological Activity | 220 |
| 6.1.3 Chirality Application as Single Isomers | 222 |
| 6.2 Resolution of Racemates | 223 |
| 6.2.1 Direct Resolution | 223 |
| 6.2.2 Indirect Resolution | 223 |
| 6.3 Reaction of Various Functional Groups | 225 |
| 6.3.1 Amines | 225 |
| 6.3.2 Carboxyls | 226 |
| 6.3.3 Hydroxyls | 227 |
| 6.3.4 Thiols and Others | 227 |
| 6.4 Derivatization for GC Analysis | 228 |
| 6.4.1 Label for Alcohols | 228 |
| 6.4.2 Label for Carboxylic Acids | 230 |
| 6.4.3 Label for Aldehydes and Ketones | 231 |
| 6.4.4 Label for Other Functional Groups | 233 |
| 6.5 Derivatization for LC Analysis | 233 |
| 6.5.1 Label for UV-VIS Detection | 235 |
| 6.5.1.1 Label for Primary and Secondary Amines | 235 |
| 6.5.1.2 Label for Carboxylic Acids | 240 |
| 6.5.1.3 Label for Alcohols | 241 |
| 6.5.1.4 Label for Other Functional Groups | 241 |


Edited by Toshimasa Toyo'oka: *Modern Derivatization Methods for Separation Sciences* © 1999 John Wiley & Sons Ltd.

6.5.2 Label for Fluorescence (FL), Laser-Induced Fluorescence (LIF) and Chemiluminescence (CL) Detection 243
6.5.2.1 Label for Primary and Secondary Amines Including Amino Acids 249
6.5.2.2 Label for Carboxylic Acids 257
6.5.2.3 Label for Alcohols 262
6.5.2.4 Label for Aldehydes and Ketones 266
6.5.2.5 Label for Thiols 266
6.5.2.6 Label for Other Functional Groups 267
6.5.3 Derivatization of Electrochemical (EC) Detection 267
6.6 Derivatization of Capillary Electrophoresis (CE) Analysis 267
6.7 Conclusion and Further Perspective 284
References 284

# 6.1 INTRODUCTION

## 6.1.1 Fundamentals of Stereochemistry

The most common type of chiral molecule contains a tetrahedral carbon atom attached to four different groups. The carbon atom is the asymetric center of the molecule. Such a molecule can exist as a pair of different stereoisomeric compounds which are non-superimposable mirror images of each other. The mirror images exist for objects with three dimensions. These look like the relation between right hand and left hand. As shown in Fig. 6.1, lactic acid is a chiral molecule and existing as a pair of mirror images. Such a relationship of stereoisomers is called enantiomer. Although a large number of chiral molecules comprise an asymmetric carbon, another class chiral molecules such as chiral axis compounds is

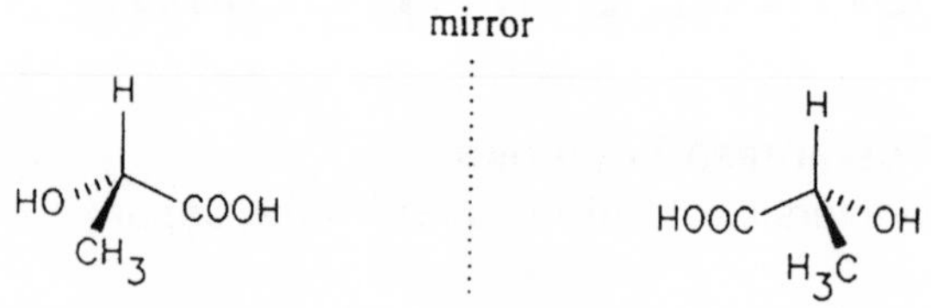

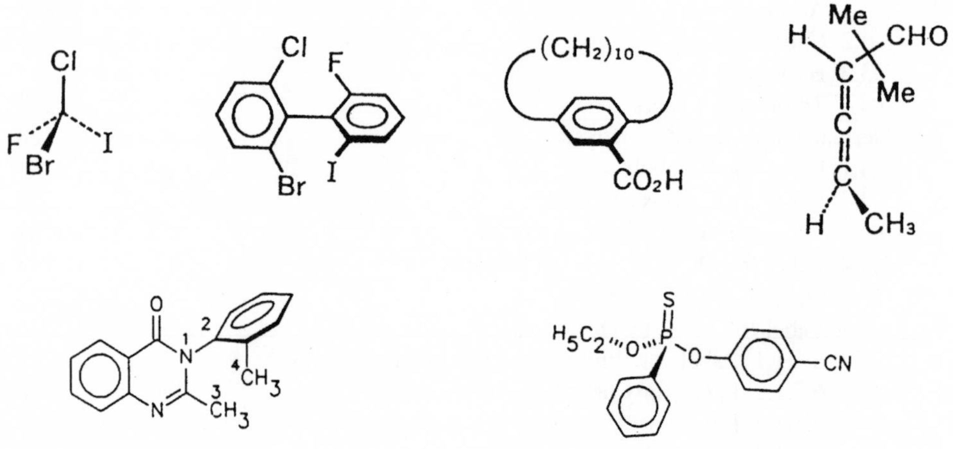

**Fig. 6.1.** Enantiomers of lactic acid.



**Fig. 6.2.** Examples of chiral molecules.

also known. In this type of chiral molecule called an atropisomer, the asymmetry is due to hindered rotation. Compounds possessing phosphorus (P) and sulfur (*S*) atoms may also possibly be chiral molecules. These examples are shown in Fig. 6.2.

In the case of compounds containing more than two asymetric carbons that are chiral centers, the number of stereoisomers depends on the stereogenic carbon ($n$) and is defined as $2^n$. When the number ($n$) of asymmetric carbons in the compounds such as ephedrine and pseudoephedrine is 2, for example, the possible stereostructures are $2^2 = 4$ (Fig. 6.3).

The stereoisomers are mainly classified into two, enatiomer and diastereomer, as shown in Fig. 6.4. The stereoisomers bearing a mirror image relationship to each other are enantiomers; whereas the stereoisomers that are not in the relationship are called diastereomers (Fig. 6.4). Therefore, achiral *cis* and *trans* geometric isomers are also included in the definition. Although both ephedrine and pseudoephedrine exist as enantiomeric pairs, (1*R*,2*S*)-ephedrine and (1*S*,2*S*)-pseudoephedrine, or (1*S*,2*R*)-ephedrine and (1*R*,2*R*)-pseudoephedrine have a non-mirror-image relationship to each other. Such diastereomers with only one asymmetric center inverted among two or more are called epimers; for example, ephedrine and pseudoephedrin.

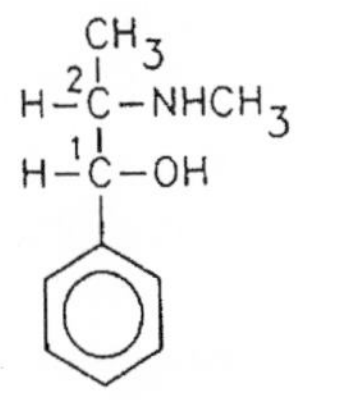

(*1R,2S*)-Ephedrine (*1S,2R*)-Ephedrine

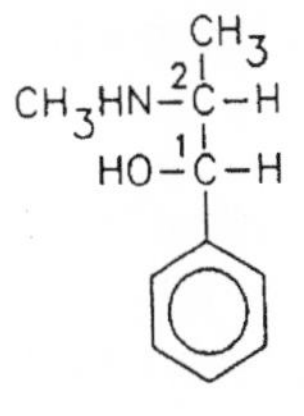

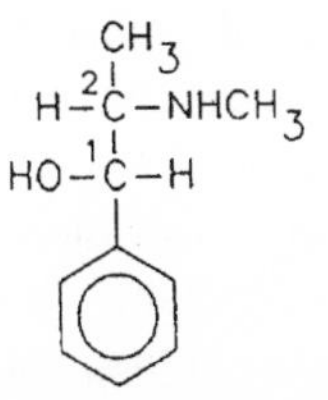

(*1S,2S*)-Pseudoephedrine (*1R,2R*)-Pseudoephedrine

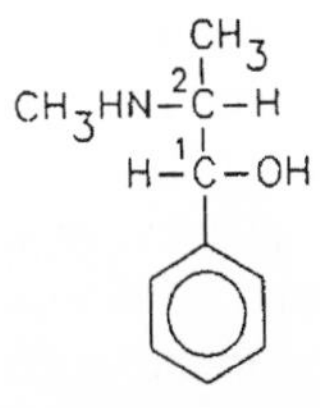

**Fig. 6.3.** Fischer projections of ephedrine and pseudoephedrine.

The physical and chemical properties such as melting point and solubility usually differ both in diastereomers. The differences in the properties have been used for the resolution of enantiomers for long time periods. On the other hand, same physico-chemical properties are observed in a pair of enantiomers. The rotation of plane-polarized light is the only specific distinguishing physical property of enantiomers. The enantiomers existing as mirror image molecules rotate plane-polarized light to an equal but opposite extent. Therefore, the racemic mixture, i.e. a 50/50% mixture of both enantiomers, does not rotate plane-polarized light the same as do achiral compounds. The direction and the degree of rotation are measured with a polarimeter, and denoted by (*d*)/(*l*) (or (+)/(−)) and a numerical value. Since the rotation depends on the concentration, solvent, temperature, cell length and wavelength of light used, the magnitude of each enantiomer is usually expressed as its specific rotation [$\alpha$]. However, there is no correlation between the direction of rotation and the notation of absolute configuration. In other words, the actual arrangement of the substituents

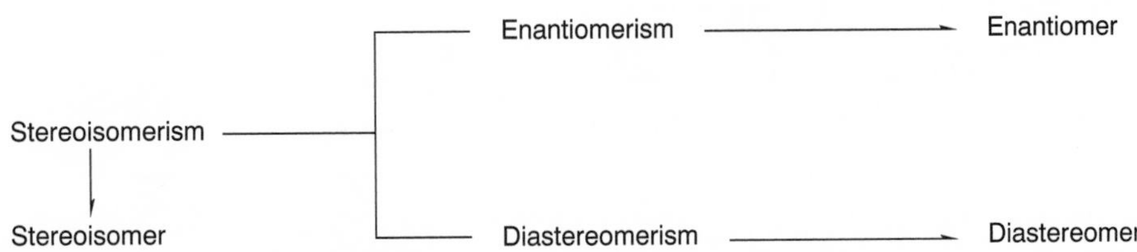


**Fig. 6.4.** Classification of stereoisomers.

around the stereogenic center (asymmetric carbon) is not defined from the rotation, (*d*)(+) and/or (*l*)(−). In the early stages, the configuration of all chiral molecules was chemically related to D(+)-glyceraldehyde or L(−)-glyceraldehyde. D(+)-glyceraldehyde is the enantiomer with the position of the OH substituent on the right-hand side of the stereogenic center in the Fischer convention [1]. This has been widely used in the series of sugar and $\alpha$-amino acid. The D and L definition by the Fischer convention can be confused with the sign of rotation (*d*) and (*l*). Consequently, the Cahn–Ingold–Prelog convention [2,3] known as the sequence rule is currently used to denote the absolute configuration. When the four substituents are represented as W, X, Y and Z, priority is essentially established by the atomic number of the atoms closest to the asymmetric carbon. In the case of the same atomic number, the priority is dependent on the next attached atoms, until a difference is established. Double bonds count as two single bonds. The priority of some substituents are listed in Fig. 6.5. For the assignment of *R* or *S* configuration, the stereogenic center is viewed with the substituent of lowest priority (W) pointed away from the observer. When the sequence of decreasing priority, X > Y > Z, is clockwise, the configuration is defined by *R*. The *S* configuration is counterclockwise (Fig. 6.6). In sugar and amino acid chemistry, the D and L labels as derived from the Fischer convention are still used and can be confused with the designation of the optical rotation *d* and *l*.

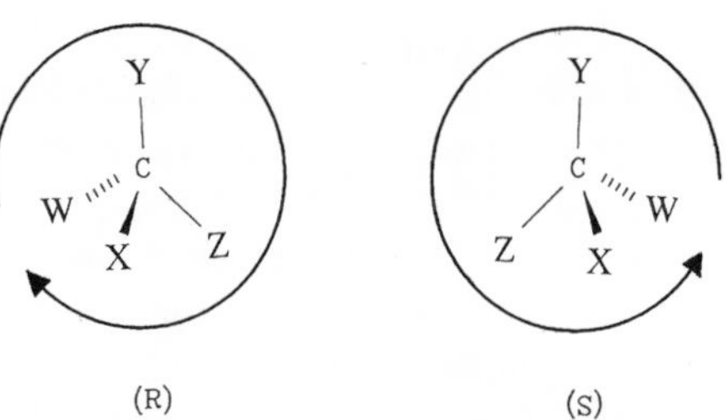


**Fig. 6.6.** Expression by Cahn-Ingold-Prelog convention.

## 6.1.2 Chirality and Biological Activity

A pair of enantiomers does not have identical chemical and physical properties. In the chiral environment, however, it is possible to exhibit the different reactivities in each enantiomer. Since the molecular components of living organisms such as enzymes and receptors keep their chirality, greater differences are usually observed for the activities of the enantiomers in a biological system. The difference is currently understood as a logical conclusion. The phenomena derived from chiral molecules can be equally applied to all bioactive substances such as drugs, agrochemicals, natural products, flavors, fragrances and food additives. Figure 6.7 shows typical examples of the differences between each enantiomer of racemates.

It is important to realize that enantioselectivity plays an important role not only in the pharmacodynamic phase involving the interaction of bioactive agents with macromolecules (enzymes and receptors) in the target organs but also in the pharmacokinetic phase involving ADME (absorption, distribution, metabolic conversion and excretion)

$Cl > S > F > O > N > C > H$

$-CH_2Cl > -CH_2OH > -CH_2CH_3 > -CH_3$

$-C(CH_3)_3 > -CH(CH_3)_2 > -CH_2CH_3 > -CH_3$

$-COOCH_3 > -COOH > -CONH_2 > COCH_3 > -CH{=}O > -CH_2OH$

$-C{=}N > -C_6H_5 > -C{=}CH > -CH{=}CH_2 > -CH_2CH_3$

$-C(CH_3)_2CH_2CH_3 > -C(CH_3){=}CH_2 > -CH(CH_3)CH_2CH_3$

$-CH{=}CH_2 > -CH(CH_3)_2$

**Fig. 6.5.** Priority of some substituents.

| Name | | Enantiomer | Effect | | Enantiomer | Effect |
|---|---|---|---|---|---|---|
| Dopa | S | 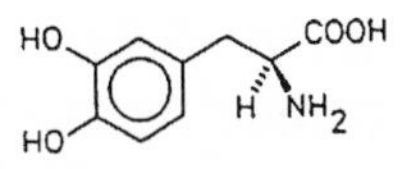 | anti-Perkinson | R | 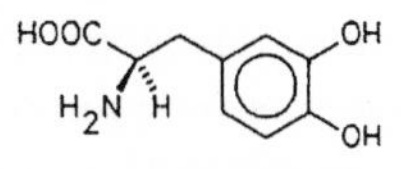 | side-effect |
| Ketamine | S | 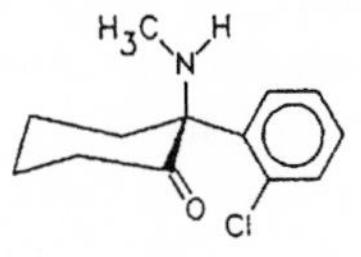 | anaesthetic | R | 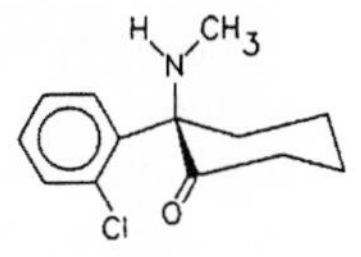 | hallucinogen |
| Penicillamine | S | 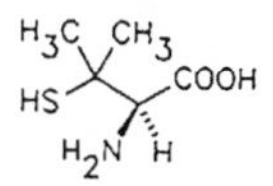 | antiarthritic | R | 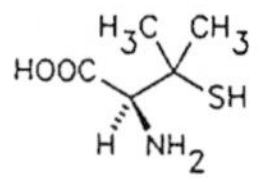 | mutagen |
| Thalidomide | S | 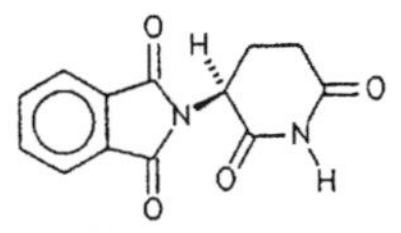 | teratogen | R | 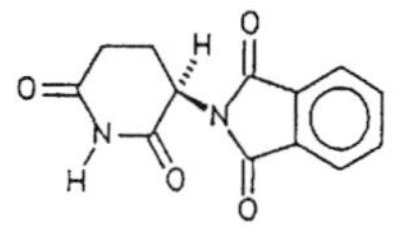 | sedative |
| Limonen | S | 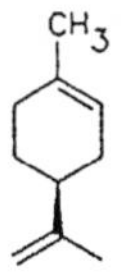 | lemon odor | R | 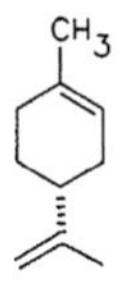 | orange odor |
| Carvone | S | 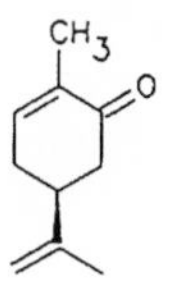 | caraway | R | 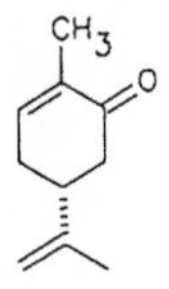 | spearmint |
| Aspartame | SS | 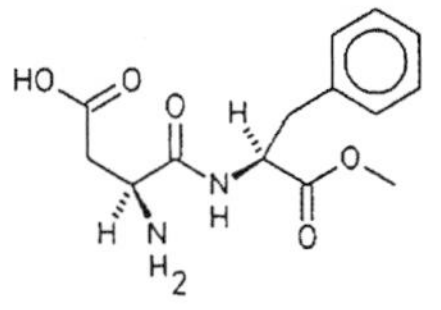 | sweet | RR | 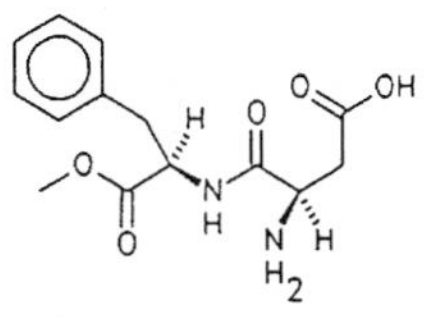 | bitter |

**Fig. 6.7.** Examples of different biological effects in each enantiomer.

of the drug. The different pharmacodynamic and pharmacokinetics of eutomer (isomer with higher affinity) and distomer (isomer with lower affinity) in the racemate leads to a variety of effects. For example, the distomer sometimes exhibits undesirable side-effects such as mutagenecity of (*R*)-penicillamine and teratogen of (*S*)-thalidomide. The distomer, however, displays no serious side-effects in some cases, such as $\beta$-blockers. The $\alpha$-arylpropionic acids like ibuprofen are all chiral molecules and (*S*)-enantiomers are responsible for the desired therapeutic effect (treatment of rheumatoid arthritis). Since the (*R*)-isomers undergo metabolic inversion of configuration with racemate to afford the active (*S*)-isomers, most of this type of drug, except for naproxen, has been administrated as racemates.

### 6.1.3 Chirality Application as Single Isomers

The majority of naturally occurring medicines have chiral molecules and almost all that exist in nature are marketed as the single enantiomer. In spite of the inherent difference with mirror images, a number of synthetic chiral drugs are still sold as racemates, instead of chiral molecules. According to research from 1990 (Fig. 6.8), approximately

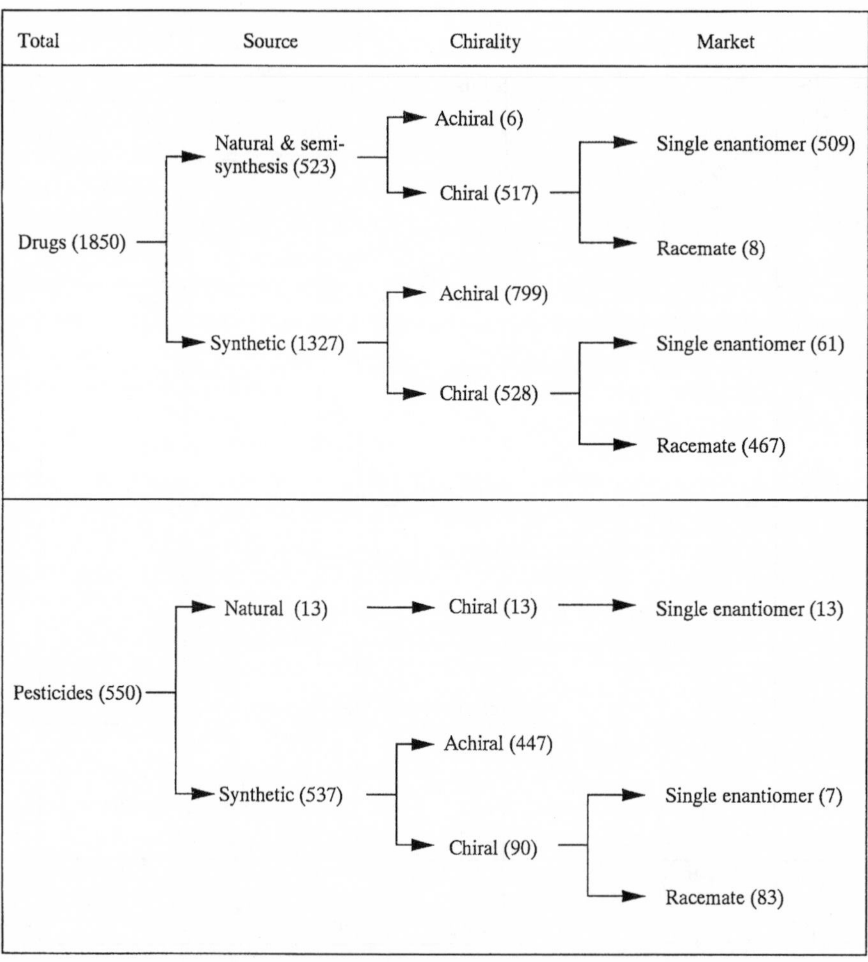


**Fig. 6.8.** Distribution of single enantiomer and racemate on the market.

40% of the drugs developed from organic synthesis possess chiral centers in the structure, but only 12% are marketed as single enantiomers [4,5]. The Food and Drug Administrations in USA (FDA) has a flexible policy on chiral drugs, and will continue to accept racemates as new chemical entities [6]. This guideline is good news for the pharmaceutical industry. Pharmaceutical companies have continued to develop chiral drugs as racemates. When the racemates are developed as drugs, however, the risk and benefit of candidate chemicals must be examined in detail. If it is feasible to prepare a single enantiomer by 'chirotechnology' at low cost, when many problems seen with racemates disappear [7,8].

## 6.2 RESOLUTION OF RACEMATES

Chiral separation is equally important to chiral synthesis, for the purposes of investigation of the differences in biological effects such as pharmacological and toxicological properties, and the establishment of specifications for drug marketing. High-performance liquid chromatography (HPLC) has been a powerful tool for chiral analysis over the last two decades. The principles are divided into two categories; one is based upon direct separation using a chiral stationary phase (CSP), and the other relys on diastereomer formation with a suitable chiral derivatization reagent. A third technique is the method of chiral ligand exchange chromatography utilizing mobile phase additives such as the amino acid enantiomer and metal ion. The separation is based upon the chilate formation between the analyte enantiomer and mobile phase additives. Limited racemates such as amino acids are separated by the method.

### 6.2.1 Direct Resolution

A number of enantioseparations have been achieved with the direct methods that employ CSP columns containing immobilized chiral selectors. Commercially available CSP columns are listed in Table 6.1. These are classified into cavity phase (cyclodextrin $\alpha$, $\beta$, $\gamma$, etc), helical phase (cellulose and amylose esters), affinity phase ($\alpha$1-acid glycoprotein and ovomucoid, etc), $\phi$-donor (smichiral columns), $\phi$-accepter (pirkle type columns), and ligand exchange. The stereo chemical resolution between the pirkle column and a pair of enantiomers is explained by the three-point chiral interaction model [9]. Separation by liquid chromatography proceeds in four steps: (1) association of enantiomers with a chiral selector on a CSP column; (2) chiral recognition based on thermodynamic differentiation of the diastereomeric complexation; (3) dissociation of enantiomers from the chiral selector; (4) repetition of the above three steps in the CSP column. The separation mode is based on the difference due to the stabilities of the diastereomeric complexes formed between the stationary phase and each enantiomer in the flow system. The characteristics of the direct method using a CSP column are listed in Table 6.2. The theoretical plate and the detection sensitivity are not always enough in sample analysis. Since there is no derivatization step for the resolution, no racemization occurs during the reaction with a chiral tagging reagent. Consequently, the direct method using the CSP column is recommended for trace analysis of antipode enantiomer in the main component, such as the determination of optical purity in bulk drug as in quality control. However, it is difficult to select the best column for the resolution of each racemate, because the separation is highly influenced by the interaction between CSP and the enantiomer. The elution order of each enantiomer (*R*- or *S*- form) is dependent upon the CSP column used, and cannot be changed easily. It is an another disadvantage that CSP columns are generally expensive.

### 6.2.2 Indirect Resolution

The indirect method involving a derivatization step with a chiral tagging reagent is an efficient technique for separation of many enantiomers. The characteristics of the indirect method by HPLC are also described in Table 6.2. Separation is based upon diastereomer formation. This indirect derivatization method is suitable for trace analysis of enantiomers in biological samples, such as blood

**Table 6.1.** Commercially available chiral stationary phase (CSP) column

| Chiral selector | Name of phase | Company |
|---|---|---|
| Cyclodextrin $\alpha$, $\beta$, $\gamma$ | Cyclobond I, II, III | Astec |
| Cyclodextrin $\beta$ | Chiral CD BR | YMC |
| Cyclodextrin $\beta$ | ChiraDex | Merck |
| Cyclodextrin $\beta$ | ES-CD | Shinwa |
| Cyclodextrin $\beta$ phenyl | ES-PhCD | Shinwa |
| (*R*)-*N*-(3,5-Dinitrobenzoyl)phenylglycine | Bakerbond Chiral Ionic DNBPG | Baker |
| | Bakerbond Chiral Covalent DNBPG | Baker |
| | Ionic D-phenylglycine | Regis |
| | Covalent D-phenylglycine | Regis |
| | SUMICHIRAL OA-2000 | Sumitomo |
| (*S*)-*N*-(3,5-Dinitrobenzoyl)phenylglycine | Covalent L-phenylglycine | Regis |
| (*S*)-*N*-(3,5-Dinitrobenzoyl)leucine | Bakerbond Chiral Covalent DNBLeu | Baker |
| | Ionic L-leucine | Regis |
| | Covalent L-leucine | Regis |
| *N*-(3,5-Dinitrobenzoyl)-(*R*)-1-naphthylglycine | SUMICHIRAL OA-2500 | Sumitomo |
| *N*-(3,5-Dinitrophenylaminocarbonyl)-L-valine | SUMICHIRAL OA-3100 | Sumitomo |
| *N*-(3,5-Dinitrophenylaminocarbonyl)-L-tert.-leucine | SUMICHIRAL OA-3200 | Sumitomo |
| *N*-(3,5-Dinitrophenylaminocarbonyl)-D-phenylglycine | SUMICHIRAL OA-3300 | Sumitomo |
| *N*-[(*S*)-1-($\alpha$-naphtyl)ethylaminocarbonyl]-L-valine | SUMICHIRAL OA-4000 | Sumitomo |
| *N*-[(*R*)-1-($\alpha$-naphtyl)ethylaminocarbonyl]-L-valine | SUMICHIRAL OA-4100 | Sumitomo |
| *N*-[(*S*)-1-($\alpha$-naphtyl)ethylaminocarbonyl]-L-proline | SUMICHIRAL OA-4400 | Sumitomo |
| *N*-[(*R*)-1-($\alpha$-naphtyl)ethylaminocarbonyl]-L-proline | SUMICHIRAL OA-4500 | Sumitomo |
| *N*-[(*S*)-1-($\alpha$-naphtyl)ethylaminocarbonyl]-L-tert.-leucine | SUMICHIRAL OA-4600 | Sumitomo |
| *N*-[(*R*)-1-($\alpha$-naphtyl)ethylaminocarbonyl]-L-tert.-leucine | SUMICHIRAL OA-4700 | Sumitomo |
| *N*-[(*S*)-1-($\alpha$-naphtyl)ethylaminocarbonyl]-(*S*)-indoline-2-carboxylic acid | SUMICHIRAL OA-4800 | Sumitomo |
| *N*-[(*R*)-1-($\alpha$-naphtyl)ethylaminocarbonyl]-(*S*)-indoline-2-carboxylic acid | SUMICHIRAL OA-4900 | Sumitomo |
| *N*,*S*-Dioctyl-D-penicillamine | SUMICHIRAL OA-5000 | Sumitomo |
| (*R*)-2-amino-1,1-bis(2-butoxy-5-tert.-butylphenyl)phenylpropanol-1 | SUMICHIRAL OA-5500 | Sumitomo |
| L-Tartaric acid mono-(*R*)-1-($\alpha$-naphtyl)ethylamide | SUMICHIRAL OA-6000 | Sumitomo |
| L-tartaric acid mono-L-valine-(*S*)-($\alpha$-naphtyl)ethylamide | SUMICHIRAL OA-6100 | Sumitomo |
| Cellulose ester derivatives | Chiracel OA, OB, OB-H, OJ, OK, CA-1 | Daicel |
| Cellulose carbamate derivatives | Chiracel OC, OD, OD-H, OF, OG | Daicel |
| Cellulose tribenzyl ether | Chiral OE | Daicel |
| Amylose carbamate derivatives | Chiracel AD, AS | Daicel |
| Poly(triphenylmethyl methacrylate) | Chiracel TO(+) | Daicel |
| Poly(2-pyridyldiphenylmethyl methacrylate) | Chiracel OP(+) | Daicel |
| 18-Crown-6 | Crownpak CR | Daicel |
| $\alpha$1-Acid glycoprotein | EnantioPac | LKB |
| Poly(*N*-acryloylphenylalanine ethyl ester) | ChiraSpher | Merck |
| Bovine serum albumin (BSA) | | Aspec |
| Acid protein | Chiral AGP | Daicel |
| Ovomucoid | ES-OVM | Shinwa |
| Ovomucoid | Enantio OVM | Tosoh |
| Chiral polymer | TSK gel Enantio-Li | Tosoh |

**Table 6.2.** Characterization of direct and indirect resolution by HPLC

| Method | Advantages | Disadvantages |
|---|---|---|
| Direct | (1) No derivatization needs except to enhance detectability. | (1) Universal stationary phase which resolves all racemates is not available. |
| | (2) Racemization is usually negligible. | (2) Generally theoretical plate is low. |
| | (3) Recovery of enantiomers is more easy than indirect method, because of direct elution from column. | (3) Control of elution order is difficult. |
| | (4) Enantiomers have essentially the same detector response. | (4) High cost of chiral stationary phase column. |
| | (5) Resolution of racemates which have no reactive functional group are possible. | |
| Indirect | (1) Ordinary achiral column such as ODS is possible to use for resolution of almost racemates. | (1) Analytes have a reactive functional group in the structure. |
| | (2) Elution order can be controlled by choosing of enantiomer of chiral reagent. | (2) Chiral tagging reagents have to have 100% optical purity. |
| | (3) Sensitive detection is possible with a suitable reagent possessing high detector response. | (3) Recovery of enantiomers is usually difficult. |
| | (4) It can be adopted for both analytical and preparative work. | (4) Resulting diastereomers may have different detector response. |
| | | (5) Need to consider both stabilities of reagent itself and the derivatives. |

and urine, because of the option of coupling with highly sensitive reagents which have the high molar absorptivity ($\varepsilon$) or the high fluorescence quantum yield ($\phi$). There are various reagents that provide ultraviolet-visible (UV-VIS), fluorescence (FL) and electrochemical (EC) labels for gas chromatography (GC), liquid chromatography (LC) and capillary electrophoresis (CE). This chapter focuses on indirect resolution using a chiral derivatization reagent. Enantiospecific separations of racemic compounds such as drugs and biologically active compounds using chiral derivatization reagents are described in the following sections. The aim of the chapter is to list the chiral reagents with the properties of chiral reagents and their application.

## 6.3 REACTIONS OF VARIOUS FUNCTIONAL GROUPS

The labelling with chiral derivatization reagent is carried our with the reaction of a reactive functional groups in substrates, e.g. amines (primary and secondary), carboxyl, carbonyl, hydroxyls (alcohol and phenol) and thiol. Many organic reactions are adopted for the labelling of various functional groups in the substrate. As described in the previous section, it is impossible to label the compounds without these functional groups.

### 6.3.1 Amines

Among the functional groups, the tagging reactions of primary and secondary amines have been extensively investigated and a large number of articles have been reported in various journals. The major types of reaction for chiral amines involving amino acids and aminoalcohols are based on the formations of amide, carbamate, urea and isourea (Fig. 6.9). The formation reactions of diastereomeric amides are widely used for the resolution of various primary amines. The reactions with acid chloride and chloroformate reagents rapidly proceed to produce corresponding amides and carbamates, respectively. The acid halogenides such as *N*-substituted prolyl chloride

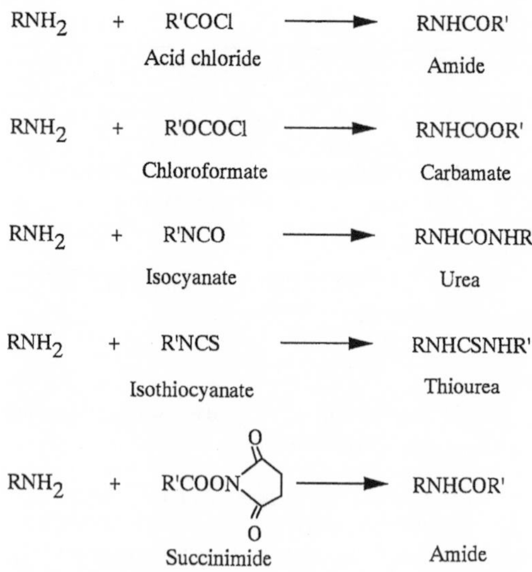


**Fig. 6.9.** Derivatization reactions for diastereomer formation of optically active amines.

is a good label because the amide produced from acylation occurs under mild conditions at room temperature or in an ice bath. Since hydrolysis of the reagents proceeds easily with water in sample solution, the contamination of water is the medium should be definitely avoided. The chiral drugs possessing the –COOH group, such as naproxen and benzoxaprofen, are converted to the corresponding –COCl and used as the resolving reagents for the chiral amines. Another amide formation is the reaction with *N*-succinimidyl ester. These types of reagent are fairly stable and can be used in aqueous media. Therefore, the main application is as a reagent for staining biological specimens such as cells and organs. Chloroformate derivatives quantitatively label the primary and secondary amines to produce corresponding carbamates at room temperature. The most important label of this type is chiral 1-(9-fluorenyl)ethylchloroformate (FLEC) and is used for amines, especially for the resolution of amino acids by HPLC. Although the labelling with isocyanates and isothiocyanates is slower than those with the above halogeno reagents, they are suitable for the resolution of racemate due to the introduction of polar functional groups of urea and isourea. The 2,3,4,6-tetra-*O*-acetyl-$\beta$-D-glucopyranosyl isothiocyanate (GITC) bearing reagents which have an isothiocyanate group as a reactive site, are one of the most popular labels for amines to produce thiourea derivatized in the presence of a base catalyst. The thiourea diastereomers are readily separated with a reversed-phase ODS column in spite of the long distance between two chiral centers. The derivatization reactions using isocyanate reagents need rather drastic conditions at high temperature and long time. The reaction of orthophtaladehyde (OPA) and a chiral thiol (e.g. *N*-acetyl-L-cysteine and tert.-butyloxycarbonyl (BOC)-L-cysteine) are also important labels for chiral amines, especially in amino acid enantiomers. The reaction is often completed within 5 min at room temperature. However, the resulting isoindole derivatives are usually reported as unstable compounds. To solve the disadvantage, post-column labeling following separation with an analytical column is adopted for the determination of primary amines.

## 6.3.2 Carboxyls

Carboxyl compounds are an important analyte the same as amines, because a wide range of drugs and chemicals of biological importance possess a carboxylic acid in their structure. The tagging is historically carried out with ester formation (Fig. 6.10). The esteration with a chiral alcohol requires extremely drastic conditions such as high temperature and a long reaction period in a strong acid solution. The reaction proceeds smoothly with activation of carboxylic acid to acid chloride using

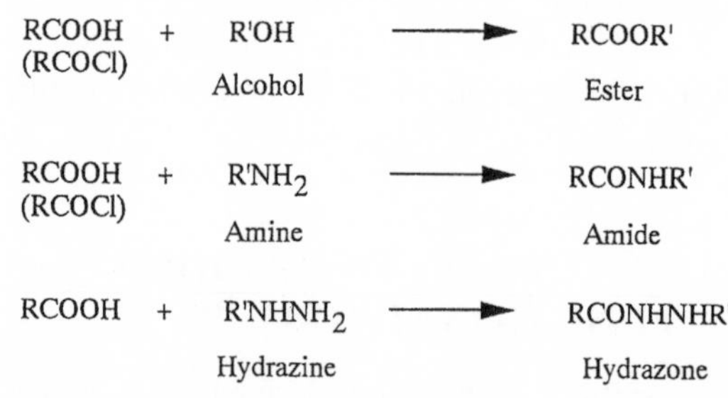


**Fig. 6.10.** Derivatization reactions for diastereomer formation of optically active carboxyls.

pentachlorophosphine ($PCl_5$) or thionyl chloride ($SO_2Cl_2$). However, the racemization during these reactions should be considered in the case of trace determination of the antipode enantiomer in an excess amount of enantiomer. Chiral alcohols are another type of labelling reagent of carboxyls for GC analysis in order to produce the diastereomer ester. Aliphatic primary and secondary amines are used as the resolving reagent for carboxylic acid labelling to yield diastereomeric amides. The amide formation reaction using a chiral amine reagent in the presence of an activation agent such as *N*,*N*-dialkylcarbodiimides usually proceeds under relatively mild conditions at room temperature. In general, the amide diastereomers are more readily separated by chromatography than the corresponding esters, particularly by the HPLC column. The separation seems to depend on the case of hydrogen bonding with the stationary phase. Consequently, a chiral amine reagent is recommended for the tagging of carboxylic acids in terms of reactivity and resolution. A number of papers using this type of reagent have been published in various journals. Many chiral amines are commercially available with high optical purity.

## 6.3.3 Hydroxyls

Hydroxyl compounds are classified into two types, alcohols and phenols. The phenol is easier to label, because of the liberation of the ionization form $Ph{-}O^-$ in a slightly alkaline medium. However, the derivatization of alcoholic compounds is fairly difficult due to low reactivity based upon a high pKa value. Esteration with acids and acid chloride, carbonation with chloroformates, and carbamation with isocyanates are adopted as the main labelling reactions of alcohols (Fig. 6.11). Esteration with acid chloride reagents and carbonation with chloroformate reagents seem to be suitable for labelling because of the good reactivity of the $-COCl$ group in the reagent. Since the isocyanate reacts with alcohols to produce carbamate diastereomers at elevated temperatures, it should be racemization during the derivatization reaction should be ensured. The

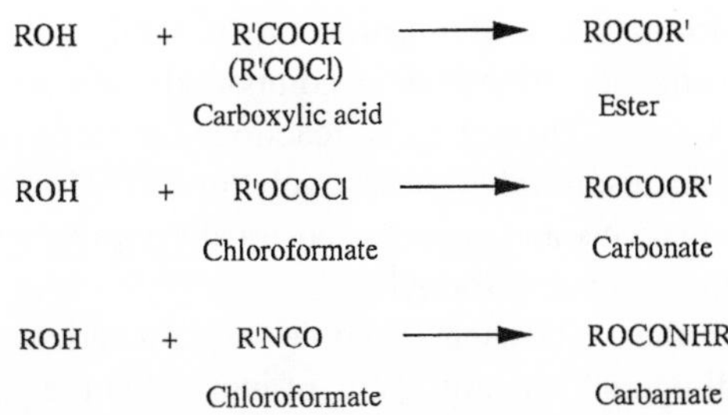


**Fig. 6.11.** Derivatization reactions for diastereomer formation of optically active hydroxyls.

ester formation with acids is also performed with severe reaction conditions such as in a strong basic medium. Therefore, the esters are normally prepared after activation with carboxylic acid to form the chloride or anhydride compound. Since the labelling for alcohols competes with the hydrolysis of the reagent, the reaction should be performed in anhydrous solvents under the protection of moisture. Furthermore, reaction conditions should be optimized in detail in each sample.

## 6.3.4 Thiols and Others

The thiol functional group ($-SH$) is highly reactive with electrophiles. However, only a few examples are reported for labelling due to low stability of thiol compounds (Fig. 6.12). Possible reagents are of isothiocyanate type which is used for the tagging of amines, and produces corresponding dithiocarbamates. Other separation of chiral thiols is performed by HPLC after pre-column reaction with OPA and a chiral amine.

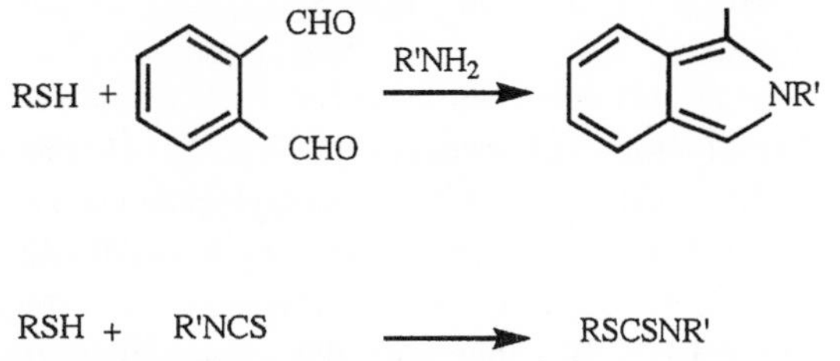

RSH + (CHO, CHO) —$R'NH_2$→ (SR, NR')

RSH + R'NCS → RSCSNR'

**Fig. 6.12.** Derivatization reactions for diastereomer formation of optically active thiols.

Various types of reaction are possible for the derivatization of carbonyl compounds, aldehydes and ketones. However, the reactions for chromatographic resolution are relatively few. Hydrazine-type reagents are mainly used for the HPLC resolution of chiral carbonyls.

In the following sections, characterization, handling and versatility, etc. of individual reagents are described, together with typical examples.

## 6.4 DERIVATIZATION FOR GC ANALYSIS

GC is one of the effective tools for the separation of volatile substances owing to its excellent separatability and detectability, especially in capillary column. Chiral resolution of various compounds was carried out by the GC method prior to 1990. The chiral compounds analyzed mainly by GC are hydroxyls (alcohols and phenols) and carbonyls (aldehydes and ketones). Since only volatile compounds are adopted for GC, the reagent which yields highly volatile derivatives is essential for GC analysis. An ideal reagent is one where the resulting derivative possesses high detectability such as good electron capturing properties. Since the separation by GC is essentially carried out at high temperature, the stability of the derivatives is important when choosing the reagent.

### 6.4.1 Label for Alcohols

The chiral reagents for hydroxyls, chiral carboxylic acid compounds, e.g. *R*(+)-transchrysanthemic acid (2) [10–12], drimanoic acid (3) [10], (*S*)-acetoxypropionic acid (1) [13–16], 3$\beta$-acetoxy-$\Delta^5$-etienic acid (4) [17], *S*(+)-phenylpropionic acid (7) [18], mandelic acid (D(−) and L(+)) (8) [19], *S*(−)-*N*-(trifluoroacetyl)proline (12) [12] and *S*(−)-$\alpha$-methoxy-$\alpha$-(trifluoromethyl)phenylacetic acid (MTPA, Mosher's reagent) (11) [20–22], are widely used to yield diastereomeric esters (Fig. 6.13). To increase the reactivity, the reagents are converted with chlorination to the corresponding acid chloride prior to reaction with alcoholic compounds. Care should be taken when these reagents are used the tagging, because certain acid chlorides can racemize under extreme acidic or basic medium and/or at elevated temperatures. Fluorine probes such as (*S*)(−)-*N*-(trifluoroacetyl)prolyl chloride and *S*(−)-$\alpha$-methoxy-$\alpha$-(trifluoromethyl)phenylacetyl chrolide, are excellence in terms of volatility and sensitivity. The reaction of alcohols with isocyanate reagents such as phenylethyl isocyanate (PEIC) (13) [23–25] and napthylethyl isocyanate (NEIC) (14) [26–28] are relatively slower than amines, and require a long time being heated at high temperatures. However, racemization of resulting carbamates is negligible in spite of severe derivatization conditions. Another type of reagent is chiral chloroformates, for example, (−)-menthyl chloroformate (10) [29], which yield carbonates under mild conditions in an alkaline medium. Therefore, the reagent is suitable for acid and/or heat labile alcohols. Some typical examples using these reagents are described in the following examples.

**Example 1.** Tagging of alcohols with chrysanthemic acid (2) [10].

*R*(+)-trans-chrysanthemic acid (2 mg) in dry toluene (100 µl) is converted to the acid chloride form with redistilled thionyl chloride (200 µl) at 60 °C for 1 hr. After removing excess thionyl chloride under a stream of nitrogen, alcohols (1 mg) and 3 molar equivalents of the tagging reagent in dry toluene are reacted in a sealed tube at 40 °C for 1–2 h. The products are directly analysed by GC without further purification. The separation is carried out on non-polar (1% SE-30) and low polarity (1% OV-17) columns. A chromatogram of the separation is show in Fig. 6.14.

**Example 2.** Tagging of 1,3-dialkylglycerol ethers with *S*(−)-1-phenylethyl isocyanate (PEIC) (13) or *S*(−)-1-(1-naphtyl)ethyl isothiocyanate (NEIC) (14) [24].

The secondary alcohols (10 µl) are reacted with *S*(−)-PEIC or *S*(−)-NEIC (20 µl) in a screw-cap vial at 60 °C for 1 h. The excess reagent

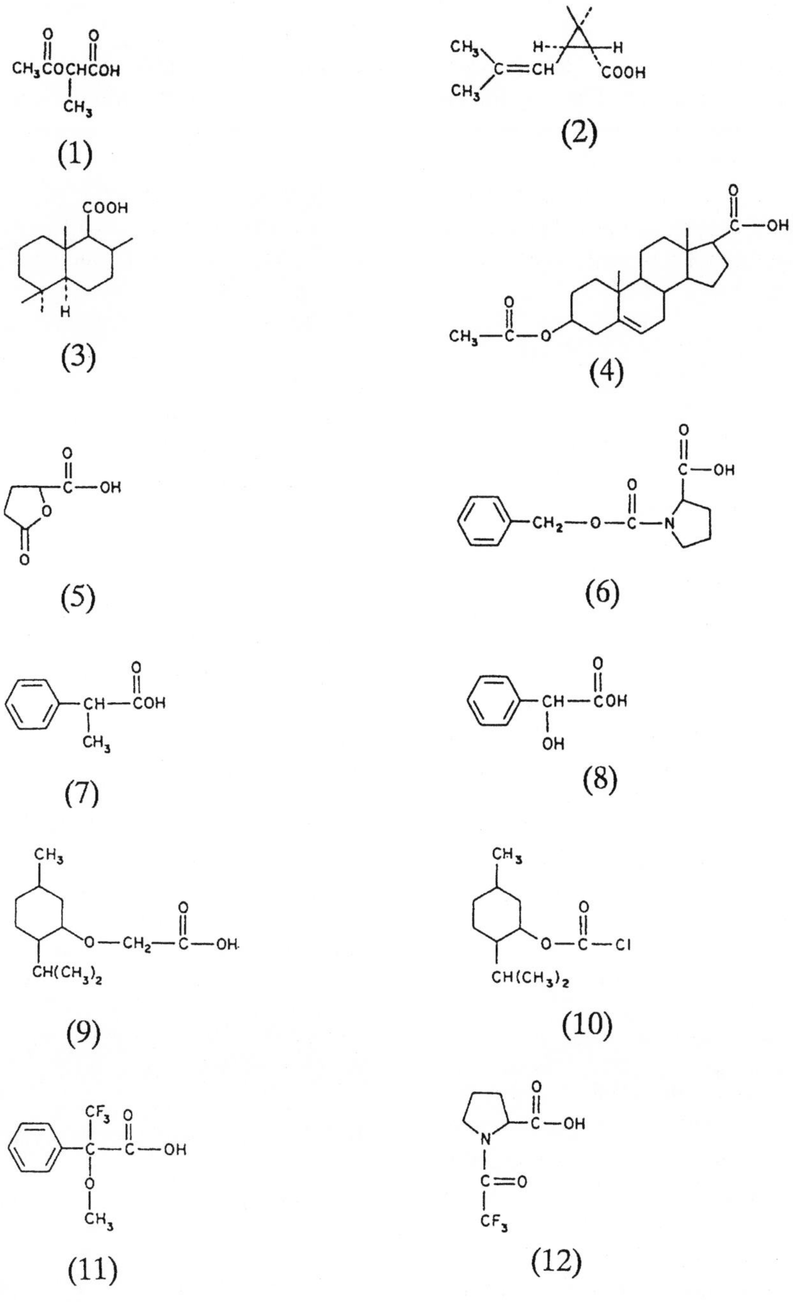


**Fig. 6.13.** Chiral tagging reagents for hydroxyl compounds by GC: (1) (*S*)-acetoxypropionic acid (AP); (2) *R*(+)-trans-chrysanthemic acid; (3) Drimanoic acid; (4) 3$\beta$-acetoxy-$\Delta^5$-etienic acid; (5) *S*-tetrahydro-5-oxo-2-furancarboxylic acid (TOF); (6) Carbobenzyloxy-L-proline; (7) *S*(+)-phenylpropionic acid; (8) D(−)/L(+)-mandelic acid; (9) (−)-menthenyloxyacetic acid; (10) (−)-menthyl chloroformate; (11) *S*(−)-$\alpha$-methoxy-$\alpha$-(trifluoromethyl)-phenylacetic acid (MTPA, Mosher's reagent); (12) *S*(−)-*N*-(trifluoroacetyl)proline; (13) *R*(+)/*S*(−)-phenylethyl isocyanate (PEIC); (14) *R*(+)/*S*(−)-naphtylethyl isocyanate (NEIC).

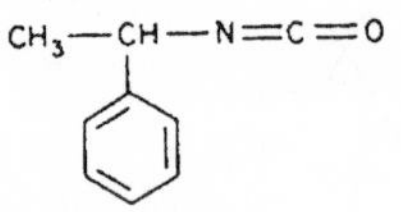

$CH_3—CH—N=C=O$ (13)

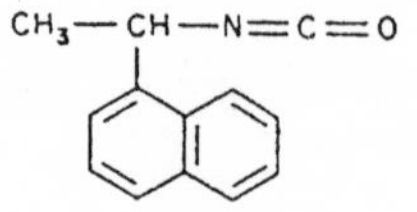

$CH_3—CH—N=C=O$ (14)

Fig. 6.13. (*continued*).

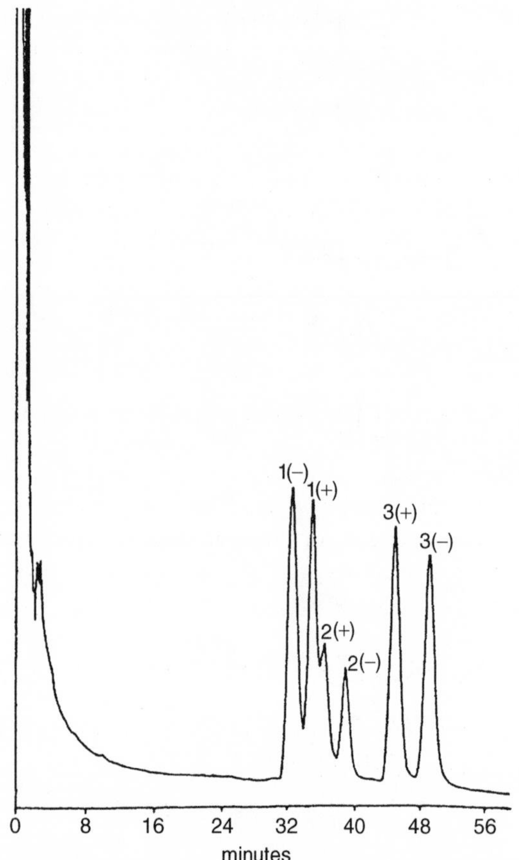


**Fig. 6.14.** GC separation of (±)-fenchyl alcohol (1); (±)-neomenthol (2); and (±)-menthol (3). [Reproduced from ref. 10, p. 898, Fig. 4.].

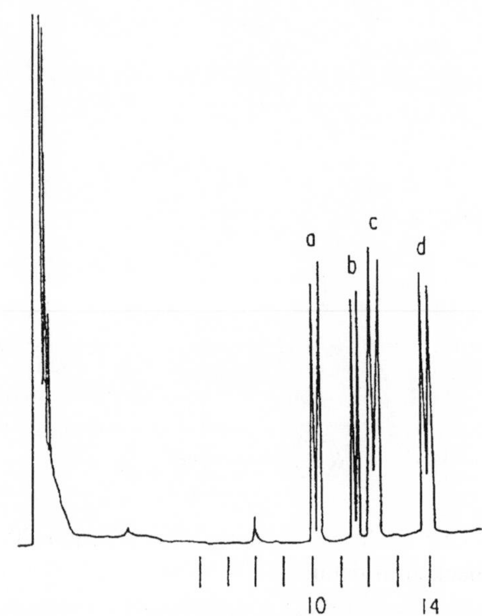


**Fig. 6.15.** GC Separations of $\alpha$-methylbenzyl carbamates obtained from 1,3-dialkylglycerol esters. [Reproduced from ref. 24, p. 212, Fig. 6.].

is discharged with 0.1 ml methanol for 15 min. The diluted solution of the resulting carbamates is analysed with GC (column, Supelco SPB-1 or SP-2340). An example of the separation of the 1-methylbenzyl carbamates is given in Fig. 6.15.

The separation factors ($\alpha$) are greater for the naphthyl derivatives than for 1-methylbenzyl carbamates. However, the greater retentions of the naphthyl derivatives limit the usefulness.

## 6.4.2 Label for Carboxylic Acids

Ester formation reactions using alcoholic reagents are adopted for the labelling of carboxylic acid. Various chiral alcohols such as 2-butanol (15) [30,31], menthol (16) [32–37] and $\alpha$-methylbenzyl alcohol (17) [31], can possibly be used for the purpose (Fig. 6.16). Since the ester formation reaction of carboxylic acid with alcohol generally requires drastic conditions, the reaction is carried out after activation of carboxylic acid to the corresponding acid chloride. Therefore, the

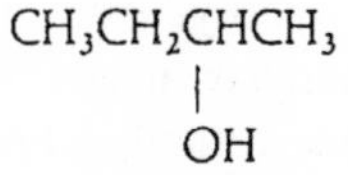

**Fig. 6.16.** Chiral tagging reagents for carboxylic acids by GC. (15) *R*(−)/*S*(+)-2-butanol; (16) (−)/(+)-menthol; (17) α-methylbenzyl alcohol; (18) *R*(−)/*S*(+)-amphetamine; (19) *O*-(−)-menthyl-*N*,*N*-diisopropylisourea (MDI).

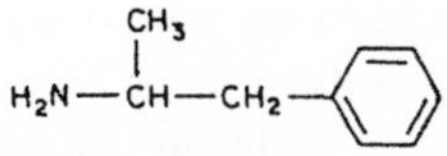

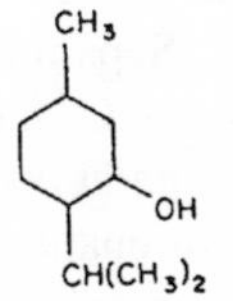

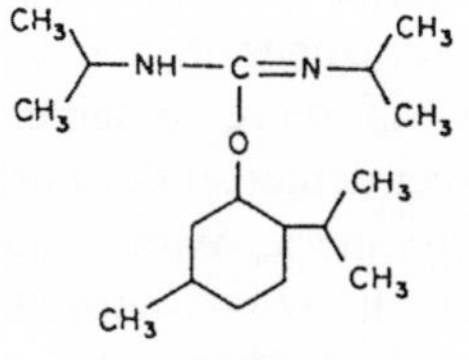

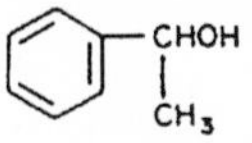

racemization during the convertion reaction should be checked in detail.

*O*-(−)-menthyl-*N*,*N*-diisopropylisourea (19) [38] reacts with carboxylic acid in the presence of base catalyst to produce diastereomeric menthyl esters. According to the report, the reagent is stable for up to four years, and no racemization occurs in its reaction with carboxylic acid.

Chiral amines are another important label for carboxylic acid, and a number of chiral amine reagents have been developed. A typical example is described later.

**Example 3.** Tagging of carboxylic acids with *O*-(−)-menthyl-*N*,*N*-diisopropylisourea (MDI) (19) [38].

To the acids dissolved in 400 μl of tetrahydrofuran/triethylamine (9:1) is added 10 μl of (−)-MDI. Then the reaction mixture is heated at 100 °C for 16 h. The solution is analysed by capillary gas chromatography (SP-2100 wall-coated open tubular glass capillary column).

As depicted in Fig. 6.17, a pair of diastereomeric esters and *N*,*N*-diisopropylurea are produced with the derivatization reaction. The disadvantage is the long time required for heating in order to complete of the reaction. The separation of resulting amides are suitable not by GC but by HPLC method.

### 6.4.3 Label for Aldehydes and Ketones

The separation of racemic carbonyls (aldehydes and ketones) has been facilitated by a wide variety of derivatizations. However, the derivatizations for chromatographic resolution are fewer. Many carbonyls have been resolved by the GC method, and HPLC is scarce. Chiral reagents for chromatographic resolution, (+)-2,2,2-trifluoro-1-phenylethyl hydrazine (24) [39,40], L-cysteine methyl ester (20) [41], 2-amino-oxy-4-methylvaleric acid [42], *O*-(−)-bornylhydroxyammonium chloride (22) [43,44],

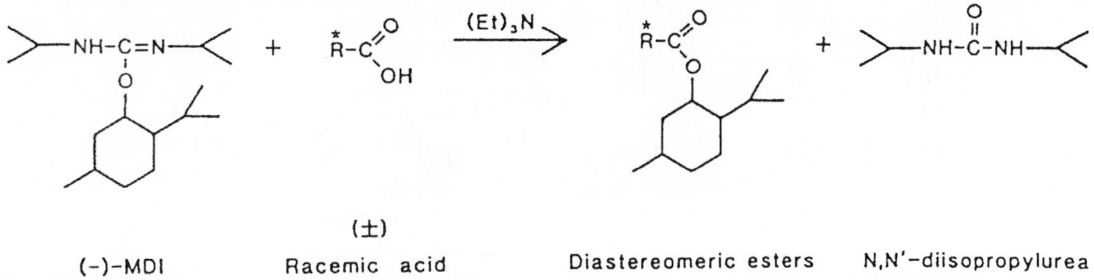

**Fig. 6.17.** Reaction of carboxylic acid with *O*-(−)-menthyl-*N*,*N*-diisopropylisourea (MDI). [Reproduced from ref. 38, p. 162.].

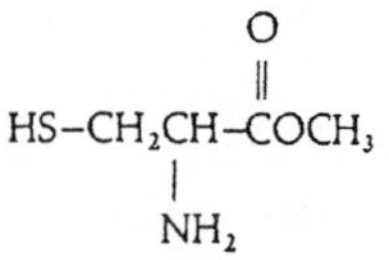

$HS{-}CH_2CH(NH_2){-}COCH_3$ (with C=O)

(20)

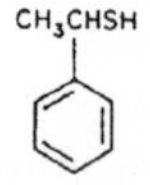

$CH_3CHSH$ (phenyl)

(21)

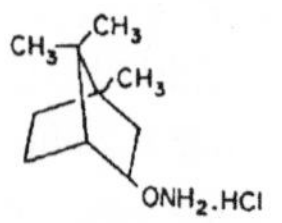

$CH_3$, $CH_3$, $CH_3$, $ONH_2.HCl$

(22)

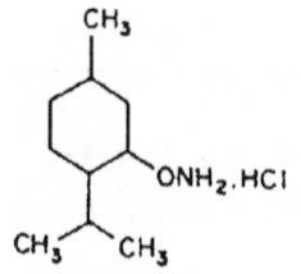

$CH_3$, $ONH_2.HCl$, $CH_3$, $CH_3$

(23)

$CF_3{-}CH(NHNH_2){-}$(phenyl)

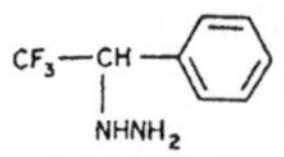

(24)

**Fig. 6.18.** Chiral tagging reagents for carbonyls by GC. (20) L-Cysteine methyl ester; (21) (+)-1-phenylethanethiol; (22) *O*-(−)-bornylhydroxyammonium chloride; (23) *O*-(−)-menthylhydroxyammonium chloride; (24) (+)-2,2,2-trifluoro-1-phenylethylhydrazine.

etc. have been reported (Fig. 6.18). The diastereomeric thiazolidines derived from the reaction of nine aldoses with L-cysteine methyl ester (20) are resolved by GC. The analysis needs silylation with hexamethyldisilazane-trimethylchlorosilane (HMDS-TMCS) to volatile the derivatives. The hydrazine type reagents such as (+)-2,2,2-trifluoro-1-phenylethyl hydrazine (24) yield the hydrazones in acidic medium. Separation by HPLC is more suitable than the GC method. Schweer demonstrates fairly good separation of the diastereomeric oximes derived from enantiomeric sugars with *O*-(−)-bornylhydroxyammonium chloride (22) [45] or *O*-(−)-menthylhydroxyammonium chloride (23) [45,46].

**Example 4.** Tagging of sugars with *O*-(−)-bornylhydroxyammonium chloride (22) [45].

Sugar or carbohydrate (about 0.5 mg) is added to *O*-(−)-bornylhydroxyammonium chloride (4 mg) and sodium acetate (3 mg) in 0.1 ml of water. The mixture is allowed to stand at 80 °C for 1 h. The GC column is a 50 m long capillary column wall-coated with OV-225 (WGA, Griesheim, G.F.R.). A GC separation of the derivatives of carbohydrates is shown in Fig. 6.19.

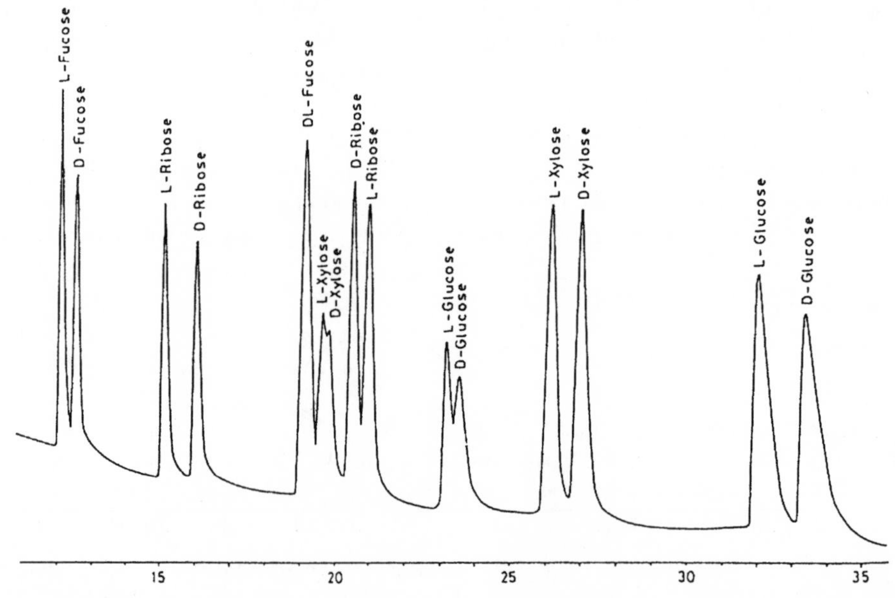


**Fig. 6.19.** GC separation of enantiomers of carbohydrates as (−)-bornyloxime pertrifluoroacetates. [Reproduced from ref. 45, p. 166, Fig. 2.].

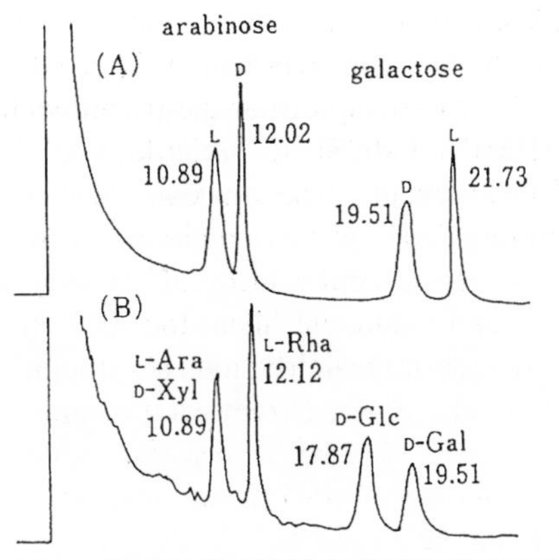


**Fig. 6.20.** Separation of enantiomers of arabinose and galactose as trimethylsilylated ethers of thiazolidine derivatives (chromatogram A); Trimethylsilylated thiazolidine derivatives of component monosaccharides of a *Thladiantha* saponin (Chromatogram B). [Reproduced from ref. 41, p. 504, Fig. 3.].

**Example 5.** Tagging of aldose enantiomers with L-cysteine methyl ester (20) [41].

The mixture of sugars (0.04 M) and L-cysteine methyl ester hydrochloride (0.06 M) in pyridine (200 μl) is warmed at 60 °C for 1 h, and then trimethylsilylation with HMDS-TMCS reagent is performed for another 30 min at the same temperature. After centrifuge, the supernatant (1 μl) is subjected to GLC column (G-SCOT Silicone OV-17 on Silanox; 0.3 mm i.d. × 50 m).

The derivatives of the aldoses having R-configuration at C-1 eluted before those of the *S*-configuration (Fig. 6.20).

### 6.4.4 Label for Other Functional Groups

Enantiodifferentiation of lactones occurs in natural products. Although direct resolution using capillary GC is applicable, indirect derivatization is more usual for the separation of the lactones. The reactions are mainly ketal formation with chiral diols such as 2,3-butanediol [47], or ring opening with 1-phenylethyl isocyanate (PEIC) (13) [48], acetoxypropionic acid (1) [49,50] or teterahydro-5-oxo-2-furancarboxylic acid (5) [51], etc. In general, severe conditions are needed for the cleavage.

Derivatization of amino alcohols and amino acids is also performed with the reagents for amino functional group (primary and secondary).

## 6.5 DERIVATIZATION FOR LC ANALYSIS

Liquid chromatography (LC), especially high-performance liquid chromatography (HPLC), is an important technique in separation science, as is high-performance capillary electrophoresis (HPCE) described later. In HPLC analysis, the separation is carried out on the column, and the sensitivity and improved selectivity can be achieved in the detection step. Therefore, highly sensitive and selective detection techniques such as fluorometry and chemiluminometry are adopted for the determination of trace substances. Although mass spectrometry (MS) can provide excellent selectivity and sensitivity, the combination with HPLC is not yet widely usable because of the high cost of the systems. On the other hand, the other detectors, i.e. ultraviolet-visible (UV-VIS), fluorescence (FL), electrochemical (EC) and chemiluminescence (CL), are popular for use in the determination of various items of biological importance.

A number of UV labels have been currently applied to the tagging of various functional groups. The sensitivity of the derivatives is not good enough in some real samples. To solve the problem, various types of fluorescence labels have been developed instead of UV-VIS labels. The basical reactions are similar to those in the labelling of UV-VIS, FL, CL and EC [52–54]. If a new type of reaction is developed, various modifications of the labelling properties would provide new reagents for UV-VIS, FL, CL and EC detections. An HPLC chromatographic run carried out at around room temperature is suitable for chiral resolution, because the possible racemization observed during the separation on column is negligible, as compared with GC analysis.

With indirect methods involving a derivatization step, the choice of the tagging reagent is of the greatest importance for the resolution of chiral molecules. A pair of mirror image enantiomers is labelled with a chiral derivatization reagent to generate corresponding two diastereomers. The separation is performed with the differences from physicochemical properties (e.g. stereochemistry and stability) with achiral stationary phase. Therefore, the selection of the reagent determines the accuracy, precision and repeatability of the quantitative analysis. Furthermore, the presence of two or more reactive functional groups within the analyte complicates the tagging reaction. Several important points worthy of consideration are listed below: (i) The optical purity of the reagent should be good enough (preferably 100%) the same as chemical purity. Since the opposite enantiomer contaminated in the reagent used also produces a corresponding diastereomer, it is obvious that incorrect determination results are observed when impure reagents is used. (ii) The degree of recemization during the labelling reaction and storage of the reagent itself is another important consideration for quantitative determination. Besides, the chemical stability of the resulting diastereomers also influences the results obtained. Good stability at least one day is required in many sample analyse because autoanalysis over night is run as usual. (iii) It is important that the reactivity of the reagent to each enantiomer and the fluorescence properties (wavelengths and intensity) of the resulting derivatives are essentially the same. When those are exactly the same, the curves of time pattern of the reaction with the reagent are superimposable. (iv) The reagent possess specificity for the target functional group and quantitatively labels the substrate under mild conditions. The resulting diastereomers exhibit an adequate detector response for sample analysis. (v) Another important point is the solubility of the reagent whether it is freely soluble in water and/or misible in an aqueous solvent such as alcohol and acetonitrile, since many bioactive chiral molecules are dissolved in aqueous solution. Items (iv) and (v) are similar considerations to those used in achiral quantitative analysis with an achiral reagent. (vi) Practically, both enantiomers of the reagent should be commercially available or easily obtained by simple synthesis, because the elution order can be controlled by the selection of the reagent enantiomer. This is necessary when the determination of a trace enantiomer is required in the presence of a large amount of the antipode enantiomer. However, few derivatization reagents are as commercially available as both enantiomers.

The elution order and degree of separation of the diastereomer derived from each reagent is not easily predicted with the achiral stationary phase of a ordinary column such as octadesyl silica (ODS). It is influenced by the distance between two asymetric centers of substrate and the reagent, and the distance should be minimized for good separation. The conformational rigidity around the chiral centers is another important factor in separation. A resolving reagent is recommended in which the free rotation near the asymetric center in the substrate is disturbed by the formation of the diastereomer. Since there is no obvious rule concerning the separation of both diastereomers, the structural difference should also be considered when choosing the reagent.

A number of optical isomers have been developed for HPLC analysis of chiral molecules having various functional groups. Chiral primary and secondary amines are easy to label with carboxyls, chloroformates, isocyanates and isothiocyanates to derive corresponding amides, carbamates, ureas and thioureas, respectively. Racemic carboxylic acids are usually labeled with chiral primary amine in the presence of an activation reagent (e.g. EDC and DPDS/TPP). The reaction proceeds under mild conditions at room temperature. Ester formation with chiral alcohol is also adopted for the derivatization of carboxyls. Since the reaction conditions are generally drastic, the possibility of racemization should be monitored during the reaction. The labelling of alcoholic OH is a most difficult functional group, because the reaction competes with water in the reaction medium. For this purpose, acid chloride-type reagents are mainly used in anhydrous solvents such as chloroform and benzene. A variety of organic reactions is possible for the carbonyl compounds, aldehydes and ketones. Few labels are reported for the carbonyl enantiomers by liquid chromatography. Those are amine and hydrazine type reagents to produce corresponding nitriles and hydrazones. Since the resulting C=N structure is unstable, a reducing agent such as sodium borohydride ($Na_2BO_4$) is usually added in the reaction medium. There is no effective reagent for the tagging of tertial amines, lactones, alkene (C=C), and alkine (C≡C). It is necessary to develop the reagents for these functional groups.

In this section, labelling reagents to UV-VIS, FL and EC detection by HPLC are classified into the types of reactions for each functional group. Some typical examples using the chiral reagents are described, together with the chromatograms.

## 6.5.1 Label for UV-VIS Detection

Tagging of substances with reagents that afford the structures absorbing UV or VIS regions is the most popular means of derivatization, because almost all laboratories possess a UV-VIS detector and analysts are experiencing manipulation. The reagents usually have large molar absorptivity ($\varepsilon$). Of course, it is ideal that the reagent quantitatively reacts with an analyte to produce a sole diastereomer without racemization. Furthermore, both the reagent and the resulting derivative must be stable. There are many undesirable substances in samples, which absorb in UV-VIS regions. Since interfere of impurities absorbing the detection wavelength is considered in real samples, especially in complex matrices such as biological specimens, the reagents possessing a visible band are preferable in terms of selectivity. However, selectivity may be solved by use of efficient columns having high theoretical plates such as GC capillary columns, in the near future. Hence selectivity problems using UV-label would be largely overcome. In the following sections, characterization and use of the chiral UV-VIS reagents for various functional groups are summarized, and some typical examples are given.

### 6.5.1.1 Label for Primary and Secondary Amines

Although a number of labels for amino functional groups have been developed and applied to real sample analysis, chiral reagents which permit the enantioseparations are fewer. Chiral UV-labels are shown in Fig. 6.21. The reagents bearing isothiocyanates (e.g. 2,3,4,6-tetra-*O*-acetyl-$\beta$-D-glucopyranosyl isothiocyanate (GITC) (26) [55–60], 2,3,4-tri-*O*-acetyl-$\alpha$-D-arabinopyranosyl isothiocyanate (AITC) (27) [61–64] and *R*(+)-1-phenylethyl isothiocyanate (PEIT) (25) [65]), isocyanates (e.g. *R*(+)-1-phenylethyl isocyanate (PEIC) (13) [66–71] and (*R*)-1-naphtylethyl isocyanate (NEIC) (14) [72–74]), acid *N*-succinimidyl esters (e.g. 1-$\alpha$-methoxy-$\alpha$-methyl-naphthaleneacetic acid *N*-succinimidyl ester (30) [75] and (*S*)-2-methoxy-2-phenylacetic acid *N*-succinimidyl ester (29) [76]), chloroformates (e.g. (−)-menthylchloroformate (10) [77–82] and (*S*)-tert.-butyl 3-(chloroformoxy)butyrate (31) [83]) and carboxylic acids and its acid chlorides (e.g. *N*-trifluoroacetyl-1-prolylchloride [84], (−)-$\alpha$-methoxy-$\alpha$-(trifluoromethyl)phenylacetyl chloride [61], (−)-camphanic acid chloride [85], 1-[(4-nitrophenyl)sulfonyl]prolyl chloride (32) [86,87] and *S*(+)-benzoxaprofen chloride (34) [88–90]) are important labels for amines. FDNP derivatives [91] such as 1-fluoro-2,4-dinitrophenyl-5-L-alanine amide (Marfey's reagent) (36) [92–94] are also used for the tagging of amino acids and peptides. $\beta$-blockers are labelled with tert.-butoxycarbonyl-L-leucine anhydride [95] in a basic medium. One advantage of the chiral reagent is that the separation of the enantiomers is possible with removal of the tert.-butoxycarbonyl group by treatment with trifluoroacetic acid (TFA). Various *O*,*O*-disubstituted (*R*,*R*)-tartaric acids [96] such as (*R*,*R*)-*O*,*O*-diacetyl tartaric acid anhydride (DATAAN) (37) [97] and (*R*,*R*)-*O*,*O*-dibenzoyl tartaric acid anhydride (DBTAAN) (38), which are synthesized by Lindner *et al.*, react with alkanolamines, e.g. propranolol, in aprotic media containing trichloroacetic acid to form tartaric acid monoesters. Some typical examples using these reagents are described here.

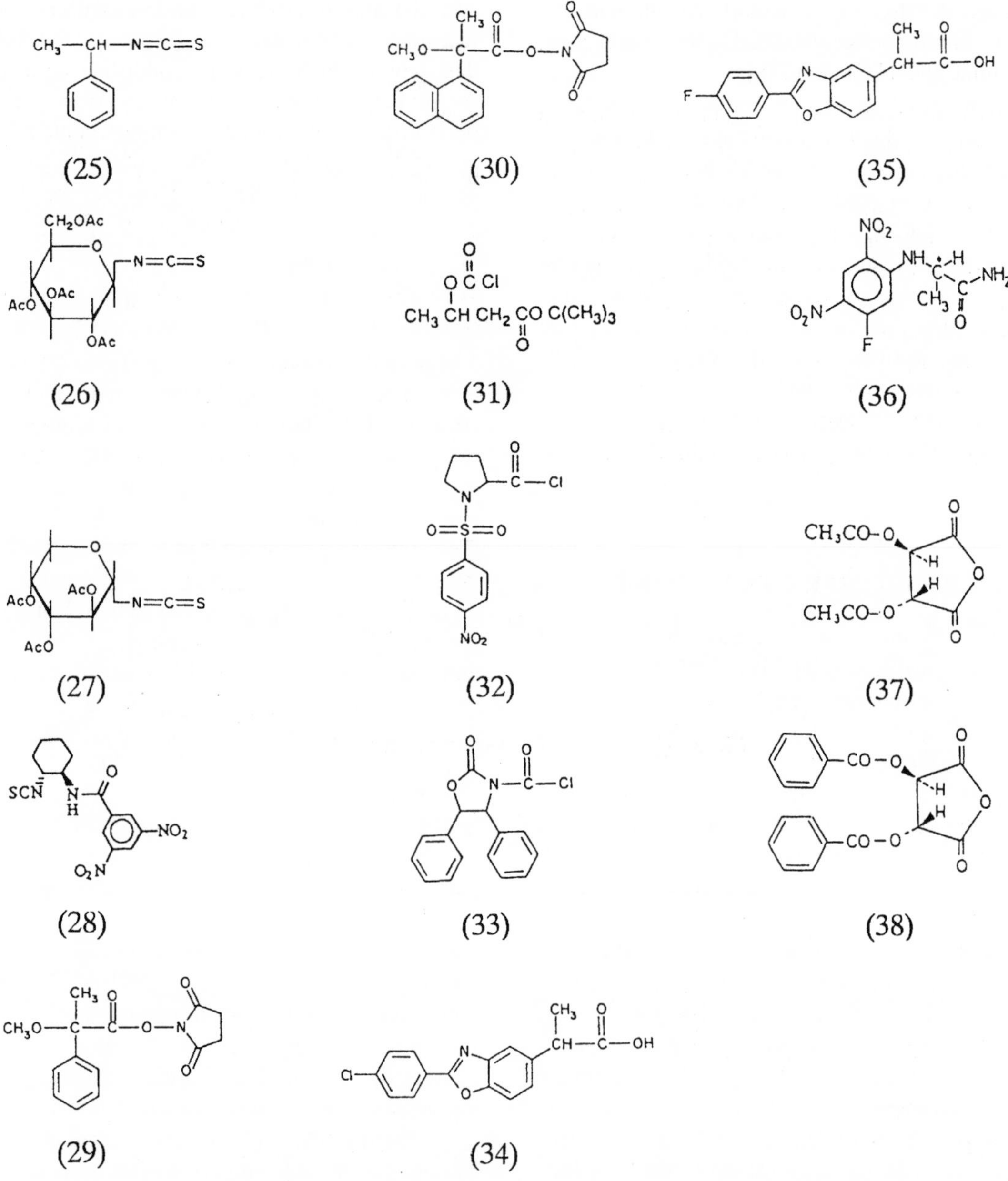


**Fig. 6.21.** Chiral UV-VIS labels for amines by HPLC: (25) *R*(+)-1-phenylethyl isothiocyanate (PEIT); (26) 2,3,4,6-tetra-*O*-acetyl-*β*-D-glucopyranosyl isothiocyanate (GITC); (27) 2,3,4-tri-*O*-acetyl-*α*-D-arabinopyranosyl isothiocyanate (AITC); (28) (*R*,*R*)-*N*-[(2-isothiocyanato)cyclohexyl]-(3,5)-dinitrobenzoylamide (DDITC); (29) *S*-2-methoxy-2-phenylacetic acid *N*-succinimidyl ester; (30) 1-*α*-methoxy-*α*-methylnaphthaleneacetic acid *N*-succinimidyl ester; (31) *S*-tert.-butyl 3-(chloroformoxy)butyrate; (32) 1-[(4-nitrophenyl)sulfonyl]prolyl chloride; (33) cis-4,5-Diphenyl-2-oxazolidone-3-carbamyl chloride; (34) *S*(+)-2-(*p*-Chlorophenyl)-*α*-methylbenzoxazole-5-acetic acid chloride (benzoxaprofen); (35) *S*(+)-(*p*-Fluorophenyl)-*α*-methylbenzoxazole-5-acetic acid chloride (flunoxaprofen); (36) 1-fluoro-2,4-dinitrophenyl-5-L-alanine amide (Marfey's reagent); (37) (*R*,*R*)-*O*,*O*-diacetyl tartaric acid anhydride (DATAAN); (38) (*R*,*R*)-*O*,*O*-dibenzoyl tartaric acid anhydride (DBTAAN).

**Example 1.** Tagging of amino alcohol with *N*-[(2-isothiocyanato)cyclohexyl]-(3,5)-dinitrobenzoylamide (DDITC) (28) [98].

To 0.01 mM of amino alcohol in 0.5 ml acetonitrile is added DDITC in 0.5 ml acetonitrile. After derivatization at 60 °C for 120 min, L-proline is added to the reaction solution to quench the excess reagent. Then the reaction mixture is placed in the oven for another 30 min. An aliquot, diluted with mobile phase and neutralized with acetic acid is injected onto the HPLC column. A chromatogram of the separation of the derivatives is shown in Fig. 6.22.

The DDITC derivatives are well separated with reversed-phase HPLC. The separation factors ($\alpha$) and the resolutions (Rs) of the diastereomeric thioureas were higher than those of the GITC derivatives which are well established as a label for primary and secondary amines. However, the derivatization of secondary amines containing a tertiary butyl group at $\alpha$-position of amino function is impossible with DDITC. That may be due to steric hindrance of the valky group.

**Example 2.** Tagging of peptide hydrolysate with Marfey's reagent (36) [92].

The hydrolysate prepared from peptide (2.5 μmol) is dissolved in 100 μl of 0.5 M sodium bicarbonate solution. Then the mixture of the solution and 200 μl of 1% Marfey's reagent in acetone is

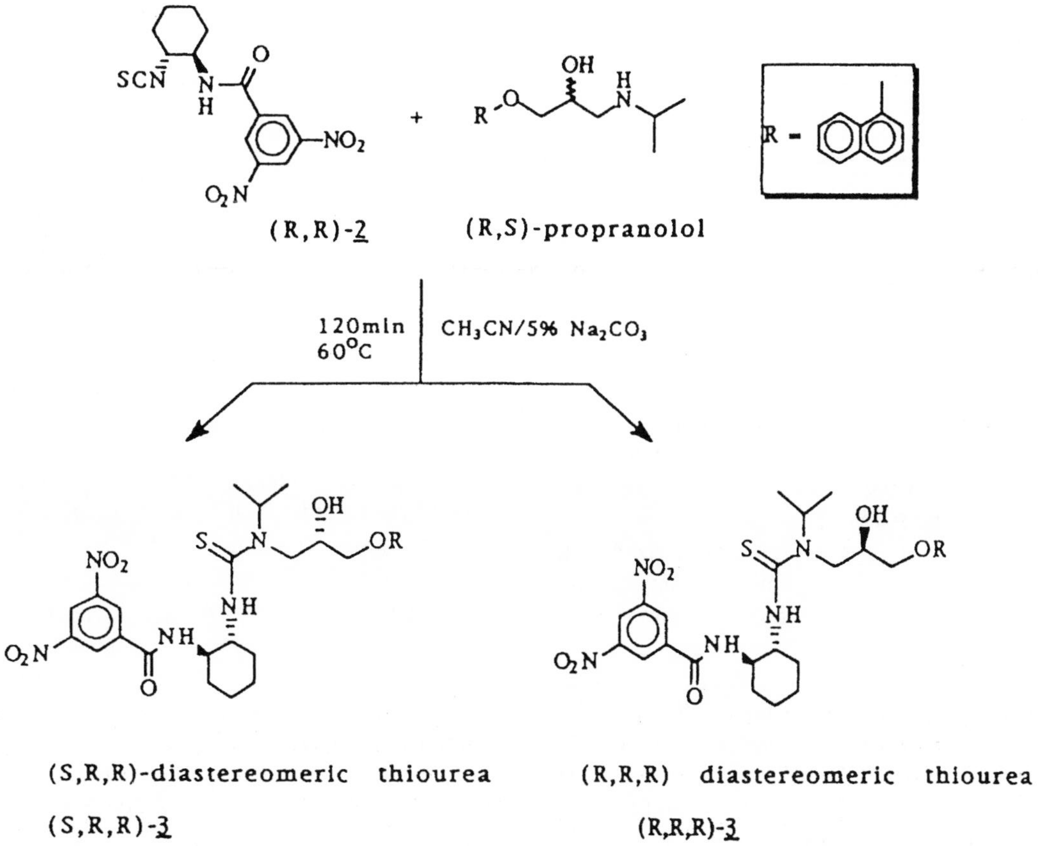

(R,R)-2 + (R,S)-propranolol (R = 1-naphthyl)

120 min, 60 °C; $CH_3CN$/5% $Na_2CO_3$

(S,R,R)-diastereomeric thiourea (S,R,R)-3

(R,R,R) diastereomeric thiourea (R,R,R)-3

**Fig. 6.22.** HPLC separation of (*R*,*S*)-metoprolol derivatized as (*R*,*R*)-DDITC- and GITC-thioureas. Column: Hypersil ODS (125 × 4 mm, i.d., 5 μm); mobile phase, acetonitrile/20 mM ammonium acetate (55:45); flow rate, 1 ml/min; detection, UV at 254 nm. [Reproduced from ref. 98, p. 41, Fig. 5.].

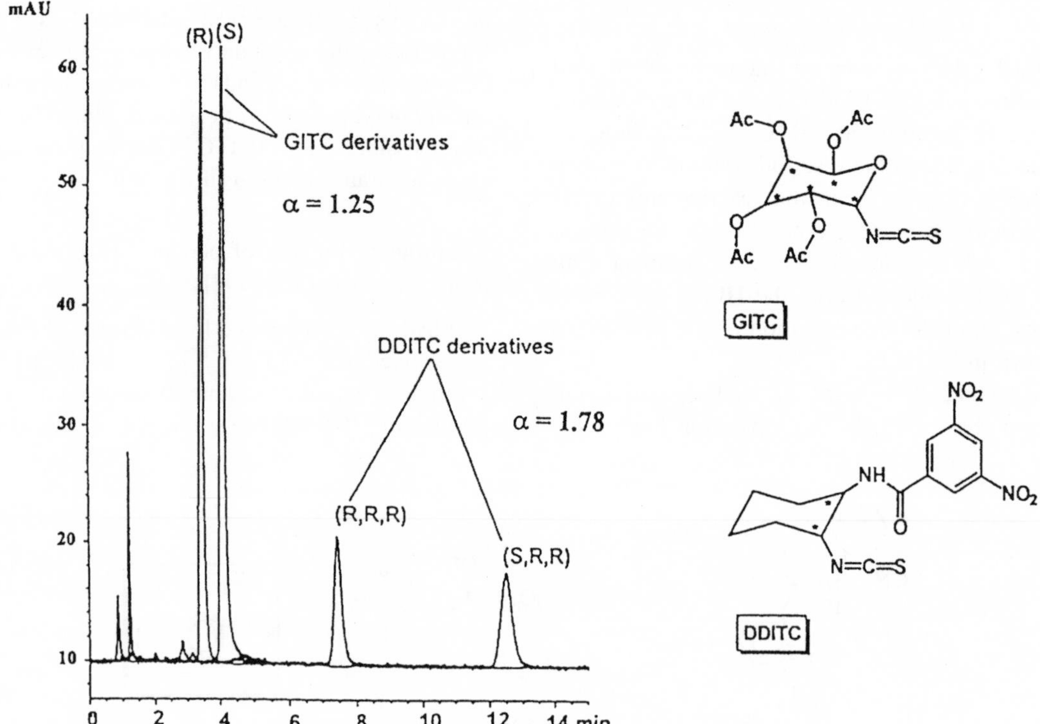


**Fig. 6.22.** (*continued*).

incubated at 40 °C for 90 min. After cooling, 25 μl of 2 M HCl are added to the reaction solution and diluted 15–20 fold in methanol or eluent. An aliquot is injected onto the HPLC column. A chromatogram of the separation of the derivatives is shown in Fig. 6.23.

Brückner and Gah [91] synthesized the analogues of Marfey's reagents, in which alanine is replaced with other chiral $\alpha$-amino acids and their derivatives. The reagents, representative of the general structures such as FDNP-Val-NHR (R=H, tert.-butyl, chiral aralkyl, phenyl, *p*-nitrophenyl), FDNP-Val-OR (R=H, $CH_3$, tert.-butyl), FDNP-(Ala)$n$-$NH_2$ ($n = 1,2$), made possible the separation of D/L-amino acids as diastereomers. The difference of the retention times in HPLC are explained from the observations of the Corey–Pauling–Koltun (CPK) space-filling molecular models.

**Example 3.** Separation of encainide (antiarrhythmic drug) and its major metabolites with (−)-menthyl chloroformate (MCF) (10) [77].

To the sample containing (±)-encainide or (±)-*N*-demethylencainide (NDE) (10 ng 5 $\mu$g), dissolved in acetonitrile containing *N*,*N*-diisopropylethylamine (50 μl, 10:90 v/v), is added (−)-MCF in acetonitrile (50 μl, 10:90 v/v). After heating at 60 °C for 2 h, the acetonitrile is evaporated and dissolved in the mobile phase. A chromatogram obtained from rat liver microsomal incubation is show in Fig. 6.24.

In general, (−)-menthyl chloroformate has been used as a chiral reagent for primary and secondary amines. Prakash *et al.* demonstrated that the reagent also reacts with the tertiary amino functional group to produce carbamate diastereomer. The resulting derivatives are efficiently separated

arylation in $NaHCO_3$/acetone
90 min, 40 °C
mol. ratio: 1.4–1.6:1

1-fluoro-2,4-dinitrophenyl-5-L-alanine amide

DL

LL + LD → separation

Reaction of DL-amino acids with Marfey's reagent.

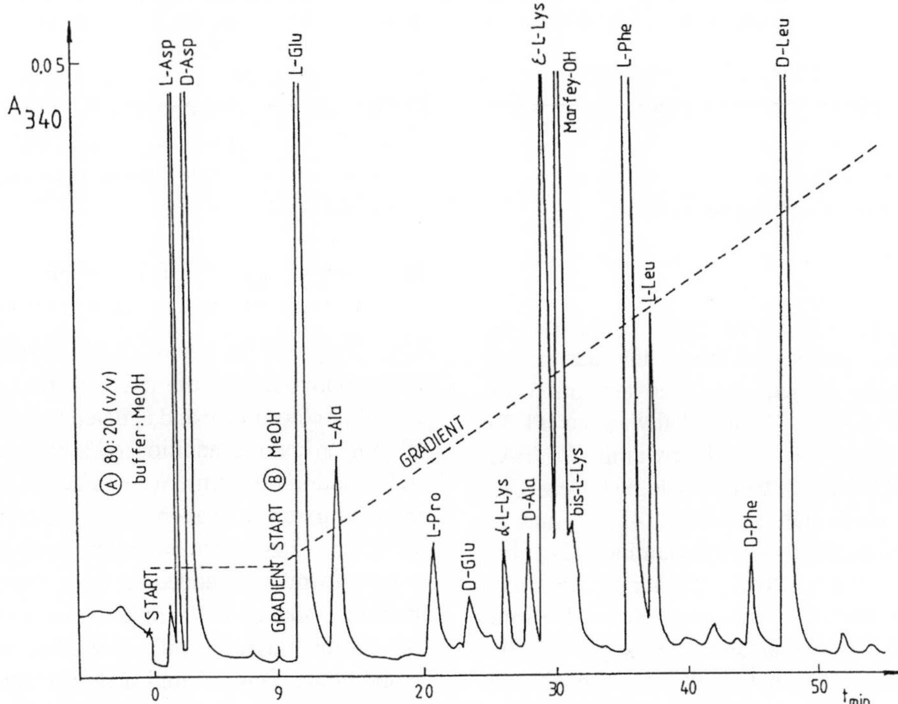


**Fig. 6.23.** HPLC separation of amino acids labelled with Marfey's reagent by gradient elution. Column: Hypersil ODS-6 (125 × 4 mm, i.d., 5 μm); eluents: A, 0.02 M sodium acetate buffer (pH 4)/methanol (4:1), flow rate, 1 ml/min; B, methanol; linear gradient atfer 9 min of 1.5 ml (B)/min. Detection, 340 nm; sample amount, 50–600 pmol of L- and D-amino acids/10 μl injection. [Reproduced from ref. 92, p. 120, Fig. 2].

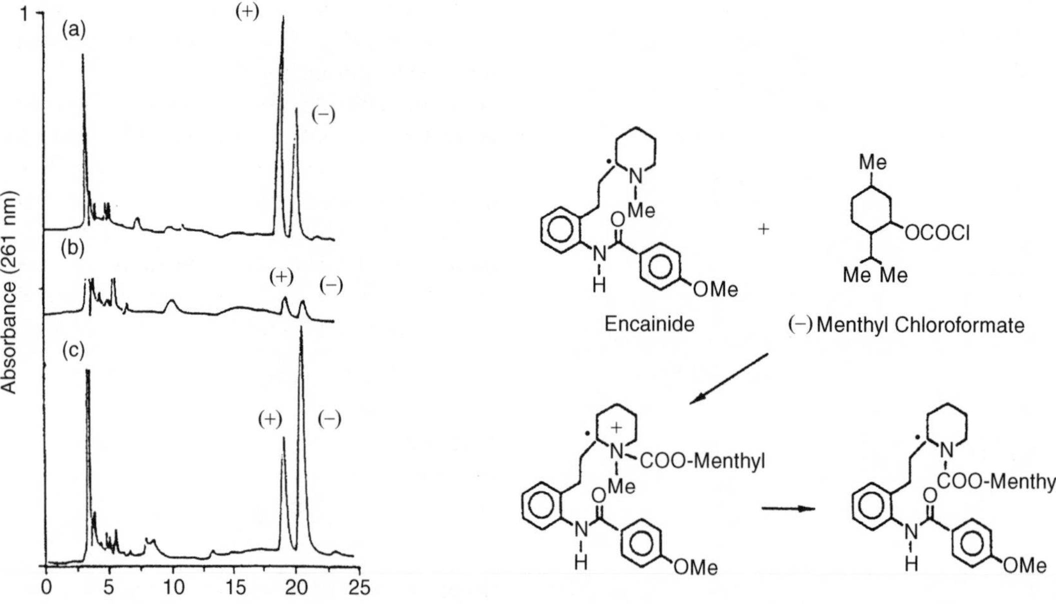


**Fig. 6.24.** Chromatograms of carbamate diastereomers of encainide and metabolites isolated from microsomal incubation of rat liver: (a) encainide; (b) NDE and (c) ODE. [Reproduced from ref. 77, p. 332, Fig. 6.].

with normal-phase chromatography on a silica column.

### 6.5.1.2 Label for Carboxylic Acids

A wide variety of amines possessing $-NH_2$ and −NH can possibly be used for the tagging of carboxylic acids. Since ester formation by direct tagging is fairly difficult, the derivatization is usually carried out after activation of carboxylic acid. Chiral 1-phenylethylamine (PEA) (39) [99–109], 1-(1-naphtyl)ethylamine (NEA) (41) [110–113] and their analogues, e.g. $\alpha$-methyl-4-nitrobenzylamine (40) [114,115], and 1-(4-dimethylamino-1-naphtyl)ethylamine (42) [116, 117] are the most popular reagents (Fig. 6.25). The diastereomers derived from the reagents of naphthalene moiety can be used to determine not only UV but also FL detection. However, the fluorescence properties not the choice for real sample analysis because of the relatively short wavelengths of excitation and emission. Other chiral amine reagents, L-alanine-$\beta$-naphtylamide (44), L-phenylalanine-$\beta$-naphtylamide [118], and L-leucinamide (43) [119–127] have been reported. Some typical examples using these reagents are described later.

**Example 4.** Tagging of arylpropionic acid with *S*(−)-1-phenylethylamine (PEA) (39) [106].

The arylpropionic acid (0.05–0.5 µmole) and 1,1-carbonyldiimidazole (52 mg) are dissolved in 2.3 ml of chloroform and agitated mechanically for 10 min. After the addition of acetic acid (10 µl) and agitation (10 min), *S*(−)-PEA (80 µl) is added and the mixture is shaken for 25 min. After 30 min reaction, 0.2 N ammonia solution (3 ml) and *n*-hexane (5 ml) are added to the solution and the mixture is shaken for 15 min. After centrifugation, the organic layer (4 ml) is washed with 0.2 N hydrochloric acid (3 ml) and centrifuged. Then 3 ml of the organic layer is evaporated under reduced pressure and the residue is dissolved in 1 ml of the mobile phase. In Table 6.3 is shown the separation of same arylpropionic acids after labelling with *S*(−)-PEA.

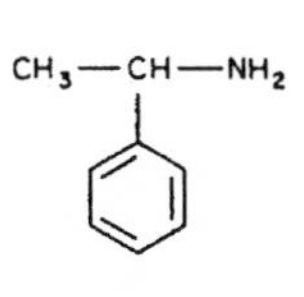

(39) 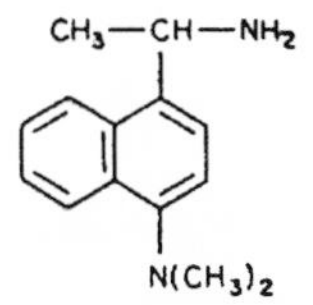 (42)

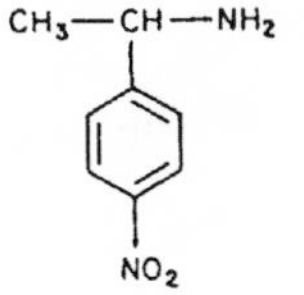

(40) 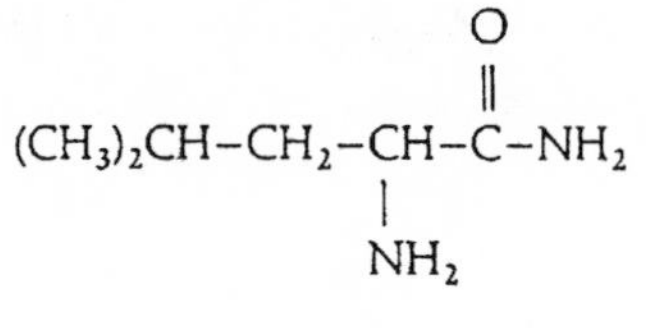 (43)

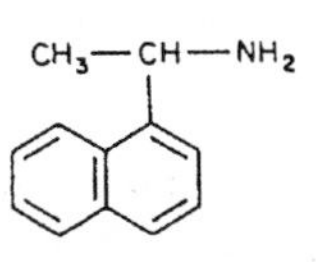

(41) 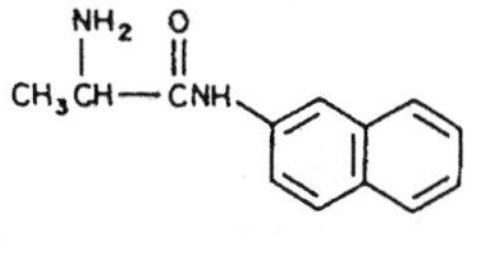 (44)

**Fig. 6.25.** Chiral UV labels for carboxyls by HPLC. (39) $S(-)/R(+)$-1-phenylethylamine (PEA); (40) $(R)$-$\alpha$-methyl-4-nitrobenzylamine; (41) $S(-)$-1-(1-naphtyl)-ethylamine (NEA); (42) (+)/(−)-1-(4-dimethylamino-1-naphtyl)ethylamine; (43) L-leucinamide; (44) L-alanine-$\beta$-naphtylamide.

### 6.5.1.3 Label for Alcohols

As described in Section 6.4.1, $\alpha$-methoxyl-$\alpha$-(trifluoromethyl)phenylacetyl chloride (MTPA) (11) is a good label of hydroxyl compounds for GC analysis, because the derivatives are highly volatile and have excellent electron capturing properties. The derivatives of PAH dihydrodiols are resolved not only by GC analysis, but also by normal-phase LC [128–131]. Asymmetric alcohols formed from anthracene 1,2-oides are also labelled with a chiral reagent, $(S)$-tetrahydro-5-oxo-2-furancarboxylic acid (TOF) (5) [22] or (−)-menthenyloxyacetic acid (MOA) (9) [132–134] (Fig. 6.13). Doolittle and Heath [22] compare the separation factor ($\alpha$) of the enantiomeric esters derived from MPTA, TOF and $(S)$-acetoxypropionic acid (AP) (1) by GC and HPLC. The results conclude that both chromatographies are similar; TOF is most suitable for acetylenic, olefinic and aromatic alcohols, whereas AP and MPTA are recommended for aliphatic alcohols and alcoholic lactones, respectively. Carbobenzoyl-L-proline is used for the formation of diastereomeric esters [135]. The determinations of warfarin and its metabolites in plasma and urine with carbobenzoyl-L-proline are shown in the following example.

**Example 5.** Determinations of warfarin and its metabolite by indirect derivatization with carbobenzoyl-L-proline (6) [135].

Carbobenzoyl-L-proline (2 mg/10 μl acetonitrile), imidazole (10 μg/10 μl acetonitrile) and dicyclo-hexylcarbodiimide (2 mg/10 μl acetonitrile) are added to the urine (or plasma) extracts in the tapered culture tubes. After 2 h reaction at ambient temperature, a 10 μl volume of ethyl acetate is added to the mixture and analysed by normal-phase HPLC with ethyl acetate-hexane-methanol-acetic acid (25:74.75:0.25:0.3) as the eluent. The silica column outlet is fitted through the polytef tubing (0.030 mm, i.d.) to a stainless steel tee, through which the post-column reagent (50% $n$-butylamine in methanol; flow rate, 0.4 ml/min) is mixed with the mobile phase. A separation chromatogram of warfarin and its metabolites in urine extract is shown in Fig. 6.26.

Although the determination of the diastereomeric ester of warfarin is possible with UV detection at 313 nm, the sensitivity is insufficient to measure the minute amounts in biological samples. In this paper, therefore, the detection by fluorescence after post-column aminolysis with $n$-butylamine is performed. The limit of determination for the enantiomer is improved to 50–100 ng levels.

### 6.5.1.4 Label for Other Functional Groups

Numerous chiral compounds are synthesized as UV labels for various functional groups. However,

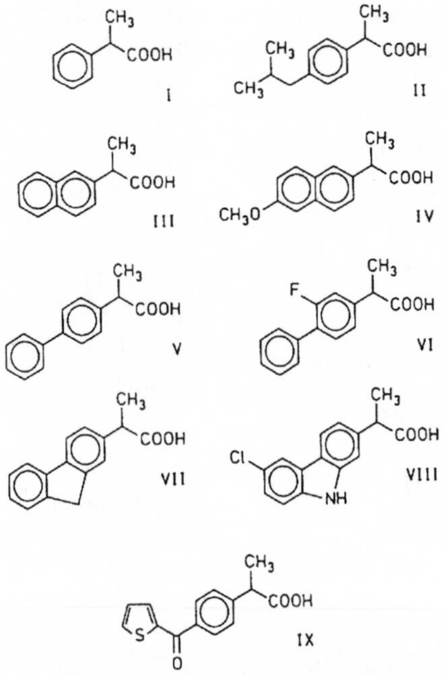

**Table 6.3.** HPLC resolution of the enantiomers of 2-arylpropionates after derivatization with $S(-)$-1-phenylethylamine (39). LiChrosorb Si 60 (10 μm), 250 mm column; mobile phase, isopropanol/cyclohexane (7.5:92.5, v/v); flow-rate, 2 ml.min

| No. | Compound | Retention time (min)* | | Resolution factor |
|---|---|---|---|---|
| | | 1st peak[†] | 2nd peak[†] | |
| I | 2-phenylpropionic acid | 2.94 | 3.81 | 2.47 |
| II | Ibuprofen | 2.40 | 3.13 | 1.46 |
| III | 2-(2-naphtyl)propionic acid | 2.86 | 3.92 | 2.91 |
| IV | Naproxen | 3.31 | 4.39 | 2.35 |
| V | 2-(4-biphenyl)propionic acid | 2.81 | 3.86 | 2.29 |
| VI | Flurbiprofen | 3.04 | 4.48 | 3.30 |
| VII | Cicloprofen | 2.84 | 3.88 | 2.74 |
| VIII | Carprofen | 3.53 | 5.18 | 2.88 |
| IX | Suprofen | 4.92 | 7.38 | 3.94 |

*Mean variation ± 0.02 min. [†]For compounds I, IV, VI and IX the (*R*-acid; *S*-amine) derivative corresponds to the first peak and the (*S*-acid; *S*-amine) derivative to the second peak. For compounds II, III, V, VII and VIII the same sequence is highly probable but was not investigated. [Reproduced from ref. 106, p. 400, Table. 1]

only a few labels for thiols and carbonyls (aldehyde and ketone) are developed for liquid chromatographic resolution. The resolution of chiral thiols by HPLC is successfully carried out with GITC (25) which is a good reagent for primary and secondary amines. Generally, the resulting dithiocarbamates are less stable than thiocarbamoyl derivatives. Recently Lindner *et al.* [98] synthesized a new NCS type reagent (DDITC) (28) for amine and thiol labeling (Fig. 6.21). The separation of the resulting dithiocarbamates derived from the reagent is much better than those with GITC (25). For the derivatization of aldehydes and ketones, the chiral compounds having hydrazino ($-NHNH_2$) and primary amino ($-NH_2$) groups are usable, and produce the corresponding hydrazones and oximes [40,42,136,137].

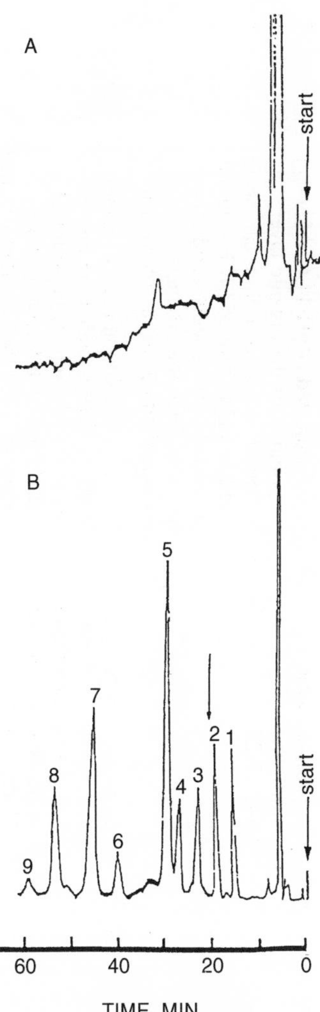


**Fig. 6.26.** Chromatogram resulting from extraction and derivatization with carbobenzyloxy-L-proline of a control urine sample (A), and urine sample (0.5 ml) obtained from a subject between 4 and 5 days after oral administration of *RS*-warfarin (1.5 mg/kg) (B): (1) and (2) internal standards; (3) (*SR*)-alcohol; (4) (*SS*)-alcohol; (5) (*RS*)-alcohol; 6-(*S*)-hydroxywarfarin; (7) 7-(*S*)-hydroxywarfarin; (8) 6-(*R*)-hydroxywarfarin; (9) 7-(*R*)-hydroxywarfarin; For (A), a sensitivity setting of 0.2 μA; for (B), the arrow indicates a change of setting from 1 to 0.2 μA. [Reproduced from ref. 135, p. 1395, Fig. 5.].

Table 6.4 shows the chiral UV-VIS labels and some typical examples using these reagents by HPLC.

### 6.5.2 Label for Fluorescence (FL), Laser-Induced Fluorescence (LIF) and Chemiluminescence (CL) Detection

As fluorometry is both sensitive and selective, a large number of papers concerning fluorescence tagging have been reported. However, the fluorescence properties of the substances tend to be greatly affected by temperature, viscosity of the solvent, pH of the medium, and the contamination of halide ions such as $Cl^-$ and $Br^-$. It should be also noted that undesirable fluorescence materials which contaminated the tested samples, especially in biological specimens, affect the determination. The FL label is the most effective for determinations in biological specimens, in terms of sensitivity and/or selectivity. However, different types of FL labelling reagents have been developed for the enantioseparations of drugs and biologically important materials.

Fluorogenic reactions can be classified into the following two types: (i) fluorescence generation, and (ii) fluorescence labelling. Pre-column and post-column derivatizations fall into two separate categories due to the reaction timing before or after the separation. The pre-column method is used more frequently than the post-column derivatization for sensitive chiral separation by HPLC. It is ideal if the chiral derivatization reagents possess the following characteristics: (1) The reagent has high purity chemically and optically. (2) The reagent is stable and does not racemize during storage. (3) Racemization is negligible under the derivatization reaction conditions. (4) The reaction rates and yields are essentially the same for each enantiomer, this is generally not the case for many of the reagents reported. (5) The resulting diastereomers exhibit an adequate detector response for sample analysis. These items are of general importance for all chiral tagging reagents, not only for FL but also UV-VIS.

The fluorescence detection in HPLC provides excellent sensitivity and selectivity. In real sample analysis, however, it is observed that the sensitivity is not good enough for trace level analysis. To solve this, laser-induced fluorescence (LIF)

**Table 6.4A.** UV-VIS chiral derivatization reagents for amines

| Reagent | Enantiomer | Wavelengths (nm) | HPLC condition | Sample and treatment | Reference |
|---|---|---|---|---|---|
| (−)-α-Methoxy-α-(trifluoromethyl)-phenylacetyl chloride (MTPA-Cl) | Amphetamine | 220 or 254 | RP ODS (250 × 4.6 mm, i.d., 5μm); methanol/water (60:40) for 20 min, followed by a gradient elution to 100% methanol at 40 min; 1–2 ml/min. | To 0.1 mg amphetamine dissolved in 0.2 ml of methylene chloride is added 0.1 ml of the reagent (0.5 mmol/ml in methylene chloride and 50μl of pyridine. The solution is heated at 70 °C for 30 min. | [61] |
| (*R*)-1-Phenylethyl isocyanate (PEIC) | Tertiary amines | 254 | LiChrosorb Si 60 or Spherisorb silica gel (250 × 4.6 mm, i.d., 5 μm); *n*-hexane/ethyl acetate (or isopropanol); 1.0 ml/min. | To the secondary amines obtained from *N*-dealkylation reaction is added a five-fold excess (*R*)-PEIC in methylene chloride and allowed to stand at RT for 30 min. | [71] |
| (+)-1-(1-Naphtylethyl isocyanate (NEIC) | Amino acids | 222 | Octadecyl Si 100 Polyol (250 × 4.6 mm, i.d.); acetonitrile/50 mM ammonium acetate (pH 6.2) (or 0.1 M acetic acid); 1 ml/min. | 0.1 ml of the reagent is added with 0.05 ml of amino acid (lower than 5 mM) and sodium borate buffer (pH 9.0). After mixing, the mixture is stood for 5 min at RT. | [74] |
| (−)-Menthyl chloroformate | Atenolol, β-Blockers | 195 | Partisil 5 ODS3 (100 × 4.6 mm, i.d., 5 μm); water/aceto-nitrile/methanol (43:35:22); 1.2 ml/min. | The drug and IS are extracted with ethylacetate (4 ml) from plasma (1 ml) adjusted at pH 11.6. After evaporation of the organic layer, 0.2 ml of saturated sodium carbonate solution and 0.2 ml of MCF solution are added to the residues and mixed for 30 s. Then the product is extracted with 2 ml of chloroform. | [78] |
| 4-Nitrophenylsulfonyl-L-prolyl chloride | Amphetamine, ephedrine | 254 | Zorbax-Sil or Supelcosil LC-Si (70 × 2.1 mm, i.d., 5 μm); chloroform/*n*-heptane; 1.4 ml/min; Zorbax ODS (150 × 4.6 mm, i.d., 5 μm); methanol/water; 1.5 ml/min. | The reagent solution (11–13 mmol) in 40 ml THF is added the mixture of amine (13–15 mmol) in 40 ml TFA and 150 ml of 10% potassium carbonate. Then the solution is heated at approximately 50 °C for 3 h, and maintained at pH 8 or above. | [86] |

| | | | | | |
|---|---|---|---|---|---|
| *S*-Flunoxaprofen chloride | Amino acid esters | 305 | Zorbax-Sil (250 × 4.6 mm, i.d., 5 µm); *n*-hexane/dichloromethane/ethanol/*n*-propanol; 1.5 ml/min. | To amino acids esterified with 3M HCl in 2-propanol is added *S*-flunoxaprofen chloride (at least 3-fold more excess) in anhydrous dichloromethane or ethyl acetate, together with 20 mg of anhydrous sodium carbonate, and then heated at 40 °C for 1 h. | [245] |
| (*S*)-2-Methoxy-2-phenylacetic acid *N*-succinimidyl ester (SMPA) | Primary amines | 254 | Rainin silica gel (250 × 4.6 mm, i.d., 5 µm); hexane/ethyl acetate (4:1); 1.0 ml/min. | The reagent (2–10 mg) is reacted with amines (1 mg) in 1 ml of TFA and 1 ml of water for 15 min on a steam bath. | [76] |
| (−)-α-Methoxy-α-methyl-1-(or-2-) naphthaleneacetic acid | Amino acids | 280 | µPorasil or µBondapak C18 (300 × 3.9 mm, i.d., 5 µm); cyclohexane/ethyl acetate or methanol/water; 1 ml/min. | To amino acid methyl esters (ca. 0.1 mg) in pyridine (0.2 ml) are added the reagent and *N*,*N*-dicyclohexyl-carbodiimide (ca. 2 mg). The mixture is reacted at room temperature for 30 min. | [246] |
| (−)-1,7-Dimethyl-7-norbornyl isothiocyanate or (+)-Neomenthyl isothiocyanate | Amino acid methyl esters | 243 | µPorasil (300 × 3.9 mm, i.d.); cyclohexane/ethyl acetate; 1.0 ml/min. | Amino acid methyl esters (ca. 0.1 mg), reagent (ca. 2mg) and sodium acetate (ca. 2mg) added in 0.2 ml of acetonitrile are allowed to stand at 37 °C for 1 hr. | [247] |
| Marfey's reagent ant its analogues | Amino acids | 340 | Spherisorb ODS 2 (250 × 4 mm, i.d., 5 µm); gradient elutions; 1 ml/min. | 25 µl (5 µmol) of DL-amino acid standard solution, 65 µl (5 µmol) of 1M $NaHCO_3$, and 50 µl (5 µmol) of reagent solution are heated at 40 °C for 1 hr. | [91] |
| (*R*,*R*)-*O*,*O*-Diacetyltartaric acid anhydride (DATAAN) | β-Blockers | 254 | Spherisorb ODS (250 × 4.6 mm, i.d., 5 µm); 2% acetic acid (pH 3.7)/methanol (50:50); 1.5 ml/min. | β-Blockers (ng-mg) dissolved in dry aprotic solvent (e.g. dichloroethane, tetrahydrofuran, acetone) and excess of TCA are added to DATAAN (at least 3-fold excess). Then the mixture is reacted at 50 °C for several hours. | [96] |
| 2,3,4,6-Tetra-*O*-acetyl-β-D-gluco-pyranosyl isothiocyanate (GITC) | Pindolol | 258 | TSK-gel ODS 120A (250 × 4.6 mm, i.d., 5 µm); acetonitrile/10 mM phosphate buffer (pH 3.4); 0.9 ml/min. | To the residues of diethylether extracts of plasma or supernatant from heart or lung homogenate alkalinized are added 50 µl of 2% GITC in acetonitrile containing 64 ng of 1-nitronaphthalene (IS). At the termination of the reaction, 10 µl of 0.2% hydrazine in acetonitrile are added to discard the unreacted GITC. | [56] |

*continued overleaf*

**Table 6.4A.** *(continued)*

| Reagent | Enantiomer | Wavelengths (nm) | HPLC condition | Sample and treatment | Reference |
|---|---|---|---|---|---|
| 2,3,4-Tetra-*O*-acetyl-α'-D-arabino-pyranosyl isothiocyanate (AITC) | Amphetamines, | 220,254 | RP18 (250 × 4.5 mm, i.d., 5 μm); methanol/water (55:45); 1.0 ml/min. | To amphetamine dissolved in 50 μl of methylene chloride is added 10% molar excess AITC (50 μl, 7.6 μmol/ml in acetonitrile or methylene chloride containing 0.2% TEA). The solution is allowed to stand at room temperature for 1 hr. | [61] |
| (*R*)-1-Phenylethyl isothiocyanate (PEI) | Propranolol, ephedrine, amino acids | 254 | Ultraspher ODS (150 × 4.6 mm, i.d., 5 μm); acetonitrile (or Methanol)/20 mM monobasic ammonium phosphate; 1.0 ml/min. | Sample and reagent (2 mg) dissolved in 0.2 ml acetonitrile are reacted for 30 min. The solution is diluted 50 times with mobile phase. | [65] |
| (*R,R*)-*N*-[(2-Isothiocyanato)-cyclohexyl]-(3,5)-dinitro-benzoylamide (DDITC) | β-Blockers | 254 | Lichrospher 60 RP select B 125 × 4 mm, i.d., 5 μm); acetonitrile/20 mM ammonium acetate (pH 7.3) (55:45); 1 ml/min. | To amino compound dissolved in 0.5 ml of acetonitrile is added 0.05 mM DDITC in 5 ml acetonitrile. Then the mixture is heated at 60 °C for 120 min. | [98] |
| (−)-1-α-Methoxy-α-methyl naphalene acetic acid *N*-succinimidyl ester | 2,5-Dimethoxy-4-methylam-phetamine | 280 | μPorasil (300 × 3.9 mm, i.d.); cyclohexane/ethyl acetate (3:1); 1.0 ml/min. | Plasma samples (0.5–2 ml) neutralized are passed through Sep-pak C18 cartridge. After washing with water (3 ml) and 10% ethanol (2 ml), the analyte is eluted with 1% methylamine/80% ethanol (4 ml). The residues evaporated is redissolved in 90% ethanol (0.5 ml) and applied to CM-LH-20 column (0.1 g, 10 × 6 mm, i.d.). After removal of neutral and acidic materials, the fraction eluted with 6% methylamine/90% ethanol (3 ml) is collected and dried. The residue is heated with the reagent (0.2 mg) in pyridine (0.2 ml) at 60 °C for 2 h. Then the derivatives are extracted, dried and redissolved in cyclohexane/ethyl acetate (3:1) (0.15 ml). | [248] |

**Table 6.4B.** UV-VIS chiral derivatization reagents for carboxyls and their application

| Reagent | Enantiomer | Wavelengths (nm) | HPLC condition | Sample and treatment | Reference |
|---|---|---|---|---|---|
| L-Leucinamide | Arylpropionic acids | 254,280 and 309 | Zorbax ODS (250 × 4.6 mm, i.d., 5μm); acetonitrile/10 mM phosphate buffer (pH 6.5) (55:45); 1.2 ml/min. | To 5 μg of drug is added 0.1 ml of TEA in dried acetonitrile (50 mM) and 50 μl of ethyl chloroformate in dried acetonitrile (60 mM). After 2 min, 50 μl of the reagent (1M) and TEA (1M) in methanol are added and reacted for 3 min, and then the reaction is stoped with the addition of 0.2 ml of 0.25 M HCl. | [119] |
| *S*(−)/*R*(+)-1-Phenylethyl-amine (PEA) | Prostaglandin | 214 | Zorbax Sil (250 × 4.6 mm, i.d., 5 μm); hexane/dioxane/water (80:20:0.16); 2 ml/min. | The sample is converted to the mixed anhydride at −20 °C using 0.4 μl isobutylchloroformate in the presence of an equimolar amount of TEA. After 30 min, 0.7 μl of (*S*)(−)-PEA is added and allowed to stand at RT for 15 min. | [105] |
| L-Alanine *β*-naphthylamide | Carboxylic acids | 254 | Zorbax Sil (150 × 4.6 mm, i.d., 5 μm); chloroform (or *n*-hexane)/methanol; 1 ml/min. | The sample (2 mg) and the reagent (20 mg), dissolved in 50 μl methanol, and *N*,*N*-dicyclohexylcarbodiimide in 2 ml chloroform (5 mg/min) are allowed to stand at RT for 1 h. | [118] |
| *S*(−)-1-(1-Naphtyl)ethylamine (NEA) | Ibuprofen | 254 | Hypersil (250 × 4.6 mm, i.d., 10 μm); hexane/ethyl acetate (4:1); 3.2 ml/min. | To the ibuprofen and internal standard, extracted from plasma, are added 1-hydroxybenzotriazole, 1-(3-dimethyl-aminopropyl)-3-ethyl-carbodiimide hydrochloride and the reagent (0.1 mg of each as 0.1 ml of 1 mg/ml in dichloromethane). The mixture is allowed to stand at RT for 2 h. | [111] |

*continued overleaf*

**Table 6.4B.** *(continued)*

| Reagent | Enantiomer | Wavelengths (nm) | HPLC condition | Sample and treatment | Reference |
|---|---|---|---|---|---|
| *S*(−)/*R*(+)-α-Methyl-4-nitro-benzylamine | Isoprenoid acids | 254 | Silica-gel (500 × 4 mm, i.d., 10 μm); THF/*n*-heptane (2:8); 1.5 ml/min. | The sample converted to the acid chloride by refluxing with oxalyl chloride is reacted with optically pure reagent. | [115] |
| (1*R*,2*R*)-(−)-1-(4-Nitrophenyl)-2-amino-1,3-propanediol | Carboxylic acids | 220 | Partisil 10 ODS (250 × 4.6 mm, i.d.); acetonitril/methanol/water/85% phosphoric acids; 1.3 ml/min. | The mixture of carboxylic acid (ca. 50 μmol), the reagent (50 μmol) and DCC (50 mmol) in 3 ml of methanol is allowed to stand for 15 min. | [249] |
| D/L-*O*-(4-Nitrobenzyl)tyro-sine methyl ester | *N*-Protected amino acids | 270 | LiChrosorb Si 60-10 (250 × 4.6 mm, i.d., 5 μm); gradient elution, hexane/2-propanol; 2 ml/min. | Sample (ca. 0.01 mmol) is dissolved in 1 ml of the reagent enantiomer and 1 ml of dicyclohexylcarbodiimide. The mixture is allowed to stand at RT for 1 h. | [250] |

detection is adopted. Today, various laser sources such as Ar-ion, He-Cd ion and semiconductor laser are commercially available. The minimum detectable concentrations with LIF are typically one to five orders of magnitude lower than those of FL and UV detection. The excitation wavelengths which can possibly be used are limited by each laser source selected, which is an important disadvantage of LIF detection. However, it may be improved with the development of various laser sources and tagging reagents matched to the wavelengths.

It is well known that some fluorescent compounds emit light during chemical reactions without the need of optical excitation with lamps such as a xenon arc. Since the flicker noise based on the lamp is negligible, extremely high sensitivity is theoretically obtained by this method. Indeed, trace analysis at attomole levels has been achieved with the technique in the reaction of CL reagents such as luminol, lucigeninm and the combination of oxalates and hydrogenperoxide ($H_2O_2$). In the HPLC-CL detection system, the chemiluminescence reagents are mixed with eluate after column separation and reacted just before the detector, because the CL reaction is very fast and the light generated immediately disappears after a few seconds. The CL method can also be applied to the detection of distereomers derived from fluorescent chiral tagging reagents.

Enantioseparations of chiral molecules possessing various functional groups as the reactive site are described in the following sections. Only the tagging reaction with fluorescent chiral reagents are listed in here. In other words, the examples using fluorescent reagent or fluorescence generation are described in the following sections. Even though the analytes are carried out with fluorescence detection, the reagents without fluorescence itself and a fluorescence generator are not covered in this section.

#### 6.5.2.1 Label for Primary and Secondary Amines Including Amino Acids

Many biochemically important compounds, such as biogenic amines, amino acids and drugs, have at least one amino functional group in the structure. Therefore, the resolution of chiral amines is one of importance. *o*-Phtalaldehyde (OPA) (45) has been widely used for the fluormetric determination of amines in the presence of thiol such as 2-mercaptoethanol (2-ME). The derivatization reaction essentially requires the thiol compound to produce the corresponding isoindole. The OPA method is extended to enantioseparation of chiral primary amines and amino acids. When a chiral mercaptan, instead of the achiral thiol, is used together with OPA for the labelling of racemic amines, the resulting diastereomers are easily separated with achiral phase chromatography. *N*-Acetyl-L-cysteine (NAC) (47) [138–149], *N*-tert.-butyloxycarbonyl-L-cysteine (BOC-C) (46) [142,143,150], *N*-acetyl-D-penicillamine (NAP) (48) [140–143], 1-thio-*β*-D-glucose (TG) (49) [141,151], 2,3,4,6-tetra-*O*-acetyl-1-thio-*β*-D-glucopyranoside (TATG) (50) [141,151] and *N*-isobutylyl-L(or D)-cysteine [152] have been used as the chiral mercaptans of the OPA method for the resolution of primary amine enantiomers (Fig. 6.27). The naturally occurring compounds like L-cysteine derivatives are expected to be ideal as the chiral mercaptan because they are available in optically pure form. Actually, the purity of BOC-C was calculated to be 99.8%, judging from the apparent D-amino acid diastereomers obtained from the reaction with L-amino acids. The diastereomers of 21 pairs of amino acids derived from OPA/BOC-C can be resolved by reversed-phase chromatography with linear gradient elution. However, proline does not react with OPA/BOC-C owing to the lack of a primary amino group, and cysteine forms only a weakly fluorescent derivative. On the other hand, this reaction is important for the determination of lysine enantiomers. The L-enantiomers elute before the corresponding D-antipode with all of the amino acids tested. Although the separation of amino acid enantiomers is also obtained with other chiral mercaptans, such as NAC and NAP, the elution order depends upon the hydrophobility of amino acids. The detection wavelengths of the derivatives are around 345 nm and 445 nm, respectively, and the minimum detectable amount is at the sub-pmol level. Nimura and Kinoshita

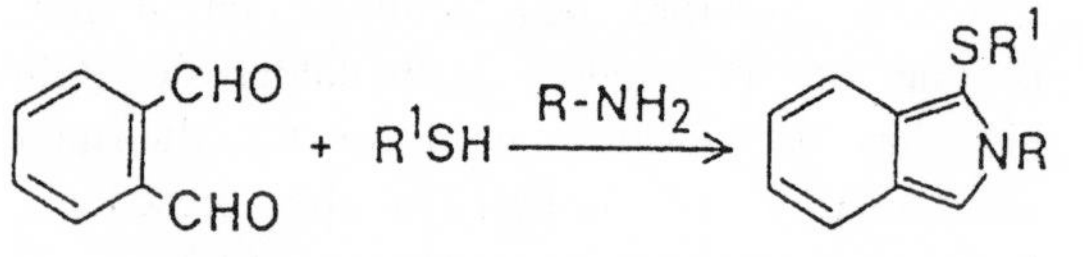

**Fig. 6.27.** Structures and reaction of OPA/chiral thiols. (45) *o*-phtalaldehyde (OPA); (46) *N*-tert.-butyloxycarbonyl-L-cysteine (BOC-C); (47) *N*-acetyl-L-cysteine (NAC); (48) *N*-acetyl-D-penicillamine (NAP); (49) 1-thio-$\beta$-D-glucose (TG); (50) 2,3,4,6-tetra-*O*-acetyl-1-thio-$\beta$-D-glucopyranoside (TATG).

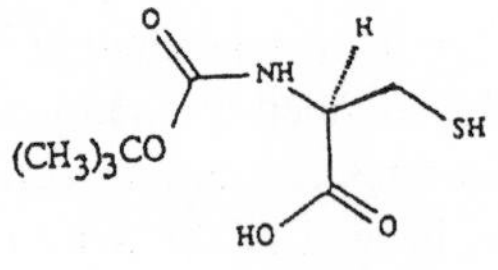

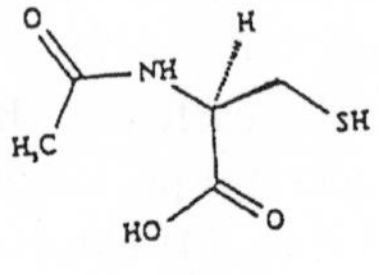

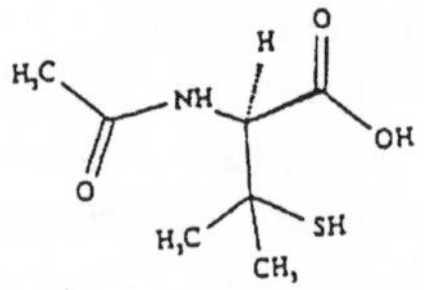

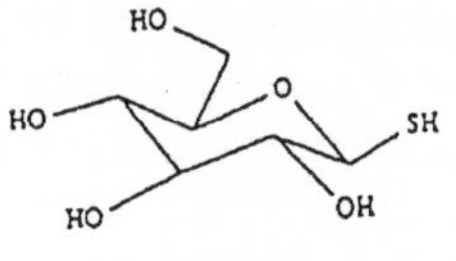

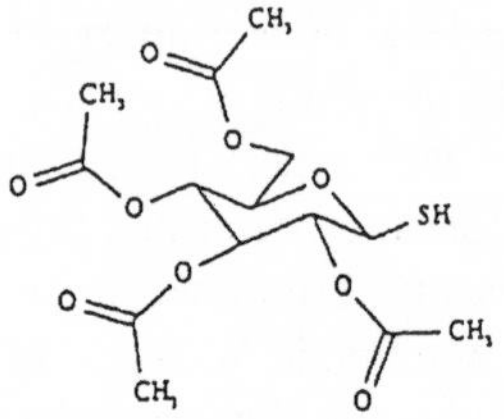

[144] obtained adequate separation of the common protein amino acid enantiomers within 70 min in a single chromatographic run by a gradient elution. Those reagents may be given some preference as they are commercially available. Numerous amines, e.g. biogenic amines (norepinephirine and norephedrine), amino acids, amino alcohols and primary-amine drugs have been resolved with combination of OPA and a chiral mercaptan. $\omega$-*N*-Oxalyl diamino acids [143], lombricine [142], tranylcypromine [145], and baclofen [146] were analysed by pre-column labeling with OPA/chiral thiols. Desai and Gal [141] report the enantioseparations of various primary amines and drugs [($\pm$)-*p*-chloroamphetamine (PCA), ($-$)-amphetamine (AMP), 3-amino-1-(4-hydroxyphenyl)butane (AHB), ($\pm$)-1-methyl-3-phenylpropylamine (APB), ($\pm$)-*p*-hydroxyamphetamine (HAM), rimantadine (RIM), tocainide (TOC), and mexiletine (MEX)]. The resolution (Rs) values depend upon the type of chiral thiols and the primary amines. Relatively good separations were obtained by use of TG or TATG. Brückner *et al.* [153] systematically investigate the chromatographic resolvability of D/L-amino acid diastereomers derived from OPA together with chiral *N*-acetylated cysteines. From their results, *N*-isobutylyl-L-cysteine (IBLC) and *N*-isobutylyl-D-cysteine (IBDC) are selected as good chiral thiols for OPA. A 41-component standard containing seventeen protein amino acids was satisfactorily resolved by the fully automated pre-column derivatization technique. The present method is applied to the determination of D-amino acids in various fields of biosamples such as bacteria, microfungi, plants and vertebrates.

**Example 1.** Enantioseparations of D/L-amino acids labelled with OPA/IBLC or OPA/IBDC [153].

To human urine (400 μl) provided by a healthy volunteer is added 30% (w/v %) aqueous 5-sulphosalicylic acid (50 μl) and 0.52 mM L-*homo*-Arg (50 μl). The mixture is centrifuged at 6000 xg and 2 μl aliquots of the supernatants are subjected to the HPLC system. The chromatograms obtained from standard amino acids and human urine are shown in Fig. 6.28 and 6.29, respectively.

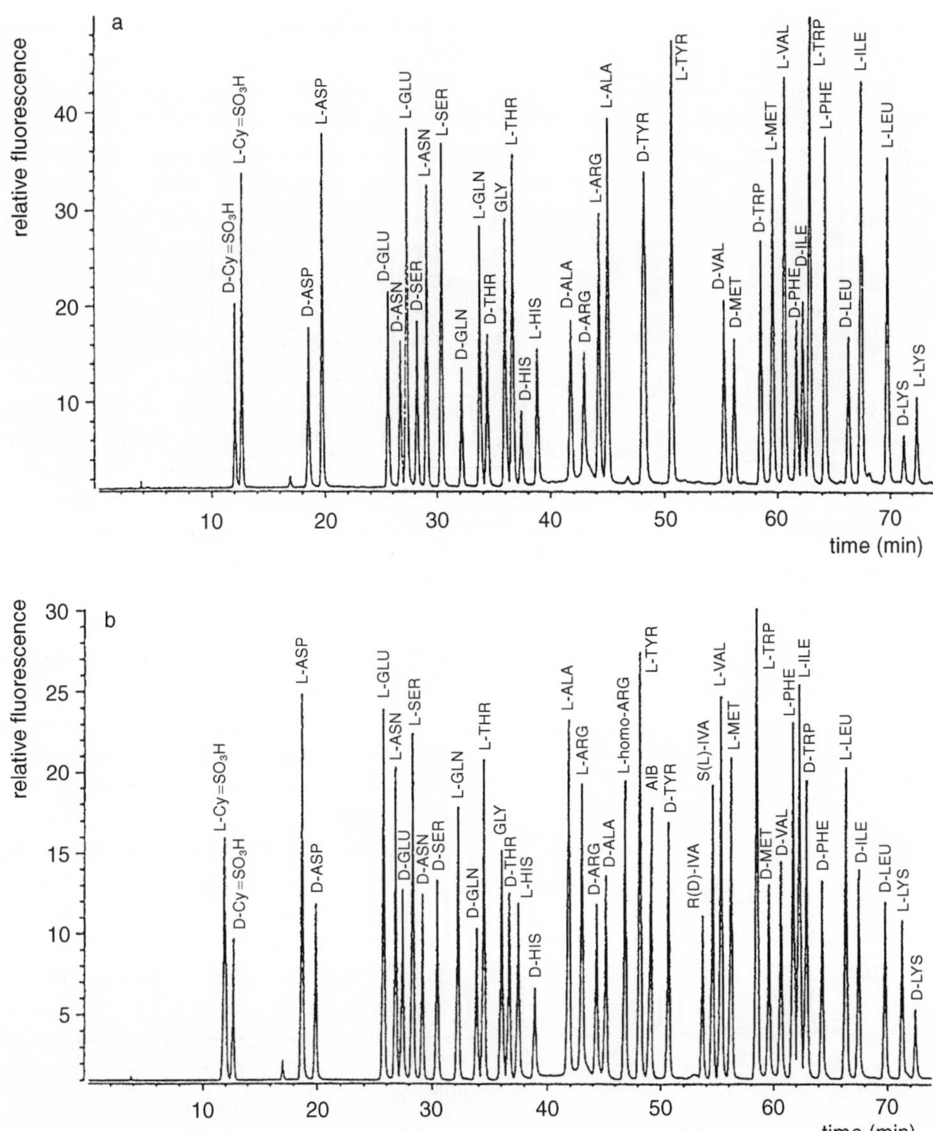


**Fig. 6.28.** Elution profiles of standard amino acids (100 pmol L-amino acids and 50 pmol D-amino acids and glycine) derived from (a) OPA/IBDC and (b) OPA/IBLC. [Reproduced from ref. 153, p. 263, Fig. 1.].

Replacement of OPA/IBLC with OPA/IBDC leads to a reversal in the elution order of the derivatives of D/L-amino acids.

Einarsson *et al.* [154] developed (+)-1-(9-fluorenyl)ethyl chloroformate [(+)-FLEC] as a fluorescent chiral tagging reagent for primary and secondary amines involving amino acids. The reaction conditions at room temperature in basic solution are mild and the resulting fluorescent products are stable. The rates of reaction with aromatic amino acids are faster than those with acidic amino acids, i.e. aspartic and glutamic acids. Good separation is obtained with increased hydrophobicity of the derivatives. The bulky and

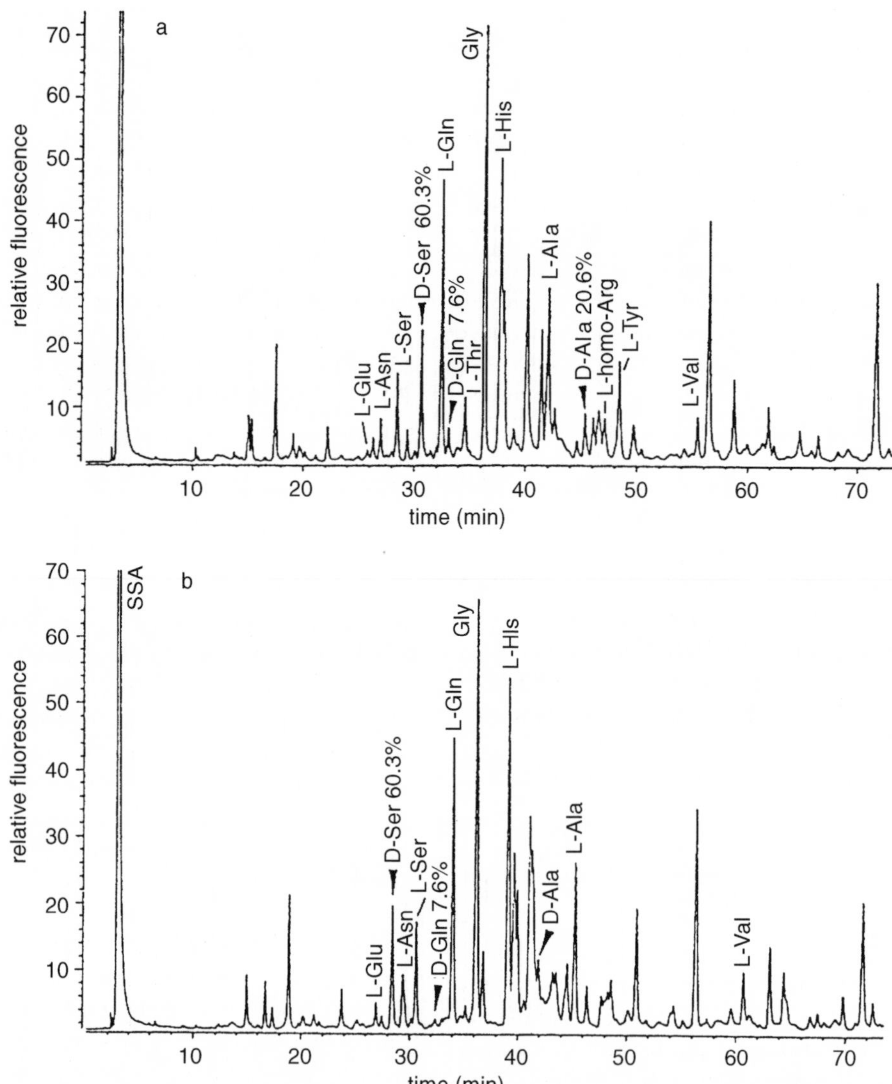


**Fig. 6.29.** Elution profiles of amino acids from human male urine derived from (a) OPA/IBLC and (b) OPA/IBDC. [Reproduced from ref. 153, p. 271, Fig. 10.].

planar fluorene moiety in the reagent structure seems to play an important role in the resolvability of the diastereomers. The D-enantiomers eluted consistently before L-enantiomers of all of the amino acids. The elution order is a distinct advantage because the L-form is the dominant component in most amino acid samples. Another advantage is the thermal stability of the carbamates derived from (+)-FLEC. The enantioseparations of pharmaceuticals in synthetic mixtures and biological fluids is equally important to the resolution of amino acids and biogenic amines. Enantioseparations of chiral amines using FLEC are carried out with (+)-enantiomer in most of the reports. In the case of separation of biogenic amines and drugs, however, the elution of the main component may be faster than the trace amount of the antipode. Since (−)-FLEC shows similar optical purity and reactivity, it can also be used as readily as (+)-FLEC for the chiral separation of amine enantiomer with a reversal of the elution order [155]. Therefore, trace quantities of the D- or L-enantiomer may be determined with a switch in the elution order by selecting the appropriate enantiomer of FLEC reagent. The following cardiovascular drugs, i.e. atenolol [156], propranolol, metoprolol, $\alpha$-hydroxymethoprolol, tocainide, (RS)-2-[(RS)-$\alpha$-(2-ethoxyphenoxy)benzyl]morpholine methanesulfonate (reboxetine) [157], (RS)-1-methyl-8-[morpholin-2-yl)methoxy]-1,2,3,4-tetrahydroquinoline [158], are separated by reversed-phase HPLC. The sensitivity of methamphetamine derivative with fluorescence detection is ca. 200 times higher than that of UV detection at 254 mm [159]. Simultaneous assays of the diastereomers of D- and L-carnitines derived from (+)-FLEC are performed, not only with HPLC but also with capillary zone electrophoresis (CZE) [155,160]. Chou *et al.* [161] synthesized solid-phase reagents, i.e. FMOC-L-proline and FMOC-L-phenylalanine, which have 9-fluorenylmethyl moiety, and applied then to the separation of a pair of methylbenzylamine enantiomers.

**Example 2.** Separation of amino acid enantiomers after derivatization with (+)-FLEC [154].

For derivatization of primary and secondary amino acids, 0.5 ml of FLEC reagent (15 mM in acetone/acetonitrile, 3/1) is added to a 0.4 ml sample solution dissolved in 0.1 ml of borate buffer (1 M pH 6.85). After 4 min reaction, the mixture is extracted with pentane to remove excess reagent. The aqueous phase is injected onto the HPLC column.

For selective derivatization of secondary amino acids, 0.1 ml of OPA reagent (50 mg OPA and 25 µl mercaptoethanol in acetonitrile) is added to a 0.9 ml sample solution dissolved in 0.1 ml of borate buffer (0.8 M pH 9.5), and allowed to stand for 30 s. Then 0.1 ml iodoacetamide (140 mg/ml in acetonitrile) is added and 30 s later 0.3 ml of FLEC (5 mM in acetone) is added. After 2 min reaction, the solution is extracted with pentane and the organic phases are discarded. The aqueous phase is injected onto the HPLC column. Figures 6.30 and 6.31 are chromatograms obtained from standard solutions of amino acids and secondary amino acids.

FLEC and OPA/optically active thiol are categorized as reagents (fluorescence generaters) that yield fluorescence products. However, chiral derivatization using a fluorescent reagent is also used as an indirect method for the enantioseparations of chiral molecules. The reagents having the some naphthalene structure as the fluorophore, e.g. 1-(1-naphthyl)ethyl isocyanate [*R*(−)- or *S*(+)-NEIC] (14) [162–168] and (−)-$\alpha$-methoxy-$\alpha$-methyl-1- (60) or 2-naphthaleneacetic acids (MMNA) (61) [169], have been used for the chiral separation of serotonin reuptake inhibitor (fluoxetine) and its desmethyl metabolite (norfluoxetine) [165], calcium antagonist prenylamine [*N*-(3,3-diphenylpropyl)-*N*-($\alpha$-methylphenethylamine)] [72], and $\beta$-adrenoceptor blocking agents, such as propranolol [166], betaxolol [163] and nadolol [162,164]. Although these chiral compounds are determined after tagging with *R*(−)-NEIC, the opposite enantiomer [*S*(+)-NEIC] is also used for the separation of both enantiomers of acebutolol [73]. This tagging is based upon the reaction of a primary or secondary amine with an isocyanate moiety in the reagent. The isocyanate type reagent also reacts with hydroxyls and thiols. Compared with amines, however, the reaction with alcohols requires rather drastic conditions. Although thiol compounds react easily with the reagent, the resulting diastereomer is not stable in many samples. It is advantageous that both enantiomers of NEIC are commercially available in pure forms. The resulting derivatives, like those of phenylethylisocyanate (PEIC) (13) [170], are sometimes determined by UV detection. The fluorometric detection based on naphthyl moiety of NEIC is more sensitive. Since the excitation and emission of the amine derivatives are in short the wavelength region (less than 330 nm), there may be interference with native substances in the sample fluids, which absorb or emit light in the same range.

Toyo'oka *et al.* have developed fluorescent chiral tagging reagents with the benzofurazan

FLEC

Derivatization reaction of D/L-amino acids with FLEC.

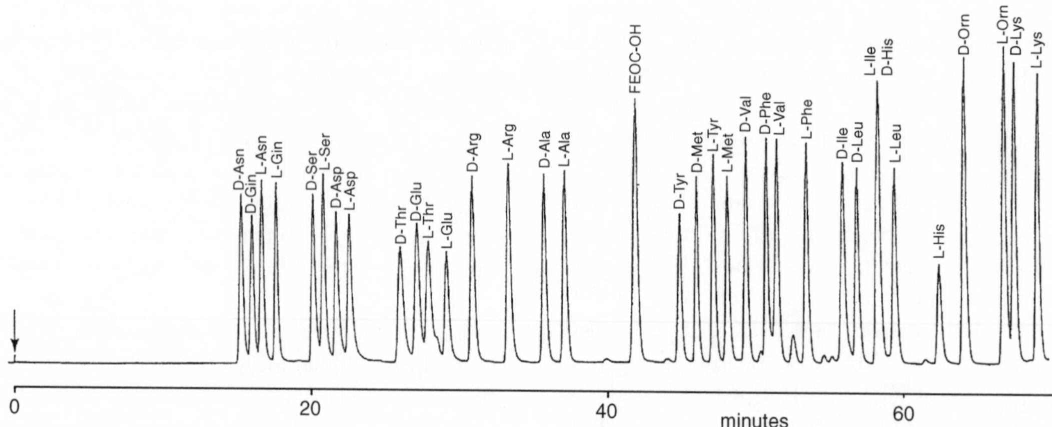


**Fig. 6.30.** Separation of standard L- and D-amino acids derived from (+)-FLEC. [Reproduced from ref. 154, p. 1194, Fig. 5.].

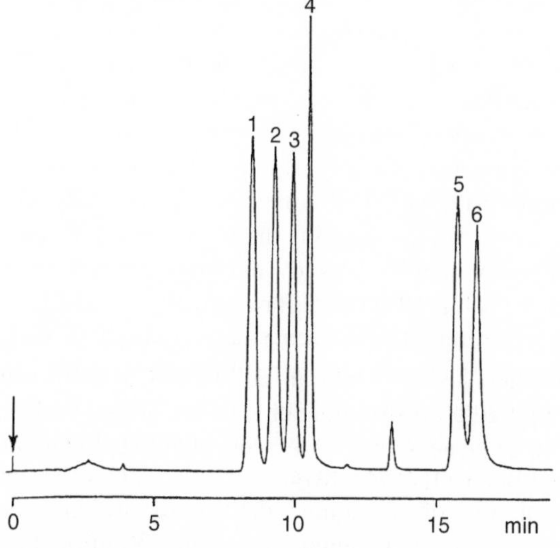


**Fig. 6.31.** Separation of proline derivatives obtained from racemic FLEC reagents. Peaks: (1) trans-4-hydroxy-L-proline; (2) trans-4-hydroxy-D-proline; (3) cis-4-hydroxy-L-proline; (4) cis-4-hydroxy-D-proline; (5) L-proline; and (6) D-proline. [Reproduced from ref. 154, p. 1193, Fig. 2.].

(2,1,3-benzoxadiazole) structure to permit enantioseparation of various types of racemates involving amino, carboxyl, carbonyl and hydroxyl functional groups (Fig. 6.32). While working on the development of chiral tagging reagents, optically active fluorescent 'Edman-type' reagents, i.e. 4-(3-isothiocyanatopyrrolidin-1-yl)-7-(*N*,*N*-dimethylaminosulfonyl)-2,1,3-benzoxadiazole [*S*(+)- and *R*(−)-DBD-PyNCS] (59) and 4-(3-isothiocyanatopyrrolidin-1-yl)-7-nitro-2,1,3-benzoxadiazole [*S*(+)- and *R*(−)-NBD-PyNCS] (58) are synthesized [171]. The usefulness of the resolution of various racemic amines including $\beta$-blockers has been demonstrated with the reagents. The thiocarbamoyl derivatives obtained from aromatic and basic amino acid enantiomers were well separated by reversed-phase chromatography [172,173]. The proposed method using these chiral reagents is applied to the examination of stereo chemical purities of synthetic peptides [174]. Various bioactive peptides consisting of L-amino acids have been separated from reptiles and insects. Therefore,

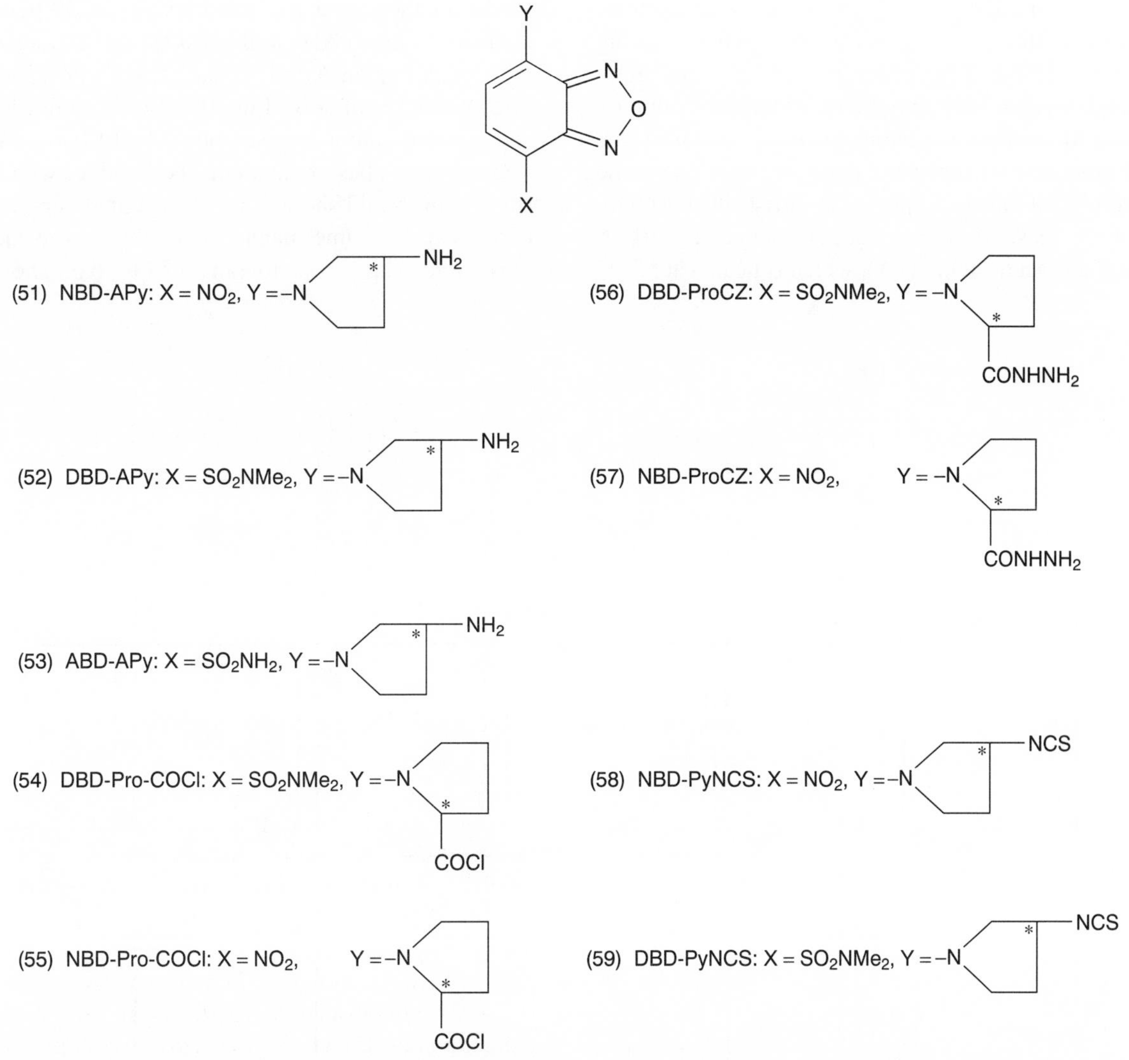

**Fig. 6.32.** Chiral tagging reagents bearing benzofurazan moiety as the fluorophore.

it is firmly believed that only L-amino acids participate in the functions of higher animals. However, some peptides (e.g. Achatin I and $\omega$-agatoxin-TK) comprising D-amino acid(s) in the sequences have been discovered in the ganglia of snails and the venom of spiders. Consequently, chiral sequence analysis of peptides and proteins, in which L-amino acids in the sequence are converted to D-amino acids, is another conceivable application for these reagents as part of a chiral Edman-degradation technique. A proposed method using a chiral DBD-PyNCS (59) was applied to the discrimination of D/L-amino acids in the sequences of [D-Ala$^2$]-leucine-enkepharin and deltolphin II [175,176].

**Example 3.** Chiral sequential analysis of [D-Ala$^2$]-leucine-enkepharin utilizing *R*(−)-DBD-PyNCS (59) [175].

To a 100 μl of peptide (0.1 nM) in water/acetonitrile/pyridine (2/1/1, v/v/v %) is added 100 μl of *R*(−)-DBD-PyNCS (10 mM) in water/acetonitrile/pyridine (2/1/1, v/v/v %). After heating at 65 °C for 1 hr under protection from light, any excess reagents are extracted out by 400 μl heptane/dichloromethane (7:3). The same extraction is repeated three times. Then the combined aqueous phase is dried in a stream of $N_2$ gas. To the residue is added 100 μl of trifluoroacetic acid (TFA), which is heated at 55 °C for 1 min, then dried in a stream of $N_2$ gas. 50 μl of distilled water is added to the residue and extracted twice with heptane/ethyl acetate (1:5, v/v %). The dried extract is dissolved in 100 μl of acetonitrile, and injected onto a reversed-phase HPLC column. The aqueous phase containing the peptide, which is one amino acid shorter than the original peptide, is dried in the same manner and subjected to the next cycle. The chromatograms of Fig. 6.33 show

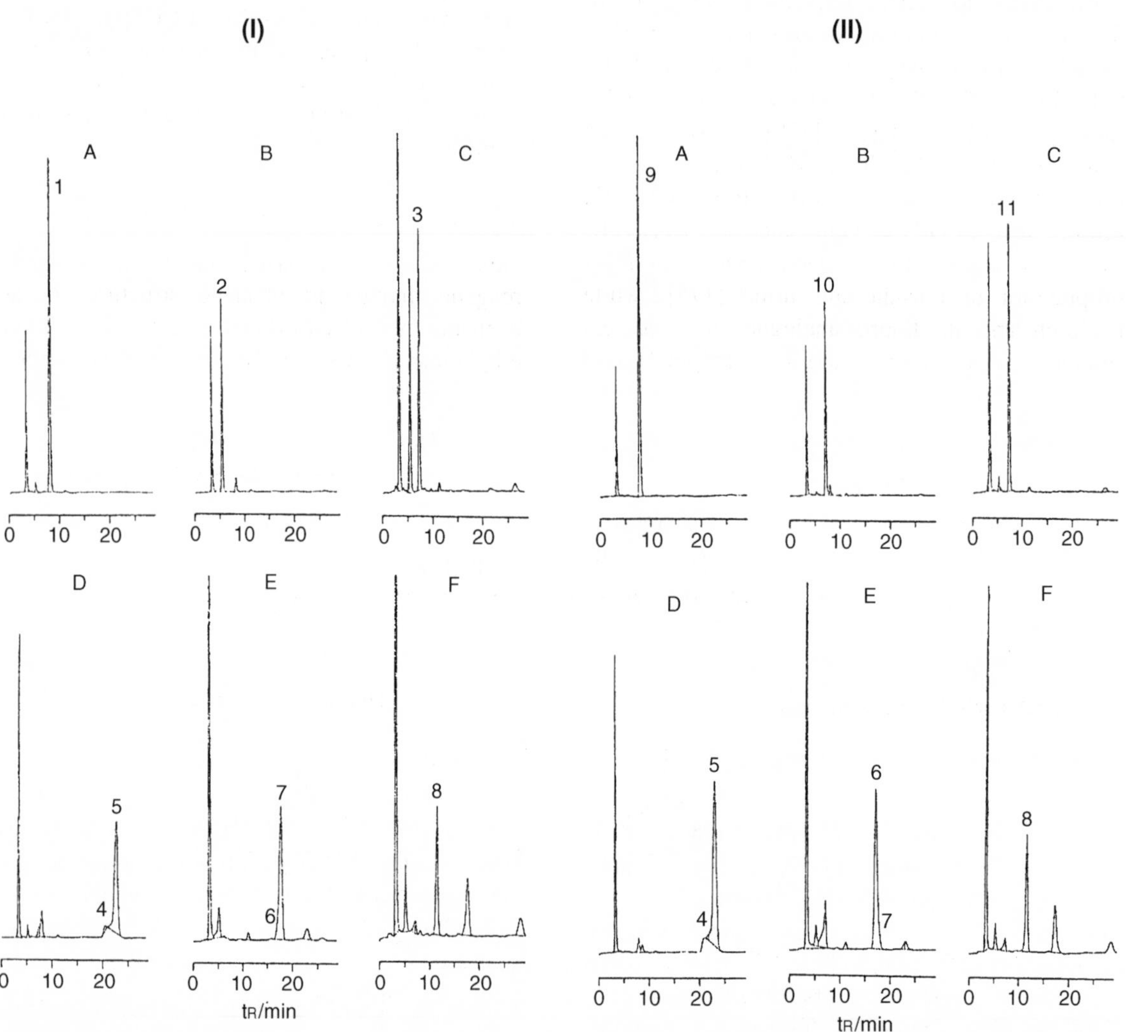


**Fig. 6.33.** Chiral sequential analysis of leucine-enkepharines using *R*(−)-DBD-PyNCS: (I), L-Tyr-L-Ala-Gly-L-Phe-L-Leu; (II), L-Tyr-D-Ala-Gly-L-Phe-L-Leu. Chromatograms: A, B and C, thiocarbamoyl derivatives obtained from each cycle; D, E and F, thiohydantoin derivatives obtained from each cycle. Peaks: (1), (2), (3), (9), (10) and (11), labelled peptides: (4) D-Tyr; (5) L-Tyr; (6) D-Ala; (7) L-Ala; (8) Gly. [Reproduced from ref. 175, p. 781, Fig. 2.].

the separation of thiohydantoin derivatives in each cycle of [D-Ala$^2$]-leucine-enkepharin and [L-Ala$^2$]-leucine-enkepharin.

The reactions of labelling, cleavage, cyclization and conversion seem to proceed as those of Edman degradation using phenylisothiocyanate (PITC). A possible reaction scheme of the proposed procedure with *R*(−)-DBD-PyNCS is shown in Fig. 6.34.

As a chiral coupling component for amines, the 2-arylpropionic acid derivatives that are introduced as an acyl chloride or an isocyanate moiety include: *S*(+)-flunoxaprofen chloride [*S*(+)-FLOP-Cl] (71) [177,178], *S*(+)-benzoxaprofen chloride [*S*(+)-BZOP-Cl] (70) [179], *S*(+)-naproxen chloride [*S*(+)-NAP-Cl] (68) [179], or *S*(−)-flunoxaprofen isocyanate [*S*(−)-FLOPIC] (62) [180]. *S*(+)-FLOP-Cl (71) and *S*(−)-FLOPIC (62) were applied to the determination of *R*/*S*-propranolol in plasma and urine [178], while baclofen and its fluoro analogue in biological samples [urine, plasma and cerebrospinal fluid (CSF)] were assayed with *S*(+)-BZOP-Cl (70) and *S*(+)-NAP-Cl (68) [179]. Amine enantiomers can be resolved with a chiral reagent that has a carboxyl group as the reactive functional group [169]. Accordingly, *N*-[4-(6-methoxy-2-benzoxazolyl)]benzoyl-L-phenylalanine (BOX-L-Phe) (64) and *N*-[4-(6-methoxy-2-benzoxazolyl)]-benzoyl-L-proline (BOX-L-Pro) (63), with the $NH_2$ group converted to COOH, were synthesized to permit separation of racemic amines [181]. 2-Methyl-2-$\beta$-naphthyl-1,3-benzodiazole-4-carboxylic acid [*S*(+)-MNB-COOH] (66) is successfully applied to the enantioseparations of amino acids by reversed-phase HPLC [182]. Figure 6.35 illustrates the structures of the above reagents.

#### 6.5.2.2 Label for Carboxylic Acids

Goto *et al.* synthesized chiral derivatization reagents having naphthalene structure, *l*- and *d*-enantiomers of 1-(4-dimethylamino-1-naphthyl)-ethylamine (DANE) (42), for carboxyls such as

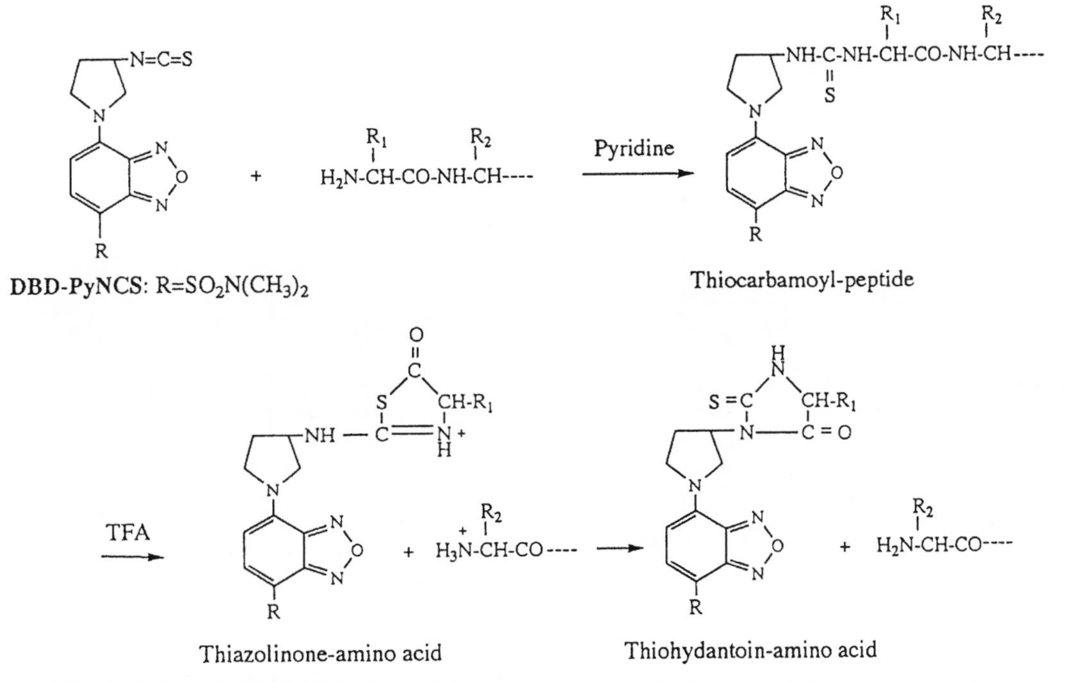

DBD-PyNCS: R=$SO_2N(CH_3)_2$

Pyridine

Thiocarbamoyl-peptide

TFA

Thiazolinone-amino acid

Thiohydantoin-amino acid

**Fig. 6.34.** Labelling of peptide with *R*(−)-DBD-PyNCS, and subsequent cleavage and cyclization reactions. [Reproduced from ref. 175, p. 780, Fig. 1.].

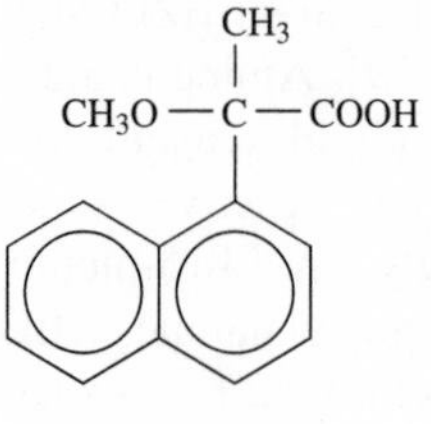

(60) 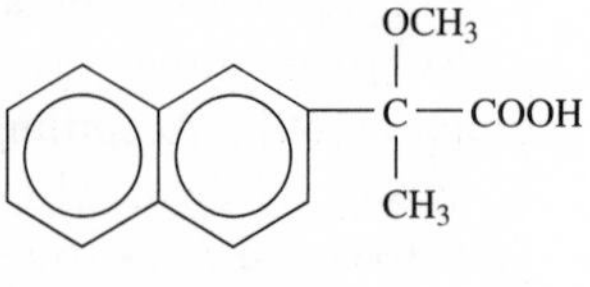 (61) 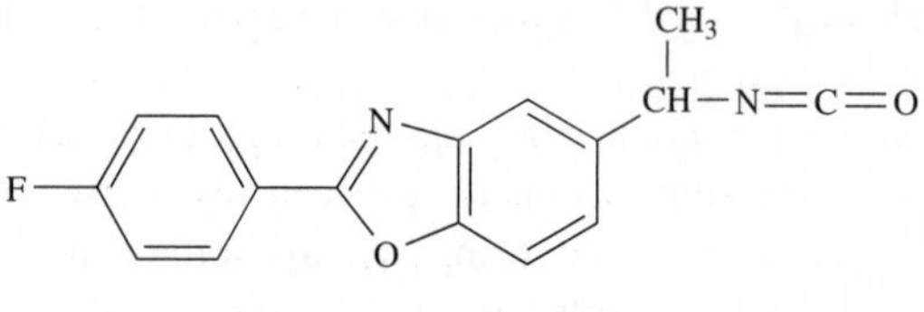 (62)

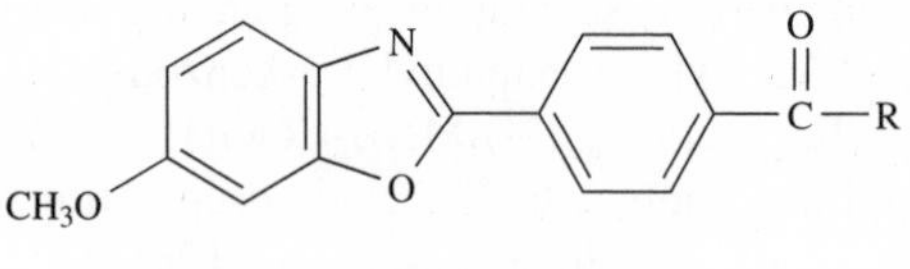

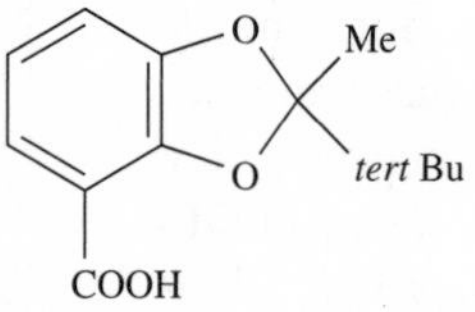

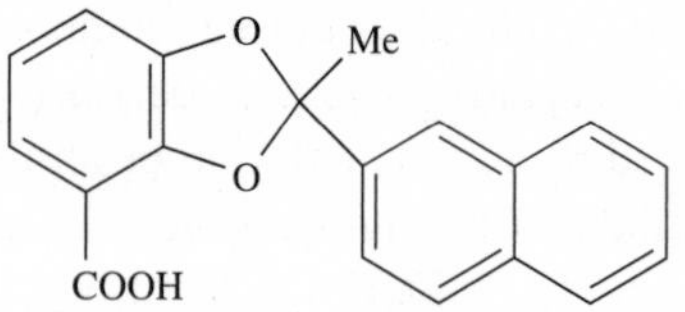

(63) BOX-L-Pro : R 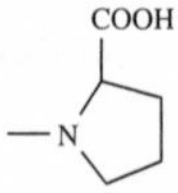 = (65) (66)

(64) BOX-L-Phe : R = —NH−CH−COOH

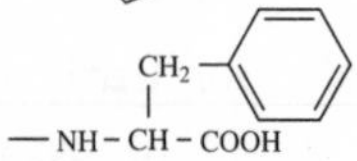

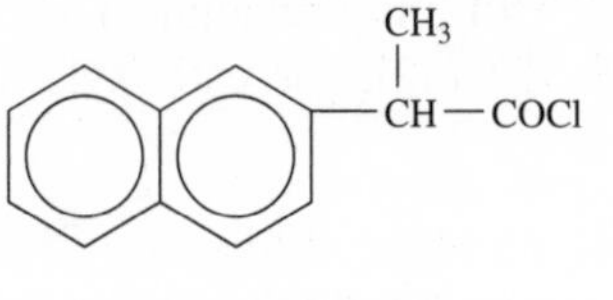

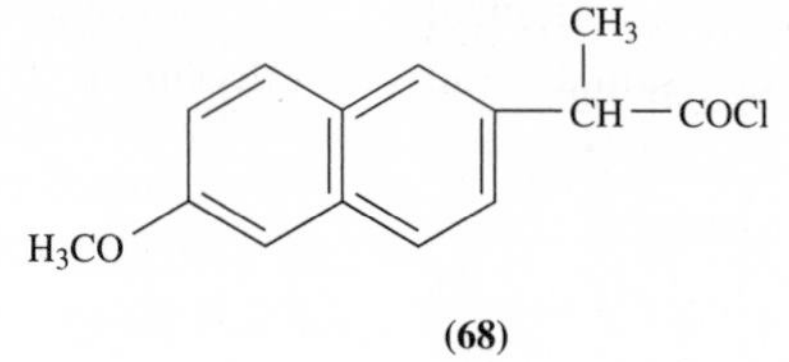

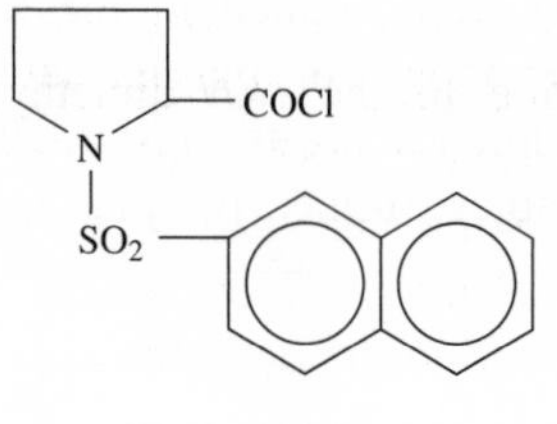

(67) (68) (69)

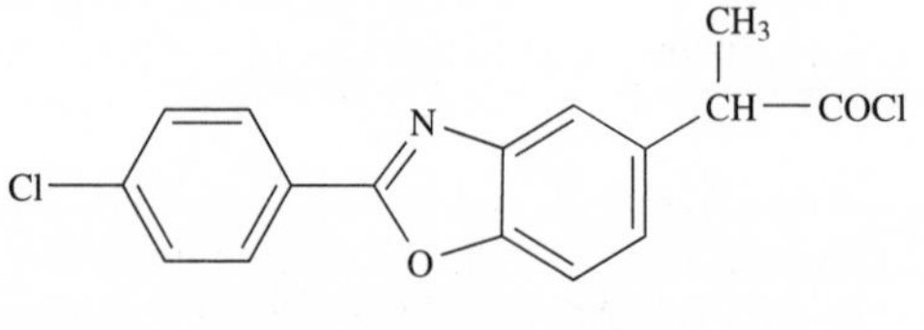

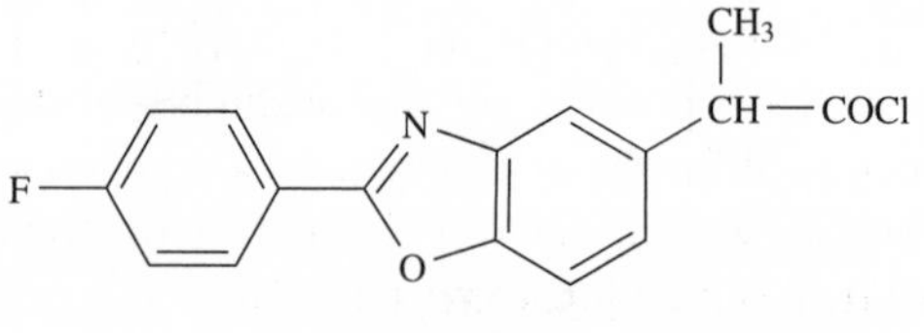

(70) (71)

**Fig. 6.35.** Chiral FL Labels for amines by HPLC: (60) (−)-$\alpha$-methoxy-$\alpha$-methyl-1-naphthaleneacetic acids (MM(1)NA); (61) (−)-$\alpha$-methoxy-$\alpha$-methyl-2-naphthaleneacetic acids (MM(2)NA); (62) *S*(−)-flunoxaprofen isocyanate [*S*(−)-FLOPIC]; (63) *N*-[4-(6-methoxy-2-benzoxazolyl)]benzoyl-L-proline (BOX-L-Pro); (64) *N*-[4-(6-methoxy-2-benzoxazolyl)]benzoyl-L-phenylalanine (BOX-L-Phe); (65) 2-tert.-butyl-2-methyl-1,3-benzodioxole-4-carboxylic acid [*R*(−)- and *S*(+)-TBMB-COOH]; (66) 2-Methyl-2-$\beta$-naphthyl-1,3-benzodiazole-4-carboxylic acid [*S*(+)-MNB-COOH]; (67) D-2-(2-Naphtyl)propionyl chloride; (68) *S*(+)-naproxen chloride [*S*(+)-NAP-Cl]; (69) *S*(−)-*N*-(1-Naphtylsulfonyl)-2-pyrrolidenecarbonyl chloride; (70) *S*(+)-benzoxaprofen chloride [*S*(+)-BZOP-Cl]; (71) *S*(+)-flunoxaprofen chloride [*S*(+)-FLOP-Cl].

free fatty acids and pharmaceuticals (Fig. 6.25). Condensation reactions with the chiral amines is carried out in the presence of activation agents such as 1-ethyl-3-(3-dimethylaminopropyl)carbodiimide (EDC) and pyridine [183]. The reagents are used as precolumn derivatization for the enantioseparations of naproxen. The resulting amide diastereomers are separated more efficiently by normal-phase rather than by reversed-phase chromatography. In the case of *N*-acetylamino acids, the resolution (Rs) values increased with carbon number of alkyl residue at $\alpha$-position [184]. Aromatic amino acids (e.g. phenylalanine) and anti-inflammatory drugs (e.g. naproxen, ibuprofen and indoprofen) provided good resolution (Rs: more than 3.0) [184]. Even with the naphthalene fluorophore, the maximal excitation and emission wavelengths shifted to the relatively long wavelength region (ex. 320 nm; em. 395–420 nm) owing to the *N*,*N*-dimethylamino group at the 4-position [183–186]. The sensitivity depends upon the naphthalene fluorophore and detection of 0.1 ng is achieved with naproxen. DANE was applied to the determination of loxoprofen and its alcohol metabolites in urine [186] and rat plasma [185], and naproxen in human serum [183]. The DANE analogue, 1-(1-naphthyl)ethylamine [*S*(−)- and *R*(+)-NEA], which has no dimethylamino group at the 4-position on the naphthalene ring, has been used for the determination of compounds of pharmacological interest such as leukotriene $D_4$ antagonist in human plasma [187] and non-steroidal anti-inflammatory drugs [188,189]. The derivatives labelled with DANE or NEA are detectable with fluorometry or absorptimetry.

Two highly sensitive, chiral derivatization reagents, 1-(1-anthryl) and 1-(2-anthryl)ethylamines (*l*- and *d*-enantiomers) (AEA) (73), have also been developed for carboxylic acids by Goto *et al.* [190] (Fig. 6.36). Although the derivatization conditions and detection wavelengths are similar to those of DANE and NEA, the sensitivity (ca. 100 fmol) is much better than that of DANE and NEA. This is due to large fluorescence quantum yield ($\phi$) of the anthracene moiety in the reagent. No racemization during the tagging reaction with AEA occurred, even after prolonged reaction.

Carboxylic acids such as *N*-acetylamino acids and anti-inflammatory drugs (e.g. naproxen and ibuprofen) react with 4-(3-aminopyrrolidin-1-yl)-7-(*N*,*N*-dimethylaminosulfonyl)-2,1,3-benzoxadiazole [*S*(+)- and *R*(−)-DBD-APy] (52) or 4-(3-aminopyrrolidin-1-yl)-7-nitro-2,1,3-benzoxadiazole [*S*(+)- and *R*(−)-DBD-APy] (51) at room temperature in the presence of 2,2-dipyridyl disulfide (DPDS) and triphenylphosphine (TPP) as the activation agents [191–193] (Fig. 6.32). The resulting diastereomers are completely separated by both reversed-phase (Rs = 1.62–6.96) and normal-phase (Rs = 2.58–7.60) chromatography [193]. Detection at the 20–50 fmol levels with a conventional fluorescence detector is possible. The diastereomers derived from NBD-APy provide sensitive detection not only with conventional fluorescence detection but also with argon-ion (488 nm) laser-induced fluorescence detection (LIF) [192,193]. On the other hand, the DBD-APy derivatives are also amenable to the peroxyoxalate chemiluminescence (CL) method [191,193]. The chromatographic detection limits are at the sub-fmol levels with these methods.

**Example 4.** Separation of anti-inflammetry drug derivatives obtained from *R*(−)-DBD-APy (52) [193].

Enantiomers of carboxylic acids (ca. 1mg each) are reacted at room temperature with 0.25 mM *R*(−)-DBD-APy in 0.5 ml of acetonitrile in the presence of 2,2-dipyridyl disulfide (DPDS) and triphenylphosphine (TPP) (2.5 mM each) as activation reagents. After 2 h reaction, 5 µl aliquots are injected onto a reversed-phase (Inertsil ODS-2) and a normal-phase (Inertsil SIL) columns. Figure 6.37 shows the reversed-phase chromatogram obtained from anti-inflammatory drugs and *R*(−)-DBD-APy. The separations of antiinflammatory drugs and *N*-acetylamino acids after derivatization with *S*(+)-DBD-PyNCS are listed in Table 6.5.

The diastereomers are also investigated with Ar-ion laser-induced detection at 488 nm. The minimum detectable levels of DBD-APy (52),

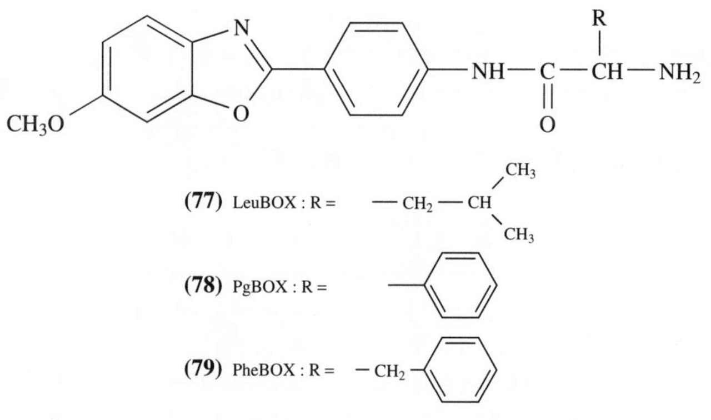

(72)

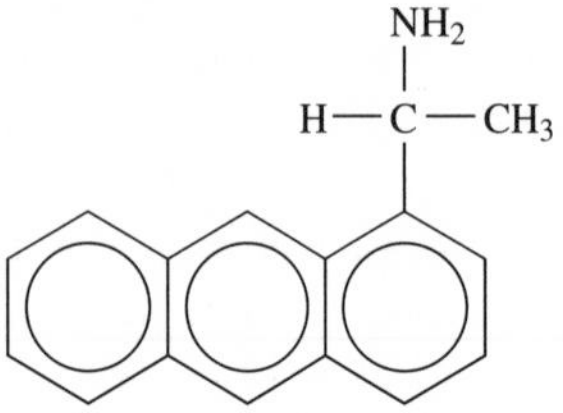

(73)

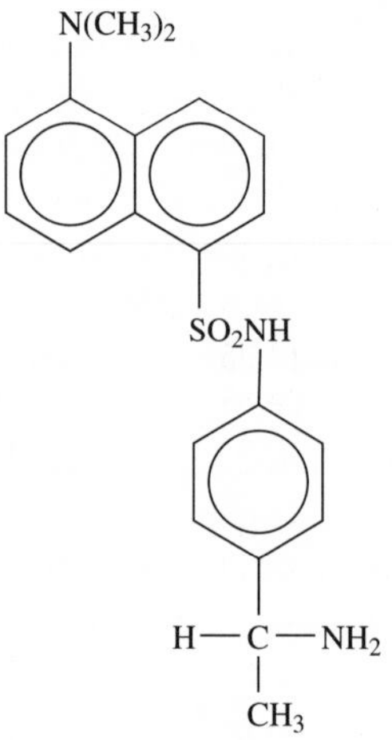

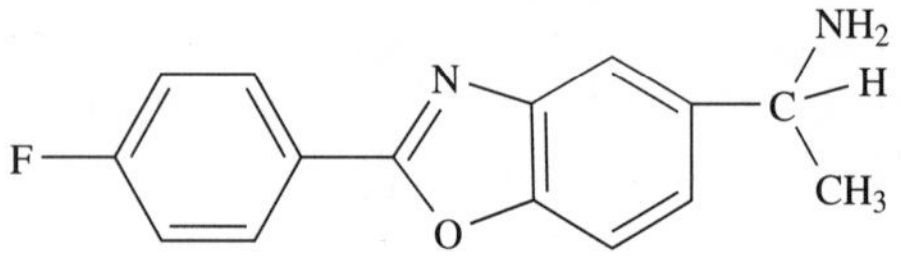

(74)

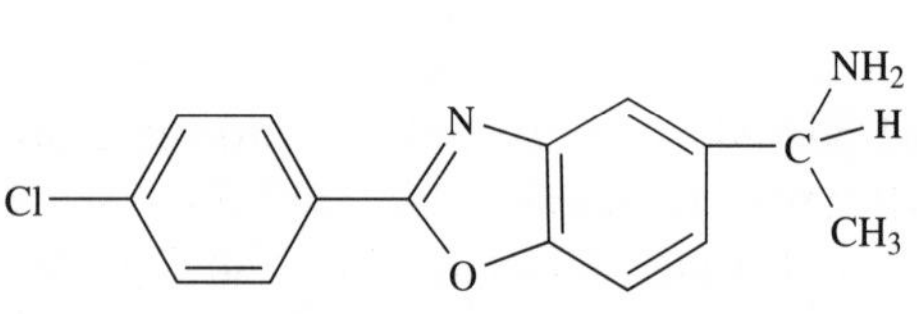

(75)

(77) LeuBOX : R =

(78) PgBOX : R =

(79) PheBOX : R =

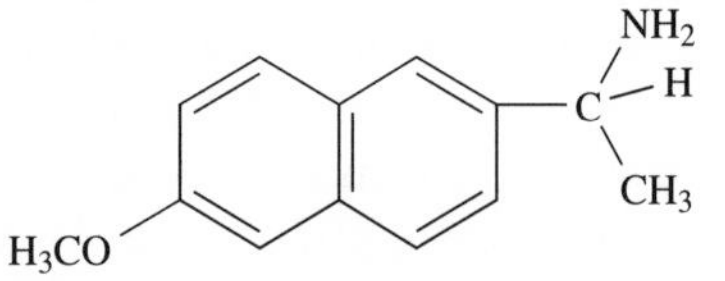

(80)

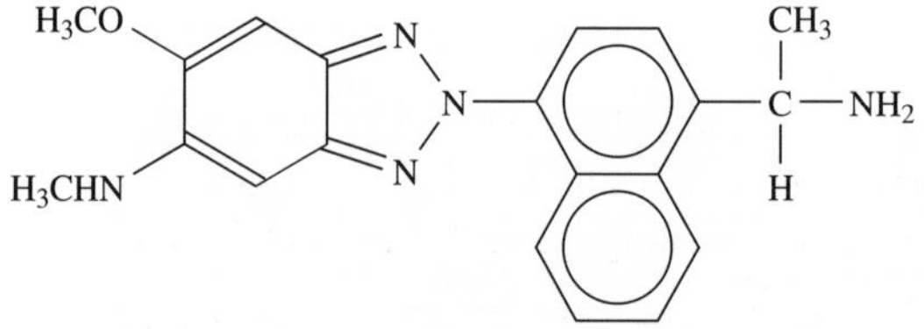

(81)

**Fig. 6.36.** Chiral FL Labels for carboxylic acids by HPLC: (72) *S*(+)-6-Methoxy-*α*-methyl-2-naphthaleneethylamine (NAPA); (73) 1-(1-anthryl)ethylamines (AEA); (74) *S*(+)-(*p*-Fluorophenyl)-*α*-methyl-5-benzoxazoleethylamine (*S*-FLOPA); (75) *S*(+)-(*p*-Chlorophenyl)-*α*-methyl-5-benzoxazoleethylamine (*S*-BOPA); (76) (−)-2-[4-(1-aminoethyl)-phenyl]-6-methoxybenzoxazole [(−)-APMB]; (77) 2-[4-(L-leucyl)aminophenyl]-6-methoxybenzoxazole (L-LeuBOX); (78) 2-[4-(D-phenylglycyl)aminophenyl]-6-methoxybenzoxazole (D-PgBOX); (79) 2-[4-(L-phenylalanyl)aminophenyl]-6-methoxybenzoxazole (L-PheBOX); (80) 1-(4-dansylaminophenyl)ethylamine (*d*- and *l*-DAPEA); (81) *S*(−)-2-[4-(1-aminoethyl)naphthyl]-6-methoxy-2H-benzotriazolyl-5-amine [*S*(−)-ANBT]; (82) *S*(+)-1-methyl-2-(2,3-naphthalimido)ethyl trifluoromethanesulfonate [*S*(+)-MNE-OTf].

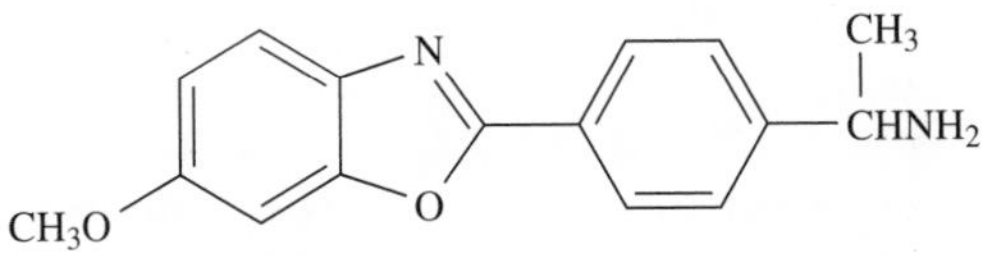

(76)

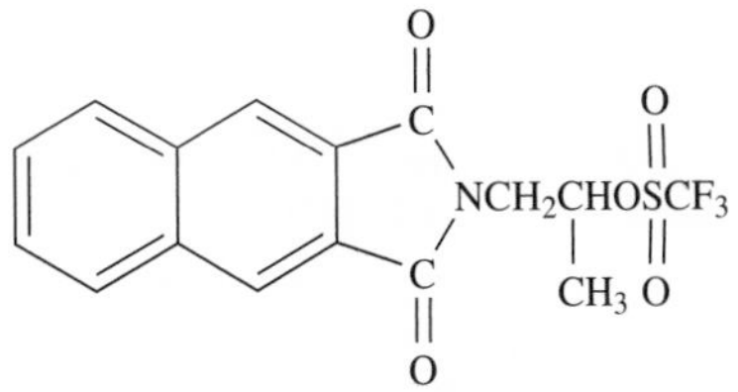

(82)

**Fig. 6.36.** *(continued)*.

**Table 6.5.** LC separation of diastereomers derived from D-DBD-APy (52) by reversed-phase chromatography

| Carboxylic acid | D-Isomer | | L-Isomer | | $\alpha$ | Rs | Eluent |
|---|---|---|---|---|---|---|---|
| | Retention time (min) | $k'$ | Retention time (min) | $k'$ | | | |
| Ibuprofen | 45.32 | 9.21 | 18.49 | 11.33 | 1.23 | 3.53 | E |
| Ibuprofen | 26.22 | 16.48 | 32.73 | 20.82 | 1.26 | 4.92 | D |
| Naproxen | 11.32 | 6.55 | 13.35 | 7.90 | 1.21 | 2.70 | D |
| Naproxen | 19.49 | 11.99 | 24.02 | 15.01 | 1.25 | 4.42 | C |
| Loxoprofen | 8.17 | 4.45 | 9.10 | 5.07 | 1.14 | 1.63 | D |
| Loxoprofen | 13.25 | 7.83 | 15.29 | 9.19 | 1.17 | 2.55 | C |
| *N*-Acetyl-Trp | 24.88 | 15.59 | 19.34 | 11.90 | 1.31 | 5.04 | B |
| *N*-Acetyl-Phe | 26.45 | 16.63 | 18.79 | 11.53 | 1.44 | 6.96 | B |
| *N*-Acetyl-Tyr | 8.79 | 4.86 | 7.47 | 3.98 | 1.22 | 2.29 | B |
| *N*-Acetyl-Tyr | 18.03 | 11.02 | 14.31 | 8.54 | 1.29 | 4.25 | A |
| *N*-Acetyl-Leu | 21.47 | 13.31 | 18.82 | 11.54 | 1.15 | 2.72 | B |
| *N*-Acetyl-Val | 11.83 | 6.88 | 11.03 | 6.35 | 1.08 | 1.18 | B |
| *N*-Acetyl-Val | 25.07 | 15.71 | 23.13 | 14.42 | 1.09 | 1.62 | A |
| *N*-Acetyl-Met | 12.80 | 7.54 | 11.68 | 6.79 | 1.11 | 1.55 | B |
| *N*-Acetyl-Met | 27.65 | 17.44 | 24.69 | 15.46 | 1.13 | 2.33 | A |

Column, Inertsil ODS-2 (150 × 4.6 mm, i.d., 5 μm) at 40 °C; eluent, (A) water/$CH_3CN$ (75:25), (B) water/$CH_3CN$ (70:30), (C) water/$CH_3CN$ (55:45), (D) water/$CH_3CN$ (50:50), (E) water/$CH_3CN$ (45:55); flow rate, 1.0 ml/min; fluorescence detection, excitation at 470 nm, emission at 580 nm. [Reproduced from ref. 193, p. 78, Table 4.]

ABD-APy (53) and NBD-APy (51), which are the chiral benzofurazan bearing reagents for carboxylic acid labelling, were 11, 29 and 3 fmol, respectively [192,193] (Fig. 6.38).

Fluorescent chiral amine derivatives, FLOPA (74), BOPA (75) or NAPA (72) were synthesized from the corresponding enantiomer of 2-arylpropionic acid (flunoxaprofen, benoxaprofen or naproxen) by the following sequence: formation of the acyl azide, conversion to the isocyanate, and finally, to the amine (Curtius degradation) [194] (Fig. 6.36). Racemic anti-inflammatory drugs (e.g. flunoxaprofen, benoxaprofen, flubiprofen, and ibuprofen) were successfully separated as diastereomeric derivatives with *S*(+)-FLOPA by normal-phase and reversed-phase chromatography. *S*(+)-FLOPA is applied to the determination of (±)-$\alpha$-phenylcyclopentylacetic acid [(±)-PCA] in human plasma and urine [195]. Kondo *et al.* report (−)-2-[4-(1-aminoethyl)phenyl]-6-methoxybenzoxazole [(−)-APMB] (76), 2-[4-(L-leucyl)aminophenyl]-6-methoxybenzoxazole (L-LeuBOX) (77), 2-[4-(L-phenylalanyl)amino-phenyl]-6-methoxybenzoxazole (L-PheBOX) (79), and 2-[4-(D-phenylglycyl)aminophenyl]-6-methoxybenzoxazole (D-PgBOX) (78) as 2-phenylbenzoxazole derivatives for carboxylic acids [196,197] (Fig. 6.36). *R*/*S*-Ibuprofens in rat plasma after oral administration were determined with (−)-

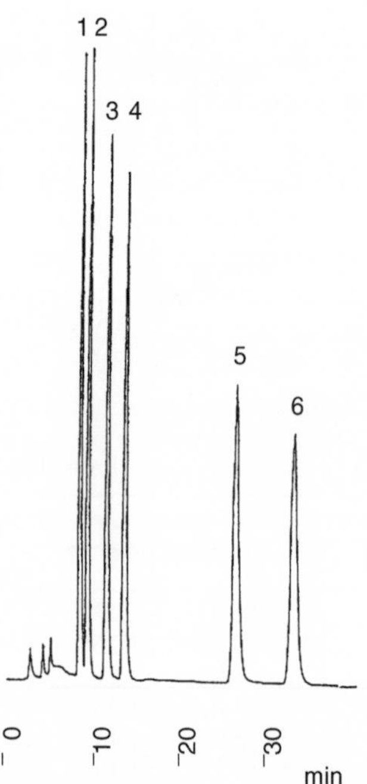


**Fig. 6.37.** Reversed-phase chromatogram of anti-inflammatory drugs derived from $R(-)$-DBD-PyNCS. Peaks: (1) D-loxoprofen; (2) L-loxoprofen; (3) D-naproxen; (4) L-naproxen; (5) D-ibuprofen; (6) L-ibuprofen. [Reproduced from ref. 193, p79, Fig. 5.].

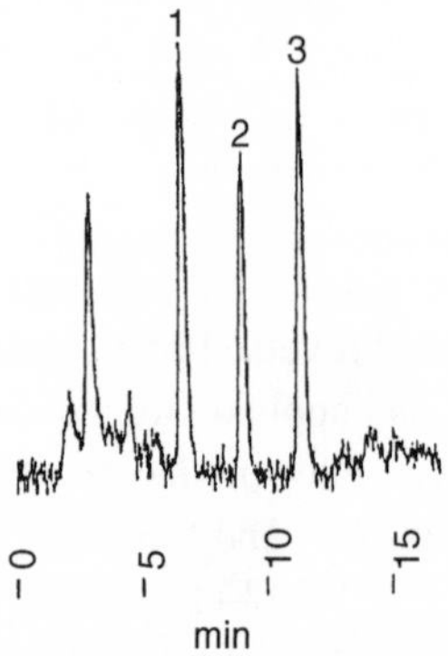


**Fig. 6.38.** Chromatogram of naproxen derivatives obtained from benzofurazan reagents with Ar-ion laser-induced fluorescence detection. Peaks: (1) ABD-APy-naproxen (40 fmol); (2) NBD-APy-naproxen (3 fmol); and (3) DBD-APy-L-naproxen (10 fmol). Column: TSK-gel PTH Pak (250 × 2.0 mm, i.d., 5 μm) at 40 °C; mobile phase, water/acetonitrile (40:60); flow rate, 0.2 ml/min; detection, Ar-ion at 488 nm (excitation), intereference filter at 540 ± 20 nm (emission). [Reproduced from ref. 192, p. 360, Fig. 2.].

APMB [198]. Since these reagents possess an amino functional group as the reactive site, the tagging conditions and the detection methods are essentially the same for all of the reagents. The detection limit of (−)-APMB, a derivative of 2-phenylpropionic acid, is 10 fmol at a signal-to-noise ratio of 3 [196]. The high sensitivity with fluorometry is almost same with DBD-APy derivative [199].

It is well known that dansylated amino acids give excellent sensitivity because of optimum fluorescence. Exploiting this important property, Iwaki *et al.* [200] synthesized dansyl-amine derivative, 1-(4-dansylaminophenyl)ethylamine (*d*- and *l*-DAPEA) (80) for carboxylic acid enantiomers (Fig. 6.36). Similarly, a chiral tagging reagent having dansyl structure, i.e. 1-(5-dimethylamino-1-naphthalenesulfonyl)-(*S*)-3-aminopyrrolidine [(*S*)-DNS-APy], is synthesized and used for the enantioseparation of carboxylic acids such as anti-inflammatory drugs [201]. For the resolution of racemic carboxylic acids, Yasaka *et al.* developed the chiral reagent involving powerful alkylating ability, *S*(+)-1-methyl-2-(2,3-naphthalimido)ethyl trifluoromethanesulfonate [*S*(+)-MNE-OTf] (82) [202] (Fig. 6.36). The reagent with carboxylic acids requires anhydrous potassium carbonate ($K_2CO_3$) and 18-crown-6 in acetonitrile. Benzotriazole derivative [*S*(−)-2-[4-(1-aminoethyl)naphthyl]-6-methoxy-2H-benzotriazolyl-5-amine] [*S*-(−)-ANBT] (81) is also reported as a tagging reagent for carboxylic acid enantiomers [203] (Fig. 6.36).

### 6.5.2.3 Label for Alcohols

Shimizu *et al.* [204,205] synthesized two chiral tagging reagents, *S*(−)-*N*-(1-naphthylsulfonyl)-2-pyrrolidinecarbonyl chloride (NSPC) and *d*-2-(2-naphtyl)propionyl chloride (NPC), containing the −COCl group as a reactive group toward alcohol enantiomers (Fig. 6.35). The reaction of alcohol with these reagents proceeds in an aprotic solvent, such as chloroform, in the presence of base catalyst (e.g. triethylamine) under mild condition at room temperature. Since carbonyl chloride type reagents decompose readily in aqueous reaction media, considerable care should be exercised to avoid moisture. Four isomers (*dl*-cis and *dl*-trans) of deacetyl diltiazem labelled with NPSC are completely separated by normal-phase chromatography. The resulting derivatives are monitored at 254 nm with a UV detector. However, fluorescence detection based on the naphthalene moiety is probably effective.

A new type of chiral tagging reagent, (−)- and (+)-2-methyl-1,1′-binaphthalene-2′-carbonyl cyanides (Methyl-BNCC) (83), with a binaphthalene moiety as the fluorophore and carbonyl cyanide as a reactive group toward the hydroxyl function, has been developed by Goto *et al.* [206] (Fig. 6.39). The tagging reaction progresses favorably with heating at 60 °C in the presence of 0.01% quinuclidine as a base. The diastereomeric esters derived from the pairs of $\beta$-hydroxy acids and $\beta$-blockers are efficiently resolved by normal-phase chromatography using organic solvents such as *n*-hexane and ethyl acetate as the eluents [207]. The detection wavelengths of 340 nm (excitation) and 420 nm (emission) are typical of the naphthalene structure. Although sub-pmol detection of alcoholic compounds is possible with this derivatization method, the sensitivity might not be adequate for quantitation of enantiomeric alcohols at trace amounts in biological fluids. To overcome the sensitivity, Goto *et al.* [208] also prepare the highly sensitive and similar tagging reagent, (aS)-2′-methoxy-1,1′-binaphthalene-2-carbonyl cyanide (Methoxy-BNCC) (84), which has a methoxy group substituted for the methyl group at the 2′-position. The limit of detection was improved at the 100 fmol level with the reagent. The enhancement in the sensitivity is due to substitution of a methoxy group, being an electron-donating group

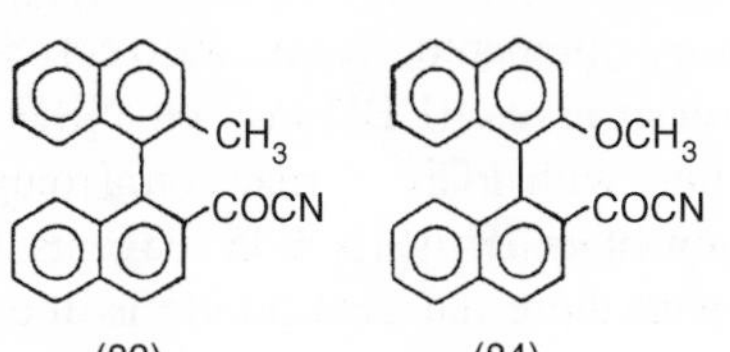

$CH_3$ COCN (83) $OCH_3$ COCN (84)

**Fig. 6.39.** Chiral axis reagents for alcohols by HPLC: (83) (aS)-2′-methyl-1,1′-binaphthalene-2-carbonyl cyanide (methyl-BNCC); (84) (aS)-2′-methoxy-1,1′-binaphthalene-2-carbonyl cyanide (methoxy-BNCC).

on the aromatic nucleus. The optical purities of the chiral axis reagents (Methoxy-BNCC, 99.5%; Methyl-BNCC, 99.0%) are adequate for analysis of biological samples. Many methods for the resolution of enantiomeric $\beta$-adrenergic blocking agents involve the derivatization with a chiral reagent through the secondary amino functional group. However, the drugs with the tert.-butylamino structure (e.g. bucumolol, carteolol) are relatively less reactive than those with the iso-propylamino structure (e.g. atenolol, propranolol) owing to steric hindrance. Therefore, the quantitative determination of tert.-butyl type compounds by tagging of the amino group is sometimes difficult, whereas the reactive site for Methoxy-BNCC or Methyl-BNCC which is the secondary alcoholic group on $\beta$-blockers, there is equal reactivity with all of the drugs. The method using Methoxy-BNCC is applied to the determination of racemic penbutolol sulfate in dog plasma [209].

**Example 5.** Determination of penbutolol enantiomers in plasma using Methoxy-BNCC (84) [209].

The plasma sample involving (*R*)-bufuranol (5 ng, internal standard) is diluted with 4 ml of 0.5 M potassium phosphate buffer (pH 7.0) and passed through a Sep-pak $C_{18}$ cartridge. After washing with 5-ml water and 5-ml of 30% ethanol successively, $\beta$-blockers are eluted with 5 ml of 0.1% methylamine in ethanol. The eluate dried is mixed with 0.2 mg Methoxy-BNCC in 0.1% quiniclidine in acetonitrile (0.1 ml), and heated at 60 °C for 20 min. Any excess reagent unreacted is decomposed with 50 μl of methanol, and the reaction mixture is evaporated under an $N_2$ stream. The residues obtained are redissolved in 1 ml of 90% ethanol and applied to a CM-LH-20 column (18 mm × 6 mm, i.d., 100 mg). After washing with 90% ethanol to remove any acidic compounds, the labelled products are eluted with 5 ml of 0.1 M methylamine in 90% ethanol. The dried eluate is redissolved in the mobile phase and injected into the HPLC column. The reaction scheme of the derivatization and the chromatograms obtained from plasma samples are shown in Fig. 6.40 and Fig. 6.41, respectively.

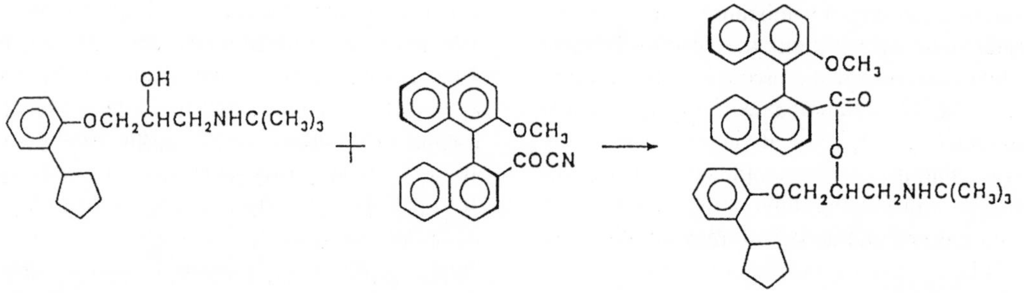

**Fig. 6.40.** Derivatization reaction of penbutolol with (aS)-2′-methoxy-1,1′-binaphthalene-2-carbonyl cyanide (methoxy-BNCC). [Reproduced from ref. 209, p. 724, Fig. 1.].

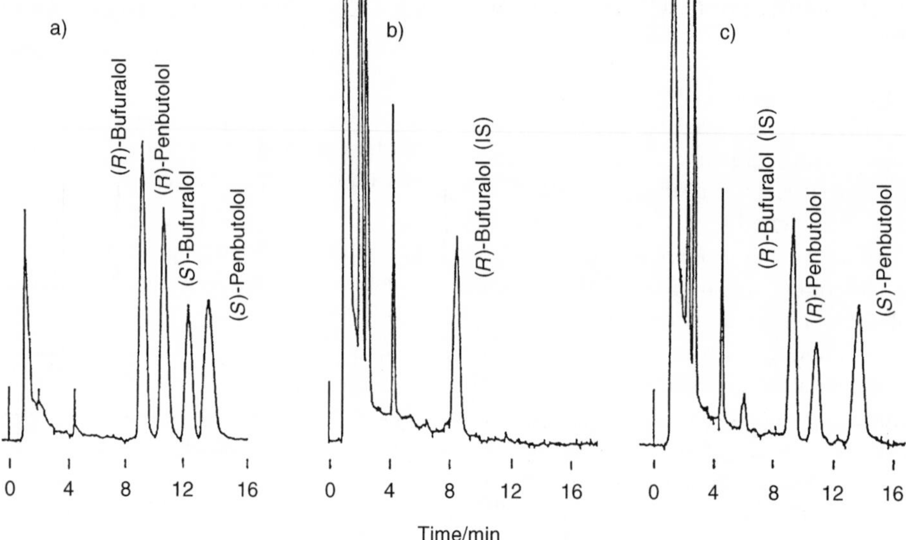


**Fig. 6.41.** Chromatograms of penbutolol enantiomers in plasma after derivatization with methoxy-BNCC. Chromatograms: a) standard penbutolol enantiomers; b) control plasma spiked with (*R*)-bufuralol (10 ng/ml, IS); c) plasma obtained from a dog at 2 hrs after oral administration of racemic penbutolol sulfate (670 μg/kg). Column: Cosmosil 5SL; mobile phase, 0.005% triethylamine in hexane/ethyl acetate (6:1); flow rate, 2 ml/min; detection, 405 nm (excitation at 290 nm). [Reproduced from ref. 209, p. 725, Fig. 4.].

Alcoholic compounds are difficult to label due to limited reactivity and the relatively poor stability of the reagents [168,207,209]. 4-(2-Chloroformylpyrrolidin-1-yl)-7-(*N*,*N*-dimethylaminosulfonyl)-2,1,3-benzoxadiazole [*S*(−)- and *R*(+)-DBD-Pro-COCl] (54) and 4-(2-chloroformylpyrrolidin-1-yl)-7-nitro-2,1,3-benzoxadiazole [*S*(−)- and *R*(+)-NBD-Pro-COCl] (55) have been developed as the chiral reagents toward hydroxyl enantiomers [210–213] (Fig. 6.32). These reagents are readily prepared from the corresponding proline derivatives (DBD-Pro or NBD-Pro) by chlorination with $PCl_5$. These chiral reagents, as with many other derivatization reagents for alcohols, depend upon an acyl halide as the reactive functional group. The acyl halide group exhibits excellent reactivity with hydroxyls in the presence of an HCl scavenger such as pyridine. Anhydrous organic solvents such as benzene are required as the reaction medium, because these reagents are sensitive to hydrolysis by moisture. The reagents also react with amines, and the reaction rate is faster than that toward the alcohols. The racemic alcohols (e.g. 2-heptanol and 1-phenylethanol) and amines[(e.g. 1-phenylethylamine and 1-(1-naphtyl)ethylamine] were completely resolved by normal-phase chromatography, while the Rs values obtained by reversed-phase method were relatively small [211].

**Example 6.** Separation of aliphatic alcohols with NBD-Pro-COCl (55) [211].

The alcohol enantiomers are reacted at 80 °C with 1 mM $S(-)$-NBD-Pro-COCl in 1 ml of benzene in the presence of 1% pyridine. After 1 h reaction, 5 µl aliquots are injected into a reversed-phase (Inertsil ODS-80A) and a normal-phase (Inertsil-SIL) columns. The eluents for reversed-phase and normal-phase chromatography are water/acetonitrile and $n$-hexane/ethyl acetate, respectively. The chromatograms obtained from alcohols are shown in Fig. 6.42.

The primary and secondary amines also react with the reagent to produce corresponding amide diastereomers. The conditions are milder than those with alcohols. The resolution (Rs) values by normal-phase chromatography are larger than those by reversed-phase chromatography.

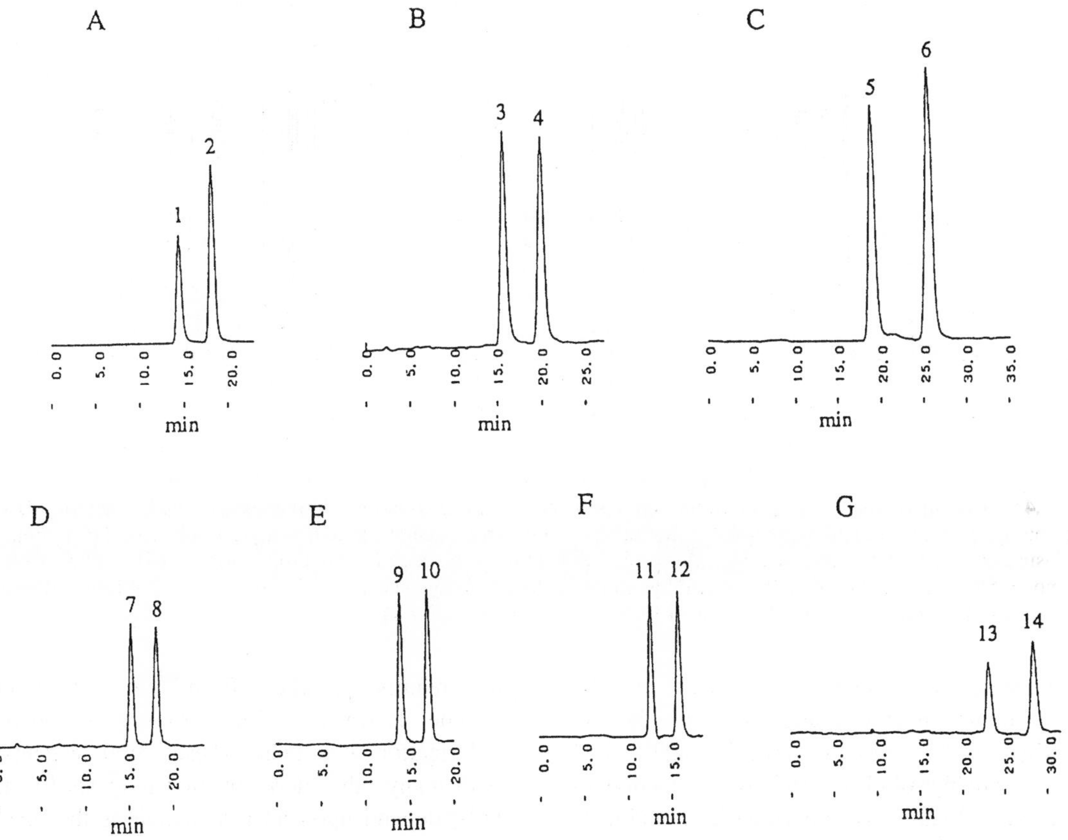


**Fig. 6.42.** Normal-phase chromatograms of racemic amines obtained from derivatization with $S(-)$-NBD-Pro-COCl. Peaks: 1, $S(-)$-1-naphtylethylamine (NEA); 2, $R(+)$-NEA; 3, $S(+)$-cyclohexylethlamine (CEA); 4, $R(-)$-CEA; 5, $S(-)$-1-phenylethylamine (PEA); 6, $R(+)$-PEA; 7, $S(+)$-2-hexanol; 8, $R(-)$-2-hexanol; 9, $S(+)$-2-heptanol; 10, $R(-)$-2-heptanol; 11, $S(+)$-2-nonanol; 12, $R(-)$-2-nonanol; 13, $S(-)$-1-phenylethomol; 14, $R(+)$-1-phenylethanol. [Reproduced from ref. 211, p. 86, Fig. 4.].

When $S(-)$-NBD-Pro-COCl is selected as the derivatization reagent, the diastereomers corresponding to the $S$-configurations are eluted before than those from R-configurations; while opposite elution orders are identified by the use of $R(+)$-NBD-Pro-COCl (Fig. 6.42).

2-Tert.-butyl-2-methyl-1,3-benzodioxole-4-carboxylic acid [$R(-)$- and $S(+)$-TBMB-COOH] (65) was developed to determine D,L-configurations of monosaccharides such as xylose, arabinose, glucose, galactose, fucose, and mannose [214]. The HPLC method involves a coupling reaction with $R(-)$-TBMB-COOH and per-$O$-acetyl pyranosyl bromides. The determination of the absolute configuration and optical purity of di-$O$-acylglycerols is carried out in these determinations with $S(+)$-TBMB-COOH [215] by normal-phase chromatography. The series of the 1,3-benzodioxole type reagents have recently been developed by Nishida *et al.* [216].

### 6.5.2.4 Label for Aldehydes and Ketones

As the reagents for chiral ketones, optically active 4-(2-carbazoylpyrrolidin-1-yl)-7-($N$,$N$-dimethylaminosulfonyl)-2,1,3-benzoxadiazole [$S(-)$- and $R(+)$-DBD-ProCZ] (56) and 4-(2-carbazoylpyrrolidin-1-yl)-7-nitro-2,1,3-benzoxadiazole [$S(-)$- and $R(+)$-NBD-ProCZ] (57) were synthesized from the reaction of DBD-Pro-COCl (54) or NBD-Pro-COCl (55) with $NH_2NH_2$ [217] (Fig. 6.32). The tagging reactions toward aldehydes and ketones proceed under mild conditions at 65 °C for 10 min in the presence of trichloroacetic acid (TCA). The separation is not adequate for racemic ketones (i.e. 2-phenylcyclohexanone, 2-phenylcycloheptanone, and 1-decalone).

### 6.5.2.5 Label for Thiols

OPA has been widely used as a label for chiral amino acids in the presence of a chiral mercaptan such as $N$-acetyl-L-cysteine (NAC) and $N$-acetyl-D-penicillamine (NAP). Moreover, combinations of OPA and a chiral $\alpha$-amino acids permit the resolution of racemic thiols by HPLC [218]. Of all of the L-amino acids tested as the chiral amine, L-valine gave the best selectivity for racemic thiols such as cysteine, 3-mercapto-1,2-propanediol, and 1-mercapto-2-propanol. Chiral DBD-PyNCS (59) reported as the label of amines involving amino acids fluorescence reagent also reacts with chiral thiols, and yields the dithiocarbamate diastereomers. Therefore, the resolution of chiral thiols is possible utilizing this reagent. The limit of detection on reversed-phase chromatography is at the sub-pmol level.

**Example 7.** Derivatization of tiopronin with $R(-)$DBD-PyNCS (59) and its separation by reversed-phase chromatography [219].

Place 10 µl of 12 mM $R(-)$DBD-PyNCS in acetonitrile and 20 µl of tiopronin enantiomers (30 µM each) in 2 mM EDTA.2Na containing 1% pyridine. After leaving to stand for 40 min, 10 µl of the mixture is injected into the column. The reaction scheme and a typical chromatogram obtained from tiopronin are shown in Fig. 6.43 and Fig. 6.44, respectively.

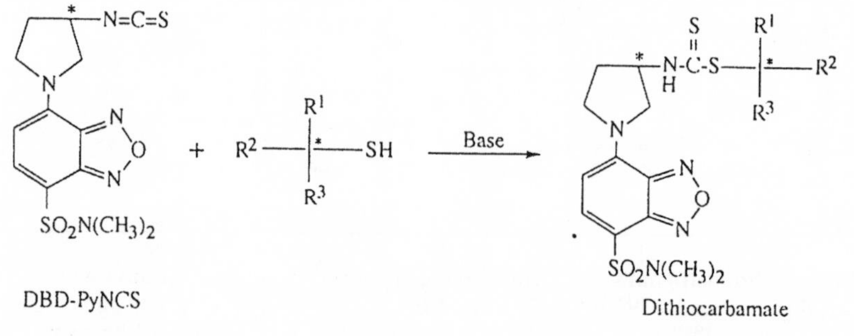

N=C=S
N
$SO_2N(CH_3)_2$
DBD-PyNCS
$R^1$ $R^2$ SH $R^3$
Base
S N-C-S H $R^1$ $R^2$ $R^3$
$SO_2N(CH_3)_2$
Dithiocarbamate

**Fig. 6.43.** Derivatization reaction of racemic thiols with $R(-)$-DBD-PyNCS. [Reproduced from ref. 219, p. 114, Fig. 2.].

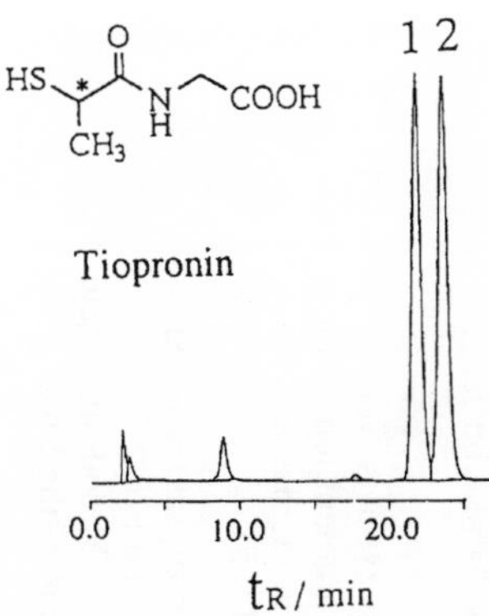


**Fig. 6.44.** Chromatogram of dithiocarbamate derivatives of racemic tiopronin obtained from the reaction with *R*(−)-DBD-PyNCS. Peaks: 1, (*R*)-tiopronin; 2, (*S*)-tiopronin. [Reproduced from ref. 219, p. 115, Fig. 3.].

#### 6.5.2.6 Label for Other Functional Groups

There is no effective fluorescence chiral tagging reagent for other functional groups such as lactones and olefine. Therefore, it is recommended to develop the reagents for these groups which are selective and sensitive.

Table 6.6 shows the fluorescent chiral labelling reagents and their application using HPLC.

### 6.5.3 Derivatization for Electrochemical (EC) Detection

A large number of chiral derivatization reagents for UV-VIS and FL detection are reported and commercially available. In contrast, the documentation of the reagents for electrochemical determination is fairly limited. These reagents are essentially active compounds to oxidation and/or reduction. Shimada *et al.* [220] developed chiral ferrocene compounds, i.e. *R*(−)-1-ferrocenylethylamine (85) and *S*(+)-1-ferrocenylpropylamine (86), for the labelling of non-steroidal inflammatory drugs such as naproxen and ibuprofen (Fig. 6.45). The Rs values of the resulting diastereomers are listed in Table 6.7. Aminotetralin stereoisomers in plasma labelled with *R*(+)-1-phenylethyl isocyanate (PEIC) (13) are determined at nanogram level with an electrochemical detector, operating in the oxidative

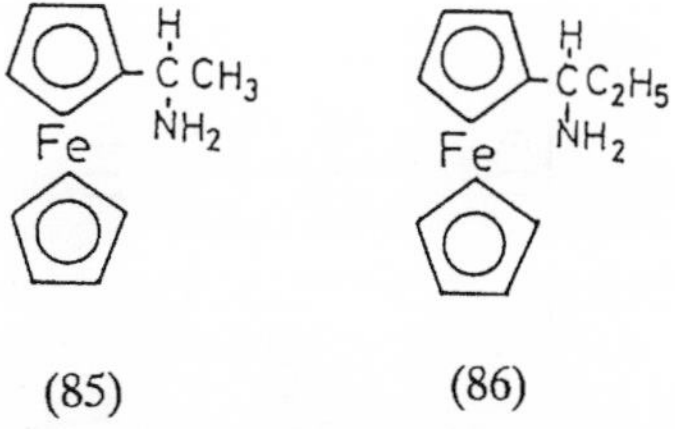


**Fig. 6.45.** Chiral derivatization reagents for EC detection: (85) *R*(−)-ferrocenylethylamine; (86) *S*(+)-ferrocenylpropylamine.

mode. Of course, the diastereomers labelled with these reagents are possible to determine with UV-VIS detector, if the reagents exhibit absorption at UV-VIS regions. *R*(−)-2-amino-1-propanol [221], (−)-heptaflurobutyrylthioproryl chloride [222] and (*S*)-1-(1-naphtyl)ethyl isothiocyanate (NEIC) (14) [223] are used for the determination of gossypol in plasma, methylphenidate and salsolinol, respectively.

**Example 1.** EC detection of aminotetralin using *R*(+)-1-phenylethyl isocyanate (PEIC) (13) [224].

The residue obtained from plasma extraction and evaporation is dissolved in acetonitrile/0.05% sodium bicarbonate (8/2, v/v %) solution (pH 7.5), and then 10 μl of 0.4% PEIC in acetonitrile is added. After reaction at $25 \pm 2$ °C for 16 h, 100 μl of 0.001 M HCl is added to the mixture. 50 μl aliquots are injected into the HPLC system. The applied potential of the screen electrode (D1), sample electrode (D2) and the potential of the guard cell are set at 0.45 V, 0.70 V and 0.9 V, respectively. Column, Spherisorb ODS-2 ($250 \times 4.6$ mm, i.d., 5 μm); Mobile phase, acetonitrile/50 mM sodium acetate (46/54, v/v %, pH 7.5). The typical chromatograms obtained from EC detection are shown in Fig. 6.46.

## 6.6 DERIVATIZATION FOR CAPILLARY ELECTROPHORESIS (CE) ANALYSIS

Chiral resolution is currently performed by chromatographic techniques including HPLC and GC.

**Table 6.6A.** Fluorescent chiral derivatization reagents for amines and their application

| Reagent | Enantiomer | Catalyst | Wavelengths (nm) | | HPLC condition | Sample and treatment | Reference |
|---|---|---|---|---|---|---|---|
| | | | ex. | em. | | | |
| OPA/BOC-C | Amino acids | | 344 | 443 | Spheri 5, RP-8 (5 μm, 220 × 4.6 mm) | Automated pre-column derivatization at room temperature. | [150] |
| OPA/BOC-C,/NAC or /NAP | Amino alcohols, peptide hydrolysate | | 344 | 443 | Hypersil ODS (3 μm, 120 × 4 mm) | Reacted at RT for 10 min in ethanol/borate buffer (pH 10). | [140] |
| OPA/BOC-C,/NAC, /NAP,/TG or /TATG | AMP, HAM, PCA, AHB, MEX, TOC, RIM, APB | | 338 | 425 | Nova-Pak ODS (4 μm, 150 × 3.9 mm) | Reacted at RT for 5 min in methanol/borate buffer (pH 9.5). | [141] |
| OPA/NAC | Tranylcy-promine | | 344 | 442 | Zorbax ODS (5 μm, 250 × 4.6 mm) | Human plasma or urine extract was reacted at RT for 5 min in ethanol/borate buffer (pH 10). | [145] |
| OPA/NAC | Baclofen | | 340 | 460 | Chrompack Cp-Spher $C_8$, 25 cm long | Reacted at 80 °C for 25 min in methanol/borate buffer (pH 10.4). | [146] |
| OPA/NAC | D,L-threo-3,4-di-hydroxy-phenylserine | | 344 | 443 | MicroSpher $C_{18}$ (3 μm, 100 × 4.6 mm) | Human plasma or urine extract was reacted at RT for 10 min in methanol/borate buffer (pH 10.0). | [139] |
| OPA/NAC | Lombricine | | | | Spherisorb ODS II "EXCEL" (5 μm, 250 × 4.6 mm) | Earthworm extracts were reacted at ambient temperature for 5 min in methanol/borate buffer (pH 9.5). | [142] |
| OPA/NAC | Amino acids | | 360 | 405 | Develosil ODS-5 (200 × 6 mm) | Protein hydrolysate was reacted at RT for 2 min in methanol/0.1 M sodium borate. Post-column method was also applied. | [144] |
| OPA/BOC-C or /NAC | *ω*-*N*-Oxalyl diamino acids | | 344 | 443 | Spherisorb ODS II "EXCEL" (5 μm, 250 × 4.6 mm) | Reacted at ambient temperature for 5 min in methanol/borate buffer (pH 8.2). | [143] |

| | | | | | | |
|---|---|---|---|---|---|---|
| OPA/NAC | MEX | 350 | 445 | Apex ODS (5 µm, 250 × 4.6 mm) | Human plasma extract was reacted at RT for 2 min in methanol/0.1 M sodium borate. | [138] |
| OPA/TG | $\alpha$-Amino acids | 360 | 420 | LiChrosorb RP-8 (5 µm, 250 × 4 mm) | Mouse serum treated with 4-fluoroglutamic acid was reacted at RT for 1 min in methanol/borate buffer (pH 10.4). | [151] |
| (+)-FLEC | Amino acids, 4-hydroxy-Proline | 260 | filter | Spherisorb octyl (5 µm, 250 × 4.6 mm) | Reacted at RT for 4 min in borate buffer (pH 6.85) and acetone/acetonitrile (3:1). | [154] |
| (+)-FLEC | *S* 12024 | 260 | 310 | Spherisorb cyanopropyl (3 µm, 150 × 4.6 mm) | Human plasma extract was reacted at RT for 1 hr in 0.01% triethylamine containing water/acetonitrile (4:5). | [158] |
| (+)-FLEC | Reboxetine | 260 | 315 | LiChroCART Supersphere 60 RP-8 (4 µm, 250 × 4.6 mm) | Human plasma extract was reacted at RT for 5 min in acetonitrile/borate buffer (pH 8) | [157] |
| (+)-FLEC | Methampheth-amine, ephedrine, pseudoe-phedrine | 295 | 315 | $5C_{18}$-AR (5 µm, 250 × 3.9 mm) | Reacted at RT for 30 min in pH 12/dichloromethane | [159] |
| (+)-FLEC | Atenolol | 227 | 310 | Microspher C18 (3 µm, 100 × 4.6 mm) | Rat plasma extract was reacted at RT for 30 min in acetone/borate buffer (pH 8.5). | [156] |
| (+)/(−)-FLEC | Carnitine | 260 | 310 | RP-18 column (5 µm, 240 × 4.6 mm) | Reacted at 45° for 60 min in acetone/carbonate buffer (pH 10.4). | [155] |
| *R*(−)-NEIC | Fluoxetine, norfluoxetine | 285 | 313 | Jones Apex Silica column (5 µm, 150 × 4.6 mm) | Human plasma extract was reacted at 60 °C for 30 min in alkaline *n*-hexane. | [165] |
| *S*(+)-NEIC | Propranolol | 225 | 280 | Whatman Partisil 5 Silica column, 250 mm long | Rat blood extract was reacted at RT for 1 min in alkaline chloroform/hexane (1:1). | [166] |
| *R*(−)-NEIC | Nadolol | 230 | 330 | Beckman ODS (5 µm, 250 × 4.6 mm) | Human plasma extract was reacted at RT for 1 hr in alkaline 1,2-dimethoxyethane. | [162] |

*continued overleaf*

**Table 6.6A.** *(continued)*

| Reagent | Enantiomer | Catalyst | Wavelengths (nm) | | HPLC condition | Sample and treatment | Reference |
|---|---|---|---|---|---|---|---|
| | | | ex. | em. | | | |
| *R*(−)-NEIC | Nadolol | | 285 | 340 | YMC-AM-303 ODS (5 μm, 250 × 4.6 mm) | Dog plasma extract was reacted at 45 °C for 5 min in alkaline methanol. | [164] |
| *R*(−)-NEIC | Prenylamine | | 285 | 333 | Nova Pak Resolve $C_{18}$ (4 μm, 150 × 309 mm) | Human plasma and urine extract was reacted at 80 °C for 10 min and then RT for 15 hr in chloroform/DMF. | [72] |
| *R*(−)-NEIC | Betaxolol | | 285 | 330 | Hypersil ODS (3 μm, 150 × 4.6 mm) | Alkaline extract of human blood was reacted at RT for 1.5 hr in methanol. | [163] |
| *S*(+)-NEIC | Acebutolol | | 220 | 389 | Whatman Partisil Silica column 5 μm, 25 cm long (5 μm, 150 × 4.6 mm) | Alkaline extract was reacted at RT for 30 s in chloroform containing pyridine. | [73] |
| DBD-Pro-COCl | Amines | | 450 | 560 | Inertsil SIL (5 μm, 150 × 4.6 mm) | Reacted at 50 °C for 90 min in benzene containing pyridine. | [212] |
| NBD-Pro-COCl | Amines | | 485 | 530 | Inertsil ODS-80A or Inertsil SIL (5 μm, 150 × 4.6 mm) | Reacted at 50 °C (amines) or 80 °C (alcohol) for 1 hr in benzene containing pyridine. | [211] |
| DBD-PyNCS | Amines, *β*-Blockers | | 460 | 540 | Inertsil ODS-80A (5 μm, 150 × 4.6 mm) | Reacted at 55 °C for 10 min in acetonitrile containing triethylamine. | [171,172] |
| NBD-PyNCS | Amines, *β*-Blockers | | 490 | 530 | Inertsil ODS-80A (5 μm, 150 × 4.6 mm) | Reacted at 55 °C for 10 min in acetonitrile containing triethylamine. | [171,172] |
| DBD-PyNCS | Amino acids, peptides | | 460 | 540 | Inertsil ODS-80A (5 μm, 150 × 4.6 mm) | Reacted at 60 °C for 10 min in acetonitrile | [172,174] |
| NBD-PyNCS | Amino acids, peptides | | 490 | 530 | Inertsil ODS-80A (5 μm, 150 × 4.6 mm) | Reacted at 60 °C for 10 min in acetonitrile | [172,174] |
| *S*(+)-FLOP-Cl | Amino acids, peptides | $Na_2CO_3$ | 305 | 355 | Zorbax Sil column (5 μm, 250 × 4.6 mm) | Reacted at 40 °C for 60 min in dichloromethane. | [177] |

| | | | | | | | |
|---|---|---|---|---|---|---|---|
| *S*(+)-FLOP-Cl | Propranolol | TEA | 305 | 355 | Ultrasphere ODS (5 μm, 250 × 4.6 mm) | Alkaline extract of human plasma was reacted at ambient temperature (overnight) in dichloromethane. | [178] |
| *S*(+)-FLOPIC | Propranolol | TEA | 305 | 355 | Nova Pak $C_{18}$ (4 μm, 150 × 3.9 mm) | Alkaline extract of human plasma was reacted at ambient temperature for 30 min in dichloromethane. | [178] |
| *S*(+)-BZOP-Cl | Baclofen, its fluoro analogue | $Na_2CO_3$ | 313 | 365 | Zorbax Sil silica gel (5 μm, 250 × 4.6 mm) | Urine, plasma, and CSF samples obtained from liquid-solid extraction were reacted at RT for 60 min in ethyl acetate solution. | [179] |
| *S*(+)-NAP-Cl | Baclofen, its fluoro analogue | $Na_2CO_3$ | 335 | 365 | Zorbax Sil silica gel (5 μm, 250 × 4.6 mm) | Urine, plasma, and CSF samples obtained from liquid-solid extraction were reacted at RT for 60 min in ethylacetate solution. | [179] |
| *S*(+)-MNB-COOH | Amino acids | | 310 | 380 | Capcell Pak ODS SG-120 (5 μm, 250 × 4.6 mm) | Amino acids were reacted with *S*(+)-MNB-COCl, converted from the reagent, at RT for 30 min in lithium carbonate buffer (pH 9.0). | [182] |

Abbreviations: RT, room temperature; TEA, Triethylamine; CSF, Cerebrospinal fluid.

**Table 6.6B.** Fluorescent chiral derivatization reagents for carboxyls and their application

| Reagent | Enantiomer | Catalyst | Wavelengths (nm) | | HPLC condition | Sample and treatment | Reference |
|---|---|---|---|---|---|---|---|
| | | | Ex. | Em. | | | |
| *S*(+)/*R*(−)-DANE | Amino acids, ibuprofen, indoprofen, naproxen | EDC | 320 | 395 | $\mu$Porasil (300 × 2.5 mm) | Reacted at RT for 3 h in pyridine or dichloromethane containing tri-*n*-butylamine. | [184] |
| *S*(+)-DANE | Loxoprofen, its metabolites | HOBT + DCC | 313 | 420 | $\mu$Porasil (10 μm, 300 × 3.9 mm) | Reacted at RT for 1 h in dichloromethane/pyridine (10:1). | [186] |
| *S*(+)-DANE | Loxoprofen | HOBT + DCC | 313 | 420 | $\mu$Porasil (10 μm, 300 × 3.9 mm) | Rat plasma extract was reacted at RT for 1 h in dichloromethane/ pyridine. | [185] |
| *R*(−)-DANE | Naproxen | HOBT + EDC | 320 | 410 | $\mu$Porasil (300 × 2.5 mm) | Human serum extract was reacted at 4 °C for 45 min in dichloromethane/ pyridine. | [183] |
| *S*(−)-NEA | MK-571 | TEA | 350 | 410 | Supelcosil R-Phenylurea column (5 μm, 250 × 4.6 mm) | Human plasma extract was reacted at RT for 10 min in acetonitrile. | [187] |
| AEA | *N*-acetyl amino acids, naproxen | EDC + HOBT | 260 | 400 | Spherical Silica column (5 μm, 150 × 4, 6 mm) 150 × 4.6 mm) | Reacted at RT for 1 hr (*N*-acetyl amino acid) in pyridine, or at 40 °C (naproxen). for 20 min. | [190] |
| DBD-APy, or NBD-APy | *N*-acetyl amino acids, anti-inflammatory drugs | DPDS/TPP or DEPC | 470 | 580 | Inertsil ODS-2 or Inertsil SIL (5 μm, 150 × 4.6 mm) | Reacted at RT for 4 h in acetonitrile. | [199] |
| NBD-APy | *N*-acetyl amino acids, anti-inflammatory drugs | DPDS/TPP | 488 | 520< (LIF detection) | Inertsil ODS-2 or Inertsil SIL (5 μm, 150 × 4.6 mm) | Reacted at RT for 4 h in acetonitrile. | [193] |

| | | | | | | | |
|---|---|---|---|---|---|---|---|
| DBD-APy | *N*-Acetyl amino acids, anti-inflammatory drugs | DPDS/TPP | Chemiluminescence | | Inertsil ODS-2 (5 μm, 150 × 4.6 mm) | Reacted at RT for 4 h in acetonitrile. | [191] |
| *S*(+)-FLOPA | Ibuprofen | EDC + HOBT | 305 | 355 | Zorbax Sil column (5 μm, 250 × 4.6 mm) | Human urine extract was reacted at ambient temperature for 2 h in dichloromethane containing pyridine. | [194] |
| *S*(+)-FLOPA | α-Phenylcyclopentylacetic acid | EDC | 305 | 355 | Zorbax Sil column (5 μm, 250 × 4.6 mm) | Human plasma and urine extract was reacted at ambient temperature for 2 h in dichloromethane containing pyridine. | [195] |
| (−)-APMB | Ibuprofen, naproxen, flurbiprofen | DPDS + TPP | 320 | 380 | TSK gel ODS-80TM (5 μm, 150 × 4.6 mm) | Reacted at RT for 20 min in dichloromethane. | [196] |
| (−)-APMB | Ibuprofen | DPDS + TPP | 320 | 380 | TSK gel ODS-80TM (5 μm, 150 × 4.6 mm) | Rat plasma extract Reacte was reacted at RT for 5 min in dichloromethane solution. | [198] |
| L-LeuBOX, L-PheBOX, or D-PgBOX | Z-Amino acids, Ibuprofen, naproxen, flurbiprofen | DPDS + TPP | 330 | 375 | TSK gel Silica-60 (5 μm, 150 × 4.6 mm) | Reacted at RT for 5 min in dichloromethane. | [197] |
| BOX-L-Phe, or BOX-L-Pro | NEA, PEA, CHEA | DPDS + TPP | 325 or 325 | 432,403 | YMCJ'sphere ODS (4 μm, 150 × 4.6 mm) or TSK gel Silica-60 (5 μm, 150 × 4.6 mm) | Reacted at RT for 5 min in dichloromethane. | [181] |

*continued overleaf*

**Table 6.6B.** *(continued)*

| Reagent | Enantiomer | Catalyst | Wavelengths (nm) | | HPLC condition | Sample and treatment | Reference |
|---|---|---|---|---|---|---|---|
| | | | Ex. | Em. | | | |
| L/D-DEPEA | Ibuprofen, flurbiprofen, pranoprofen, phenoprofen, naproxen | DPDS + TPP | 338 | 535 | TSK gel ODS-80TM (5 μm, 150 × 4.6 mm) | Reacted at RT for 3 h in acetonitrile. | [200] |
| *S*(+)-MNE-OTf | α-Methoxy-phenylacetic acid | 18-Crown-6 + $K_2CO_3$ | 259 | 394 | TSK gel ODS-80TM (5 μm, 150 × 4.6 mm) | Reacted at RT for 1 h in acetonitrile. | [202] |
| *S*(−)-ANBT | *S*-8666 | BEPT + MDPP | 355 | 480 | Nucleosil $5C_{18}$ (250 × 4.6 mm) | Reacted at RT for 2 h in acetonitrile. | [203] |

Abbreviations: RT, room temperature; BEPT, 2-bromo-1-ethylpyridinium tetrafluoroborate; CHEA, 1-Cyclohexylethylamine; DCC, dicyclohexylcarbodiimide; DEPC, diethyl phosphorocyanidate; DPDS, 2,2′-Dipyridyl disulfide; EDC, 1-ethyl-3-(3-dimethylaminopropyl)-carbodiimide; HOBT, 1-hydroxybenzotriazole; MDPP, 9-methyl-3,4,-dihydro-2*H*-pyrido[1,2-*a*]pyrimidin-2-one; NEA, 1-(1-Naphtyl)ethylamine; PEA, 1-phenylethylamine; TEA, triethylamine; TPP, Triphenylphosphine.

**Table 6.6C.** Fluorescent chiral derivatization reagents for hydroxyls, carbonyls and thiols and their application

| Reagent | Enantiomer | Catalyst | Wavelengths (nm) | | HPLC condition | Sample and treatment | Reference |
|---|---|---|---|---|---|---|---|
| | | | Ex. | Em. | | | |
| Methyl-BNCC | Hydroxyls | TEA | 342 | 420 | Cosmosil 5SL (5 μm, 150 × 4.6 mm) | Reacted at 60 °C for 20 min in acetonitrile. | [206] |
| (+)-Methyl-BNCC | β-Hydroxy acids | TEA | 342 | 420 | Cosmosil 5SL (5 μm, 150 × 4.6 mm) | Reacted at 60 °C for 20 min in acetonitrile. | [207] |
| (+)-Methyl-BNCC | Propranolol, penbutolol | Quinuclidine | 342 | 420 | Spherical Silica column (5 μm, 150 × 4.6 mm) | Reacted at 60 °C for 20 min in acetonitrile. | [207] |
| (aS)-Methoxy-BNCC | Propranolol, penbutolol, bufuralol | Quinuclidine | 330 | 420 | Cosmoil 5SL (5 μm, 150 × 4.6 mm) | Reacted at 60 °C for 20 min in acetonitrile. | [208] |
| (aS)-Methoxy-BNCC | Penbutolol | Quinuclidine | 290 | 405 | Cosmosil 5SL (5 μm, 150 × 4.6 mm) | Dog plasma extract was reacted at 60 °C for 20 min. | [209] |
| DBD-Pro-COCl | Alcohols | | 450 | 560 | Inertsil SIL (5 μm, 150 × 4.6 mm) | Reacted at 80 °C for 4 h in benzene containing pyridine. | [210] |
| NBD-Pro-COCl | Alcohols | | 485 | 530 | Inertsil ODS-80A or Inertsil SIL (5 μm, 150 × 4.6 mm) | Reacted at 50 °C (amines) or 80 °C (alcohol) for 1 h in benzene containing pyridine. | [211] |
| DBD-ProCZ | Ketones | | 450 | 540 | Inertsil ODS-80A or Inertsil SIL (5 μm, 150 × 4.6 mm) | Reacted at 65 °C for 10 min in acetonitrile/water containing trichloroacetic acid. | [217] |
| NBD-ProCZ | Ketones | | 490 | 530 | Inertsil ODS-80A or Inertsil SIL (5 μm, 150 × 4.6 mm) | Reacted at 65 °C for 10 min in acetonitrile/water containing trichloroacetic acid. | [217] |
| *S*(+)-TBMB-COOH | Acylglycerols | | 310 | 370 | Deverosil 60-3 Silica gel (5 μm, 50 × 4.6 mm) | Silylated glycerols were reacted with TBMB-COCl converted from TBMB-COOH | [215] |
| *R*(−)-TBMB-COOH | Mono-saccharides | | 310 | 370 | ODS column (150 × 4.6 mm) | Reducing sugars treated with acetic anhydride/$HClO_4$ and HBr/AcOH were reacted with TBMB-COOH/$KHCO_3$ in acetone at 60 °C for 1 h. | [216] |
| OPA/L-Valine | Cysteine, homo-cysteine, 3-MPD, 1-MPR | | 345 | 440 | Capcell Pak $C_{18}$ AG120 (5 μm, 150 × 4.6 mm) | Reacted at RT for 1 min in borate-phosphate buffer (pH 9.0). | [218] |

Abbreviations: RT, room temperature; 3-MPD, 3-mercapto-1,2-propanediol; 1-MPR, 1-mercapto-2-propanol; TEA, Triethylamine.

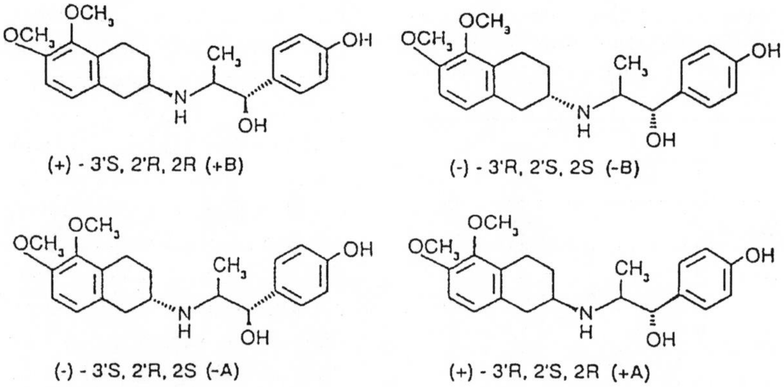

(+) - 3'S, 2'R, 2R (+B)

(-) - 3'R, 2'S, 2S (-B)

(-) - 3'S, 2'R, 2S (-A)

(+) - 3'R, 2'S, 2R (+A)

Structures of the stereoisomers of I.

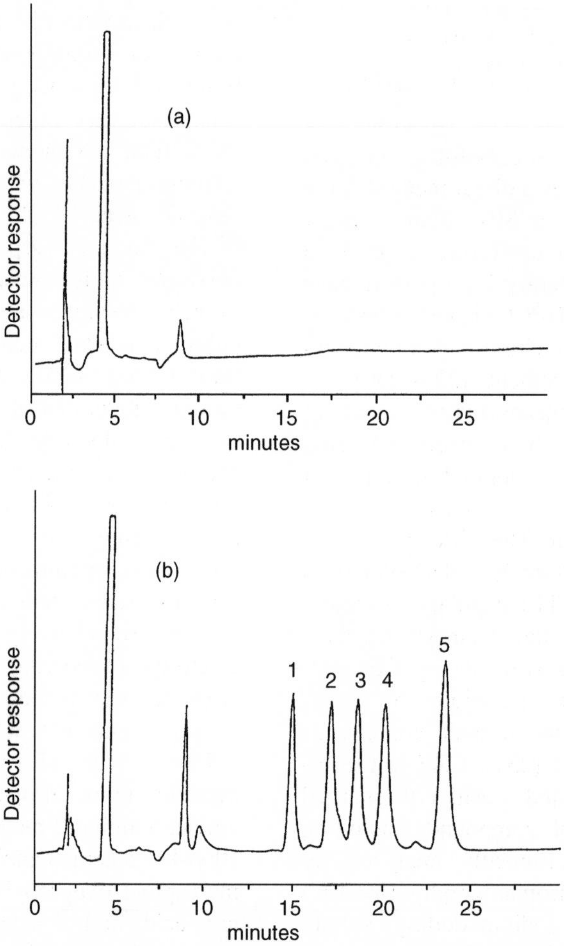

**Fig. 6.46.** Chromatograms of aminotetralin in plasma extracts derived from $R(+)$-PEIC. Chromatograms: (a) blank sample; (b) sample spiked with I (400 ng/ml). [Reproduced from ref. 224, p. 100, Fig. 6.].

**Table 6.7.** HPLC separation of diastereomeric amides derived from *N*-acetylamino acids with *R*(−)-ferrocenylethylamine (85) and *S*(+)-ferrocenylpropylamine (86)

| *N*-acetylamino acid | | (85) | | (86) | |
|---|---|---|---|---|---|
| | | k′ | Rs | k′ | Rs |
| Alanine | D | 13.2[a] | | 28.8[c] | |
| | L | 12.0 | 1.20 | 32.0 | 1.50 |
| Valine | D | 5.4[b] | | 6.4[d] | |
| | L | 6.4 | 1.33 | 5.2 | 1.70 |
| Leucine | D | 9.6[b] | | 5.8[d] | |
| | L | 12.4 | 2.40 | 4.6 | 1.50 |
| Phenylalanine | D | 12.4[b] | | 7.4[d] | |
| | L | 16.4 | 2.67 | 5.6 | 1.80 |

Conditions: mobile phase, tetrahydrofuran/1.5% AcONa (pH 5.0), [a] (1:4), [b] (1:2), [c] (1:4), [d] (2:3). Flow rate was set at 0.5 ml/min ($t_0 = 2.5$ min) except for c) (1 ml/min, $t_0 = 1.25$ min). [Reproduced from ref. 220, p. 3166, Table 1.]

Almost all racemates are successfully separated and determined by HPLC as a direct method using a chiral stationary phase (CSP) column and/or indirect method involving derivatization using a chiral tagging reagent. Recently, high-performance capillary electrophoresis (HPCE) has also become another important tool for chiral separation with the advance of HPCE instruments [225–230].

HPCE is an excellent means for separation of multi-components with its high speed and separatability. The technique is applied in different fields of research such as biological, pharmaceutical and environmental. The different separation modes, named capillary gel electrophoresis (CGE), isotachophoresis (ITP), capillary zone electrophoresis (CZE) and micellar electrokinetic chromatography (MEKC), are available by CE analysis. HPCE is a popular technique for separating various inorganic and organic compounds involving chiral molecules [231]. CZE is recommended for negatively and positively charged compounds. A number of compounds including neutral substances are efficiently analyzed by micellar electrokinetic chromatography (MEKC) using detergents such as sodium dodesyl sulfate (SDS) and cholesterols [232–234]. Among these modes, CZE and MEKC are mainly used for the separation of a pair of enantiomers. One approach for the CE separation of enantiomers involves the use of buffer additives such as cyclodextrine derivatives, chiral surfactants or cyclodextrine plus chiral surfactant. The other method for the CE analysis of optical isomers is classified into the MEKC separation of diastereomers resulting from the reaction of enantiomers with chiral tagging reagents. Although chiral molecules are also resolved with the MEKC mode, the addition of a chiral selector such as $\alpha$-, $\beta$- and $\gamma$-cyclodextrines (CDs) is required for effective separation. The separation of two enantiomers is a difficult task in CE analysis because they possess similar physico-chemical properties, unless a chiral environment is used in order to selectively modify their electrophoretic mobilities. Only a minute amount of chiral selector used for chiral separation is usually added to the background electrolyte. Therefore, very expensive chiral selectors can be used. Table 6.8 shows some typical examples for enantioseparation after derivatization with achiral reagents using HPCE.

The indirect separation method involving derivatization is mainly applied in CE for the enantiometric resolution of amino acids using different chiral reagents, such as GITC, OPA/a chiral thiol, OPA/a chiral amine, (+)-*O*,*O*′-dibenzoyl-L-tartaric anhydride (DBTAAN) [235], (+)-diacetyl-L-tartaric anhydride (DATAAN) [236], FLEC and Marfey's reagents. The examples of diastereomeric resolution by HPCE are shown in Table 6.9. CE analysis using indirect chiral derivatization is still rarely used for chiral separation, because the enantiomeric derivatizations obtained from an achiral reagent are relatively easy to separate with the electrolytes involving a chiral selector and micel. However, the method is essentially applicable to all chiral molecules which are resolved by HPLC.

Leroy *et al.* [237] evaluated the separation of racemic amino drugs ($\alpha$-methylbenzeneethanolamine, 6-amino-2-methyl-2-heptanol and 1-aminoethyl-benzenemethanol) and thiol drugs [*N*-(2-mercapto-1-oxopropyl)glycine, 2-mercaptopropionic acid and *N*-acetyl-3-mercaptovaline] after indirect derivatization by reversed-phase HPLC and CE. According to the reactive group of the analytes, OPA/homochiral thiol (i.e. *N*-acetyl-L-cysteine or *N*-acetyl-D-penicillamine) for chiral

**Table 6.8.** HPCE separation of chiral compounds labeled with achiral reagents

| Reagent | Enantiomer | CE mode | Chiral selector | Detection | Conditions (BGE and capillary) | Sample and treatment | Reference |
|---|---|---|---|---|---|---|---|
| PITC | Amino acids, peptides | MEKC | TM-$\beta$-CD gigitonin $\beta$-escin | UV269nm | 50 mM sodium phosphate (pH 3.0), 50 mM SDS and 25 mM chiral selector; uncoated capillary (50 cm × 50 μm, i.d., 30 cm effective length); −15kV; 26 °C. | Samples (0.1 ∼ 1 mg) in 0.1 ml dimethylallylamine-propanol-water (1.7:30:20) or metanol-water-triethylamine (7:1:1) and PITC (5–10 μl) are heated at 55 °C for 30 min. After remove the solvent, 12.5–25% TFA in 0.1 ml water or 1 M HCl is added and heated at 55 °C for 40 min or at 80 °C for 5 min. | [251] |
| ANDSA | Phenoxy acid herbicides | CZE | $\beta$-CD, TM-$\beta$-CD | UV230 nm, FL420 nm (He–Cd laser ex. 325 nm) | 25 mM sodium phosphate, 600 mM borate (pH 5.0), containing 10 mM chiral selector; uncoated capillary (57 cm × 50 μm, i.d., 50 cm effective length); 20 kV; 30 °C. | 5 μl EDAC ($5 \times 10^{-5}$ M) pH 5.0 solution is added to the mixture of 5 μl sample ($5 \times 10^{-5}$ M) in acetonitrile and 10 μl ANDSA ($5 \times 10^{-5}$ M) in 50 mM phosphate buffer (pH 3.0), then stirred for 2.0 h at RT. | [252] |
| AEOC | Amino acids | MEKC | $\beta$-CD, $\gamma$-CD | UV256 nm, | 50 mM phosphate (pH 7.5), 40 mM SDS, 1M urea, 15% IPA, 45 mM $\beta$-CD; 50 mM phosphate (pH 7.5), 40 mM SDS, 15% IPA, 10 mM $\gamma$-CD; capillary (67 cm × 25 μ m, i.d., 46 cm to detector); 30 kV; 25 °C. | 0.2 ml of 10 mM AEOC and 0.2 ml of 2 mM amino acid in 0.2 M borate buffer (pH 9.0) react at RT for 2 min. | [253] |
| FMOC | Peptides | MEKC | Vancomycin | UV254 nm, | 25 nM HEPES (pH 7.6), 25 mM SDS, 1 mM vancomycin; capillary (68 cm × 25 μm, i.d., 46 cm to detector); 25 kV; 25 °C. | Sample (0.1–0.2 mM) reacts with FMOC. | [254] |

| | | | | | | | |
|---|---|---|---|---|---|---|---|
| FMOC | Amino acids | MEKC | $\beta$-CD, $\gamma$-CD | UV256 nm, | 50 mM phosphate (pH 7.5), 50 mM SDS, 12 mM $\beta$-CD (or $\gamma$-CD), 15% 15% IPA; capillary (67 cm × 25 μm, i.d., 45 cm to detector); 25 kV; 25 °C. | 0.2 ml of 10 mM AEOC and 0.2 ml of 2 mM amino acid in 0.2 M borate buffer (pH 9.0) react at RT for 2 min. | [254] |
| AQC | Amino acids | MEKC | (*R*) or (*S*)-DDCV | UV254 nm, | 25 mM $Na_2HPO_4/Na_2B_4O_7$ (pH 9.0), and 0.1 M (*R*)- or (*S*)-DDCV; capillary (60 cm × 50 μm, i.d., 52.5 cm to detector); 16 kV; 30 °C. | 0.1 ml of amino acid (0.25 mg/ml) in 0.7 ml borate buffer (0.2 M, pH 8.8) is voltexed with 0.1 ml of AQC (3 mg) in acetonitrile. | [255] |
| DNS-Cl | Amino acids | CZE | $\beta$-CD | UV254 nm, | *N*-methylformamide containing | 0.1 ml of amino acid (0.5 mM) | [256] |
| | | | | | 10 mM NaCl and 25–200 mM $\beta$-CD capillary (58.5 cm × 50 μm, i.d., 50 cm to detector); 30 kV; 25 °C. | in 0.1 M borate buffer (pH 9.5) reacts with 0.1 ml DNS-Cl (8 mM) in acetonitrile at RT for 1 h. | [257] |
| DNS-Cl | 2-methyltaurine | MEKC | $\beta$-CD and $\gamma$-CD | UV254 nm, | 0.1 M borate buffer (pH 8.6) prepared with 0.618 boric acid in water-metanol (8:2) and 2 M NaOH, 0.1 M SDS, 60 mM $\beta$-CD and 10 mM $\gamma$-CD; capillary (80 cm × 75 μm, i.d., 71 cm effective length); 20 kV; 25 °C. | To 1 ml of 2-methyltaurine of DNS-Cl in acetonitrile (1 mg/ml) is added 2 ml aminosulfonic acid solution (0.3 mg/ml) in 40 mM sodium carbonate buffer (adjusted to pH 9.5 with 2 M HCl) at RT for 1 h. | [258] |
| DNS-Cl | Amino acids | MEKC | DDA | UV275 nm, | 2 mM copper acetate, 4 mM DDA, 50 mM SDS, 20 mM ammonium acetate, and 5% glycerol (pH 7.0); capillary (97.5 cm × 50 μm, i.d., 64 cm to detector); 0.3 kV/cm; RT. | DNS-Amino acids | [259] |

Abbreviations: ANDSA, 7-Aminonaphthalene-1,3-disulfonic acid; AEOC, 2-(9-Anthryl)ethyl chloroformate; AQC, 6-aminoquinoyl-*N*-hydroxysuccinimidyl carbamate; BGE, Background electrolyte; DNS-Cl, 4-Dimethylaminobenzene-4′-sulfonyl chloride; DDCV, *N*-dodecoxycarbonylvaline; DDA, *N*,*N*-di-decyl-D-alanine; FMOC, 9-Fluorenylmethyl chloroformate; TM-$\beta$-CD, 2,3,6-Tri-*O*-methyl-$\beta$-cyclodextrin.

**Table 6.9.** Diastereomer separation by HPCE and their applications

| Reagent | Enantiomer | CE type | Detection | Conditions (BGE and capillary) | Sample and treatment | Reference |
|---|---|---|---|---|---|---|
| DBTAAN | Amino acids $\alpha$-hydroxy acids | CZE | UV233 nm | 0.025 M phosphate buffer (pH 7.0) or 0.025 M phosphate buffer (pH 5.8), 3% PVP; fused silica capillary (56 cm × 100 μm, i.d., 39 cm to detector); coated with 1% 3-(tri-methoxysilyl)propyl-methacrylate in 50% aqueous acetone and 3% acrylamide in water containing 0.04% TEMED and 0.05% ammonium persulfate; 12 kV. | 0.25 mmol sample in acetonitrile (dry solvent) and 0.5 mmol reagent reacted at 50 °C for 20 h. | [260] |
| DBTAAN | Amino acids | CZE | UV233 nm | 0.025 M phosphate buffer 3% PVP; capillary (56 cm × 100 μm, i.d., 39 cm to detector), coated with 1% 3-(tri-methoxysilyl)propyl-methacrylate in ethanol and 4% acrylamide in water containing 0.1% TEMED and 0.1% ammonium persulfate; 12 kV. | 0.25 mmol sample in acetonitrile (dry solvent) and 0.5 mmol reagent reacted at 50 °C for 20 h. | [235] |
| DATAAN | D/L-Tryptophan | CZE | UV233 nm | 0.025 M phosphate buffer (pH 6.35) 6% PVP; capillary (54 cm × 100 μm, i.d., 38 cm to detector), coated with 1% 3-(tri-methoxysilyl)propyl-methacrylate in ethanol and 4% acrylamide in water containing 0.1% TEMED and 0.1% ammonium persulfate; 10 kV; 24 °C. | 10 mM tryptophan in 25 ml acetonitrile (dry solvent) and 20 mM reagent reacted at 50 °C for 20 h. | [236] |
| Marfey's reagent | Amino acids | MEKC (SDS) | UV214nm or 340 nm | 0.05 M ammonium phosphate buffer (pH 3.3) or 0.1 M sodium borate buffer (pH 8.5); capillary (57 cm × 75 μm, i.d., 50 cm effective length); 10 kV or 12 kV; 25 °C. | 0.05 M amino acid or peptide in 50 μl water and 1% reagent in 70 μl acetonitrile (or acetone) reacted at 35 °C for 90 min (or 60 °C for 15 min). | [241] |
| Marfey's reagent | Amphetamine | MEKC | UV340 nm | 80% aqueous solution of 5 mM sodium | amphetamine (A), 4-hydroxy- | [241] |

| | | | | | | |
|---|---|---|---|---|---|---|
| | related drugs | (SDS) | | borate (pH 9.0), 0.1 M SDS, 20% methanol; non-coated capillary (48.5 cm × 50 µm, i.d., 40 cm effective length); 30 kV; 40 °C. | amphetamine (HA), 3,4-methylene-dioxymethamphetamine (MDMA) and 3,4-methylenedioxyethamphetamine (MEDA), labelled with Marfey's reagent. | [242] |
| GITC | Amphetamine, methamphetamine and their hydroxy-phenylethylamine precursors | MEKC (SDS) | UV210 nm | 0.01 M phosphate-borate buffer (pH 9.0) containing 0.1 M SDS; capillary (48 cm × 50 µm, i.d., 26 cm effective length); 20 kV; 40 °C. | Phenylethylamines (0.4 mg) in 0.1 ml of 50% aqueous acetonitrile containing 0.2% triethylamine and 0.1 ml of 1.28% GITC-acetonitrile solution reacted at RT for 15 min. | [238] |
| FLEC | Amino acids | MEKC (SDS) | UV200nm LIF (KrF) ex. 248 nm | 5 mM sodium borate (pH 9.2), 25 mM SDS and 10% acetonitrile; capillary (65 cm × 50 µm, i.d., 60 cm effective length); 20 kV; RT. | 10 µl of amino acid (~0.14 mg/ml of each), 10 µl of 0.2 M borate buffer (pH 8.0) and 10 µl of FLEC (18 mM in acetone) reacted at RT for 4 min. | [239] |
| FLEC | Carnitine | CZE | UV210 nm, 214 nm and 254 nm | 50 mM phosphate (pH 2.6), 20 mM tetrabutylammonium bromide; capillary (67 cm × 75 µm, i.d., 60 cm effective length); 20 kV, or 25 kV; RT. | 30 µl of carnitine in 30 µl carbonate buffer (50 mM, pH 10.4) and 80 µl FLEC (1 mM in acetone) reacted at 45 °C for 60 min. | [240] |
| OPA/Valine | *N*-acetyl-cysteine | CZE | UV214 nm, | 155 mM borate buffer (pH 8.98) | 5 ml of 10 mM *N*-acetylcysteine, | [261] |
| | | | | containing 5% PEG; capillary (37 cm × 50 µm, i.d., 30 cm effective length); 25 kV, at 25 °C. | 5 ml of 10.1 mM OPA in 133 mM borate buffer (pH 10.4) and 5 ml of 10.1 mM L-valine in 0.01 M HCl are mixed and used for analysis. | [262] |
| OPA/chiral thiol | Amino drugs | MEKC | UV335 nm, | 0.1 M borate buffer (pH 9.5) containing 50 mM SDS; capillary (70 cm × 75 µm, i.d.); 20 kV, at 25 °C. | 0.1 ml analyte (0.1 ~ 0.001 M) in 0.01 M HCl, 0.5 ml of 0.1 M borate buffer (pH 9.5), 0.2 ml of chiral thiol (2.5 mM) in 0.01 M HCl and 0.2 ml of OPA (5 mM) in ethanol reacted at 20 °C for 10 min. | [237] |

amines and OPA/chiral amine (i.e. [(−)-(1*R*,2*S*)-norephedrine, L-phenylalanine, L-tyrosine or 3-hydroxy-L-tyrosine] for chiral thiols are studied as the chiral labelling reagents. For racemic amine drugs, the isoindole diastereomers are separated well by the techniques of MEKC and MEKC with $\beta$-CD; whereas some racemic thiols are resolved by CZE using $\beta$-CD. Although gradient elution would be needed in HPLC for the separation of diastereomeric derivatives, CE allows the separation with different polarity under constant migration conditions. GITC shown to be an excellent UV label for the analysis of primary and secondary amines by HPLC is utilized in CE analysis for the separation of amphetamine, methamphetamine and their related compounds [238]. FLEC [239,240] and Marfey's [241–243] reagents are important labels as a chiral reagent for indirect resolution of racemates not only in HPLC but also CE analysis. Enantioseparations of amino acids are achieved indirectly by MEKC after pre-column labelling with (+)-FLEC [236]. In the resulting separated diastereomers, the L-enantiomers migrated faster than the corresponding D-isomers. Opposite migrations are obtained with use of (−)-FLEC reagent. When laser-induced fluorescence at 248 nm (KrF) is used for detection, the limit of detection is improved to nM levels which is about two orders of magnitude more sensitive than UV detection at 200 nm. The chiral separations are also performed with a combination of chiral and achiral ion pairing reagents, such as (*R*)(−)-camphersulfonic acid, sodium cyclamate and quinine, and CDs [244].

**Example 1.** CE separation of amino acids labelled with FLEC [239].

To 10 µl of amino acids mixture (−0.14 mg/ml each) is added 10 µl of 200 mM borate buffer (pH 8.0) and 10 µl of (+)-FLEC of (+)- or (−)-FLEC in acetone (4 mM). The mixture is left to stand for 4 min, and then diluted with deionized water before injection into the CE system. The separations are performed at 20 °C and 20 kV with 50 µm × 67 cm fused-silica capillaries (60 cm to detector). The samples are injected into the capillary by applying pressure (0.5 psi) for 5 s. LIF excitation is provided by a compact, pulsed laser operating at 248 nm. An electropherogram of the MEKC separation of (+)-FLEC derivatives of amino acids is shown in Fig. 6.47. The comparison between UV and LIF detection is shown in the electroherograms of Fig. 6.48.

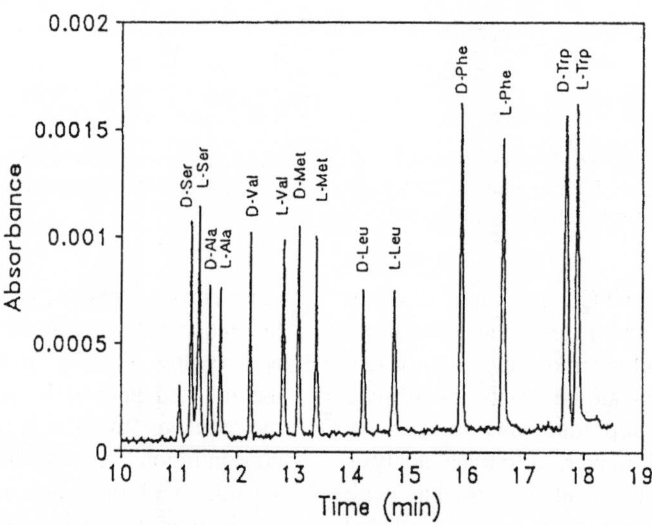


**Fig. 6.47.** MEKC separation of racemic amino acids derived from (−)-FLEC. Buffer, 10 mM sodium borate (pH 9.2), 25 mM SDS, and 10% acetonitrile. [Reproduced from ref. 239, p. 508, Fig. 7.].

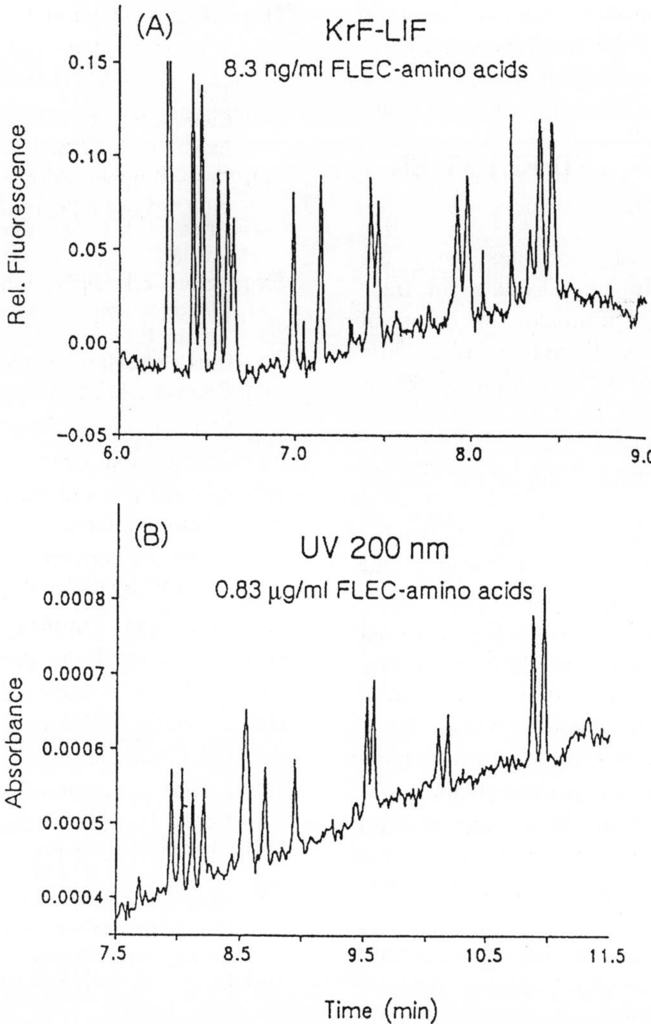


**Fig. 6.48.** Comparison of LIF and UV detection of (+)-FLEC derivatized amino acids. Electropherograms: (A) KrF-LIF detection; capillary, 65 cm × 50 mm; voltage, 20 kV; (B) UV detection at 200 nm; capillary, 67 cm × 50 mm; voltage, 20 kV. [Reproduced from ref. 239, p. 508, Fig. 9.].

The advantages of HPCE are rapid analysis of multi-components with excellent resolution. The identification of chiral molecules with direct and indirect methods can be easily established with relative good reproducibility and at a low running cost. Moreover, ultratrace analysis at attomole–zeptomole level is possible with sensitive detection modes such as laser-induced fluorescence (LIF) and mass spectrometry (MS). The equilibration of the capillary is faster than the other chromatographic techniques like HPLC, and the requested volume of background electrolyte is very low, usually in the $\mu$l-ml range. Since the injected volume is extremely low (nl level), the aqueous sample of low volume and high concentration is suitable for CE analysis. Although separation of racemates is possible with HPCE, application to real samples such as plasma and urine is still limited due to interference of separation and the sticking of endogenous components to the capillary column. However, the method of CE will be used incresingly for enantioseparation in various samples.

## 6.7 CONCLUSION AND FURTHER PERSPECTIVE

The direct resolution of racemates using the CSP column is usually not adequate for trace analysis in biological specimens, in terms of sensitivity and selectivity. Experimentally, chiral resolution with indirect derivatization is not as easy as the direct method using the CSP column. In spite of many problems associated with the indirect method, i.e. optical purity of the reagent, stability of the reagent, possibility of racemization during the tagging reaction and supply of the reagent; the good sensitivity and selectivity of the indirect method using a fluorescent chiral derivatization reagent are attractive points for the determination of chiral molecules in real samples. Since the optical purity of chiral derivatization reagents is generally less than 99%, a strict assay of the trace quantity of enantiomer in a large amount of antipode is relatively difficult using the indirect method. Thus, the indirect method is recommended for analysis such as metabolic studies in biological specimens, because the % CV is usually in the acceptable range of error. Since the resolution of racemates are largely influenced by the selection of the chiral tagging reagent, derivatization conditions and chromatographic conditions, information concerning the chiral reagent and derivatization conditions which affect the reaction rate, yield and racemization should be considered. The chiral resolution of important compounds such as drugs and biogenic amines has mainly been carried out by HPLC. However, HPCE may be also used as an effective tool for chiral analysis. Furthermore, the technique of capillary electrochromatography (CEC) is worthy of note for the separation and detection of more hydrophobic chiral compounds than the compounds suited to CE analysis. The CEC may well become a powerful technique in chiral resolution as in occupies the middle position between HPLC and HPCE.

## REFERENCES


[1] Fischer, E. (1919) *Ber. Dtsch. Chem. Ges.*, **524**, 129.

[2] Cahn, R.S., Ingold, C.K. and Prelog, V. (1956) *Experientia*, **12**, 81.

[3] Cahn, R.S., Ingold, C.K. and Prelog, V. (1966) *Angew. Chem. Int. Ed. Engl.*, **5**, 385.

[4] Ariens, E.J. (1990) *Chem. & Eng. News*, March 19, 38.

[5] Ariens, E.J. (1988) *Stereoselectivity of Pesticides. Biological and Chemical Problems*, pp 39–108. Ariens, E.J., Van Resen, J.J.S. and Welling W. (Eds.) Elsevier, Amsterdam.

[6] Borman, S. (1992) *Chem. & Eng. News*, June 15, 5.

[7] Stinson, S.C. (1992) *Chem. & Eng. News*, September 28, 46.

[8] Sheldon, R.A. (1993) *Chirotechnology*, Marcel Dekker Inc. New York.

[9] Pirkle, W.H., Hyun, M.H. and Bank, B. (1984) *J. Chromatogr.*, **316**, 585.

[10] Brooks, C.J.W., Gilbert, M.T. and Gilbert, J.D. (1973) *Anal. Chem.*, **45**, 896.

[11] Burden, R.S., Deas, A.H.B. and Clarke, T. (1987) *J. Chromatogr.*, **391**, 273.

[12] Attygalle, A.B., Morgan, E.D., Evershed, R.P. and Rowland, S.J. (1983) *J. Chromatogr.*, **260**, 411.

[13] Gil-Av, E., Charles-Sigler, R., Fischer, G. and Nurok, D. (1966) *J. Gas Chromatogr.*, **4**, 51.

[14] Charles, R., Fischer, G. and Gil-Av, E. (1963) *Isr. J. Chem.*, **1**, 234.

[15] Mosandl, A., Gessner, M., Gunther, C., Deger, W. and Singer, G. (1987) *J. High Resolut. Chromatogr. Chromatogr. Commun.*, **10**, 67.

[16] Gil-Av, E. and Nurok, D. (1962) *Proc. Chem. Soc.*, 146.

[17] Anders, M.W. and Copper, M.J.. (1971) *Anal. Chem.*, **43**, 1093.

[18] Hammarstrom, S. and Hamberg, M. (1973) *Anal. Biochem.*, **52**, 169.

[19] Cross, J.M., Putney, B.F. and Bernstein, J. (1970) *J. Chromatogr. Sci.*, **8**, 679.

[20] Dale, J.A., Dull, D.L. and Mosher, H.S. (1969) *J. Org. Chem.*, **34**, 2543.

[21] Michelsen, P. and Odham, G. (1985) *J. Chromatogr.*, **331**, 295.

[22] Doolittle, R.E. and Health, R.R. (1984) *J. Org. Chem.*, **49**, 5041.

[23] Pereira, W., Bacon, V.A., Patton, W. and Halpern, B. (1970) *Anal. Lett.*, **3**, 23.

[24] Sonnet, P.E., Piotrowski, E.G. and Boswell, R.T. (1988) *J. Chromatogr.*, **436**, 205.

[25] Gaydou, E.M. and Randriamiharisoa, R.P. (1987) *J. Chromatogr.*, **396**, 378.

[26] Pirkle, W.H. and Hoekstra, M.S. (1974) *J. Org. Chem.*, **39**, 3904.
[27] Pirkle, W.H. and Boeder, C.W. (1978) *J. Org. Chem.*, **43**, 1950.
[28] Yamazaki, Y. and Maeda, H. (1986) *Agric. Biol. Chem.*, **50**, 79.
[29] Westley, J.W. and Halpern, B. (1968) *J. Org. Chem.*, **33**, 3978.
[30] Kamerling, J.P., Duran, M., Gerwig, G.R., Ketting, D., Bruinvis, L., Vliegenthart, J.F.G. and Wadman, S.K. (1981) *J. Chromatogr.*, **222**, 276.
[31] Kaneda, T. (1986) *J. Chromatogr.*, **366**, 217.
[32] Ackman, R.G., Cox, R.E., Eglinton, G., Hooper, S.N. and Maxwell, J.R. (1972) *J. Chromatogr. Sci.*, **10**, 392.
[33] Chapman, R.A. and Harris, C.R. (1979) *J. Chromatogr.*, **174**, 369.
[34] Jiang, M. and Soderlund, D.M. (1982) *J. Chromatogr.*, **248**, 143.
[35] Rowland, S.J., Larcher, A.V., Alexander, R. and Kagi, R.I. (1984) *J. Chromatogr.*, **312**, 395.
[36] Kamerling, J.P., Gerwig, G.R., Vliegenthart, J.F.G. (1977) *J. Chromatogr.*, **143**, 117.
[37] Horiba, M., Kitahara, H., Takahashi, K., Yamamoto, S., Murano, A. and Oi, N. (1979) *Agric. Biol. Chem.*, **43**, 2311.
[38] Ballard, K.D., Eller, T.D. and Knapp, D.R. (1983) *J. Chromatogr.*, **275**, 161.
[39] Halpern, B. *Derivatization for Chromatogrraphy*, (1977) pp 457–499, Blau, K. and King, G.S. (Eds.) Heyden, London.
[40] Pereira, W.E., Salomon, M. and Halpern, B. (1971) *Aust. J. Chem.*, **24**, 1103.
[41] Hara, S., Okabe, H. and Mihashi, K. (1987) *Chem. Pharm. Bull.*, **35**, 501.
[42] Pappo, R., Collins, P. and Jung, C. (1973) *Tetrahedron Lett.*, **12**, 943.
[43] Schweer, H. (1983) *J. Chromatogr.*, **259**, 164.
[44] Theilacker, W. and Ebke, K. (1956) *Angew. Chem.*, **68**, 303.
[45] Schweer, H. (1982) *J. Chromatogr.*, **243**, 149.
[46] Schweer, H. (1983) *Carbohydr. Res.*, **116**, 139.
[47] Saucy, G., Borer, R., Trullinger, D.P., Jones, J.B. and Lok, K.P. (1977) *J. Org. Chem.*, **42**, 3206.
[48] Engel, K.H., Albrecht, W. and Heidlas, J. (1990) *J. Agric. Food Chem.*, **38**, 244.
[49] Mosandl, A., Gessner, M., Gunther, C., Deger, W. and Singer, G. (1987) *J. High Resolut. Chromatogr. Chromatogr. Commun.*, **10**, 67.
[50] Deger, W., Gessner, M., Gunther, C., Singer, G. and Mosandl, A. (1988) *J. Agric. Food Chem.*, **36**, 1260.
[51] Gessner, M., Deger, W. and Mosandl, A. (1988) *Int. Food Res. Technol.*, **186**, 417.
[52] Blau, K. and Halket, J. (Eds.) *Handbook of Derivatization for Chromatogrraphy*, 2nd Edition (1993) John Wiley & Sons, Chichester.
[53] Frei, R.W. and Zeck, K. (Eds.) *Selective Sample Handling and Detection in HPLC, Part A and B*, (1988) Elsevier, Amsterdam.
[54] Lingeman, H. and Underberg, W. (Eds.) *Detection-orientated Derivatization Techniques in LC* (1990) Marcel Dekker, New York.
[55] Nimura, N., Ogura, H. and Kinoshita, T. (1980) *J. Chromatogr.*, **202**, 375.
[56] Hasegawa, R., Murai-Kushiya, M., Komuro, T. and Kimura, T. (1989) *J. Chromatogr.*, **494**, 381.
[57] Demian, I. and Gripshover, D.F. (1989) *J. Chromatogr.*, **466**, 415.
[58] Allgire, J.F., Juenge, E.C., Damo, C.P., Sullivan, G.M. and Kirchhoefer, R.D. (1985) *J. Chromatogr.*, **325**, 249.
[59] Nimura, N., Toyama, A. and Kinoshita, T. (1984) *J. Chromatogr.*, **316**, 547.
[60] Eisenberg, E.J., Patterson, W.R. and Kahn, G.C. (1989) *J. Chromatogr.*, **493**, 105.
[61] Miller, K.J., Gal, J. and Ames, M.M. (1984) *J. Chromatogr.*, **307**, 335.
[62] Sedman, A.J. and Gal, J. (1983) *J. Chromatogr.*, **278**, 199.
[63] Noggle, F.T. and Clark, C.R. (1986) *J. Forensic Sci.*, **31**, 732.
[64] Noggle, F.T., DeRuiter, J. and Clark, C.R. (1986) *Anal. Chem.*, **58**, 1643.
[65] Gal, J. and Sedman, A.J. (1984) *J. Chromatogr.*, **314**, 275.
[66] Schaefer, H.G., Spahn, H., Lopez, L.M. and Derendorf, H. (1990) *J. Chromatogr.*, **527**, 351.
[67] Laganiere, S., Kwong, E. and Shen, D.D. (1989) *J. Chromatogr.*, **488**, 407.
[68] Gulaid, A.A., Houghton, G.W. and Boobis, A.R. (1985) *J. Chromatogr.*, **318**, 393.
[69] Chin, S.K., Hui, A.C. and Giacomini, K.M. (1989) *J. Chromatogr.*, **489**, 438.
[70] Pflugmann, G., Spahn, H. and Mutschler, E. (1987) *J. Chromatogr.*, **421**, 161.
[71] Maibaum, J. (1988) *J. Chromatogr.*, **436**, 269.
[72] Gietl, Y., Spahn, H. and Mutschler, E. (1988) *J. Chromatogr.*, **426**, 305.
[73] Piquette-Miller, M., Foster, R.T., Pasutto, F.M. and Jamali, F. (1990) *J. Chromatogr.*, **526**, 129.
[74] Dunlop, D.S. and Neidle, A. (1987) *Anal. Biochem.*, **165**, 38.
[75] Goto, J., Goto, N., Hikichi, A. and Nambara, T. (1979) *J. Liq. Chromatogr.*, **2**, 1179.
[76] Husain, P.A., Colbert, J.E., Sirimanne, S.R., Van Derveer, D.G., Herman, H.H. and May, S.W. (1989) *Anal. Biochem.*, **178**, 177.
[77] Prakash, C., Jajoo, H.K., Blair, I.A. and Mayol, R.F. (1989) *J. Chromatogr.*, **493**, 325.
[78] Mehvar, R. (1989) *J. Pharm. Sci.*, **78**, 1035.

[79] Prakash, C., Koshakji, R.P., Wood, A.J.J. and Blair, I.A. (1989) *J. Pharm. Sci.*, **78**, 771.
[80] Schmitthenner, H.F., Fedorchuk, M. and Walter, D.J. (1989) *J. Chromatogr.*, **487**, 197.
[81] Mehvar, R. (1989) *J. Chromatogr.*, **493**, 402.
[82] Li, F., Cooper, S.F. and Cote, M. (1995) *J. Chromatogr. B*, **668**, 67.
[83] Ahnoff, M., Chen, S., Green, A. and Grundevik, I. (1990) *J. Chromatogr.*, **506**, 593.
[84] Hermansson, J. and Von Bahr, C. (1980) *J. Chromatogr.*, **221**, 109.
[85] Naganuma, H., Kondo, J. and Kawahara, Y. (1990) *J. Chromatogr.*, **532**, 65.
[86] Clark, C.R. and Barksdale, J.M. (1984) *Anal. Chem.*, **56**, 958.
[87] Barksdale, J.M. and Clark, C.R. (1985) *J. Chromatogr. Sci.*, **23**, 176.
[88] Pflugmann, G., Spahn, H. and Mutschler, E. (1987) *J. Chromatogr.*, **416**, 331.
[89] Weber, H., Spahn, H., Mutschler, E. and Mohrke, W. (1984) *J. Chromatogr.*, **307**, 145.
[90] Spahn, H., Weber, H., Mutschler, E. and Mohrke, W. (1984) *J. Chromatogr.*, **310**, 167.
[91] Brückner, H. and Gah, C. (1991) *J. Chromatogr.*, **555**, 81.
[92] Szokan, G., Mezo, G. and Hudecz, F. (1988) *J. Chromatogr.*, **444**, 115.
[93] Aberhart, D.J., Cotting, J-A. and Lin, H-J. (1985) *Anal. Biochem.*, **151**, 88.
[94] Szokan, G., Mezo, G., Hudecz, F., Majer, Z., Schon, I., Nyeki, O., Szirtes, T. and Dolling, R. (1984) *J. Liq. Chromatogr.*, **12**, 2855.
[95] Guttendorf, R.J., Kostenbauder, H.B. and Wedlund, P.J. (1989) *J. Chromatogr.*, **489**, 333.
[96] Lindner, W., Leitner, C. and Uray, G. (1984) *J. Chromatogr.*, **316**, 605.
[97] Lindner, W., Rath, M., Stoschitzky, K. and Uray, G. (1989) *J. Chromatogr.*, **487**, 375.
[98] Kleidernigg, O.P., Posch, K. and Lindner, W. (1996) *J. Chromatogr. A*, **729**, 33.
[99] Rossetti, V., Lombard, A. and Buffa, M. (1986) *J. Pharm. Biomed. Anal.*, **4**, 673.
[100] Sallustio, B.C., Abas, A., Hayball, P.J., Purdie, Y.J. and Meffin, P.J. (1986) *J. Chromatogr.*, **374**, 329.
[101] Maitre, J.M., Boss, G., Testa, B. and Hostettmann, K. (1986) *J. Chromatogr.*, **356**, 341.
[102] McKay, S.W., Mallen, D.N.B., Shrubsall, P.R., Swann, B.P. and Williamson, W.R.N. (1979) *J. Chromatogr.*, **170**, 482.
[103] Rickett, F.E. (1973) *Analyst*, **98**, 687.
[104] Pedrazzini, S., Zanoboni-Muciaccia, W., Sacchi, C. and Forgione, A. (1987) *J. Chromatogr.*, **415**, 214.
[105] Clark, C.P., Snider, B.G. and Bowman, P.B. (1987) *J. Chromatogr.*, **408**, 275.
[106] Maitre, J., Boss, G. and Testa, B. (1984) *J. Chromatogr.*, **299**, 397.
[107] Jiang, M. and Soderlund, D.M. (1982) *J. Chromatogr.*, **248**, 143.
[108] Sioufi, A., Colussi, D., Marfil, F. and Dubois, J.P. (1987) *J. Chromatogr.*, **414**, 131.
[109] Knadler, M.P. and Hall, S.D. (1989) *J. Chromatogr.*, **494**, 173.
[110] Hutt, A.J., Fournel, S. and Caldwell, J. (1986) *J. Chromatogr.*, **378**, 409.
[111] Averginos, A. and Hutt, A.J. (1987) *J. Chromatogr.*, **415**, 75.
[112] Lau, Y.Y. (1996) *J. Liq. Chromatogr. & Rel. Technol.*, **19**, 2143.
[113] Tsina, I., Tam, Y.L., Boyd, A., Rocha, C., Massey, I. and Tarnowski, T. (1996) *J. Pharm. Biomed. Anal.*, **15**, 403.
[114] Valentine, D., Chan, K.K., Scott, C.G., Johnson, K.K., Toth, K. and Saucy, G. (1976) *J. Org. Chem.*, **41**, 62.
[115] Scott, C.G., Petrin, M.J. and McCorkle, T. (1976) *J. Chromatogr.*, **125**, 157.
[116] Nagashima, H., Tanaka, Y., Watanabe, N., Hayashi, R. and Kawada, K. (1984) *Chem. Pharm. Bull.*, **32**, 251.
[117] Goto, J., Goto, N. and Nambara, T. (1982) *J. Chromatogr.*, **239**, 559.
[118] Fujimoto, Y., Ishii, K., Nishi, H., Tsumagari, N., Kakimoto, T. and Shimizu, R. (1987) *J. Chromatogr.*, **402**, 344.
[119] Spahn, H. (1987) *J. Chromatogr.*, **423**, 334.
[120] Lehr, K.H. and Damm, P. (1988) *J. Chromatogr.*, **425**, 153.
[121] Bjorkman, S. (1987) *J. Chromatogr.*, **414**, 465.
[122] Foster, R.T. and Jamali, F. (1987) *J. Chromatogr.*, **416**, 388.
[123] Jamali, F., Pasutto, F.M. and Lemko, C. (1988) *J. Liq. Chromatogr.*, **12**, 1835.
[124] Berry, B.W. and Jamali, F. (1988) *Pharm. Res.*, **5**, 123.
[125] Mehvar, R. and Jamali, F. (1988) *Pharm. Res.*, **5**, 53.
[126] Mehvar, R., Jamali, F. and Pasutto, F.M. (1988) *J. Chromatogr.*, **425**, 135.
[127] Spahn, H., Spahn, I., Pflugmann, G., Mutshler, E. and Benet, L.Z. (1988) *J. Chromatogr.*, **433**, 331.
[128] Balani, S.K., van Bladeren, P.J., Cassidy, E.S., Boyd, D.R. and Jerina, D.M. (1987) *J. Org. Chem.*, **52**, 137.
[129] van Bladeren, P.J., Sayer, J.M., Ryan, D.E., Thomas, P.E., Levin, W. and Jerina, D.M. (1985) *J. Biol. Chem.*, **260**, 10226.
[130] Armstrong, R.N., Kedzierski, B., Levin, W. and Jerina, D.M. (1981) *J. Biol. Chem.*, **256**, 4726.
[131] Duke, C.C. and Holder, G.M. (1988) *J. Chromatogr.*, **430**, 53.

[132] Lee, H. and Harvey, R.G. (1984) *J. Org. Chem.*, **49**, 1114.
[133] Yagi, H. and Jerina, D.M. (1982) *J. Am. Chem. Soc.*, **104**, 4026.
[134] Harvey, R.G. and Cho, H. (1977) *Anal. Biochem.*, **80**, 540.
[135] Banfield, C. and Rowland, M. (1984) *J. Pharm. Sci.*, **73**, 1392.
[136] Oshima, R., Yamaguchi, Y. and Kumanotani, J. (1982) *Carbohydr. Res.*, **107**, 169.
[137] Oshima, R., Kumanotani, J. and Watanabe, C. (1983) *J. Chromatogr.*, **259**, 159.
[138] Abolfathi, Z., Belanger, P.-M., Gilbert, M., Rouleau, J.R. and Turgeon, J. (1992) *J. Chromatogr.*, **579**, 366.
[139] Boomsma, F., Van Der Hoorn, F.A.J., Man In't Veld, A.J. and Schalekamp, M.A.D.H. (1988) *J. Chromatogr.*, **427**, 219.
[140] Buck, R.H. and Krummen, K. (1987) *J. Chromatogr.*, **387**, 255.
[141] Desai, D.M. and Gal, J. (1993) *J. Chromatogr.*, **629**, 215.
[142] Euerby, M.R., Partridge, L.Z. and Rajani, P. (1988) *J. Chromatogr.*, **447**, 392.
[143] Euerby, M.R., Nunn, P.B. and Partridge, L.Z. (1989) *J. Chromatogr.*, **466**, 407.
[144] Nimura, N. and Kinoshita, T. (1986) *J. Chromatogr.*, **352**, 169.
[145] Spahn-Langguth, H., Hahn, G., Mutschler, E., Mohrke, W. and Langguth, P. (1992) *J. Chromatogr.*, **584**, 229.
[146] Wuis, E.W., Beneken Kolmer, E.W.J., Van Beijsterveldt, L.E.C., Burgers, R.C.M., Vree, T.B. and Van Der Kleyn, E. (1987) *J. Chromatogr.*, **415**, 419.
[147] Nimura, N., Iwaki, K. and Kinoshita, T. (1987) *J. Chromatogr.*, **402**, 387.
[148] Aswad, D.W. (1984) *Anal. Biochem.*, **137**, 405.
[149] Maurs, M., Trigalo, F. and Azerad, R. (1988) *J. Chromatogr.*, **440**, 209.
[150] Buck, R.H. and Krummen, K. (1984) *J. Chromatogr.*, **315**, 279.
[151] Jegorov, A., Triska, J., Trnka, T. and Cerny, M. (1988) *J. Chromatogr.*, **434**, 417.
[152] Brückner, H., Wittner, R. and Godel, H. (1991) *Chromatographia*, **32**, 383.
[153] Brückner, H., Haasmann, S., Langer, M., Westhauser, T., Wittner, R. and Godel, H. (1994) *J. Chromatogr. A*, **666**, 259.
[154] Einarsson, S., Josefsson, B., Moller, P. and Sanchez, D. (1987). Anal. Chem. **59**, 1191.
[155] Vogt, C., Georgi, A. and Werner, G. (1995) *Chromatographia* **40**, 287.
[156] Rosseel, M.T., Vermeulen, A.M. and Belpaire, F.M. (1991) *J. Chromatogr.*, **568**, 239.
[157] Frigerio, E., Pianezzola, E. and Benedetti, S. (1994) *J. Chromatogr. A*, **660**, 351.
[158] Boursier-Neyret, C., Baune, A., Klippert, P., Castagne, I. and Sauveur, C. (1993) *J. Pharm. Biomed. Anal.*, **11**, 1161.
[159] Chen, Y.-P., Hsu, M.-C. and Chien, C.S. (1994) *J. Chromatogr. A*, **672**, 135.
[160] De Witt, P., Deias, R., Muck, S., Galletti, B., Meloni, D., Celletti, P. and Marzo, A. (1994) *J. Chromatogr. B*, **657**, 67.
[161] Chou, T-Y., Gao, C-X., Grinberg, N. and Krull, I.S. (1989) *Anal. Chem*, **61**, 1548.
[162] Belas, F.J., Phillips, M.A., Srinivas, N.R., Barbhaiya, R.H. and Blair, I.A. (1995) *Biomed. Chromatogr.*, **9**, 140.
[163] Darmon, A. and Thenot, J.P. (1986) *J. Chromatogr.*, **374**, 321.
[164] Hoshino, M., Yajima, K., Suzuki, Y. and Okahira, A. (1994) *J. Chromatogr. B* **661**, 281.
[165] Peyton, A.L., Carpenter, R. and Rutkowski, K. (1991) *Pharm. Res.*, **8**, 1528.
[166] Piquette-Miller, M. and Jamali, F. (1993) *Pharm. Res.*, **10**, 294.
[167] Pirkle, W.H. and Hoekstra, M.S. (1974) *J. Org. Chem.*, **39**, 3904.
[168] Sakaki, K. and Hirata, H. (1991) *J. Chromatogr.*, **585**, 117.
[169] Goto, J., Hasegawa, M., Nakamura, S., Shimada, K. and Nambara, T. (1978) *J. Chromatogr.*, **152**, 413.
[170] Pham-Huy, C., Sahui-Gnassi, A., Saada, V., Gramond, J.P., Galons, H., Ellouk-Achard, S., Levresse, V., Fompeydie, D. and Claude, J.R. (1994) *J. Pharm. Biomed. Anal.*, **12**, 1189.
[171] Toyo'oka, T. and Liu, Y.-M. (1995) *Analyst*, **120**, 385.
[172] Liu, Y.-M., Miao, J.-R. and Toyo'oka, T. (1995) *Anal. Chim. Acta*, **314**, 169.
[173] Toyo'oka, T. and Liu, Y.-M. (1995) *J. Chromatogr. A*, **689**, 23.
[174] Liu, Y.-M. and Toyo'oka, T. (1995) *Chromatographia*, **40**, 645.
[175] Toyo'oka, T., Suzuki, T., Watanabe, T. and Liu, Y.-M. (1996) *Anal. Sci.*, **12**, 779.
[176] Suzuki, T., Watanabe, T. and Toyo'oka, T., (1997) *Anal. Chim. Acta*, **352**, 357.
[177] Langguth, P., Spahn, H. and Merkle, H.-P. (1990) *J. Chromatogr.*, **528**, 55.
[178] Spahn-Langguth, H., Podkowik, B., Stahl, E., Martin, E. and Mutschler, E. (1991) *J. Anal. Toxicol.*, **15**, 209.
[179] Spahn, H., Krau$\beta$, D. and Mutschler, E. (1988) *Pharm. Res.*, **5**, 107.
[180] Martin, E., Quinke, K., Spahn, H. and Mutschler, E. (1989) *Chirality* **1**, 223.
[181] Kondo, J., Imaoka, T., Suzuki, N., Kawasaki, T., Nakanishi, A. and Kawahara, Y. (1994) *Anal. Sci.*, **10**, 697.

[182] Ito, E., Nishida, Y., Horie, H., Ohrui, H. and Meguro, H. (1995) *Bunseki Kagaku* **44**, 739.
[183] Goto, J., Goto, N. and Nambara, T. (1982) *J. Chromatogr.*, **239**, 559.
[184] Goto, J., Goto, N., Hikichi, A., Nishimaki, T. and Nambara, T. (1980) *Anal. Chim. Acta*, **120**, 187.
[185] Nagashima, H., Tanaka, Y., Watanabe, H., Hayashi, R. and Kawada, K. (1984) *Chem. Pharm. Bull.*, **32**, 251.
[186] Nagashima, H., Tanaka, Y. and Hayashi, R. (1985) *J. Chromatogr.*, **345**, 373.
[187] Robinett, R.S.R. and Hsieh, J.Y.-K. (1991) *J. Chromatogr.*, **570**, 157.
[188] Hutt, A.J., Fournel, S. and Caldwell, J. (1986) *J. Chromatogr.*, **378**, 409.
[189] Avgerinos, A. and Hutt, A.J. (1987) *J. Chromatogr.*, **415**, 75.
[190] Goto, J., Ito, M., Katsuki, S., Saito, N. and Nambara, T. (1986) *J. Liq. Chromatogr.*, **9**, 683.
[191] Toyo'oka, T., Ishibashi, M. and Terao, T. (1992) *J. Chromatogr.*, **627**, 75.
[192] Toyo'oka, T., Ishibashi, M. and Terao, T. (1992) *J. Chromatogr.*, **625**, 357.
[193] Toyo'oka, T., Ishibashi, M. and Terao, T. (1993) *Anal. Chim. Acta* **278**, 71.
[194] Spahn, H. and Langguth, P. (1990) *Pharm. Res.*, **7**, 1262.
[195] Liebmann, B., Mayer, S., Mutschler, E. and Spahn-Langguth, H. (1992) *Arzneim. Forsch.*, **42**, 1354.
[196] Kondo, J., Imaoka, T., Kawasaki, T., Nakanishi, A. and Kawahara, Y. (1993) *J. Chromatogr.*, **645**, 75.
[197] Kondo, J., Suzuki, N., Imaoka, T., Kawasaki, T., Nakanishi, A. and Kawahara, Y. (1994) *Anal. Sci.*, **10**, 17.
[198] Kondo, J., Suzuki, N., Naganuma, H., Imaoka, T., Kawasaki, T., Nakanishi, A. and Kawahara, Y. (1994) *Biomed. Chromatogr.*, **8**, 170.
[199] Toyo'oka, T., Ishibashi, M. and Terao, T. (1992) *Analyst* **117**, 727.
[200] Iwaki, K., Bunrin, T., Kameda, Y. and Yamazaki, M. (1994) *J. Chromatogr. A*, **662**, 87.
[201] Al-Kindy, S., Santa, T., Fukushima, T., Homma, H. and Imai, K. (1997) *Biomed. Chromatogr.*, **11**, 137.
[202] Yasaka, Y., Matsumoto, T. and Tanaka, M. (1995) *Anal. Sci.*, **11**, 295.
[203] Narita, S. and Kitagawa, T. (1989) *Anal. Sci.*, **5**, 361.
[204] Shimizu, R., Ishii, K., Tsumagari, N., Tanigawa, M., Matsumoto, M. and Harrison, I.T. (1982) *J. Chromatogr.*, **253**, 101.
[205] Shimizu, R., Kakimoto, T., Ishii, K., Fujimoto, Y., Nishi, H. and Tsumagari, N. (1986) *J. Chromatogr.*, **357**, 119.
[206] Goto, J., Goto, N. and Nambara, T. (1982) *Chem. Pharm. Bull.*, **30**, 4597.
[207] Goto, J., Goto, N., Shao, G., Ito, M., Hongo-Ishikawa, A., Nakamura, S. and Nambara, T. (1990) *Anal. Sci.*, **6**, 261.
[208] Goto, J., Shao, G., Fukasawa, M., Nambara, T. and Miyano, S. (1991) *Anal. Sci.*, **7**, 645.
[209] Goto, J., Shao, G., Ito, M., Kuriki, T. and Nambara, T. (1991) *Anal. Sci.*, **7**, 723.
[210] Toyo'oka, T., Ishibashi, M., Terao, T. and Imai, K. (1993) *Analyst*, **118**, 759.
[211] Toyo'oka, T., Liu, Y.-M., Hanioka, N., Jinno, H., Ando, M. and Imai, K. (1994) *J. Chromatogr. A*, **675**, 79.
[212] Toyo'oka, T., Liu, Y.-M., Jinno, H., Hanioka, N., Ando, M. and Imai, K. (1994) *Biomed. Chromatogr.*, **8**, 85.
[213] Toyo'oka, T., Liu, Y.-M., Hanioka, N., Jinno, H. and Ando, M. (1994) *Anal. Chim. Acta*, **285**, 343.
[214] Nishida, Y., Bai, C., Ohrui, H. and Meguro, H. (1994) *J. Carbohydrate Chem.*, **13**, 1003.
[215] Kim, J.-H., Uzawa, H., Nishida, Y., Ohrui, H. and Meguro, H. (1994) *J. Chromatogr. A*, **677**, 35.
[216] Nishida, Y., Ito, E., Abe, M., Ohrui, H. and Meguro, H. (1995) *Anal. Sci.*, **11**, 213.
[217] Toyo'oka, T. and Liu, Y.-M. (1994) *Anal. Proc.*, **31**, 265.
[218] Sano, A., Takitani, S. and Nakamura, H. (1995) *Anal. Sci.*, **11**, 299.
[219] Jin, D., Takehana, K. and Toyo'oka, T. (1997) *Anal. Sci.*, **13**, 113.
[220] Shimada, K., Haniuda, E., Oe, T. and Nambara, T. (1987) *J. Liq. Chromatogr.*, **10**, 3161.
[221] Wu, D-F., Reidenberg, M.M. and Drayer, D.E. (1988) *J. Chromatogr.*, **433**, 141.
[222] Lim, H.K. (1985) Master Thesis, Univ. Saskatchewan, Saskatoon, Canada.
[223] Pianezzola, E., Bellotti, V., Fontana, E., Moro, E., Gal, J. and Desai, D.M. (1989) *J. Chromatogr.*, **495**, 205.
[224] Rondelli, I., Mariotti, F., Acerbi, D., Redenti, E., Amari, G. and Ventura, P. (1993) *J. Chromatogr.*, **612**, 95.
[225] Fanali, S., Cristalli, M., Vespalec, R. and Bocek, P. (1994) *Advances in Electrophoresis*, p 3, Chrambach, A., Dunn, M.J. and Radola, B.J. (Eds.) VCH, Weinheim.
[226] Snopek, J., Jelinek, I. and Smolkova-keulemansova, E. (1991) *New Trends in Cyclodextrins and Derivatives*, p 483, Duchene, D. (Ed.) Edition de Sante, Paris.
[227] Li, S. and Purdy, W.C. (1992) *Chem. Rev.*, **92**, 1457.
[228] Okafo, G.N. and Camilleri, P. (1993) *Capillary Electrophoresis*, p 163, Camilleri, P. (Ed.) CRC Press, Boca Raton, FL.

[229] Otuka, K. and Terabe, S. (1993) *Trends Anal. Chem.*, **12**, 125.
[230] Fanali, S. (1993) *Capillary Electrophoresis Technology*, p 731, Guzman, N.A. (Ed.) Marcel Dekker, New York.
[231] Fanali, S. (1996) *J. Chromatogr. A*, **735**, 77.
[232] Michaelsen, S., Møller, P. and Søresen, H. (1994) *J. Chromatogr. A*, **680**, 299.
[233] Otsuka, K., Karuhara, K., Higashimori, M. and Terabe, S. (1994) *J. Chromatogr. A*, **680**, 317.
[234] Terabe, S., Otsuka, K., Ichikawa, K., Tsuchiya, A. and Ando, T. (1984) *Anal. Chem.*, **56**, 111.
[235] Schutzner, W., Fanali, S., Rizzi, A. and Kenndler, E. (1996) *J. Chromatogr. A*, **719**, 411.
[236] Schutzner, W., Fanali, S., Rizzi, A. and Kenndler, E. (1993) *J. Chromatogr.*, **639**, 375.
[237] Leroy, P., Bellucci, L. and Nicolas, A. (1995) *Chirality*, **7**, 235.
[238] Lurie, I.S. (1992) *J. Chromatogr.*, **605**, 269.
[239] Chan, K.C., Muschik, G.M. and Issaq, H.J. (1995) *Electrophoresis*, **16**, 504.
[240] Vogt, C., Georgi, A. and Werner, G. (1995) *Chromatographia*, **40**, 287.
[241] Tran, A., Blanc, T. and Leopold, E.J. (1990) *J. Chromatogr.*, **516**, 241.
[242] Cladrowa-Runge, S., Hirz, R., Kenndler, E. and Rizzi, A. (1995) *J. Chromatogr. A*, **710**, 339.
[243] Smyth, A.X. and Smyth, M.R. (1996) *Anal. Lett.*, **29**, 991.
[244] Bunke, A. and Jira, T. (1996) *Pharmazie*, **51**, 479.
[245] Langguth, P., Spahn, H. and Merkle, H-P. (1990) *J. Chromatogr.*, **528**, 55.
[246] Goto, J., Hasegawa, M., Nakamura, S., Shimada, K. and Nambara, T. (1978) *J. Chromatogr.*, **152**, 413.
[247] Nambara, T., Ikegawa, S., Hasegawa, M. and Goto, J. (1978) *Anal. Chim. Acta*, **101**, 111.
[248] Goto, J., Goto, N., Hikichi, A. and Nambara, T. (1979) *J. Liq. Chromatogr.*, **2**, 1179.
[249] Ladanyi, L., Sztruhar, I., Slegel, P. and Vereczekey-Donath, G. (1987) *Chromatographia*, **24**, 477.
[250] Gorog, S., Herenyi, B. and Low, M. (1986) *J. Chromatogr.*, **353**, 417.
[251] Kurosu, Y., Murayama, K., Shindo, N., Shisa, Y. and Ishioka, N. (1996) *J. Chromatogr. A*, **752**, 279.
[252] Mechref, Y. and Rassi, Z.E. (1996) *Anal. Chem.*, **68**, 1771.
[253] Wan, H., Engstrom, A. and Blomberg, L.G. (1996) *J. Chromatogr. A*, **731**, 283.
[254] Wan, H. and Blomberg, L.G. (1996) *J. Microcolumn Separations*, **8**, 339.
[255] Swartz, M.E., Mazzeo, J.R., Grover, E.R. and Brown, P.R. (1995) *Anal. Biochem.*, **231**, 65.
[256] Valko, I.E., Siren, H. and Riekkola, M-L. (1996) *J. Chromatogr. A*, **737**, 263.
[257] Volko, I.E., Siren, H. and Riekkola, M-L. (1996) *Chromatographia*, **43**, 242.
[258] Anselmi, S., Braghiroli, D., Di Bella, M., Schmid, M.G., Wintersteiger, R. and Gubitz, G. (1996) *J. Chromatogr. B*, **681**, 83.
[259] Sundin, N.G., Dowling, T.M., Grinberg, N. and Bicker, G. (1996) *J. Microcolumn Separations*, **8**, 323.
[260] Schutzner, W., Caponecchi, G., Fanali, S., Rizzi, A. and Kenndler, E. (1994) *Electrophoresis*, **15**, 769.
[261] Dette, C. and Watzig, H. (1994) *Electrophoresis*, **15**, 763.
[262] Dette, C., Watzig, H. and Aigner, A. (1994) *Pharmazie*, **49**, 245.

# Index


Achatin I 255
acesulfame K 22, 28
acetic anhydride 33, 67
3$\beta$-acetoxy-$\Delta^5$-etienic acid 228
(*S*)-acetoxypropionic acid (AP) 241
acetyl-CoA 93, 94
*N*-acetyl-L-cysteine (NAC) 21, 39, 43, 103, 226, 249, 266
*N*-acetylneuraminic acid (NANA) 182
*N*-acetyl-D-penicillamine (NAP) 249, 266
acridinium ester 168
(*N*-9-acridinyl)bromoacetamide (Br-AA) 123, 125, 131
*N*-(9-acridinyl) maleimide (NAM) 18
adamantyl-1,2-dioxetane derivative 168
adenine and its nucleos(t)ides 183
ADME 220
adsorbent method 52, 53
aflatoxins (aflatoxin $B_1$, $B_2$, $G_1$, $G_2$) 15, 35, 38
$\omega$-agatoxin-TK 255
L-alanine-$\beta$-naphtylamide 240
aliin 19
alprostadil 79
Aminex A-27 35
aminoalditol 89
*p*-aminobenzoic acid hydrazide 16
*N*-(4-aminobenzoyl)-L-glutamic acid (ABG) 89
(−)-2-[4-(1-aminoethyl)phenyl]-6-methoxybenzoxazole [(−)-APMB] 4, 5, 6, 261, 262
3-aminofluoranthene 180
2-amino-4-hydroxypteridine-6-carboxylic acid 26
aminomethyl phosphoric acid 60, 74, 75
5-aminonaphthalene-2-sulfonate (ANA) 17
8-aminonaphthalene-1,3,6-trisulfonic acid (ANTS) 17
2-amino-oxy-4-methylvaleric acid 231
9-aminophenanthrene (9-AP) 126
*p*-aminophenol 205
*R*(−)-2-amino-1-propanol 267
9-aminopyrene-1,4,6-trisulfonate (APTS) 17
2-aminopyridine (2-AP) 17
4-(3-aminopyrrolidin-1-yl)-7-(*N*,*N*-dimethylaminosulfonyl)-2,1,3-benzoxadiazole [*S*(+)- and *R*(−)-DBD-APy] 259
4-(3-aminopyrrolidin-1-yl)-7-nitro-2,1,3-benzoxadiazole [*S*(+)- and *R*(−)-NBD-APy] 259
6-aminoquinoline (6-AQ) 78, 91
6-aminoquinolyl-*N*-hydroxysuccinimidyl carbamate (AQC) 18, 20, 39, 76, 112
4-(aminosulfonyl)-7-fluoro-2,1,3-benzoxadiazole (ABD-F) 143
amitrole 60
amoxicillin 30
amperometric detection 192, 193, 194, 196, 211
amphetamine 171, 174
ampicillin 30
analgesics 10
analysis of residual pesticides and herbicides 59
*N*-(4-anilinophenyl)maleimide 205
2,3-(anthracenedicarboximido)ethyl trifluoromethanesulfonate (AE-OTf) 41, 129
1-anthroylnitrile (1-AN) 138
*p*-(9-anthroyloxy)phenacylbromide (9-APB) 125
1-anthrylcarbocyanide 41
9-anthryldiazomethane (ADAM) 26, 32, 40, 60, 125, 129
1-(1-anthryl) and 1-(2-anthryl)ethylamines (*l*- and *d*- enantiomers) (AEA) 259
4-(1-anthryl)-1,2,4-triazoline-3,5-dione (A-TAD) 152
anti arrhythmia agents 9
anti-cholesteremic agents 9
anti-hypertensive agents 9
anti-neoplastic agents 8
antitubercular agents 8
antiviral agents 8
arachidonic acid metabolites 178
ascorbic acid (AA) 25, 28
aspartame 28
asulam 60
ATP 185

atrazine 15
avermectins 15, 33

baclofen 198
bentazon (BEN) 35
benzimidazole 14
benzoin 119
benzoylchloride 42, 71, 84, 85, 86
3-benzoyl-2-quinoline carbaldehyde (BQCA) 105
*S*(+)-benzoxaprofen chloride [*S*(+)-BZOP-Cl] 235, 257
1-benzyl-2-chloropyridinium bromide (BCPB) 92, 93
bioluminescence (BL) 167, 185
bimanes 145
biotin 26
bis(2,6-difluorophenyl)oxalate (DFPO) 169
bis(2,4-dinitrophenyl)oxalate (DNPO) 169
bis(pentafluorophenyl)oxalate (PFPO) 169
bis(2,4,6-trichlorophenyl)oxalate (TCPO) 169, 173, 174, 175, 177, 178
bis[2-(3,6,9-trioxadecyloxycarbonyl)-4-phenyl]oxalate (TDPO) 169, 173, 174, 175, 179, 183
meso-1,2-bis(4-methoxyphenyl)-ethylenediamine (*p*-MOED) 155
Bolton and Hunter type reagent 204
boron 17
*O*-(−)-bornylhydroxyammonium chloride 231, 232
bovine serum albumin 185
α-bromo-2′-acetonaphtone (BAN) 78, 79
3-bromoacetyl-7-methoxycoumarin (Br-AMC) 123
3-bromoacetyl-6,7-methylenedioxycoumarin (Br-ADMC) 122
4-bromomethyl-7-acetoxycoumarin (Br-MAC) 122
1-bromoacetylpyrene (Br-AP) 9, 32, 123, 131
*N*-bromoacetyl-N′-(dansyl)piperazine (DNS-BAP) 123
2-bromoacetyl-6-methoxynaphthalene (Br-AMN) 123, 127, 131
9-bromomethylacridine (Br-MA) 123
4-bromomethyl-6,7-dimethoxycoumarin (Br-DMC) 27, 122
4-bromomethyl-6,7-dimethoxy-1-methyl-2(1H)quinoxalinone (Br-DMEQ) 123, 125, 129, 157
4-bromomethyl-7-methoxycoumarine (Br-MMC) 26, 122, 123, 127, 129, 157, 177
4-bromomethyl-6,7-methylenedioxycoumarin (Br-MDC) 123
4-bromomethyl-6,7-methylenedioxy-1-methyl-2(1*H*)quinoxalinone (Br-MMEQ) 123, 129
*p*-bromophenacyl bromide (BPB) 27, 78
*p*-bromophenylcarbamate (SIBr-PC) 76
2-(5-bromopyridylazo)-5-diethylaminophenol 27
butylisothiocyanate (BITC) 70
butylthiocarbamyl (BTC) 70

cadaverine 71, 76, 78
canister 53
capillary electrophoresis (CE) 65, 73, 89, 91, 94, 102, 185, 186, 187, 267
capillary electrochromatography (CEC) 284
capillary gel electrophoresis (CGE) 277
capillary zone electrophoresis (CZE) 91, 253, 277
captopril 92, 93, 94
(−)-camphanic acid chloride 235
carbamates 15
4-(2-carbazoylpyrrolidin-1-yl)-7-nitro-2,1,3-benzoxadiazole [*S*(−)- and *R*(+)-NBD-ProCZ] 266
carbobenzoyl-L-proline 241
carbofuran 15
carbohydrate 84, 89, 91, 92
3-(4-carboxybenzoyl)-2-quinoline carbaldehyde (CBQCA) 17, 105
cardiolipin 86
ceric sulfate 23
chemiluminescence (CL) 16, 243
chinomethionate (CIN) 35
chiral stationary phase (CSP) 223, 224
cholic acid 176
4-(2-chloroformylpyrrolidin-1-yl)-7-(*N*,*N*-dimethylaminosulfonyl)-2,1,3-benzoxadiazole [*S*(−)- and *R*(+)-DBD-Pro-COCl] 264, 266
4-(2-chloroformylpyrrolidin-1-yl)-7-nitro-2,1,3-benzoxadiazole [*S*(−)- and *R*(+)-NBD-Pro-COCl] 264, 266
3-chloro-4-methoxyaniline 60
9-chloromethylanthracene (9-CA) 22, 41
7-chloro-2,1,3-benzoxadiazole-4-sulfonate (SBD-Cl) 143
2-chloro-4,5-bis(*p*-*N*,*N*-dimethylaminosulfonyl)oxazole (SAOX-Cl) 114, 144
2-(5-chlorocarbonyl-2-oxazolyl)-5,6-methylenedioxybenzofuran (OMB-COCl) 136
4-(*N*-chloroformylmethyl-*N*-methyl)amino-7-*N*,*N*-dimethylaminosulfonyl-benzofurazan (DBD-COCl) 112
2-chloro-1-methylpyridinium iodide (CMPI) 92
4-chloro-7-nitro-2,1,3-benzoxadiazole (NBD-Cl) 19, 58, 112
chlorophenols 178
cholesterol 85
CIEEL mechanism 169
ciguatoxin (CTX) 41
citrulline 21
CL detection system for HPLC 179, 176
CL detection interface for CE 185, 186
clenbuterol 34
cloxacillin 30
CL reaction 168

column-switching 31, 32, 41, 43
copper(II) sulfate 26
cosmetics 94
coulometric detection 192, 193, 198
18-crown-6 27
crystal violet 28
2-cyanoacetamide (CA) 92, 150
cyanogen bromide 25
cyclamate 22, 28, 29
cyclodextrines 38, 277
1,3-cyclohexadione 23
4-(2-cyanoisoindolyl)phenyl isothiocyanate (CIPITC) 112
L-cysteine 25, 66, 67, 68, 74, 91, 92, 94, 179
L-cysteine methyl ester 231

2-dansylaminoethanol (DNS-AE) 128
1-(4-dansylaminophenyl)ethylamine (*d*- and *l*-DAPEA) 262
dansyl cadaverine (DNS-CD) 126
dansyl ethylenediamine (DNS-ED) 3
dansyl semipiperazine (DNS-PZ) 126, 130
dansyl chloride (DNS-Cl) 18, 20, 42, 61, 68, 72, 73, 107, 173, 174, 205
2-dansyl ethylchloroformate (DNS-ECF) 139
dansyl hydrazine (DNS-H) 34, 56, 57, 146, 150
dehydroascorbic acid (DHAA) 25
dermatan sulfate (DS) 179
desmosine 199, 211
destomycin A (DM) 31
dexamethasone 182
(*R*,*R*)-*O*,*O*-diacetyl tartaric acid anhydride (DATAAN) 235, 277
1,2-diarylethylenediamines 183
(*R*,*R*)-*O*,*O*-dibenzoyl tartaric acid anhydride (DBTAAN) 235, 277
1,2-diamino-4,5-dimethoxybenzene (DDB) 133, 135, 140, 148
1,2-diamino-4,5-ethylenedioxybenzene (DEB) 148
1,2-diamino-4,5-methylenedioxybenzene (DMB) 133, 148
diarrhtic shellfish poisoning toxins (DSP) 40
4-diazomethyl-7-methoxycoumarin (DAM-MC) 125, 126
dibutyltin (DBT) 45
$\alpha$-dicarbonyl compounds 182
7-(diethylamino)coumarin-3-carbohydrazide (DCCH) 177
7-(diethylamino)-3-[4-((iodoacetyl)amino)phenyl]-4-methylcoumarin (DCIA) 177, 181
1,3-diethyl-2-thiobarbituric acid (DETBA) 181
diethyldithiocarbamate (DDTC) 94, 95
diethylethoxymethylenemalonate (DEMM) 75, 76
1,2-(difluoro-1,3,5-triazinyl)benz[f]isoindolo-[1,2-b][1,3]benzothioazolidine (FBIBT) 139
dihydrostreptomycin (DSM) 30
3,4-dihydro-6,7-dimethoxy-4-methyl-3-oxoquinoxaline-2-carbonyl azide (DMEQ-$CON_3$) 137
3,4-dihydro-6,7-dimethoxy-4-methyl-3-oxoquinoxaline-2-carbonyl chloride (DMEQ-COCl) 110, 112
(1,2-dihydroxyethyl)furo[3,4-b]quinoxaline-1-one (DFQ) 25
diketohydrindylidene-diketohydrindamine 74
4-(5,6-dimethoxy-2-benzimidazol)-benzohydrazide (DMBI-BH) 127
4-(5′,6′-dimethoxybenzothiazolyl)benzoyl fluoride (BHBT-COF) 110, 111
5,6-dimethoxy-2-(4′-hydrazinocarbonylphenyl)-benzothiazole (BHBT-BH) 127
4-[2-(6,7-dimethoxy-4-methyl-3-oxo-3,4-dihydroquinoxalinyl)ethyl]-1,2,4-triazoline-3,5-dione (DMEQ-TAD) 152
4-(5′,6′-dimethoxybenzothiazolyl)phenyl isothiocyanate (BHBT-NCS) 112
6,7-dimethoxy-1-methyl-2(1H)-quinoxalinone-3-propionylcarboxylic acid hydrazide (DMEQ-PAH) 127, 129, 178
*N*-[4-(5,6-dimethoxy-2-phthalimidyl)-phenyl]maleimide (DPM) 141
*p*-dimethylaminobenzaldehyde 30
*N*-[4-(6-dimethylamino-2-benzofuranyl)-phenyl]maleimide (DBPM) 141, 179
5-dimethylaminonaphthalene-1-[*N*-(2-aminoethyl)]-sulfonamide (DNS-EDA) 153
1-(5-dimethylamino-1-naphthalenesulfonyl)-(*S*)-3-aminopyrrolidine [(*S*)-DNS-APy]
5-*N*,*N*′-dimethylaminonaphthalene-1-sulfonohydrazide (DNS-H) 179, 180
1-(4-dimethylamino-1-naphtyl)ethylamine (DANE) 240, 257
4-(*N*,*N*-dimethylamino)-1-naphthylisothiocyanate (DANITC) 112
4-(*N*,*N*-dimethylaminonaphthalene-5-sulfonylamino)-phenylisothiocyanate (DNSITC) 112
4-(*N*,*N*-dimethylaminosulfonyl)-7-(2-carbazoylpyrrolidin-1-yl)-2,1,3-benzoxadiazole (DBD-ProCZ) 129, 150, 151, 266
4-(*N*,*N*-dimethylaminosulfonyl)-7-fluoro-2,1,3-benzoxadiazole (DBD-F) 39, 112, 143, 175
4-(*N*,*N*-dimethylaminosulfonyl)-7-hydrazino-2,1,3-benzoxadiazole (DBD-H) 147, 180
4-(*N*,*N*-dimethylaminosulfonyl)-7-*N*-piperazino-2,1,3-benzoxadiazole (DBD-PZ) 129, 130
4-*N*,*N*-dimethylaminoazobenzene-4′-isothiocyanate (DABITC) 70
4-*N*,*N*-Dimethylaminoazobenzene-4′- sulfonyl chloride (Dabsyl-Cl, DABS-Cl) or 1-di-*n*-butylaminonaphthalene-5-sulfonyl chloride (Dabsyl-Cl) 73, 108
*N*,*N*-dimethylformamide 33

3,4-dimethoxyphenylglyoxal (DMPG) 156
3,5-dinitrobenzoyl chloride 29
2,4-dinitrofluorobenzene (DNFB, Sanger reagent) 18, 20, 204
2,4-dinitrophenyl (DNP) 65, 66
2,4-dinitrophenyl hydrazine (DNPH) 55, 56, 57
3,5-dinitrophenyl isocyanate 88
*N*,*N*-(dinitrophenyl)octylamine 89
dinophysistoxins (DTX-1,2,3) 40
dioctylsulfosuccinate 22, 29
1,2-dioxetanes 168
diphenoylperoxide 168
2-diphenylacetyl-1,3-indandion-1-hydrazone (DAIH) 56
1,2-diphenylethylenediamine (DPE) 116, 204
diphenyl-1-pyrenylphosphine (DPPP) 45
direct resolution 223, 225
direct sampling method 53
disuccinimido carbonate (DSC) 76
2,2′-dithiobis(1-amino-4,5-dimethoxybenzene) (DTAD) 148
5,5′-dithiobis(2-nitrobenzoic acid) (DTNB, Ellman's reagent) 93, 94
dithiothreitol 25
dodecyltin 45
dopamine 175, 183

Edman reagent 68, 70
eicosapentaenoic acid 176
electrochemical detection (ECD) 192, 195, 205
electrochemical oxidation 33, 39
electrode 192, 193, 194, 195
electrophore 200
electrospray ionization (ESI)
end-column CL detector for CE
environmental contaminants 51, 55
EPA TO-14 53
EPA TO-5 55
eosine Y 185
epinephrine 175, 183
eptam 15
erysorbic acid 28
ethacrynic acid 94
ethyl-*p*-aminobenzoate 17
1-ethyl-3-(3-dimethylaminopropyl) carbodiimide hydrochloride (EDC) 23, 81
ethylenediamine (ED) 183
ethyleneglycol 84, 85, 86

*R*(−)-1-ferrocenylethylamine 267
ferrocenyl group 200, 201, 206, 207, 208, 211
*S*(+)-1-ferrocenylpropylamine 267
FDNB:1-fluoro-2,4-dinitrobenzene (FDNB, Sanger reagent) 65, 66
firefly luciferase bioluminescence reaction 185
fluazifop-P-butyl 15
*S*(+)-flunoxaprofen chloride [*S*(+)-FLOP-Cl] 257
*S*(−)-flunoxaprofen isocyanate [*S*(−)-FLOPIC] 257
1-(9-fluorenyl)ethyl chloroformate (FLEC) 226, 251, 277
9-fluorenylmethyl chloroformate (FMOC) 18, 39, 42, 43, 58, 60, 68, 73, 74, 109, 121
1-fluoro-2,4-dinitrophenyl-5-L-alanine amide (Marfey's reagent) 67, 235, 277
4-fluoro-7-nitrobenzofurazan (NBD-F) 39, 112
fluorescamine 18, 19, 28, 30, 38, 60, 173
7-fluoro-2,1,3-benzoxadiazole-4-sulfonate (SBD-F) 143
fluorophore for PO-CL detection system
  3-aminofluoranthene 180
  4-(bromomethyl)-7-methoxycoumarin (Br-Mmc) 177
  dansyl chloride (Dns-Cl) 173, 174
  1,2-diarylethylenediamines 183
    7-(diethylamino)coumarin-3-carbohydrazide (DCCH) 177
  7-(diethylamino)-3-[4-((iodoacetyl)amino)phenyl]-4-methylcoumarin (DCIA) 177, 181
  1,3-diethyl-2-thiobarbituric acid (DETBA) 181
  6,7-dimethoxy-1-methyl-2(1H)-quinoxalinone-3-proprionylcarboxylic acid hydrazide 178
  *N*-[4-(6-dimethylamino-2-benzofuranyl)phenyl]-maleimide (DBPM) 179
  5-*N*,*N*′-dimethylaminonaphthalene-1-sulfonohydrazide (Dns-H) 179, 180
  4-(*N*,*N*-dimethylaminosulfonyl)-7-fluoro-2,1,3-benzoxadiazole (DBD-F) 175
  4-(*N*,*N*-dimethylaminosulfonyl)-7-hydrazino-2,1,3-benzoxadiazole (DBD-H) 180
  eosine Y 185
  ethylenediamine (ED) 183
  fluorescamine 105, 173
  4-fluoro-7-nitrobenzoxadiazole (NBD-F) 174, 175
  2-(4-hydrazinocarbonylphenyl)-4,5-diphenylimidazole (HCPI) 177, 178
  lissamine Rhodamine B sulfonyl chloride (laryl chloride) 178
  luminarin 1 175
  Naphthalene-2,3-dicarboxaldehyde (NDA) 121, 174, 175
  near-infrared (near-IR) fluorescent dyes 184
  pyrimido[5,4-d]pyrimidines 184
fluorescein isothiocyanate (FITC) 112, 121
2-fluoro-4,5-diphenyloxazole (DIFOX) 114
fluoropyrimidines, 14, 181
4-fluoro-3-nitrotrifluoromethylbenzene (FNBT) 66
4-fluoro-7-nitro-2,1,3-benzoxadiazole (NBD-F) 8, 112, 143
fluvoxamine 175
folacin 26
formaldehyde/trichloroacetic acid 30
fradiomycin (FM) 32
fructosedehydrogenase (FDH) 16

fruit acids 23
fumonisins (fumonisin $B_1$,$B_2$,$B_3$) 39
furazolidone 14
3-(2-furoyl)quinoline-2-carbaldehyde (FQCA) 105

GABA agonists 10
galactitol 88
gel permiation chromatography (GPC) 14, 22, 35
gentamycin (GM) 31
glucoamylase (GAM) 16
glucose 16
glucose oxidase (GOD) 16
*β*-D-glucosidase 16
glucosides 16
glutamic acid 20
glutathione (GSH) 92, 179, 199, 208, 212
glycosphingolipid 86
glyphosphate(*N*-(phosphonomethyl)glycine) 74
glyphosate 60
gossypol 212, 267
guanine and its nucleos(t)ides 183

hexacyanoferrate(III) 16, 117
high performance capillary electrophoresis (HPCE) 89, 277
high performance liquid chromatography (HPLC) 91, 92, 93, 94
histamine 43, 173
histidine (His) 20, 21
DL-homocysteine 25
hyaluronic acid (HA) 92, 179
hydrazine 78, 81
2-(5-hydrazinocarbonyl-2-furyl)-5,6-methylenedioxybenzofuran (FM-BH) 127
2-(5-hydrazinocarbonyl-2-oxazolyl)-5,6-methylenedioxybenzofuran (OM-BH) 127
2-(hydrazinocarbonylphenyl)-4,5-diphenylimidazole (HCPI) 127
2-(5-hydrazinocarbonyl-2-thienyl)-5,6-methylenedioxybenzofuran (TM-BH) 127
hydrogen peroxide ($H_2O_2$) 16, 39
hydroxyalcohol 86
hydroxycarboxylic acid 86
hydroxyeicosatetraenoate (HETE) 86
hydroxyethylamine 25
hydroxy fatty acid 86, 88
9-hydroxymethylanthracene (HMA) 128
4-hydroxymethyl-7-methoxycoumarin (HMC) 128
5-hydroxyindoles 182
hydroxylamine 78, 82, 83
hydroxyproline 19
8-hydroxyquinoline (8-HQ) 27
hygromycin B 14, 31
hypoglycin A (HG-A) 42

ibuprofen 177
immobilized antibody extraction 2
immobilized enzyme reactors 16, 24
indirect resolution 223, 225
indophenol 28
invertase (INV) 16
iodine 38
2-iodo-1-methylpyridinium chloride (IMPC) 92
isatoic anhydride 17, 87
*N*-isobutylyl-L(or D)-cysteine 249, 250
isocyanate 45, 53, 67, 84, 88, 95, 202
isoluminol 45
isotachophoresis (ITP) 277
isothiocyanate 67, 68, 70, 202
*N*-[(2-isothiocyanato)cyclohexyl]-(3,5)-dinitrobenzoylamide (DDITC) 237, 242
4-(3-isothiocyanatopyrrolidin-1-yl)-7-(*N*,*N*-dimethylaminosulfonyl)-2,1,3-benzoxadiazole [*S*(+)- and *R*(−)-DBD-PyNCS] 254
4-(3-isothiocyanatopyrrolidin-1-yl)-7-nitro-2,1,3-benzoxadiazole [*S*(+)- and *R*(−)-NBD-PyNCS] 254
ivermectin 6, 14, 32

kanamycin (KM) 31, 32
*α*-keto acids 182
kryptofix 222 32

L-lactate oxidase 24
lactic acid 23
lasalocid 32
laser-induced fluorescence (LIF) detection 2, 3, 17, 103, 116, 243
L-leucinamide 240
2-[4-(L-leucyl)aminophenyl]-6-methoxybenzoxazole (L-LeuBOX) 261
leuco malachite green (LMG) 33
liquid-liquid extraction 54
lipid peroxides 45
lophine (2,4,5-triphenylimidazole) 167
lophine derivatives 185
luminol (5-amino-2,3-dihydro-1,4-phthalazinedione) 16, 45, 168, 169, 185
luminol derivative
  *N*-(4-aminobutyl)-*N*-ethylisoluminol (ABEI) 169, 171, 176, 177
  6-aminomethylphthalhydrazide (6-AMP) 182
  4,5-diaminophthalhydrazide (DPH) 181, 182
  4-isothiocyanatophthalhydrazide 171
  6-isothiocyanobenzo[g]phthalazine-1,4-(2H,3H)-dione (IPO) 171, 172, 173
lucigenin (*N*, *N*′-dimethyl-9,9′-bisacridinium dinitrate) 168
lysozyme 27

malachite green (MG) 33
maleic anhydride 38
malondialdehyde 181
maltose 16

maltooligosaccharide 16
mandelic acid 228
maprotiline 173
matrix solid-phase dispersion (MSPD) 14, 33, 35
medroxyprogesterone acetate 180
menaquinones 26
(−)-menthyl chloroformate 228, 235
*O*-(−)-menthyl-*N*,*N*-diisopropylisourea (MDI) 231
2-mercaptoethanol (2-ME) 21, 32, 35, 38, 43, 60, 249
3-mercaptopropionic acid (3-MP) 20, 35, 43, 103
mercury(II) chloride 26
methamphetamine 171, 174
methiocarb 15
methocarbamol 88
methomyl 14
methoxaron 60
metoprolol 175
4-methoxybenzamidine (*p*-MBA) 150
*N*-[4-(6-methoxy-2-benzoxazolyl)]benzoyl-L-phenylalanine (BOX-L-Phe) 257
*N*-[4-(6-methoxy-2-benzoxazolyl)]benzoyl-L-proline (BOX-L-Pro) 257
(aS)-2′-methoxy-1,1′-binaphthalene-2-carbonyl cyanide (Methoxy-BNCC) 263
(−)-*α*-methoxy-*α*-methyl-1-(or 2)-naphthaleneacetic acids (MMNA) 253
1-*α*-methoxy-*α*-methylnaphthaleneacetic acid *N*-succinimidyl ester 235
6-methoxy-2-methylsulfonylquinoline-4-carbonyl chloride (MMSQ-COCl) 110
2-methoxy-2,4-diphenyl-3[2H]-furanone (MDF) 106
1-(*p*-methoxy)phenyl-3-methyl-5-pyrazolone (PMPMP) 91
*S*(−)-*α*-methoxy-*α*-(trifluoromethyl)phenylacetic acid (MTPA, Mosher's reagent) 228
*S*(−)-*α*-methoxy-*α*-(trifluoromethyl)phenylacetyl chrolide 228, 241
10-methyl-9-acridinium carboxylate 178
9-(*N*-methylaminomethyl)anthracene (MAMA) 45, 95
(−)- and (+)-2-methyl-1,1′-binaphthalene-2′-carbonyl cyanides (Methyl-BNCC) 263
*N*-methylcarbamates 35
methylene blue 22, 29
2-[*p*-(5,6-methylenedioxy-2H-benzotriazol-2-yl)]phenethylamine (MBPA) 126
4-(*N*-methyl-*N*-(2-hydroxyethyl)amine-7-nitrobenz-3-oxa-1,3-diazole (NBD-OH) 112
1-methylimidazole 23, 33
4-(6-methylnaphthalene-2-yl)-4-oxo-2-buteneoic acid 94
2-methyl-2-*β*-naphthyl-1,3-benzodiazole-4-carboxylic acid [*S*(+)-MNB-COOH] 257
*S*(+)-1-methyl-2-(2,3-naphthalimido)ethyl trifluoromethanesulfonate [*S*(+)-MNE-OTf] 262
*α*-methyl-4-nitrobenzylamine 240
4-(1-methyl-2-phenanthro[9,10-*d*]imidazol-2-yl)-benzohydrazide (MPIB-BH) 147
4-methylumbelliferyl tri-*N*-acethyl-*β*-chitotrioside (4-MU-$(GlcNAc)_3$) 27
methyl violet 2B 22, 28
mexiletine 174
micellar electrokinetic chromatography (MEKC) 21, 69, 73, 277
monoglycerides (MGL) 28, 29
molybdenum 26
monensin 32
monobromobimane 145
monosaccharide 86, 87, 88, 89, 92
myo-inositol 87, 88
mycotoxins 35

2,3-naphthalenedialdehyde or naphthalene-2,3-dicarboxaldehyde (NDA) 39, 105, 199
1,2-naphthoylenebenzimidazole-6-sulfonyl chloride (NBI-SOCl) 136
2-(2,3-naphthalimido)ethyl trifluoromethanesulphonate (NE-OTf) 129
1,2-naphtoquinone-4-sulfonic acid (NQS) 31, 70, 71
naphthoyl chloride 9, 86
naphthylcarbamoyl amino acid 67
2-naphthyl chloroformate (NT-COCl) 110
1-(1-naphtyl)ethylamine (NEA) 4, 10, 240
*N*-(1-naphthyl)ethylenediamine (NEDA) 126
1-(1-naphthyl)ethyl isocyanate (NEIC) 9, 10, 88, 228, 253, 267
1-naphtyl isothiocyanate (NIC) 58, 59, 67
*d*-2-(2-naphtyl)propionyl chloride (NPC) 262
*S*(−)-*N*-(1-naphthylsulfonyl)-2-pyrrolidinecarbonyl chloride (NSPC) 262
*S*(+)-naproxen chloride [*S*(+)-NAP-Cl] 257
naracin 32
neomycin (NM) 32
neosaxitoxin 39
nicarbazine 14
ninhydrin 18, 20, 31, 43, 71, 74, 75
4-nitro-7-(2-carbazoylpyrrolidin-1-yl)-2,1,3-benzoxadiazole (NBD-ProCZ)
1-[(4-nitrophenyl)sulfonyl]prolyl chloride 235
4-nitro-7-*N*-piperazino-2,1,3-benzoxadiazole (NBD-PZ) 129
Norit 25
normal phase high performance liquid chromatography 86, 88
nitric oxide (NO) 21
*p*-nitrobenzoyl chloride (PNB-Cl) 17
*O*-(4-nitrobenzyl)-*N*,*N*′-diisopropylisourea (PNBDI) 23
*p*-nitrobenzylhydroxylamine (PNB) 89
2-nitrophenylhydrazine hydrochloride (2-NPH) 22, 24, 81
*p*-nitrophenylhydrazine 32, 208
*N*-nitrosamines 15

*N*-2′-nitro-4-trifluoromethylphenyl (NTP) 66, 67
norepinephrine 175, 183

*n*-octyl-5-dithio-2-nitrobenzoic acid (ODNB) 65, 93
octyltin 45
okadaic acid (OA) 40
on-column derivatization 15, 21, 27, 43
organochlorine pesticides 14
organotin 14, 43
ornithine 21
*o*-phthalaldehyde (OPA) 10, 18, 19, 20, 21, 30, 31, 32, 38, 39, 43, 60, 68, 71, 103, 121, 198, 226, 249, 277
oxalyl chloride 168
1,1′-oxalyldiimidazole (ODI) 169
oxamyl 14
oxolinic acid 14
oxosteroids 208
oxo-steroids and oxo-bile acid ethyl esters 179

panacyl bromide 26, 78
paralytic shellfish poison toxins (PSP) 39
PCB 15
penicillins 30
pentaazapentacosane (PAPC) 72, 73
2′,3,4′,5,7-pentahydroxyflavon (molin reagent) 45
peptides 17, 21
peroxyoxalate chemiluminescence (PO-CL) 57, 168, 169, 170
phenacyl bromide 23
phenacyl bromidephenyldimethylsilyl chloride (PHB) 78
phenoxy acid herbicides 60
L-phenylalanine-*β*-naphtylamide 240
2-[4-(L-phenylalanyl)aminophenyl]-6-methoxybenzoxazole (L-PheBOX) 261
*β*-phenylethylamine 42
*S*(−)-1-phenylethylamine (PEA) 240
*R*(+)-1-phenylethyl isothiocyanate (PEIT) 235
2-[4-(D-phenylglycyl)aminophenyl]-6-methoxybenzoxazole (D-PgBOX) 261
phenylglyoxal (PGO) 117, 155, 183
phenyl isocyanate (PHI) 17, 18, 87, 88
phenylisothiocyanate (PITC) 18, 19, 42
*o*-phenylenediamine (OPDA) 25, 26, 28
phenylethylisocyanate (PEIC) 228, 267
phenylisocyanate (PIC) 67
phenyl isothiocyanate (PITC) 68, 69, 70
1- phenyl-3-methyl-5-pyrazolone (PMP) 17, 91
*S*(+)-phenylpropionic acid 228
3-phenyl-2-thiohydantoin (PTH) 68, 69
phosphatidylcholine (PC) 86
phosphoamino acid 69
phosphoserine 67, 69
phosphothreonine 69
phosphotyrosine 69
photolytic derivatization 28, 35
4-(2-phthalimidyl)benzohydrazide (PBH) 147
4-(*N*-phthalimidyl)benzenesulfonyl chloride (Phisyl-Cl) 108
3-(2-phthalimidyl)benzoyl azide (*m*-Phibyl-$N_3$) 137
4-(2-phthalimidyl)benzoyl azide (*p*-Phibyl-$N_3$) 137
phytic acid 29
phytonadione 26
platina 26
polyamines 42, 65, 66, 67, 71, 72, 76, 78
polymeric benzotriazole 76, 77, 78
polymeric 3,5-dinitrobenzoyl 76, 78
polymeric reagent 78
polyphosphates 29
post-capillary derivatization 102
post-column derivatization 67, 71, 102
post-column ion-pair extraction 22, 28, 29
potassium cyanide 25
potassium ferricyanide 25
potassium permanganate 26
Pravastatin 2
pre-capillary derivatization 102
pre-column derivatization 102
progesterone 34
propylene glycol esters of fatty acids (PGE) 29
prostaglandins (PG) 78, 79, 88, 198, 205, 206
pteroylglutamic acid 26
pulsed amperometric detection (PAD) 16
purpald, 4-amino-3-hydrazino-5-mercapto-1,2,4-triazole (AHMT) 92
putrescine 42, 66, 67, 71, 76, 78
pyranose oxidase (PyOD) 16
pyrazoxyfen (PYR) 35
pyrene-1-carbonylnitrile (PCN) 138
1-pyrenemethanol (1-PM) 128
1-pyrenyldiazomethane (PDAM) 26, 125
pyrene sulfonyl chloride (PSCL) 139
2-(1-pyrenyl)ethyl chloroformate (PE-COCl) 110
pyridinium dichromate 32, 88
pyridoxal (PL) 25
pyridoxamine (PM) 25
pyridoxine (PN) 25
*N*-2-pyridylglycamine 17
1-(2-pyridyl)-piperazine (PYP) 94
5-(4-pyridyl)-2-thiophenemethanol (PTM) 128

reversed phase high performance liquid chromatography (RP-HPLC) 66, 68, 69, 70, 71, 72, 75, 76, 78, 84, 85, 86, 87, 88, 89, 91, 92, 93, 94
Roundup 74
Ruhemann purple 74

saccharin 22, 28
salicylic acid chloride 203

salinomycin 32
saxitoxin 39
semduramicin 32
serotonin uptake inhibitors 10
sodium borohydride 26
sodium hypochlorite (NaOCl) 21
solid-phase extraction (SPE) 14, 21, 26, 28, 30, 31, 32, 33, 34, 35, 39, 40, 43, 54, 55
solid-phase microextraction (SPME) 15, 53
solid-phase reagent 76
solution trapping method 51, 52, 53
sorbic acid 17, 27
sorbitol 88
soxhlet extraction 54
spectinomycin (SPCM) 32
spermidine 42, 66, 67, 71, 72
spermine 42, 66, 67
streptomycin (SM) 30, 31
succinimido phenycarbamate 76
*N*-succinimidyl-1-fluorenylcarbamate (SIPC) 112
*N*-succinimidyl-1-naphthylcarbamate (SINC) 112
sucrose 16
sucrose esters of fatty acids (SuE) 29
sulfamethazine 173
sulfonamides 14, 15, 30
sulfosalicylic acid 29
sulfurous acid 25
supercritical fluid extraction (SFE) 14, 35, 54
supersonic extraction method 54

taurine 21
TEDLAR bag 53
tetracycline 27
2,3,4,6-tetra-*O*-acetyl-*β*-D-glucopyranosyl isothiocyanate (GITC) 9, 226, 235, 277
2,3,4,6-tetra-*O*-acetyl-1-thio-*β*-D-glucopyranoside (TATG) 249
3*α*,5*β*-tetrahydroaldosterone 182
tetrahydroboric acid 26
teterahydro-5-oxo-2-furancarboxylic acid (TOF) 233, 241
tert.-butyloxycarbonyl (BOC)-L-cysteine 226, 249
2-tert.-butyl-2-methyl-1,3-benzodioxole-4-carboxylic acid [*R*(−)- and *S*(+)-TBMB-COOH] 266
(*S*)-tert.-butyl 3-(chloroformoxy)butyrate 235
tetrodotoxin (TTX) 41
thiamine 25
thiochrome 25
thiofluor 21
1-thio-*β*-D-glucose (TG) 249
*m*-toluoyl chloride 58
*R*(+)-trans-chrysanthemic acid 228
2,3,4-tri-*O*-acetyl-a-D-arabinopyranosyl isothiocyanate (AITC) 235
1,2,4-triazole/mercuric chloride 30
tri-*n*-butyltin chloride (TBTC) 44
bis(tri-*n*-butyltin) oxide (TBTO) 44
trifluoroacetic acid (TFA) 15, 35
*S*(−)-*N*-(trifluoroacetyl)prolyl chloride 228
*S*(−)-*N*-(trifluoroacetyl)proline 228
(+)-2,2,2-trifluoro-1-phenylethyl hydrazine 231, 232
2,4,6-trinitrobenzene-1-sulfonic acid (TNBS) 67
triphenylphosphine (TPP) 81, 93, 259
tris(2,2′-bipyridyl)ruthenium(II) [$Ru(bpy)_3^{2+}$] 175, 176
tryptamine 42, 45
tyramine 42

vacuum glass bottle 53
vanilline 32
vitamin $B_1$ 25
vitamin $B_6$ 25
vitamin C 25
vitamin K 26